Scientific Computation

Editorial Board

J.-J. Chattot, Davis, CA, USA
P. Colella, Berkeley, CA, USA
R. Glowinski, Houston, TX, USA
M. Holt, Berkeley, CA, USA
Y. Hussaini, Tallahassee, FL, USA
P. Joly, Le Chesnay, France
H. B. Keller, Pasadena, CA, USA
D. I. Meiron, Pasadena, CA, USA
O. Pironneau, Paris, France
A. Quarteroni, Lausanne, Switzerland
J. Rappaz, Lausanne, Switzerland
R. Rosner, Chicago, IL, USA
J. H. Seinfeld, Pasadena, CA, USA
A. Szepessy, Stockholm, Sweden

Springer
Berlin
Heidelberg
New York
Barcelona
Hong Kong
London
Milan
Paris
Tokyo

Physics and Astronomy ONLINE LIBRARY

http://www.springer.de/phys/

Scientific Computation

A Computational Method in Plasma Physics
F. Bauer, O. Betancourt, P. Garabechan

**Implementation of Finite Element Methods
for Navier-Stokes Equations**
F. Thomasset

**Finite-Different Techniques
for Vectorized Fluid Dynamics Calculations**
Edited by D. Book

Unsteady Viscous Flows
D. P. Telionis

Computational Methods for Fluid Flow
R. Peyret, T. D. Taylor

**Computational Methods in Bifurcation
Theory and Dissipative Structures**
M. Kubicek, M. Marek

Optimal Shape Design for Elliptic Systems
O. Pironneau

The Method of Differential Approximation
Yu. I. Shokin

Computational Galerkin Methods
C. A. J. Fletcher

**Numerical Methods
for Nonlinear Variational Problems**
R. Glowinski

Numerical Methods in Fluid Dynamics
Second Edition   M. Holt

**Computer Studies of Phase Transitions
and Critical Phenomena**   O. G. Mouritsen

**Finite Element Methods
in Linear Ideal Magnetohydrodynamics**
R. Gruber, J. Rappaz

Numerical Simulation of Plasmas
Y. N. Dnestrovskii, D. P. Kostomarov

**Computational Methods for Kinetic Models
of Magnetically Confined Plasmas**
J. Killeen, G. D. Kerbel, M. C. McCoy,
A. A. Mirin

Spectral Methods in Fluid Dynamics
Second Edition   C. Canuto, M. Y. Hussaini,
A. Quarteroni, T. A. Zang

**Computational Techniques
for Fluid Dynamics 1**
Fundamental and General Techniques
Second Edition   C. A. J. Fletcher

**Computational Techniques
for Fluid Dynamics 2**
Specific Techniques
for Different Flow Categories
Second Edition   C. A. J. Fletcher

**Methods for the Localization of Singularities
in Numerical Solutions
of Gas Dynamics Problems**
E. V. Vorozhtsov, N. N. Yanenko

**Classical Orthogonal Polynomials
of a Discrete Variable**
A. F. Nikiforov, S. K. Suslov, V. B. Uvarov

**Flux Coordinates and Magnetic Filed
Structure: A Guide to a Fundamental Tool
of Plasma Theory**
W. D. D'haeseleer, W. N. G. Hitchon,
J. D. Callen, J. L. Shohet

**Monte Carlo Methods
in Boundary Value Problems**
K. K. Sabelfeld

The Least-Squares Finite Element Method
Theory and Applications in Computational
Fluid Dynamics and Electromagnetics
Bo-nan Jiang

**Computer Simulation
of Dynamic Phenomena**
M. L. Wilkins

Grid Generation Methods
V. D. Liseikin

Radiation in Enclosures
A. Mbiock, R. Weber

**Large Eddy Simulation
for Incompressible Flows**
An Introduction   P. Sagaut

Series homepage – http://www.springer.de/phys/books/sc/

Alexander S. Lipatov

The Hybrid Multiscale Simulation Technology

An Introduction
with Application to Astrophysical
and Laboratory Plasmas

With 124 Figures

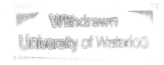

 Springer

Professor Dr. Alexander S. Lipatov
Dialogue Science – Computing Center
Russian Academy of Sciences
Vavilova St. 40, GSP-1
Moscow 117967, Russia

and

Department of Problems of Physics and Energetics
Moscow Institute of Physics and Technology
Institutsky Per. 9, Dolgoprudny
141700 Moscow Region, Russia

Library of Congress Cataloging-in-Publication Data.

Lipatov, Alexander S., 1946– .
The hybrid multiscale simulation technology: an introduction with application to
astrophysical and laboratory plasmas/Alexander S. Lipatov.
p.cm. – (Scientific computation, ISSN 1434-8322)
Includes bibliographical references and index.
ISBN 3-540-41734-6 (acid-free paper)
1. Plasma (Ionized gases)–Simulation methods. 2. Plasma astrophysics–Simulation
methods. 3. Computer simulation. I. Title. II. Series.
QC718.4.L56 2002   530.4'4–dc21   2002019987

ISSN 1434-8322
ISBN 3-540-41734-6 Springer-Verlag Berlin Heidelberg New York

This work is subject to copyright. All rights are reserved, whether the whole or part of the material is concerned, specifically the rights of translation, reprinting, reuse of illustrations, recitation, broadcasting, reproduction on microfilm or in any other way, and storage in data banks. Duplication of this publication or parts thereof is permitted only under the provisions of the German Copyright Law of September 9, 1965, in its current version, and permission for use must always be obtained from Springer-Verlag. Violations are liable for prosecution under the German Copyright Law.

Springer-Verlag Berlin Heidelberg New York
a member of BertelsmannSpringer Science+Business Media GmbH

http://www.springer.de

© Springer-Verlag Berlin Heidelberg 2002
Printed in Germany

The use of general descriptive names, registered names, trademarks, etc. in this publication does not imply, even in the absence of a specific statement, that such names are exempt from the relevant protective laws and regulations and therefore free for general use.

Typesetting by the author
Cover design: *design & production* GmbH, Heidelberg
Printed on acid-free paper   SPIN: 10788430   55/3141/tr - 5 4 3 2 1 0

To My Family

Preface

This book addresses hybrid simulation of plasmas; it is aimed at developing insight into the essence of plasma behavior. Major current applications are to astrophysical and space plasmas. Some applications are connected with active experiments in space. However, hybrid simulations are also being used to gain an understanding of basic plasma phenomena such as particle acceleration by shocks, magnetic field reconnection in neutral current sheets, generation of waves by beams, mass loading of the supersonic flow by heavy pickup ions and the dynamics of tangential discontinuities. Such simulations may be very important not only for the study of the astrophysical plasmas, but also for the study of the magnetically and inertially contained fusion plasmas, and other laboratory plasma devices.

Plasma is the fourth state of matter, consisting of electrons, ions and neutral atoms, usually at temperatures above 10^4 K. The stars and sun are plasmas; the local interstellar medium, the solar wind, magnetospheres and ionospheres of planets and comets, Van-Allen belts, etc., are all plasmas. Indeed, much of the known matter in the universe is plasma.

Current studies of physical phenomena, in particular, plasma physics, include theory, experiment and computational experiment (numerical simulation) using powerful workstations or vector/massive parallel computer systems. Use of computational experiments is very extensive because they can produce results useful to theory and experiment and because current direct investigation of astrophysical plasma–gas objects, laboratory experiments and modeling require considerable time and money. Computational experiments are very important for the design of plasma devices and space vehicles. During the last three decades, different numerical methods for plasma simulation have been developed. Among the most popular simulation techniques are particle simulation methods and methods of directly solving the Vlasov–Maxwell equations. While the direct solution of the Vlasov equation needs to use traditional numerical techniques, which are widely used, for example, in hydrodynamics, particle simulations need to use a wide range of numerical methods: integration of particle trajectories, the *Monte-Carlo* technique, the weighting and smoothing technique, and integration of the Maxwell equations. Although, many different methods have been developed for multiscale plasma simulation (from electrostatic explicit models to electromagnetic full particle implicit models), the available computer resources dictate a strict specifications of the numerical models. This circumstance results in the development of so-called *rational* or realistic models, which describe the object of our research well, and which provide an acceptable ratio of the approximation error to the computer resources. In many cases, even if we have the possibility of using a modern supercomputer, it is reasonable to exploit hybrid codes. Traditionally, it has been implied that hybrid code is code in which we use a particle description for ions and magnetohydrodynamic (MHD) equations for electrons. A large portion of this book is concerned with the numerical algorithms and application of this method to space and plasma physics. However, the hybrid codes, which are now widely used, include a wide spectrum of different types of description for different species of plasma and neutrals (ions, protons, positrons, dust grains and electrons), such as particle, gyrokinetic and orbit-averaging approaches, and electron inertia effects. These simulation techniques may be useful for the investigation of the nonlinear kinetic processes in a fully or partially ionized gas. Thus, researchers may choose the simulation model with desirable characteristics. For this reason a review of different types of model is included. This book gathers information which is valuable to simulation scientists, some of which is scattered through the published literature.

The book is organized in into two parts and twelve chapters. In Part I, I consider the different hybrid models and relevant numerical methods and algorithms. In some sections I shall present program examples which may be useful in practical simulations.

Chapter 1 classifies the physical and plasma systems and corresponding computational models. I consider the traditional numerical methods for direct integration of the Vlasov equation as well as recently designed Vlasov hybrid simulation method.

In Chap. 2, I describe the different combined approaches (e.g., particle, gyrokinetic, orbit-averaging, etc.) which can be used to study plasma systems. The models with charge exchange and photo-ionization processes are also included.

Chapter 3 gives an analysis of the different numerical schemes for updating the particle velocities and positions. I consider the direct methods and also the splitting of operator methods.

In Chap. 4, I describe the particle and the field-weighting techniques. The conservation laws for general hybrid codes are also considered. The multipole expansion method for periodic systems is also discussed.

Chapter 5 presents the analysis of numerical schemes for time integration of the electromagnetic field and electron pressure equations. The time-integration schemes for electromagnetic potential equations as well as the generalized field equations are also considered. Some important ways to design the high-order compact schemes are given.

In Chap. 6, I consider possible numerical algorithms for multiscale simulation. The general computational loops which are most useful in recent hybrid simulations are also discussed.

In Chap. 7, I discuss the generation of the particle distribution inside the computational domain and the particle flux with a given distribution at the boundaries. The different types of boundary conditions (reflection, radiation, etc.) for electric and magnetic fields are also considered.

In Part II, I consider the application of hybrid code simulation to space, astrophysical and laboratory plasmas.

In Chap. 8, I consider the simulation of collisionless shock wave structures and the particle dynamics at the shock fronts. The effect of pickup ions on the structure of the transition layer is discussed here. Special attention is given to particle acceleration processes.

Chapter 9 is devoted to the dynamics of tangential discontinuities. The stability of tangential discontinuities, wave generation in the transition layer and diffusion processes are considered. The effect of the Kelvin–Helmholtz instability is also discussed.

In Chap. 10, I present the simulation results of magnetic field reconnection processes inside neutral current sheets. The use of Ampere's and hybrid models is discussed. I consider the linear, nonlinear and explosive regimes of instability and the particle acceleration processes.

In Chap. 11, I discuss the simulation results for the generation of low-frequency electromagnetic waves by beams (or plasma clouds) that propagate in homogeneous and inhomogeneous background plasma. The mass-loading processes of the solar wind by pickup ions are also considered.

Chapter 12 is devoted to global simulation of the interaction of the solar wind with comets, nonmagnetic planets and other astrophysical objects. I discuss, in detail, the formation of the inductive magnetosphere, bow shock wave and the magnetic barrier of nonmagnetic planets. The splitting of the cometary tails is also discussed. I also consider the interaction between the interstellar local medium and the heliosphere.

I recommend *Computational Physics* by D. Potter (Wiley, 1973) as an introduction to simulation methods which may be used as the first course in the instruction of undergraduate students. The book *Computer Simulation Using Particles* by R.W. Hockney and J.W. Eastwood (McGraw–Hill International Book Company, 1981; Adam Hilder, IOP Publishing Ltd, 1988) is recommended as a complementary text in which the simulation scientists will find a mathematically sequential presentation of the particle method.

This book may be used for teaching last-year-undergraduate or first-year-graduate students in courses on numerical analysis, computational science, or computational physics. Students and scientists in plasma physics are recommended to study *Plasma Physics via Computer Simulation* by C.K. Birdsall and A.B. Langdon (1985, McGraw–Hill Book Company) and *Computational Plasma Physics* by T. Tajima (1989, Addison–Wesley Publishing Company,

Inc.). To study the methods of implicit multiscale simulation, it is very important to read *Multiple Time Scales* (ed. by J.U. Brackbill and B.I. Cohen, 1985, Academic Press. Inc.).

The work of the author has developed over the last 25 years, principally at the Space Research Institute (IKI) and later in the Dialogue Science-Computing Center of the Russian Academy of Sciences, the East–West Space Science Center of the University of Maryland, the Space and Plasma Physics Group in the Department of Physics and Astronomy of the University of Maryland at College Park, the Max Planck Institute for Extraterrestrial Physics (Berlin–Adlershoff), the Bartol Research Institute of the University of Delaware, the Max Planck Institute for Aeronomy, and the Institute for Theoretical Physics of the Technical University of Braunschweig. The simulations included in text were supported in part by the Russian Academy of Sciences (Basic Plasma Physics Projects), the VENERA-HALLEY Project, the Russian Foundation of Fundamental Science Grant 96-02-16326, the Russian Space Agency (Project Interball -IKI), the NASA ISTP Grants, the NASA Contract NASR-36811, an NSF-DOE Grant ATM 9713223, the NASA Grant NAG5-6469, the JPL NASA Contract 959167 and the NASA Delaware Space College Grant NGT5-40024, the Deutsche Forschungs-Gemeinschaft (DFG) Grant 436 RUS 17/2/00, and the Max-Planck-Gesellschaft (MPG). A portion of this book was used by the author in his lectures for undergraduate and graduate students "Numerical Methods for Space Plasma" and "Introduction to Computational Physics", which he has provided since 1973 at the Department of Problems of Physics and Energetics of the Moscow Institute of Physics and Technology. I would like to thank my advisers, Academicians Prof. Roald Z. Sagdeev and Prof. Albert A. Galeev, for collaboration since 1973. I would like to thank Academician Prof. Oleg M. Belotserkovskii for his advice and guidance during my first steps in computational physics. I would like to thank Dr. T. Bagdonat, Dr. K. Baumgärtel, Priv.-Doz. Dr. J. Buechner, Dr. C.L. Chang, Dr. Yu. Krasheninnikova, Dr. A.V. Lobachev, Prof. G.G. Lominadge, Dr. A.A. Malgichev, Dr. G.M. Milikh, Prof. U. Motschmann, Dr. H.R. Müller, Prof. N.F. Ness, Prof. K. Papadopoulos, Dr. H.L. Pauls, Dr. A.S. Sharma, Prof. K. Sauer, Prof. V.D. Shapiro, Prof. V.I. Shevchenko, Dr. I.N. Syrovatskii, Dr. A.L. Taktakishvili, Prof. L.L. Vanyan, Prof. G.P. Zank and Prof. L.M. Zelenyi, with whom I have collaborated at different times in the last 25 years. I would like to thank Prof. S.I. Anisimov, Prof. M. Ashour-Abdalla, Dr. E.V. Belova, Prof. D. Burgess, Prof. P.J. Cargill, Prof. L.M. Degtyarev, Prof. J. Drake, Prof. T.A. Fritz, Dr. C.C. Goodrich, Prof. G. Haerendal, Dr. M. Hesse, Prof. V.P. Korobeinikov, Dr. H. Kucharek, Dr. M. Kuznetsova, Prof. E. Marsch, Prof. H. Matsumoto, Prof. J.F. McKenzie, Dr. N. Omidi, Prof. A. Otto, Prof. M. Scholer, Prof. B.A. Tverskoi and Prof. D. Winske for fruitful discussions of simulation results which were included in this manuscript. I express my sincere thanks to the many plasma simulation physicists whose papers have provided the main source of inspiration and material for this book. I would like to thank Mrs. A. Brandt (Max Planck Institute for Aeronomy) for preparation of the electronic version of the figures for this book. I thank my wife Elena for her gracious understanding and encouragement at every stage of this project. I also thank her for assistance in computer simulation of some problems considered in this book. I thank my son Sergei for software support on our home computer.

I have been most fortunate to work with patient and professional editors and associates at Springer-Verlag, notably Prof. Wolf Beiglböck and his assistants, who suffered through our difficult task of producing camera-ready copy in Heidelberg. I would like to thank the referees for their helpful comments.

Katlenburg-Lindau, *Alexander S. Lipatov*
March 2002

Contents

Part I. Computational Models and Numerical Methods

1. **Physical Systems and Computational Models** 3
   1.1 Introduction 3
   1.2 The Basic Steps of Computational Experiments 3
   1.3 Classification of the Plasma Systems. Space and Time Scales . 6
       1.3.1 Solar Wind 6
       1.3.2 Solar-Wind–Earth Interaction 7
       1.3.3 Solar-Wind–Moon Interaction 8
       1.3.4 Solar-Wind–Venus and Solar-Wind–Mars Interaction .. 8
       1.3.5 Solar-Wind–Comet Interaction 10
       1.3.6 Solar-Wind–Heliosphere Interaction 12
       1.3.7 Collisionless Shocks and Neutral Current Layers ... 12
       1.3.8 Beams and Plasma Clouds 14
       1.3.9 Fusion Plasma 14
   1.4 Classification of the Computational Models 16
       1.4.1 Direct Solution of the Vlasov–Maxwell Equations ... 17
       1.4.2 Water-Bag Methods 20
       1.4.3 Vlasov Hybrid Simulation (VHS) Method 21
       1.4.4 Conventional Particle Models 22
   Summary 23

2. **Particle-Mesh Models** 25
   2.1 Introduction 25
   2.2 Hybrid Quasineutral Models 25
       2.2.1 Electrostatic Model 26
       2.2.2 Ampere Magnetoinductive Model 27
       2.2.3 Particle-Ion–Fluid-Electron Model 29
       2.2.4 Particle-Ion–Fluid-Ion–Fluid-Electron Model 36
       2.2.5 Particle-Ion-Element–Fluid-Electron Model 37
       2.2.6 Gyrokinetic-Ion–Fluid-Ion–Fluid-Electron Models.
             δF Method 37
       2.2.7 Guiding-Center-Ion–Fluid-Electron Model 45
       2.2.8 Particle-Electron–Fluid-Ion Model 48
   2.3 Particle Nonneutral Models 49

        2.3.1   Full Particle Models 50
        2.3.2   Particle-Ion–Guiding-Center-Electron Model 53
        2.3.3   Guiding-Center-Ion–Guiding-Center-Electron
                (Drift-Kinetic) Model 55
        2.3.4   Particle-Electron–Immobile-Ion Model 56
    2.4 Photo-ionization and Charge Exchange Processes 56
        2.4.1   Hybrid Particle-Neutral-Component–Fluid-
                Plasma Models..................................... 56
        2.4.2   Hybrid Particle-Neutral-Component–Kinetic-
                Plasma Models 62
    Summary 64
    Exercise 65

3. **Time Integration of the Particle Motion Equations** 67
    3.1 Introduction 67
    3.2 Explicit Leapfrog Method 71
    3.3 Implicit Method 71
    3.4 Operator Splitting Method 73
        3.4.1   Splitting of the Particle Motion Equations.
                Boris's Scheme 73
        3.4.2   Analytical Time Integration. Buneman's Scheme 74
        3.4.3   Time Integration with $\Omega \Delta t \gg 1$ 75
    3.5 Stability and Accuracy of the Leapfrog Schemes 76
    3.6 Implicit Time Integration. C1 and D1 Class Schemes....... 78
        3.6.1   C1 Class Scheme 78
        3.6.2   D1 Class Scheme 79
    3.7 Runge–Kutta Schemes 80
    3.8 Relativistic Particle Motion Equations 81
    Summary 82
    Exercises 82

4. **Density and Current Assignment. Force Interpolation.
    Conservation Laws**... 83
    4.1 Introduction 83
    4.2 Cloud and Assignment Function Shapes 83
    4.3 NGP, CIC and TSC Weighting 85
        4.3.1   Cloud (S) and Assignment (W) Function Hierarchy... 85
        4.3.2   Weighting in Two- and Three-Dimensional Space 87
    4.4 Force Interpolation 90
    4.5 Mass, Momentum and Energy Conservation 92
        4.5.1   Mass Conservation 92
        4.5.2   Momentum Conservation 94
        4.5.3   Energy Conservation 98
    4.6 Periodic Systems. Multipole Expansion Method 102
    Summary 103

5. **Time Integration of the Field and Electron Pressure Equations** 105
   5.1 Introduction 105
   5.2 Predictor–Corrector Methods 107
       5.2.1 The Upwind Method 108
       5.2.2 The Leapfrog Scheme 109
       5.2.3 Lax–Wendroff Scheme. Explicit Calculation
             of the Electric Field 111
       5.2.4 Implicit Calculation of the Electric Field 114
   5.3 Operator Splitting Methods 119
       5.3.1 Splitting Schemes 119
       5.3.2 Predictor–Corrector/Operator Splitting Scheme 120
   5.4 The Transportive Property 121
   5.5 High-Order Schemes 124
       5.5.1 Multipoint Stencil Schemes 124
       5.5.2 Differential Consequences
             from the Governing Equations 125
       5.5.3 Compact Schemes with Spectral-Like Resolution 125
       5.5.4 Advection and Diffusion Equations 131
       5.5.5 Maxwell's Equations 132
       5.5.6 Filtering of Spurious Oscillations 134
   5.6 Time Integration of the Equations
       for Electromagnetic Potentials 135
   5.7 Time Integration of the Generalized Field Equations 138
   5.8 Time Integration of the Electron Pressure Equation 140
   Summary 140
   Exercises 141

6. **General Loops for Hybrid Codes. Multiscale Methods** 143
   6.1 Introduction 143
   6.2 Examples of the Conventional
       Hybrid Simulation Loops 143
       6.2.1 General Predictor–Corrector Loop 143
       6.2.2 Implicit Time Integration
             of the Electromagnetic Equations 144
       6.2.3 The Moment Method 145
       6.2.4 The Richardson Extrapolation Method 147
   6.3 Multiple-Time-Scale Methods 147
       6.3.1 Electromagnetic Field Subcycling.
             Current Advanced Methods
             and Cyclic Leapfrog Schemes 148
       6.3.2 Light Ion (Electron) Subcycling 151
       6.3.3 Orbit Averaging 154
   6.4 Multiple-Space/Time-Scale Methods 156
       6.4.1 Variational Methods 157

        6.4.2　Adaptive Mesh and Particle Refinement Methods 158
    Summary 163

7. **Particle Loading and Injection. Boundary Conditions** 165
    7.1　Introduction 165
    7.2　Loading the Particles Inside the Computational Domain 165
        7.2.1　Loading Nonuniform Distributions $f_0(v), n_0(x)$ 165
        7.2.2　Loading a Maxwellian Velocity Distribution 166
        7.2.3　Loading a Ring Velocity Distribution 167
        7.2.4　Loading a Shell Velocity Distribution 173
    7.3　Particle Injection at Boundaries 174
        7.3.1　Loading a Maxwellian Velocity Distribution Flux 174
        7.3.2　Loading a Ring Velocity Distribution Flux 176
        7.3.3　Loading a Shell Velocity Distribution Flux 178
    7.4　Charge Exchange Processes 180
    7.5　Boundary Conditions for Particles
         and the Electromagnetic Field 181
        7.5.1　Plasma–Vacuum Interface 181
        7.5.2　Field Radiation and Absorption at the Boundaries.... 182
        7.5.3　Boundary Conditions at the Conducting Wall 185
    Summary 186

Part II. Applications

8. **Collisionless Shock Simulation** 189
    8.1　Introduction 189
    8.2　Collisionless Shocks Without Mass Loading 192
        8.2.1　Quasiperpendicular Shocks 192
        8.2.2　Oblique Shocks 198
        8.2.3　Quasiparallel Shocks 202
    8.3　Collisionless Shocks with Mass Loading by Heavy Ions 208
        8.3.1　Quasiperpendicular Shocks 210
        8.3.2　Oblique Shocks 215
        8.3.3　Quasiparallel Shocks 219
        8.3.4　Pickup Ion Acceleration at Shock Front. Shock Surfing 224
    Summary 236
    Exercises 236

9. **Tangential Discontinuity Simulation** 237
    9.1　Introduction 237
    9.2　Formulation of the Problem and Mathematical Model 238
    9.3　One-Dimensional Structures 239
    9.4　Two-Dimensional Structures 241

   9.4.1 Magnetic Field Oriented Perpendicular
     to the Simulation Plane 241
   9.4.2 Magnetic Field in the Simulation Plane 243
   9.4.3 Analysis of the Waves at the TD
     and the Wave–Particle Cross-Field Transport 243
   9.4.4 Dependence of the Final Thickness of TDs
     on Initial Conditions 246
   9.4.5 Dependence of the TD Width on Anomalous
     Resistivity and Numerical Viscosity................. 247
   9.4.6 The Kelvin–Helmholtz Instability at the TD 248

10. Magnetic Field Reconnection Simulation 255
 10.1 Introduction 255
 10.2 Ion Tearing Instability 256
   10.2.1 Formulation of the Problem and Mathematical Model 257
   10.2.2 Multimode Regime 260
   10.2.3 Single-Mode Regime 262
   10.2.4 Explosive Regime. Ion Acceleration 264
 10.3 Electron Effects on Reconnection 268
   10.3.1 Effects of Electron Inertia
     and Electron Pressure Anisotropy 270
   10.3.2 Effects of Anomalous Resistivity on Reconnection 275
 Summary 280
 Exercises 281

11. Beam Dynamics Simulation 283
 11.1 Introduction 283
 11.2 Cold Beam Dynamics 284
   11.2.1 One-Dimensional Models 285
   11.2.2 Two-Dimensional Models 288
 11.3 Mass Loading of the Supersonic Flow by Heavy Ions 291
   11.3.1 One-Dimensional Models 291
   11.3.2 Two-Dimensional Models 295
 11.4 Finite Size Beam (Plasma Cloud) Dynamics 301
   11.4.1 Generation of Low-Frequency Waves
     by Three-Dimensional and 2.5-Dimensional Beams
     in a Homogeneous Background 301
   11.4.2 Interaction of the 2.5-Dimensional Beam
     with Tangential Discontinuities..................... 302
 Summary 306

12. Interaction of the Solar Wind with Astrophysical Objects 309
 12.1 Introduction 309
 12.2 Interaction of the Solar Wind with Strong Comets 309
   12.2.1 Formulation of the Problem and Mathematical Model 310

          12.2.2 Structure of the Region of Mass Loading
                 by Cometary Ions 315
          12.2.3 Induced Magnetosphere, Bow Wave
                 and Magnetic Barrier 316
   12.3 Interaction of the Solar Wind with Weak Comets
        and Related Objects 320
          12.3.1 Formulation of the Problem and Mathematical Model   320
          12.3.2 Interaction of the Solar Wind with Very Weak Comets 321
          12.3.3 Interaction of the Solar Wind with Weak Comets 329
          12.3.4 Interaction of the Solar Wind with Pluto 329
   12.4 Interaction of the Solar Wind with Venus 333
          12.4.1 Formulation of the Problem and Mathematical Model   333
          12.4.2 Results and Conclusions 334
   12.5 Interaction of the Solar Wind with the Moon 337
          12.5.1 Formulation of the Problem and Mathematical Model   338
          12.5.2 Method of Solution 340
          12.5.3 Results and Conclusions 342
   12.6 Interaction of Neutral Interstellar Atoms with the Heliosphere 344
          12.6.1 Formulation of the Problem and Mathematical Model . 344
          12.6.2 Results and Conclusions 346
   Summary 353
   Exercises 353

13. **Appendix** 355
   13.1 Coordinate Form of Maxwell's Equations
        and the Electron Pressure Equations 355
          13.1.1 Cartesian Coordinates 355
          13.1.2 Cylindrical Coordinates 357
          13.1.3 Spherical Coordinates 358
   13.2 Solving One-Dimensional Difference Equations 361
          13.2.1 Three-Point Difference Equation
                 with Nonperiodic Boundary Conditions:
                 Forward-Elimination–Backward-Substitution Method . 361
          13.2.2 Three-Point Difference Equation
                 with Periodic Boundary Conditions:
                 Forward-Elimination–Backward-Substitution Method . 362
          13.2.3 Five-Point Difference Equation:
                 Forward-Elimination–Backward-Substitution Method . 363

14. **Solutions** 365

References 380

Index 401

Part I

Computational Models and Numerical Methods

1. Physical Systems and Computational Models

1.1 Introduction

One of the characteristic properties of current scientific research is the use of strong mathematical methods – the intensive use of mathematical modeling and computer simulation not only in physics, mechanics and industry, but in the other "nonphysical" sciences. What is meant by mathematical modeling? In principle, it means the determination of a behavior and the characteristics of the considered phenomenon, process or state by means of the solution of a set of equations – the mathematical model. It is important to design a discrete numerical model which may describe the characteristic properties of the considered phenomenon, while at the same time this model must be capable of producing meaningful research results.

1.2 The Basic Steps of Computational Experiments

Let us determine the main steps in computer experiments and compare the computational experiment with the physical experiment. First, we describe the mathematical model by analyzing the object of research. In physical experiments, this step corresponds to an analysis and choice of a scheme for an experiment – the choice of the elements of its construction and the device itself. Then for chosen differential or integral operators we design a numerical scheme and study its numerical stability. In the real experiment, at this step we design and prepare the experimental installation and provide time and effort for debugging. As a result we receive a tool for research (effective program or device). The mathematical model, computer plus code (program) represent a real device for an investigation of the physical phenomenon. The computer experiment is very important in the field of physics, where we have a gap between theory and experimental results.

The method of numerical modeling or computational simulation has a lot of applications in physics, chemistry, biology, medicine, economics (e.g., financial transport), etc., and in device design processes. Numerical study is very important when laboratory modeling is impossible, for example, in space physics and astrophysics investigations, or in the design of submicron electron devices. Numerical modeling may also be considered as a "theoretical" experiment which may result in the further development of theory.

Computational experiments also play a very important part in industrial applications, namely the triad *Mathematical Model (Computational Experiment)–Expert System–Automatic Design System*, which is the rational basis that permits faster progress for the design of an industrial installation and development a new technology (see, e.g., [37]). Computational models can help equipment manufacturers and process engineers understand the complex relationships among device parameters (e.g., inductively coupled plasma reactors), plasma parameters (e.g., reactor geometry, radio-frequency power, gas pressure and gas composition) and process performance (e.g., etch rate, anisotropy, uniformity, selectivity and damage) [112].

The use of methods of mathematical modeling is particularly important in the problems of mathematical physics, space and astrophysics, plasma physics and mechanics (hydrodynamics and gas dynamics). The starting point of all computational experiments is some physical phenomenon. The object of computational experiments is to obtain physically useful results that provide insight into the physical phenomenon under study. Between these two points we may identify a number of distinct design steps.

The basic steps in computational experiments are the following:

$$\begin{bmatrix} Physical & Phenomenon \\ Mathematical & Model \\ Discrete & Algebraic & Approximation \\ Numerical & Algorithm \\ Computer & Program \\ Computer & Experiment \end{bmatrix}$$

Each step introduces a set of constraints. Invariably the mathematical formulation is only an approximated description of the physical phenomenon.

The computational scientist cannot just take equations at face value, in spite of the fact that the development of the mathematical description of physical phenomena lies in the realm of the theoretician. He has to know what simplifying assumptions are made in order to identify the regime of validity of the equations and hence his simulation model.

More severe restrictions arise at the step at which the discrete algebraic approximation is designed, where the differential equations of the mathematical model are replaced by algebraic approximations in order to allow numerical solution using the available computers. At this step we have to solve the problems which are concerned with the consequences of a finite time step, discrete spatial meshes and, for particle models, the limited number of particles.

The onus is always on the numerical experimenter to demonstrate that the results from the simulation model are physically meaningful – we shall see in later chapters that numerical errors can have disastrous effects.

Impinging on the choice of discretization is the question of the numerical algorithm. Discretization replaces continuous variables of the mathematical model with arrays of values and differential equations with algebraic equations. Unless the hundreds (and sometimes thousands) of thousands of algebraic equations arising can be rapidly solved on the computer, the proposed simulation becomes impracticable. Generally, kinetic/hybrid simulations involve solving Maxwell's equations (or modified electromagnetic equations) based on self-consistent source terms (usually density [charge] and current), which in turn are generated from a distribution function or a particle distribution that represents the plasma subject to the electromagnetic fields. Usually, this process involves advancing the particles in time with a small time step, Δt, to collect the source terms that are used to solve for the fields. Once the new fields have been obtained, the particles can be moved again to obtain new source terms, and the process repeated over many time steps to follow the evolution over the desired time interval. This typical time-looping process, which was used widely by the simulation community for more than 30 years, is illustrated in the dashed region of Fig. 1.1. Before beginning to run the simulation, there are a number of basic decisions about the calculation which must be made. The first two of these decisions are (1) determining

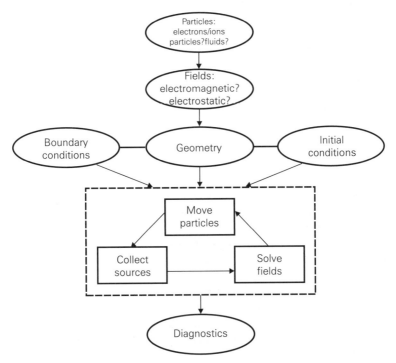

Fig. 1.1. Basic steps in kinetic particle/hybrid simulations (from [569])

which plasma or neutral species will be represented by particles or by markers (guiding centers) (the rest by fluid equations) and (2) deciding on which subset of Maxwell's equations to solve. Answering these questions determines what type of simulation code should be used. Before the actual simulation can proceed, one must also decide on the geometry of calculation (step 3) and devise initial conditions to start the problem (step 4). In addition, one must choose boundary conditions to be obeyed by the particles and the fields (step 5). These decision processes are depicted at the top of Fig. 1.1. Finally, after the simulation is finished, one analyzes the results through appropriate diagnostics (step 6), as indicated at the bottom of Fig. 1.1.

1.3 Classification of the Plasma Systems. Space and Time Scales

Plasma is the fourth state of matter, consisting of electrons, ions and neutral atoms, usually at temperatures above 10^4 K. The Sun and solar wind, the magnetosphere of planets, the induced magnetosphere of nonmagnetic planets, the coma of comets, the ionospheres of planets and comets, Van-Allen belts, the local interstellar medium (LISM), stars, etc., are all plasmas. Indeed, plasma makes up much of the known matter in the universe. The plasma phenomena play very important roles in active experiments in space: the releases of clouds in the ionosphere and in the magnetosphere; the modification of the ionosphere by electromagnetic heating; and the generation of waves by a tether in the ionosphere. Other very important objects for plasma research are the magnetically and inertially contained fusion plasmas, and other laboratory plasma devices.

1.3.1 Solar Wind

The sun emits a highly conducting plasma at supersonic speeds of about 500 km/s (sometimes the speed of the solar wind may exceed 1000 km/s). The solar wind consists mainly of electrons and protons, with an admixture of heavy ions and atoms. The local composition of the solar wind depends on the distance from the sun. Near the Earth the typical electron density, electron temperature and the interplanetary magnetic field are $n_e \approx 5\,\text{cm}^{-3}$, $T_e \approx 10^5$ K, and $B \approx 5\,\text{nT}$, respectively. At distances more than 100 AU the composition of the solar wind is determined mostly by its interaction with neutral and ionized components from the LISM. A synergy of data analysis, analytic theory and numerical simulations demonstrates that the solar wind may be considered as a turbulent magnetofluid (see, e.g., [199]). In the inner heliosphere and in undisturbed flows, the interplanetary fluctuations have preserved strong signatures of coronal processes, while in the outer heliosphere and near regions of strong shear in velocity, the interplanetary medium appears to have been stirred with the concomitant in situ generation of turbulence.

There are three types of interaction of the supersonic solar wind with planets known: a strong interaction, with the formation of an extended magnetosphere (Earth, Jupiter, etc.); a weak (Moon-like) interaction; and an intermediate interaction, when the planets or comets have a significant ionosphere and a weak magnetic field themselves (Venus, Mars, Pluto, etc., and comets).

1.3.2 Solar-Wind–Earth Interaction

As a result of the interaction of the solar wind with a strong magnetic field of the Earth's magnetosphere, a complicated structure is formed (Fig. 1.2). In front of the magnetopause – the interface which separates the solar wind plasma and magnetosphere – a strong bow shock is formed. The shocked solar wind plasma in the magnetosheath cannot easy penetrate the magnetopause, and it is mostly deflected around it. The kinetic pressure of the solar wind distorts the outer part of the geomagnetic dipole field. At the day side the solar wind compress the field, while at the night side the geomagnetic field forms a long magnetotail. Direct observation of the plasma and electromagnetic field shows that different regions of the magnetosphere are separated by different boundary layers (Fig. 1.2). A boundary layer is a region adjacent to a boundary that contains plasmas from both sides. For example, the magnetopause's boundary layer contains both magnetosheath and magnetospheric plasmas. This boundary layer has been extensively studied by spacecraft such as ISEE, INTERBALL and EQUATOR-S. In the boundary layer at the tail-ward flanks Kelvin–Helmholtz instability may play an im-

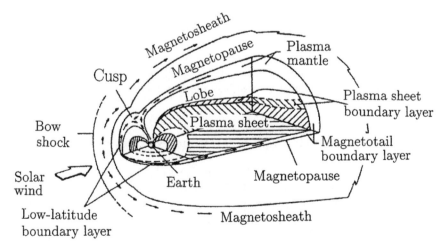

Fig. 1.2. Schematic of the magnetosphere including boundary layers (from [141])

portant role in energy transport across the magnetopause. The geomagnetic tail usually keeps a large amount of energy, and it may cause magnetic storms. The magnetic field reconnection plays an important role in the reconstruction of the tail topology and in the acceleration of particles.

1.3.3 Solar-Wind–Moon Interaction

Measurements of the structure of the interplanetary magnetic field and plasma in the vicinity of the Moon, taken by Explorer 35 [389, 489], have established that in the interaction of the solar wind with the Moon no bow shock wave is formed and there is a cavity free of plasma on the night side. Figure 1.3 shows a schematic of the solar wind flow around the Moon. The magnetic field perturbations in the lunar wake include: (a) the umbrella increase in the magnetic field in the core region of the lunar wake, (b) the penumbral decrease in the magnetic field in the region around the umbrella increase, and (c) the penumbral increase, an additional small increase in the field magnitude sometimes observed outside the penumbral decrease. A joint analysis of plasma and magnetic field data found that the penumbral increases in the magnetic field were correlated with small increases in the plasma density and possibly a small ($< 3°$) outward deflection of the plasma flow away from the wake.

According to [492], the Moon has a conducting core ($\sigma = 10^{-3} \, \Omega^{-1} \, \mathrm{m}^{-1}$) surrounded by a poorly conducting layer ($\sigma = 10^{-7} \, \Omega^{-1} \, \mathrm{m}^{-1}$). The existence of conducting layer may strongly affect the configuration of plasma flow and the magnetic field near the Moon and in the wake. Further investigation of the solar wind flow around the Moon was continued partially by the spacecrafts WIND and Lunar Prospector. This observation data allow us to further study the plasma and field in the day side environment and the dynamics processes in the plasma wake. It was shown that kinetic instabilities play an important role in wake dynamics [153]. In particular, electrostatic instability results in ion replenishment in the wake [152], whereas outside of the wake the electron wings and enhanced plasma wave activity in many modes were observed [24, 151]. The kinetic instabilities which are associated with a non-uniform distribution of ions may result in plasma density and electromagnetic perturbations in the wake.

1.3.4 Solar-Wind–Venus and Solar-Wind–Mars Interaction

The physics of the interaction of the solar wind with Venus and Mars is very different from the solar wind–Earth interaction. Even before the first space explorations of Venus and Mars provided by the VENERA and Pioneer Venus Orbiter (PVO) missions to Venus, and the Viking, MARS, and PHOBOS missions to Mars, it was supposed that the weakness of their planetary magnetic fields would have consequences for atmospheric escape. It

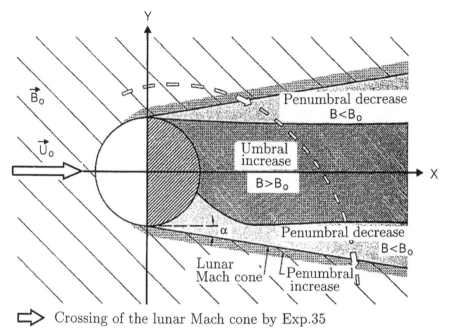

⇨ Crossing of the lunar Mach cone by Exp.35

Fig. 1.3. Schematic of the solar wind flow around the Moon (from [554])

was recognized [371, 545] that planetary ions produced in the rarefied upper atmosphere, where the solar wind plasma penetrated, could be swept away since they would become coupled to that plasma as will be described below. Figure 1.4 shows a schematic of the space environment of a weakly ionized planet. It demonstrates the close-in bow shock, the magnetosheath in which the shocked solar wind plasma flows around the ionopause and the upper atmosphere extending above the ionopause. Ions produced in the solar wind plasma, by a process such as photo-ionization, are picked up by the ambient plasma, which carries an embedded magnetic field. Because the ions gyrate around the field, they may escape into the wake, or they may re-enter the atmosphere. The direct observations the plasma environment near Mars and Venus created a lot of physical problems which must be resolved in the future. The pickup heavy ions may provide the mass loading of the solar wind and therefore change the position of the bow shock.

On the basis of magnetic field and plasma measurements in the vicinity of Mars [134, 205, 533] and Venus [135, 204, 533], the presence of collisionless bow shock waves near the planets, a weak magnetic field near Mars of $\approx 50\,\gamma$, and no appreciable field near Venus were established. The solar wind–Mars interaction is primarily atmospheric in nature; pronounced draping of the interplanetary magnetic field over the ionosphere in the north polar region, superposed with the magnetic field from crustal sources, gives rise to asym-

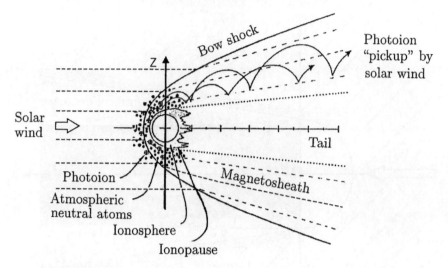

Fig. 1.4. Space environment of a weakly ionized planet showing the close-in bow shock, the magnetosheath in which the shocked solar wind plasma flows around the ionopause, and the upper atmosphere extending above the ionopause (from [354])

metries and the formation of localized Earth-like "cusps" and/or magnetic reconnection regions where plasmas from the martian ionosphere, its comet-like tail and the solar wind can intermix (see recent observations made near Mars [2, 101]). It is reasonable to suppose that even a weak magnetic field may control the mechanism of solar wind penetration into the atmosphere of Mars. However, this has not yet been resolved. The dynamics of the heavy pickup ions is very complicated. They may be strongly accelerated by the solar wind or by different gradients of plasma and field parameters near the magnetic barrier. Pickup ions may be accelerated inside the magnetotails of Venus and Mars and particularly inside the plasma sheet (a frictional pre-acceleration, a field-aligned electric field acceleration and a cross-tail electric field acceleration [276]).

1.3.5 Solar-Wind–Comet Interaction

In the case of a planet, the solar wind encounters a localized obstruction (the magnetosphere or the ionosphere of the planet), but in the case of a comet the influence of the obstruction is felt at much greater distances. This is because the gravitational field of the comet's nucleus is negligible, so that the neutral gas which evaporates from the surface of the nucleus under the influence of solar radiation expands freely into interplanetary space. At a distance of 1 AU from the Sun, i.e., at the Earth's orbit, the rate of expansion is of the order of 1 km/s, and therefore the neutral gas manages to escape to millions of kilometers from the comet's nucleus before it is ionized under the influence of

1.3 Classification of the Plasma Systems. Space and Time Scales

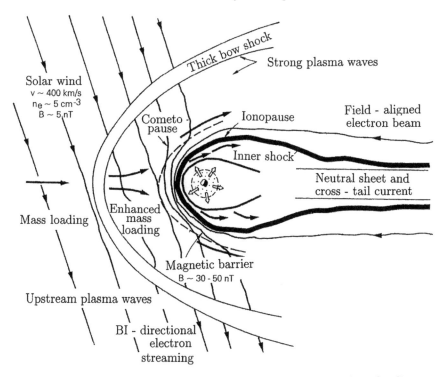

Fig. 1.5. Schematic of the solar wind flow around a comet (from[370])

the solar radiation and solar wind electrons or owing to charge exchange (the characteristic times of all these processes are of the order of 10^6 s) (Fig. 1.5).

The structure of the detached shock wave in front of comets [416] and nonmagnetic planets moving in a supersonic flow of solar wind plasma is of interest not only in view of the probes launched to Venus, Mars, Halley's comet and the Giacobini–Zinner comet, but also because of the basic differences between the ways the solar wind interacts with nonmagnetic planets and comets and with planets which have a strong magnetic field.

As is well known (see, e.g., [246]), depending on the rate of evaporation of cometary material, the physics of the gas dynamics interaction of the solar wind with a comet is analogous to the interaction of the solar wind with the Moon [314, 496, 577] (very low evaporation rate) and Venus [38, 318, 497] (high evaporation rate) or to the interaction of the interstellar gas with the solar wind [26, 547] (very high evaporation rate). The physics of the cometary atmosphere is very complicated due to the instability processes in the region of mass loading and the shock wave, and the complicated cometary tail dynamics and dust grain dynamics. In case of small comets, for example, Shoemaker–Levy 9 and Wirtanen, the kinetic effects due to the gyroradius of heavy ions and dust grains begin to play a key role. As in the case of nonmagnetic planets, the charge exchange processes may cause extremely ultraviolet (EUV)/X-ray sources in the cometary environment. Some of these processes will be consider in Part 2.

1.3.6 Solar-Wind–Heliosphere Interaction

At distances greater than 100 AU, the dynamics of the solar wind is determined mostly by the interaction with helium and other heavy neutral and ionized components coming from the LISM (Fig. 1.6). The heliopause separates the plasma from the LISM and the solar wind plasma. Figure 1.6 shows typical two shock model of the heliosphere. The bow shock is formed in the LISM, and the termination shock is formed in the solar wind.

The global interaction of the LISM with the solar wind is a fundamental problem of heliospheric physics. It requires the solution of a highly nonlinear coupled set of integro-MHD-Boltzmann equations which describe the dynamics of the ionized solar wind and the LISM together with interstellar and heliospheric neutral atoms. To the first order, the plasma and neutral atoms are coupled by resonant charge exchange, although other coupling processes do occur.

The characteristic scale of the ionized components is usually determined by the typical ion gyroradius, which is much less than the characteristic global heliospheric scales of interest. By contrast, the mean free path of neutral particles is comparable to characteristic heliospheric scales such as the distances separating the bow shock and heliopause or the heliopause and termination shock, and even the radial extent of the supersonic solar wind. Consequently, the Knudsen number, $Kn = \lambda/L$ (λ is the mean free path of neutral particles and L is a characteristic heliospheric scale), which is a measure of the distribution relaxation distance, satisfies $Kn \approx 1$. Thus, it is difficult to assume that the neutral H distribution can relax to a Maxwellian distribution; one needs ideally to solve a Boltzmann equation for the neutral component in which charge exchange and photo-ionization processes are included. The study of the interaction of the solar wind with the LISM is very important not only to understand the physics of the outer heliosphere, but also for prediction of the near-Earth orbit parameters of the solar wind and material from the LISM.

1.3.7 Collisionless Shocks and Neutral Current Layers

The dynamics of the solar wind, magnetosphere and atmosphere of planets are determined by plasma processes inside the different types of boundaries and discontinuities. Namely, inside these discontinuities – collisionless shocks, regions of mass loading by pickup ions, and current and neutral sheets – a strong transformation of the energy of the electromagnetic field and particles

1.3 Classification of the Plasma Systems. Space and Time Scales

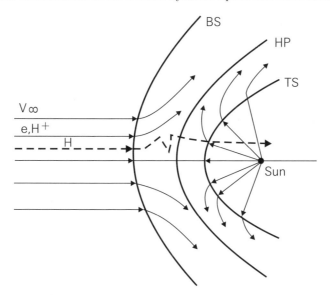

Fig. 1.6. Schematic of the solar wind interaction with LISM. BS: bow shock; HP: heliopause; TS: termination (internal) shock

takes place. Indeed the energy of the supersonic flow transforms into energy for electromagnetic waves, heating and acceleration of particles of the solar wind and pickup ions at the front of bow shocks, interplanetary shocks and the termination shock. The Kelvin–Helmholtz (K–H) instability and magnetic field reconnection may result in a strong transfer of solar wind energy and momentum across the day-side magnetopause of the Earth (and planets), creating a solar wind–Earth energy connection. The dynamical processes (magnetic field reconnection) inside the neutral sheet of the magnetotail may result in particle acceleration and may cause substorms at the Earth. The turbulence in the region of mass loading near planets and in the solar wind may strongly accelerate the pickup ions.

For this reason we shall consider the kinetic processes inside the collisionless shocks, current neutral layers and finite size beam (plasma cloud) dynamics in a background plasma. The study of heating and acceleration of particles by collisionless shocks (see Fig. 8.1) is very important for understanding the processes near planetary and cometary bow shocks. It is very important to estimate the efficiency of pickup ion and anomalous cosmic ray acceleration by the interplanetary shocks and the termination shock. The physics of shocks may also be applied for solar flares, supernovae shocks, active experiments in space and fusion research. The processes inside the shock transition layer include a wide spectrum of plasma instabilities, the interaction of particles with waves and nonstationary fine shock structures. In general, the investigation of collisionless shocks needs a multiscale approach.

14   1. Physical Systems and Computational Models

The study of magnetic field reconnection processes inside the neutral current layers is very important for an understanding of magnetotail and magnetopause dynamics, solar flares, the dynamics of the solar wind sector structures and the heliopause. The explosive regime of the magnetic field reconnection may result in a strong acceleration of particles [514, 594]. Very often the magnetic field reconnection interacts with the K–H instability, for example, at the lobe-tail flank of a magnetopause, resulting in nonlinear acceleration of both processes. Indeed, the flow accelerated by reconnection may provide the velocity shear which causes the K–H instability on the day-side current layer [469]. On the other hand, the K–H instability could modulate the current sheet, thus creating separated and pinched regions, and thereby provide the conditions for patchy reconnection at the day-side magnetopause [286]. The magnetic field reconnection processes also include the different electron and ion scales.

1.3.8 Beams and Plasma Clouds

The three-dimensional beam (or plasma cloud) dynamics in homogeneous background plasma is very important for studying the whistler and Alfvén wave generation using a tether in the ionosphere [94] and the motion of ionized clouds in the ionosphere [43] and the magnetotail. It was applied to the study of heavy ion release by the active magnetospheric particle tracer explorers (AMPTE) and the plasma environment near weak comets and asteroids. The interaction of plasma clouds from the solar wind with a magnetopause gives rise to emission of whistler waves, Alfvén waves and helicons along the geomagnetic field. It may results in a reformulation of the Chapman–Ferraro problem for the magnetosphere [95]. Pickup ions from the LISM and planetary and cometary heavy ions mass-load the solar wind accompanying the generation of Alfvén and magnetosonic waves. The generation of electromagnetic waves by beams involves again a wide spectrum of spatial and temporal scales.

As we can see from our very short discussion, the solar wind dynamics and its interaction with astrophysical objects – the LISM and the magnetospheres and atmospheres of planets – involve a cascade of coupled spatial and temporal multiscale phenomena: relatively slow-changing MHD flows and waves in the solar wind, in the LISM (MHD turbulence, magnetic helicity and vortex sheets) and inside the magnetospheres and atmospheres of planets and comets; wave–particle interactions – acceleration of ions, positrons, electrons and dust grains, generation of electromagnetic waves by beams, turbulence and anomalous transport; charge exchange and photo-ionization processes.

1.3.9 Fusion Plasma

The phenomena in laboratory plasmas are as complicated as in astrophysical plasmas. The hierarchy of structures in fusion plasmas exists in both multiple

1.3 Classification of the Plasma Systems. Space and Time Scales

time scales and multiple spatial scales. In terms of time scales, the duration or maintained time of a typical tokamak [377] fusion plasma is of the order of 1–10 s, whereas the electron cyclotron period and the electron plasma period are of the order of 1 ps. A lot of various other time scales are between these two extremes. The processes may include high-frequency electromagnetic waves (light), the electron plasma frequency and the ion plasma frequency. In a magnetized plasma the frequency of ion oscillations can be much higher, in fact, on the order of ω_{pi}, the characteristic ion plasma frequency, if the wave propagates nearly perpendicular to the magnetic field. This is because electrons are magnetized and have to follow the magnetic field line to respond to the charge separation. Thus, the ions determine the characteristic frequency.

In case of ion gyration we have a longer period, $T = 2\pi/\Omega_{\mathrm{i}}$. Electrostatic ion cyclotron waves and electromagnetic ion waves are two examples of phenomena in this regime (see, e.g., [372]). The MHD phenomena are even slower than the ion gyration. In this regime we can describe global perturbations of plasma configurations. An inhomogeneous magnetized plasma has a set of high-frequency and smaller space scale instabilities. The drift wave instability can be caused by the presence of the plasma density gradient. Although the frequency of the wave is much smaller than Ω_{i} and it might be possible for coarse-grain particle motion to occur, the kinetic effects remain essential, while the slow period of the wave and slow evolution make the wave easily couple with the transport processes that are generally slower than the drift period [372].

Collisions may result in fundamental changes in (ideal) MHD. In ideal MHD the fluid particles are connected to the magnetic field line [288], whereas with resistivity the fluid slips away from the field line. This property may result in the possibility of the magnetic field reconnection [505], by the tearing instability [165]. Such resistive MHD processes are much faster than the collisional phenomena such as the collisional transport process [234].

It is impossible to resolve multiple time and space scales of coupled phenomena in astrophysical or laboratory plasmas using a single (particle or MHD) code. However, in some cases we can use implicit codes, which allow us to investigate the interesting scales by suppression of all other time and space scales. An example of the hierarchy of time and space scales of plasma phenomena and the corresponding numerical models is illustrated in Table 1.1, which represents a modification of Table 1.2 in [509]. However, in general, we have to design so-called combined or general hybrid codes which use a multilevel description for different sorts of particles (Boltzmann/Vlasov/gyrokinetic/guiding center/hydrodynamics) to keep a set of time and space scales in the simulation. The use of inhomogeneous or adaptive grids give us the additional possibility of keeping the space multiple scales in the simulation. Examples of such approaches will be consider in Sect. 6.4. To keep multiple time scales, we can also use the orbit averaging method, light ion (electron) subcycling method, and the field subcycling method (see

Table 1.1. Hierarchy of time and space scales, and corresponding plasma processes and numerical codes

Plasma phenomena	Cyclotron plasma wave	Lower hybrid wave	MHD/kinetic Alfvén wave	Drift wave	Resistive MHD transport confinement collisions
Time scales	$(N\Omega_e)^{-1}$ $(N\Omega_i)^{-1}$	$(\Omega_i\Omega_e)^{-1}$ ω_{pi}^{-1}	$(kv_A)^{-1}$ $\geq \Omega_i^{-1}$	$\dfrac{eB}{cT_e n K_y}$	$(\omega_A^\alpha \nu^{1-\alpha})^{-1}$
Space scales	$\lambda_D, c/\omega_{pi}$	$\sqrt{\varrho_{ce}\varrho_{ci}}$		ϱ_{ci}	
Simulation methods (codes)	Explicit particle codes	Ion-scale particle quasi-neutral codes	Explicit MHD codes Conventional hybrid codes Guiding center codes Gyrokinetic codes Implicit particle codes		Implicit resistive MHD codes Transport codes

Sect. 6.3). Another way to keep as many scales as possible is to improve the accuracy of the finite-difference approximation to Maxwell's equations. There are several promising ways: spectral methods (see, e.g., [88] and references therein), compact (three-point) schemes (see, e.g., [528] and references therein), and wavelet methods (see, e.g., [538] and references therein). Some of these methods will be considered in Sect. 5.7.

In this book we shall consider mostly the collisionless systems, which are described well by the Vlasov–Maxwell equations. However, in some models we shall take into account the collision processes between ions and electrons (shocks and neutral current layers) by using the anomalous resistivity (see Sect. 2.2.3). In the case of collisions between neutrals and charged particles by means of charge exchange processes (dynamics of atoms inside the heliosphere), we use the Boltzmann simulation approach (see Sects. 2.4 and 12.6).

1.4 Classification of the Computational Models

Generally, depending on the particular problem, kinetic/hybrid simulations involve solving the Vlasov equation (or a discrete numerical model or other modifications of the Vlasov equation), or the Boltzmann equation for some species of particles, an MHD equation for the rest of the particle species and Maxwell's equations (or modified electromagnetic equations) based on

1.4 Classification of the Computational Models

self-consistent source terms (usually density/charge and current), which in turn are generated from distribution functions or particle distributions that represent the plasma subject to the electromagnetic fields.

The Vlasov–Maxwell set of equations for a collision-free plasma are as follows: An equation for distribution function of particle species α and f_α,

$$\left(\frac{\partial}{\partial t} + \boldsymbol{v} \cdot \frac{\partial}{\partial \boldsymbol{x}} + \frac{q_\alpha}{M_\alpha}(\boldsymbol{E} + \boldsymbol{v} \times \boldsymbol{B}) \cdot \frac{\partial}{\partial \boldsymbol{v}}\right) f_\alpha = 0. \tag{1.1}$$

The Maxwell equations are Faraday's law (induction equation),

$$-\frac{1}{c}\frac{\partial \boldsymbol{B}}{\partial t} = \nabla \times \boldsymbol{E}, \tag{1.2}$$

Ampere's law,

$$\nabla \times \boldsymbol{B} = \frac{4\pi}{c}\boldsymbol{J} + \frac{1}{c}\frac{\partial \boldsymbol{E}}{\partial t}, \tag{1.3}$$

and the equations for the divergency of the electric and magnetic fields,

$$\nabla \cdot \boldsymbol{E} = 4\pi\varrho \quad \text{and} \quad \nabla \cdot \boldsymbol{B} = 0. \tag{1.4}$$

In the case of collisions between atoms and charged particles, we have to use the Boltzmann equation for the distribution function (see Sect. 2.4). If we can neglect the kinetic effects, we can use the MHD equations for the ionized components. In general case we have to decide what kind of description should be used for different sorts of particles. The direct numerical solution of the Vlasov–Maxwell or Boltzmann–Maxwell equations with partial derivatives meets difficulties even if advance supercomputer systems are available. Indeed, the structure of the Vlasov equation is hyperbolic, and we can use the numerical methods for fluid dynamics. Since the Maxwell equations are linear one can use the standard algorithm to find their solution. However, very often we cannot receive an appropriate solution to the fully three-dimensional kinetic phenomenon even if modern supercomputer systems or massive parallel computers are available. An alternative way to derive some modification of the Vlasov equation is to use the particles, markers, guiding centers, etc. In this way the reduced Vlasov–Maxwell equations usually have a complex nonlinear structure. Therefore, in this case we have to develop appropriate numerical methods and again investigate the effects of finite time step, space grid, and "shot noise" effects.

1.4.1 Direct Solution of the Vlasov–Maxwell Equations

Most of the numerical simulations of the Vlasov equations have used a Lagrangian representation. The trajectories of a large number of particles are calculated, and these particle distributions are used to calculate charge and current densities. In direct Vlasov methods we consider the Eulerian representation of the Vlasov equations in phase space.

1.4.1.1 Fourier–Fourier and Fourier–Hermite Transform Methods.

The transform method is motivated partially by an attempt to eliminate partial differentiation in favor of algebraic operations in the Vlasov–Maxwell equations. Both space derivatives can be eliminated by a Fourier transform in x. Two different transformations in v have been used: the Fourier transform in v and the Hermite (Gram–Charlier) transform in v.

Let us consider a periodic plasma of length DX, with a uniform background density of ions, n_0, which is described by the Vlasov–Poisson system in one dimension [13, 14, 273].

The problem considered here is the initial-value problem for the Vlasov–Maxwell equations, i.e., given $f(x, v, 0)$, calculate $E(x, t)$ and $f(x, v, t)$ for $t > 0$. If $f(x, v, 0)$ is periodic in x initially, it will remain so indefinitely, and also E will be spatially periodic. The most difficult obstacle, numerically, is that it is in the nature of $f(x, v, t)$ to develop steep derivatives in the v-direction as time increases [21], and thus, the last term in (1.1) becomes progressively more difficult to compute accurately. Here we consider two approaches in order to solve the Vlasov–Maxwell equations, namely, the Fourier–Fourier and Fourier–Hermite transforms.

For the investigation of wave propagation in infinite and finite plasmas, it is convenient to first decompose the Vlasov equation (1.1) into a Fourier series in x [13]. As a second step we use the Fourier transform in velocity space ($v \to y$); thus we obtain the equation which contains the derivatives in time and y-space. The final equation may be solved by the method of characteristics.

In the Fourier–Hermite transform method we represent the electron distribution function as a Fourier series in x and then as a Gram–Charlier series in v. In this representation, the Vlasov equation appears as a matrix of ordinary equations. These equations may be solved by standard Runge–Kutta method.

The above codes were designed for simulation of the nonlinear damping of stable distributions, two-stream instabilities, nonlinear Landau damping, a strongly unstable counterstreaming electron plasma, etc. It is easily seen that transformed Vlasov–Maxwell equations do not form a closed system in either the Fourier index n or the Hermite index m. The nonlinear term arising from convective term in velocity space couples each Fourier mode to an infinite number of other modes, and the convective term couples each Hermite index to the one above and below. The truncation error may be estimated by comparison of the runs for different boundary indices $N(M)$.

For more detail regarding the conservation law, the boundary conditions, etc., the reader is referred to [13].

1.4.1.2 Fast Fourier–Eulerian Splitting Scheme.

According to [188, 249] the first equation of the Vlasov–Poisson system may be split in two parts; one is a free streaming equation between the impulsions,

1.4 Classification of the Computational Models

$$\frac{\partial f}{\partial t} + v\frac{\partial f}{\partial x} = 0, \tag{1.5}$$

and the other is

$$\frac{\partial f(x,v,t)}{\partial t} + E^*\frac{\partial f}{\partial v} = 0. \tag{1.6}$$

The continuous self-consistent electric field $E(x,t)$ is replaced by a succession of Dirac pulses, E^*:

$$E^*(x,t) = \sum_{p=1}^{\infty} E\left(x, (2p+1)\Delta T/2\right) \times \delta\left(t - (2p+1)\Delta T/2\right) \tag{1.7}$$

with the moments of time $t_n = n\Delta T$ and $t_{n+1/2} = (n+1/2)\Delta T$. During the impulsion time, the effect of the term $v\partial f/\partial x$ can be neglected. This equation is referred to here as the *sudden approximation*. To solve the system (1.5–1.7) for a full time step, three fractional numerical steps are needed:

(1) Solution of (1.5) for a half step. We use the forward and inverse Fourier transforms. The time integration in k-space was performed analytically.

(2) Calculation of the field (Poisson equation) and solution of (1.6) (the impulsion time). We use the forward and inverse Fourier transforms. The time integration in k-space was performed analytically.

(3) Solution of (1.5) again for another half time step with the same manner as in (1).

An extension of this scheme to the relativistic electromagnetic case may be founded in [35] and [188].

1.4.1.3 Finite-Difference Methods. In this paragraph we shall discuss the finite-difference methods for the solution of the Vlasov equation. The Vlasov equation is an hyperbolic equation; hence for the finite-difference methods perhaps the most straightforward approach is to use a $2N$-dimensional difference scheme with a rectangular mesh in phase space. Such a method was adopted for the two-dimensional problem in [270]. This type of calculation is simple but lengthy, since several thousand mesh points are required, and truncation errors will lead to some numerical diffusion even if a fourth-order-accurate difference scheme is employed [447]. In [84] a two-step Lax–Wendroff method for one-dimensional and two-dimensional problems was used. However, we can use other methods which were developed for fluid dynamics problems (see, e.g., [157]). The most simple finite-difference schemes are well known as "upstream–downstream" schemes, three-time-level leapfrog schemes, two-time-level schemes centered in space, and two-step Lax–Wendroff methods [84]. These standard techniques work well in one-dimensional and two-dimensional simulations. However, in three-dimensional simulations even the best Vlasov code is outstripped by particle simulation techniques, and we should probably use the higher order methods recently developed for fluid dynamics. The compact (three-point) schemes (see, e.g., [528]), spectral methods (see, e.g., [88]), and wavelet methods ([538], and references therein) and flux-corrected transport method (see, e.g., [58]) may be useful for time integration of the Vlasov equation as well as of Faraday's law (see Chap. 5). The finite-difference method may also be combined with the transform method. A multistep technique for the numerical solution of a two-dimensional Vlasov equation, which uses the Hermite transform in velocity space and the finite-difference approximation in coordinate space, was developed in [487]. The transformed equations are integrated at each time step first in the x-direction and then in the y-direction, and vice versa. Despite some successful one-dimensional and two-dimensional applications, the finite-difference methods need further development for three-dimensional problems. However, recent parallel computer systems will probably allow us to make fully three-dimensional direct Vlasov simulations.

1.4.2 Water-Bag Methods

The Vlasov equation describes the motion of an ideal incompressible fluid in a $2N$-dimensional phase space, where q and p are N-dimensional canonical coordinates and momentum vectors. The configuration of this phase fluid is represented by a distribution function $f(q,p)$ which evolvs with time according to the equation of particle motion. The force is obtained by solving a field equation and may be determined partly by external sources and partly by internal sources that are functionals of f itself, so that one can talk of a "self-interacting" phase fluid [47, 48].

In the one-dimensional case the fluid picture is particularly clear, since a continuous distribution $f(x,v)$ can be represented by contours in the two-dimensional (x,v) phase plane. Each contour moves with the fluid and the topology is preserved. The area enclosed between any two contours is invariant of the motion. Each point (x_i, v_i) on a contour obeys the equation of the particle motion. In this method, each contour C_j is approximated by a set of Lagrangian points $P_{j,i}$ linked to each other by straight-line segments. Adjacent points must remain sufficiently close to one another in order for the sum of the elementary trapezoidal areas to accurately approximate the invariant area within the contour.

To calculate the field equation, we use a fixed Eulerian grid and divide the phase plane into vertical strips $x_m \leq x < x_{m+1}$. To determine the electromagnetic field from the Maxwell equations, it is necessary to calculate the charge and current in each Eulerian strip m. If the area encompasses more than one Eulerian strip, then partial areas are allocated to each of the strips. Without any additional checks, this method automatically correctly processes any arbitrary contour configuration, including curves with multiple values of v for a given x, and it is exact for polygonal contours.

This method contains a weak computational instability that can be stabilized by periodically synchronizing the contours associated with the odd and even time steps [48]. During the simulation the contour topology may change and become very complicated, so for optimal approximation of contours by particles it is very useful to periodically determine the relocation of the simulation particles [48]. The advantages of the water-bag method are considerable information per unit CPU-time/memory and conceptual simplicity. Its disadvantages are that the aforementioned steep phase-space gradients develop and the boundaries of the phase-space regions rapidly become ragged. This method has demonstrated the advantageous characteristics of one-dimensional simulations; however, it needs further development for multidimensional applications.

1.4.3 Vlasov Hybrid Simulation (VHS) Method

The VHS method may be considered as a further development of the "phase fluid" approach, although this method use the elements of the particle simulation technique as we shall see below. The VHS method was developed in [397] for numerical simulation of hot collision-free plasmas. In this method a time-varying phase-space computational domain and grid are defined, and the phase fluid within the domain is filled with simulation particles. The distribution function f (or δf) is defined on the phase trajectory of each particle. At each time step Δt the value of f (or δf) is interpolated from the simulation particles onto the phase space grid. It should be noted that it may be not appropriate for all particle species to be treated by VHS. Some species may be described by fluid equations. Some components of the plasma may represent a cold plasma or cold beams which may be better described analytically or by particle in cell (PIC) codes. The phase-space domain will be a function of time as the simulation progresses [397]. For example, in a resonant wave–particle interaction problem the phase domain would cover the range of velocities resonant with the current wave field. As the wave-field spectrum changes during the simulation, so does the range of resonant velocities, and thus the phase domain moves. We assume a regular grid with constant elementary volume. The VHS technique is able to accommodate adaptive grids.

At each time step every particle is pushed according to the usual equation of motion. Each particle trajectory is continuously followed until it leaves the computational domain. Trajectories are not continuously restarted at phase space grid points as in [127]. New trajectories are continually started at the phase domain boundary. Each simulation particle is embedded in the Vlasov phase fluid and moves with it. According to Liouville's theorem, each simulation particle conserves its value of the distribution function $f(\boldsymbol{x}, \boldsymbol{v}, t)$. The values of f (or δf) are defined on the phase trajectories of the simulation particles. Thus, during the simulation the value of the distribution function is known at a large number of points in the phase domain; these points are the locations of the simulation particles.

At each time step this information is used to calculate the particle distribution function on the fixed phase space $(\boldsymbol{x}-\boldsymbol{v})$ grid and thus to estimate the plasma charge and density at the grid nodes. At each time step we have to interpolate the values of the distribution function f_i from the particles onto the fixed phase space grid, giving grid values f_{ijk}. This process of interpolation is quite different from that in the PIC and Vlasov codes.

The above technique will give a very low noise level in f_{ij}. At the grid points that do not have any simulation particles within the surrounding squares, f_{ij} has to be calculated by linearly interpolating values of f from neighboring grid points. Once a distribution function $f_{ijk...}$ is defined on the fixed phase space grid, estimates of charge ϱ and current $\boldsymbol{J}$ are easily obtained.

At the initial moment of time $t = 0$, simulation particles are given a value for the distribution function of f_0, where f_0 should be self-consistent with the presumed initial fields. At the boundaries the unperturbed initial value for the distribution function should be assigned to the inserted simulation particles.

The VHS method has significant advantages over PIC codes [236] and also over other Vlasov codes. The most significant positive features are:

(a) VHS is a very stable algorithm since no attempt is made to evaluate the derivatives of f in phase space.
(b) The technique has a very low noise level.
(c) The distribution function fine structure is handled easily.
(d) The method makes very efficient use of particles, particularly when $\delta f \ll f_0$.
(e) No numerical diffusion of the distribution function is needed in order to attain stability.
(f) The VHS technique is particularly effective when the flux of phase fluid at the boundary of the phase-space computational domain is significant. The methodology allows a dynamic population of simulation particles in which redundant particles may be discarded from the simulation and new particles inserted into the phase fluid as required.

For more details of the VHS model, the reader is referred to [397].

1.4.4 Conventional Particle Models

There are three principal types of particle simulation model (see, e.g., [236]): the particle–particle (PP) model, the particle–mesh (PM) model, and the particle–particle–particle–mesh (the PPPM or P^{3}M) model. The PP model uses the action at a distance formulation of the force law, and it is conceptually and computationally the simplest. The state of the physical system at some time t is described by the set of particle positions and velocities $\{\boldsymbol{x}_i(t), \boldsymbol{v}_i(t); i = 1, ..., N_p\}$. The time-step loop updates these values using the forces of interaction and equations of motions to obtain the state of the system at a slightly later time $t + \Delta t$.

PM model treats the force as a field quantity, approximating it on a mesh, and it uses the equations for the electromagnetic field. This algorithm is much faster than that for PP, but the force calculation is usually less accurate than in the PP method. Field quantities, which in the physical system are determined in all space, are represented approximately by values on a regular array of mesh points. Differential operators are replaced by finite-difference approximations on the mesh. The electromagnetic fields and forces at particle positions are obtained by interpolating on the array of mesh-defined values. Mesh-defined densities and currents are obtained by the opposite process of particle weighting nearly to mesh points in order to create the mesh-defined values of density and current.

The P^3M model exploits the best advantages of the PP and PM models. The PP method can be used for small systems with long-range forces or for large systems where the forces of interaction are nonzero for only a few interparticle distances. The PM method, on the other hand, is computationally fast, but can only handle smoothly varying forces. The P^3M enables large correlated systems with long-range forces to be simulated. The choice of model is dictated partly by the physics of the phenomenon under investigation and partly by consideration of computational costs.

In conventional hybrid simulations, the PM technique is most suitable, since the electromagnetic fields are described by partial derivative equations with a complicated structure. However, for some simple models – electrostatic and Ampere models (Sects. 2.2.1 and 2.2.2) – one can use the PP and P^3M techniques. For a detailed discussion the reader is referred to [51, 236, 509].

Summary

The analysis of space, astrophysical and laboratory plasmas has demonstrated that the plasma environment of these objects involves multiscale processes. To study these phenomena, we have to use (and design) so-called rational computational models which describe the major processes well. The behavior of the physical system is usually controlled by the basic dimensionless parameters, namely, Mach numbers, Reynolds numbers, plasma betas, etc. In simulation research of industrial objects it is very important to have the real values of these parameters. In scientific research we have to simulate a wide range of these parameters to determine the scale for particular physical processes. Unfortunately, very often it is impossible to provide a simulation with the real values of the major dimensionless parameters. In such a case we can use so-called non-full modeling, with the parameters being far from the exact values, and then extrapolate the solution to exact values of the input parameters.

Most of computational models considered have been developed for solution of general Vlasov–Maxwell systems. However, all these computational models (except the PP and PPPM models) may be applied successfully to kinetic simulation in the frame of different hybrid and full particle codes.

2. Particle-Mesh Models

2.1 Introduction

In real plasmas, the electrons and ions can stream along magnetic field lines, gyrate around a magnetic field line, bounce back and forth in either an electric or magnetic trap, drift and precess across the magnetic field, and collide with other particles. Each of these motions usually possesses its own characteristic frequency, and the large difference in ion and electron masses increases the range of time scales. The inclusion of self-consistent electromagnetic fields adds additional time scales for the collective oscillations of the plasma. From the point of view of numerical analysis, stiff problems commonly arise in plasma simulations because the phenomena of interest develop on relatively slow time scales, but the systems support high-frequency normal modes. The goal of using the largest possible time step to resolve the interesting physics and to minimize the computer cost, while preserving numerical stability and accuracy, makes stringent demands on the numerical methods.

Subcycling and orbit averaging successfully interleave modules for advancing the field and particle equations with different time steps. These methods differ significantly from "reduced" methods, in which high-frequency modes are removed from the governing equations, e.g., by setting a time derivative or the electron mass equal to zero. Darwin algorithms [392], which eliminate the transverse displacement and radiation fields, and electrostatic algorithms, which solve Poisson's equation and omit electromagnetic fields altogether, are popular examples of reduced methods. These models use the same time step for particle and field updating.

2.2 Hybrid Quasineutral Models

Many macroscopic problems in plasma physics are characterized by ion Larmor radii comparable to the scale lengths of the system. For these problems, and for problems involving micro-instabilities, a fluid description of the ions is inadequate, and the ions must instead be treated in a fully kinetic manner. When the frequencies of interest are low compared to the ion cyclotron frequency, the effects of high-frequency phenomena, such as electromagnetic radiation and waves associated with electron inertia, are generally negligible.

In the conventional hybrid models the field calculations generate the self-consistent electric and magnetic fields in the radiation-free limit (Darwin model). Ampere's law may be decomposed into its longitudinal (curl-free) and transverse (divergence-free) parts,

$$\nabla \times \boldsymbol{B} = \frac{4\pi}{c}\boldsymbol{J}_\mathrm{t} + \frac{1}{c}\frac{\partial \boldsymbol{E}_\mathrm{t}}{\partial t}, \tag{2.1}$$

$$0 = \frac{4\pi}{c}\boldsymbol{J}_\mathrm{l} + \frac{1}{c}\frac{\partial \boldsymbol{E}_\mathrm{l}}{\partial t}, \tag{2.2}$$

where the subscripts l and t, respectively, refer to the longitudinal and transverse parts of a vector quantity. We assume quasineutrality, setting $n_\mathrm{i} = n_\mathrm{e}$, which implies $\nabla \cdot \boldsymbol{J} = 0$. If $\boldsymbol{J}_\mathrm{l}$ vanishes at the boundaries, or if the system has periodic boundaries, then $\boldsymbol{J}_\mathrm{l} = 0$ throughout the plasma system. For the study of low-frequency phenomena, the Darwin approximation is made, that is, the transverse displacement current is neglected. Ampere's law then reduces to the simple form

$$\nabla \times \boldsymbol{B} = \frac{4\pi}{c}(\boldsymbol{J}_\mathrm{e} + \boldsymbol{J}_\mathrm{i}). \tag{2.3}$$

In the first one-dimensional and two-dimensional hybrid simulations of space and plasma phenomena, the electrons were treated as a massless fluid and ions as fully kinetic macroparticles [83, 97, 217, 227, 318, 480, 598]. The first one-dimensional [160] and multidimensional [231] hybrid codes retaining electron inertia demonstrate the possibility of simulating many phenomena which require finite electron cyclotron frequency.

During the last ten years, the fully three-dimensional hybrid codes have been developed in [66, 238, 334, 486]. In this chapter we shall describe typical hybrid models which use a different description for different species of particles – ions, dust grains, positrons and electrons.

2.2.1 Electrostatic Model

For low-frequency waves in an electrostatic problem, electrons respond to the field almost instantaneously (compared with the wave period); the ions are thereby shielded and charge separation effects become much reduced. When the phenomenon under study is at a low frequency, the high-frequency oscillations associated with the electron inertia can be neglected. Electron are adjusting quickly to the charge separation along the field line. If, on the other hand, the wave develops perpendicular or nearly perpendicular to the magnetic field direction, the quasineutrality condition may not hold even for low-frequency waves [509].

Electrons under the present condition adjust to the electric field along the field line so quickly that they behave adiabatically. This leads to the well-known Boltzmann equilibrium distribution for the electron density n_e:

2.2 Hybrid Quasineutral Models

$$n_e = n_0 \exp(e\phi/T_e), \tag{2.4}$$

where n_0 is the equilibrium density, ϕ the electric potential, and T_e the electron temperatures. Poisson's equation is

$$\nabla^2 \phi = -4\pi e(n_i - n_e), \tag{2.5}$$

the ion density $n_i = n_0 + \delta n_i$. If the nonlinearity is small, the exponential function in (2.4) may be expanded to give $1 + e\phi/T_e + O(e\phi/T_e)^2$. Substitution of (2.4) into (2.5) yields [42, 403, 558]:

$$\nabla^2 \phi = -4\pi e \delta n_i + k_{De}^2 \phi, \tag{2.6}$$

where $k_{De}^2 = 4\pi n_0 e^2 / T_e$. Equations for ion dynamics are:

$$\frac{d\boldsymbol{v}_l}{dt} = \frac{Z_i e}{M_i} \boldsymbol{E},$$

$$\frac{d\boldsymbol{x}_l}{dt} = \boldsymbol{v}_l. \tag{2.7}$$

This plasma model eliminates the high-frequency plasma oscillations; its oscillations are now ion-acoustic waves.

In the nonlinear case ($e\phi/T_e$ is not small) we have to solve the nonlinear expressions (2.4–2.5):

$$\nabla^2 \phi = -4\pi e n_i + 4\pi e n_0 \exp(e\phi/T_e). \tag{2.8}$$

The last equation may be solved by an iteration procedure. A faster convergence may be achieved by using Newton's method [236]. This model may be employed even for a magnetized plasma [403]. As long as we ignore the nonadiabatical effect of electrons, the main electron effect is considered to be the screening of charge. The condition for this regime is

$$\omega \ll k_\parallel v_e,$$

where $k_\parallel = \boldsymbol{k} \cdot \hat{\boldsymbol{b}}$, and $\hat{\boldsymbol{b}} = \boldsymbol{B}/B$.

2.2.2 Ampere Magnetoinductive Model

For the study of some two-dimensional phenomena (e.g., magnetic field reconnection processes) it may be useful to use the so-called Ampere's plasma model [131, 215, 236, 349, 351, 381, 514, 592, 594], which does not include the electrostatic field. Let us consider the two-dimensional geometry when the magnetic field lies in the simulation plane x, z. Let us suppose that the ions play an active role, i.e., the dynamics of a plasma system is determined mostly by ion-kinetic effects. In this case the electrons only provide the neutralization of ion charge, so that

$$|\nabla\phi| \ll \frac{1}{c}|\frac{\partial \mathbf{A}}{\partial t}|. \tag{2.9}$$

The above condition may be also satisfied when the electrons and ions have the same mass and the same absolute value of charge

$$m = M_i, \quad |q_i| = |q_e| = e/2. \tag{2.10}$$

Then, the magnetic field, the current, the electric field, and the vector potential may be represented as follows:

$$\mathbf{B} = (B_x, 0, B_z), \quad \mathbf{J} = (0, J_y, 0),$$
$$\mathbf{E} = (0, E_y, 0), \quad \mathbf{A} = (0, A_y, 0). \tag{2.11}$$

Using Ampere's equation and the relation between the magnetic field and the vector potential,

$$\nabla \times \mathbf{B} = \frac{4\pi}{c}\mathbf{J}; \quad \mathbf{B} = \nabla \times \mathbf{A}, \tag{2.12}$$

we obtain:

$$\triangle_{x,z} A_y = -\frac{4\pi}{c}\langle Z_i e n v_y \rangle. \tag{2.13}$$

The induction equation

$$-\frac{1}{c}\frac{\partial \mathbf{B}}{\partial t} = \nabla \times \mathbf{E} \tag{2.14}$$

allows us to determine the electric field via a vector potential,

$$E_y = -\frac{1}{c}\frac{\partial A_y}{\partial t}. \tag{2.15}$$

The equation of ion motion may be written in coordinate form:

$$\frac{d}{dt}v_x = \frac{Z_i e v_y B_z}{M_i c}, \quad \frac{d}{dt}v_z = -\frac{Z_i e v_y B_x}{M_i c}, \tag{2.16}$$

$$\frac{d}{dt}x = v_x, \quad \frac{d}{dt}z = v_z. \tag{2.17}$$

The y-component of the equation of ion motion has the integral

$$P_y = M_i v_y + \frac{Z_i e A_y}{c} = \text{const.}, \quad \frac{d}{dt}y = v_y. \tag{2.18}$$

Since the considered problem is not dependent on y, $\partial/\partial y = 0$, we do not need to update the y-coordinate of particle positions. Finally, for the vector potential we have the following equation:

$$\left(-\triangle_{x,z} + \frac{4\pi(Z_i e)^2}{M_i c^2}\langle n \rangle\right) A_y = \frac{4\pi Z_i e}{M_i c}\langle n P_y \rangle. \tag{2.19}$$

The model described above calculates explicitly the motion of heavy protons, taking into account their inertia. The effect of finite gyroradius and gyroperiod, which are absent from the MHD particle-ion-element model (Sect. 2.2.5), are correctly taken into account. This enables the model to be used near the field reversal configuration (see, e.g., Astron plasma simulation [131]). Sound and Alfvén waves are also represented. The role of the electrons, which are not explicitly represented in the model, is to maintain charge neutrality and to cancel any proton currents in the (x, z) plane [236, 349, 351]. The application of this model to simulation of the magnetic field reconnection and particle acceleration inside the current sheets will be considered in Chap. 10.

2.2.3 Particle-Ion–Fluid-Electron Model

In this model we use a kinetic description of the positively charged particles – ions, positrons or dust grains – and the electrons are considered to be a fluid. Such a type of plasma model is appropriate for the next conditions: $\lambda_i \approx \lambda_e > L$, $\varrho_{ci} \approx L$, and $\varrho_{ce} \ll L$, where $\lambda_{i(e)}$ denote free light lengths and L denotes the characteristic scale of the problem. This model may be useful for studying the global interaction of the solar wind with planets, collisionless shocks and current sheets. However, in some cases the hybrid codes may also be used for kinetic simulation of a plasma system with negative dust grains (see, e.g., [337, 563]).

2.2.3.1 Formulation via Electromagnetic Fields. The basic equations of the considered model are the ion equations of motion, the generalized Ohm's law, Faraday's induction equation and the equation for the electron pressure. The Vlasov equation for ion distribution function $f_s(t, \boldsymbol{x}, \boldsymbol{v})$ is of the form

$$\frac{\partial}{\partial t} f_s + \boldsymbol{v} \frac{\partial}{\partial \boldsymbol{x}} f_s + \frac{\boldsymbol{F}}{M_s} \frac{\partial}{\partial \boldsymbol{v}} f_s = 0. \tag{2.20}$$

The equations of ion motion are of the form

$$\frac{\mathrm{d}\boldsymbol{x}_{sl}}{\mathrm{d}t} = \boldsymbol{v}_{sl}, \quad \frac{\mathrm{d}\boldsymbol{v}_{sl}}{\mathrm{d}t} = \frac{Z_s e}{M_s} \left(\boldsymbol{E}^* + \frac{\boldsymbol{v}_{sl} \times \boldsymbol{B}}{c} \right), \quad \boldsymbol{E}^* = \boldsymbol{E} - \sigma_{\mathrm{eff}}^{-1} \boldsymbol{J}, \tag{2.21}$$

where $\boldsymbol{E}^*$ denotes the effective electric field which takes into account the collision between ions and electrons, s denotes the ion species and l is the particle index. In order to receive the transport equation for electrons (ions) one has to multiply the Vlasov equation by 1, $m(M_s)$, and $mv^2/2(M_s v^2/2)$ and integrate over velocity. In contrast to standard hybrid codes the electrons are considered to have a nonzero mass here. As a result the electron momentum transport equation (the generalized Ohm's law [63]) is

$$\frac{\partial \boldsymbol{J}_e}{\partial t} - e \nabla \cdot \mathsf{K}_e - \frac{\omega_{\mathrm{pe}}^2}{4\pi} \boldsymbol{E} + \frac{e}{mc} (\boldsymbol{J}_e \times \boldsymbol{B}) + \frac{e}{m} \boldsymbol{R}_e = 0. \tag{2.22}$$

The ion momentum transport equation for species s is

$$\frac{\partial \boldsymbol{J}_s}{\partial t} + Z_s e \nabla \cdot \mathsf{K}_s - \frac{\omega_{ps}^2}{4\pi}\boldsymbol{E} - \frac{Z_s e}{M_s c}(\boldsymbol{J}_s \times \boldsymbol{B}) - \frac{Z_s e}{M_s}\boldsymbol{R}_s = 0, \tag{2.23}$$

where ω_{pe} and ω_{ps} denote the electron and ion plasma frequencies; $\mathsf{K}_{s(e)}$ is the ion (electron) kinetic energy tensor

$$\mathsf{K}_s = \int \boldsymbol{vv} f_s \mathrm{d}^3 \boldsymbol{v}, \quad \mathsf{K}_e = \int \boldsymbol{vv} f_e \mathrm{d}^3 \boldsymbol{v}, \tag{2.24}$$

in which f_s and f_e are the ion and electron distribution functions.

Decomposition of K_e into the divergence of the electron bulk flow, the scalar pressure, and a tensor of the viscosity stress gives the following momentum equation:

$$\frac{\partial \boldsymbol{J}_e}{\partial t} + (\nabla \boldsymbol{U}_e)\boldsymbol{J}_e - \frac{e}{m}(\nabla p_e + \partial \pi^e_{\alpha\beta}/\partial x_\beta) - \frac{\omega_{pe}^2}{4\pi}\boldsymbol{E} + \frac{e}{mc}(\boldsymbol{J}_e \times \boldsymbol{B}) + \frac{e}{m}\boldsymbol{R}_e = 0, \tag{2.25}$$

where p_e is a scalar electron pressure and $\pi^e_{\alpha\beta}$ is a tensor of the viscosity stress. Here one has taken into account the transformations

$$\langle v_\alpha v_\beta \rangle = U_\alpha U_\beta + \langle v'_\alpha v'_\beta \rangle, \quad p_e = n_e m \frac{\langle v'^2_e \rangle}{3} = n_e T_e$$

and

$$\pi^e_{\alpha\beta} = n_e m \langle v'_{e\alpha} v'_{e\beta} - \frac{v'^2_e}{3}\delta_{\alpha\beta}\rangle, \tag{2.26}$$

where $v' = v - U$ is the difference between the individual particle velocity and the bulk velocity. The tensor π^e describes the anisotropic part of the electron pressure. The complete pressure tensor for a given species is

$$\mathsf{P}_{\alpha\beta} = \int m v'_\alpha v'_\beta f(t, \boldsymbol{r}, \boldsymbol{v}) \mathrm{d}\boldsymbol{v} = nm\langle v'_\alpha v'_\beta \rangle = p\delta_{\alpha\beta} + \pi_{\alpha\beta}. \tag{2.27}$$

The quantities $\boldsymbol{R}_e$ ($\boldsymbol{R}_s$) represent the mean change in momentum of the particles of a given species due to collision with all other particles

$$\boldsymbol{R} = \int m \boldsymbol{v}' C \mathrm{d}\boldsymbol{v}, \tag{2.28}$$

where C denotes the collision term in the initial Boltzmann equations.

The fourth and fifth terms of (2.25) and (2.23) describe the action of the electric and magnetic fields on the electron and ion fluid.

The energy transport equation for electrons (ions) may be written as follows:

$$\frac{\partial}{\partial t}\left(\frac{M_k n}{2}\langle v^2 \rangle\right) + \nabla\left(\frac{M_k n}{2}\langle v^2 \boldsymbol{v}\rangle\right) + Z_k en\boldsymbol{E} \cdot \boldsymbol{U} = \int \frac{M_k v^2}{2} C \mathrm{d}\boldsymbol{v}. \tag{2.29}$$

The subscript k denotes electrons ($k = \mathrm{e}$, $Z_e = -1$) and ions.

Taking into account the transformations

$$\langle \frac{v^2}{2} v_\beta \rangle = \frac{1}{2} U^2 U_\beta + U_\alpha \langle v'_\alpha v'_\beta \rangle + \frac{1}{2} \langle v'^2 \rangle U_\beta + \langle \frac{1}{2} v'^2 v'_\beta \rangle$$

$$= \left(\frac{1}{2} U^2 + \frac{5p}{2mn} \right) U_\beta + \frac{1}{mn} U_\alpha \pi_{\alpha\beta} + \langle \frac{1}{2} v'^2 v'_\beta \rangle, \quad (2.30)$$

one can reduce (2.29) to the form

$$\frac{\partial}{\partial t} \left(\frac{M_k n}{2} U_k^2 + \frac{3}{2} n T_k \right) + \frac{\partial}{\partial x_\beta} \left[\left(\frac{M_k n}{2} U_k^2 + \frac{5}{2} n T_k \right) U_{k,\beta} + (\pi^k_{\alpha\beta} \cdot U_{k\alpha}) + q^k_\beta \right]$$

$$= -Z_k e n \boldsymbol{E} \cdot \boldsymbol{U}_k + \boldsymbol{R}_k \cdot \boldsymbol{U}_k + Q_s. \quad (2.31)$$

Here one has introduced the notation

$$q^k = \int \frac{M_k v'^2}{2} \boldsymbol{v} f_k(t, \boldsymbol{r}, \boldsymbol{v}) d\boldsymbol{v} = n M_k \langle \frac{v'^2_k}{2} \boldsymbol{v}_k \rangle \quad (2.32)$$

and

$$Q_k = \int \frac{M_k v'^2}{2} C d\boldsymbol{v}. \quad (2.33)$$

The vector $\boldsymbol{q}^k$ is the flux density of heat carried by particles of a given species and represents the transport of the energy associated with the random motion in the coordinate system in which the particle gas as a whole is at rest at a given point in space [63]. The quantity Q_k is the heat generated in a gas of particles of a given species as a consequence of collisions with particles of another species [63].

The first term of (2.31) represents the variation of the total energy of particles of a given species; this consists of the kinetic energy and internal energy (per unit volume). The second term is the full flux of energy, which contains the transport of full energy with bulk velocity $\boldsymbol{U}$, the microtransport of energy and the work of the pressure force. The terms on the right side of the equation include the work of other forces and heat release.

Ampere's law in the nonradiative limit is

$$\frac{4\pi}{c} \boldsymbol{J} = \nabla \times \boldsymbol{B}, \quad (2.34)$$

and the induction (Faraday's law) equation is

$$\frac{1}{c} \frac{\partial \boldsymbol{B}}{\partial t} + \nabla \times \boldsymbol{E} = 0, \quad (2.35)$$

where the total current includes the internal and external currents

$$\boldsymbol{J} = \boldsymbol{J}_e + \boldsymbol{J}_i + \boldsymbol{J}_{\text{ext}}, \quad \boldsymbol{J}_i = \sum_{k=1}^{N_s} Z_k e n_k \boldsymbol{U}_k, \quad (2.36)$$

where $\boldsymbol{J}_{\text{ext}}$ is external current and N_s denotes the total number of particle species.

We assume the validity of the quasineutrality condition,

$$n_e = \sum_{k=1}^{N_s} Z_k n_k \equiv n. \tag{2.37}$$

One can obtain the final equation for the electric field by combining (2.22), (2.23), Ampere's law (2.34) and Faraday's law (2.35):

$$c^2 \nabla \times (\nabla \times \boldsymbol{E}) + \omega_{\text{pe}}^2 \left(1 + \sum_{k=1}^{N_s} \frac{n_k Z_k^2 m}{n_e M_k} \right) \boldsymbol{E}$$

$$+ \frac{e}{m} \left[\frac{4\pi}{c} \sum_{k=1}^{N_s} \left(1 + \frac{Z_k m}{M_k} \right) \boldsymbol{J}_k - \nabla \times \boldsymbol{B} + \frac{4\pi}{c} \boldsymbol{J}_{\text{ext}} \right] \times \boldsymbol{B}$$

$$= -\frac{4\pi e}{m} \left(\nabla p_e + \partial/x_\beta \pi_{\alpha\beta}^e \right) + 4\pi \left(\sum_{k=1}^{N_s} Z_k e \nabla \cdot \mathsf{K}_k + (\nabla \boldsymbol{U}_e) \boldsymbol{J}_e \right)$$

$$+ \frac{4\pi \partial \boldsymbol{J}_{\text{ext}}}{\partial t} - \frac{4\pi e}{m} \left(\boldsymbol{R}_e - \sum_{k=1}^{N_s} \frac{Z_k m}{M_k} \boldsymbol{R}_k \right). \tag{2.38}$$

Notice that in the limit of massless electrons this equation transforms to the standard generalized Ohm's law, which one can obtain also directly from (2.22)

$$\boldsymbol{E} = \frac{1}{enc} (\boldsymbol{J}_e \times \boldsymbol{B}) - \frac{1}{en} (\nabla p_e + \partial \pi_{\alpha\beta}^e / \partial x_\beta) + \frac{1}{en} \boldsymbol{R}_e. \tag{2.39}$$

In some cases it is convenient to eliminate the kinetic energy from the energy transport equation (2.31) for electrons by means of the equation of continuity and the equation of motion ([63], also Chap. 4). We then obtain an equation for the transport of internal electron energy (electron pressure) or the heat-balance equation:

$$\frac{3}{2} \frac{\partial p_e}{\partial t} + \frac{3}{2} \nabla (p_e \cdot \boldsymbol{U}_e) + p_e \nabla \cdot \boldsymbol{U}_e = -\pi_{\alpha\beta}^e \frac{\partial U_{e\alpha}}{\partial x_\beta} - \nabla \cdot \boldsymbol{q}_e + Q_e, \tag{2.40}$$

where $\boldsymbol{q}_e$ is

$$\boldsymbol{q}_e = nm \langle \frac{v'^2}{2} \boldsymbol{v} \rangle.$$

The transfer of momentum from ions to electrons by collisions $\boldsymbol{R} = \boldsymbol{R}_u + \boldsymbol{R}_T$ is made up of two parts: the force of friction $\boldsymbol{R}_u$ due to the existence of a relative velocity $\boldsymbol{u} = \boldsymbol{U}_e - \boldsymbol{U}_i$, and a thermal force $\boldsymbol{R}_T$, which arises by virtue of the gradient in the electron temperature. In accordance with [63], at large values of $\omega_{\text{pe}} \tau_e$ the friction force is

$$\boldsymbol{R}_u = en\left(\frac{\boldsymbol{J}_\parallel}{\sigma_\parallel} + \frac{\boldsymbol{J}_\perp}{\sigma_\perp}\right)$$

and the thermoforce is

$$\boldsymbol{R}_T = -0.71 n_e \nabla_\parallel T_e - \frac{3}{2}\frac{n_e}{\Omega_e \tau_e}\boldsymbol{e} \times \nabla T_e.$$

The heat generated in the electrons as a consequence of collisions with ions [63] is

$$Q_e = \frac{J_\parallel^2}{\sigma_\parallel} + \frac{J_\perp^2}{\sigma_\perp} + \frac{1}{en_e}\boldsymbol{J}\boldsymbol{R}_T - \frac{3m}{M}\frac{n_e}{\tau_e}(T_e - T_i). \tag{2.41}$$

If the thermoforce and the anisotropy of a frictional force are neglected, we have for the electrons

$$\boldsymbol{R}_e = -n_e m \sum_{s\neq e}\nu_{es}(\boldsymbol{U}_e - \boldsymbol{U}_s) \tag{2.42}$$

and for the ions

$$\boldsymbol{R}_s = -n_s \sum_{j\neq s}\nu_{sj}\tilde{m}_{sj}(\boldsymbol{U}_s - \boldsymbol{U}_j), \tag{2.43}$$

where

$$\tilde{m}_{sj} = \frac{m_s m_j}{m_s + m_j}, \tag{2.44}$$

in which we neglect terms of the order m/M_i; ν_{sj} is the frequency of the collision of a particle of species s with a particle of species j; and $\boldsymbol{U}_j$ is the bulk velocity of the particles of species j. Summation is made over the other species of particles.

In case where the viscosity stress tensor π^e and the heat flux $\boldsymbol{q}_e$ are neglected, a simpler equation may be used:

$$\frac{\partial p_e}{\partial t} + \boldsymbol{U}_e \cdot \nabla p_e + \gamma p_e \nabla \cdot \boldsymbol{U}_e = (\gamma - 1)\boldsymbol{J}^2/\sigma_{\text{eff}}. \tag{2.45}$$

When the effective collision frequencies are very small, one can treat the electron gas adiabatically:

$$p_e \propto n_e^\gamma, \tag{2.46}$$

where an adiabatic index $\gamma = 5/3$ is taken. In an isothermal regime one can use

$$T_e = \text{const.} \ (p_e = n_e T_e).$$

As it will be shown later, the anisotropic electron pressure may play a key role in some plasma phenomena, e.g., in the magnetic field reconnection. The hybrid simulation models which include a π tensor will be considered in Chap. 10. In the limit of zero mass of electrons, the set of equations (2.20–2.21), (2.34–2.37), (2.39) and (2.45) or (2.46) are the basic equations for conventional hybrid models.

34   2. Particle-Mesh Models

In the above model we use a phenomenological treatment of the anomalous resistivity. If a self-consistent treatment is desired, the equation of ion motion (2.21) should be modified, since it does not produce any heating. In the one-dimensional case ion heating can be included if the equations of ion motion in the direction perpendicular to the magnetic field are written as follows [214, 410]:

$$M \mathrm{d}v_x/\mathrm{d}t = eE_x + \frac{ev_y B_z}{c} + \frac{\Delta v_x M}{\Delta t}$$

and

$$M \mathrm{d}v_y/\mathrm{d}t = eE_y - \frac{ev_x B_z}{c} + \frac{\Delta v_y M}{\Delta t}. \quad (2.47)$$

The velocity corrections Δv_x and Δv_y per time step Δt are introduced so that the anomalous ion heating and the relaxation among the various ion temperature components are taken into account. The values of Δv_x and Δv_y are predicted from the ion heating rate of the particular instability. They can be implemented by use of Monte-Carlo techniques, scattering techniques, or by the simple method developed in [214] that is described below. The temperatures T_{xi} and T_{yi} are obtained from the mean square deviations of v_x and v_y at every spatial point. The temperature T_{zi} is given by

$$\frac{\partial T_{zi}}{\partial t} + u_x \frac{\partial T_{zi}}{\partial x} = Q_x + \frac{T_{xi} + T_{yi} - 2T_{zi}}{\tau_\mathrm{r}} - \frac{T_{zi} - T_\mathrm{e}}{\tau_\mathrm{ei}}, \quad (2.48)$$

where Q_z is the field-aligned anomalous heating, τ_r is the isotropization time to electromagnetic ion cyclotron instabilities (which is typically $\tau_\mathrm{r} \approx 3 - 5\Omega_\mathrm{i}^{-1}$; see [410]), and τ_ei is the electron–ion temperature equipartition time. Notice that there are as yet no theories producing anomalous values for τ_ei.

In order to compute Δv_x and Δv_y, we first calculate the ion temperature changes due to the ion heating and the relaxation at each mesh point during each time step Δt as

$$\Delta T_{xi} = \left(Q_x + \frac{T_\mathrm{e} - T_{xi}}{\tau_\mathrm{ei}} - \frac{T_{xi} - T_{zi}}{\tau_\mathrm{r}}\right) \Delta t,$$

and

$$\Delta T_{yi} = \left(Q_y + \frac{T_\mathrm{e} - T_{yi}}{\tau_\mathrm{ei}} - \frac{T_{yi} - T_{zi}}{\tau_\mathrm{r}}\right) \Delta t.$$

In general, the ratio $Q_x{:}Q_y{:}Q_z$ is proportional to $k_x^2{:}k_y^2{:}k_z^2$ of the particular instability for quasilinear theory. The coupling between T_x and T_y has already been included by the $\boldsymbol{v} \times \boldsymbol{B}$ terms of (2.47). By knowing of ΔT_{xi} and ΔT_{yi}, we can rescale at each point and time step the ion velocity deviation $(v_x^j - u_{xi})$, $(v_y^j - u_{yi})$ keeping u_{xi} and u_{yi} fixed, so that the mean velocity spreads give new correct ion temperatures at each point. Thus,

$$\Delta v_\alpha^j = \left(\frac{T_{\alpha i} + \Delta T_{\alpha i}}{T_{\alpha i}}\right)^{1/2} \leftarrow (v_\alpha^j - u_{\alpha i}).$$

The reader must refer to [214] for details. A numerical method for solving (2.35–2.39) or (2.35–2.38) will be discussed in Sects. 5.2–5.5.

2.2.3.2 Formulation via Electromagnetic Potentials. The essence of the Darwin model is to neglect only the inductive or solenoidal part of the displacement current. We also split the electric field into longitudinal and transverse parts, denoted by E_l and E_t, respectively, defined by the requirements $\nabla \times E_l = 0$ and $\nabla \cdot E_t = 0$. In the case when the plasma configuration is connected to an external current circuit, we have to make the longitudinal current follow Ampere's law. The resulting field equations are

$$\nabla \cdot E_l = 4\pi e(n_i - n_e), \quad \nabla \times E_l = 0, \tag{2.49}$$

$$\nabla \times E_t = -\frac{1}{c}\frac{\partial B}{\partial t}, \quad \nabla \cdot E_t = 0, \tag{2.50}$$

$$\nabla \times B = \frac{4\pi J}{c} + \frac{1}{c}\frac{\partial E_l}{\partial t}, \quad \nabla \cdot B = 0. \tag{2.51}$$

The field equations thus differ by only one term from the exact Maxwell equations – specifically, in the omission of the time derivative of the transverse electric field. However, this simple alteration significantly changes their character; the approximate Maxwell equations, when coupled by the particle equations through the source terms, are elliptic in type rather than hyperbolic. This ellipticity is of considerable computational significance. When solving the full Maxwell equations the time derivatives may be used to advance the field, while the divergence equations serve as initial conditions which are preserved in time because of the charge continuity equation. Working in the Coulomb gauge, (2.49–2.51) may be expressed in terms of the scalar and vector potentials, ϕ and A, respectively, as in [231]:

$$\nabla^2 \phi = -4\pi e(n_i - n_e),$$

$$\nabla^2 A = -\frac{4\pi}{c} J_t, \tag{2.52}$$

$$\nabla^2 E_t = \frac{4\pi}{c^2} \dot{J}_t, \tag{2.53}$$

where

$$B = \nabla \times A \tag{2.54}$$

and

$$E_l = -\nabla \phi. \tag{2.55}$$

In this quasineutral case a method for determining ϕ can be found by summing the electron and ion momentum equations (2.22) and (2.23). Note that we do not take into account the anomalous resistivity. The resulting equation is

$$(\dot{J}_i + \dot{J}_e) = D + \mu(E_l + E_t) + \xi \times B, \tag{2.56}$$

where

$$\mu = \frac{\omega_{pe}^2 + \omega_{pi}^2}{4\pi},$$

$$\xi = \frac{e}{c}\left(\frac{Z_i \mathbf{J}_i}{M_i} - \frac{\mathbf{J}_e}{m}\right),$$

$$\mathbf{D} = (-Z_i e \nabla \cdot \mathsf{K}_i + e \nabla \cdot \mathsf{K}_e),$$

and $\mathsf{K}_{i(e)}$ is the ion (electron) kinetic energy tensor (2.24). Taking the divergence of both sides of (2.56) and solving for $\mu \mathbf{E}_l$ gives

$$\nabla \cdot (\mu \nabla \phi) = \nabla \cdot (\mathbf{D} + \mu \mathbf{E}_t + \boldsymbol{\xi} \times \mathbf{B}). \qquad (2.57)$$

On left side of this equation we assumed that the variation time of a longitudinal electric field is much smaller than the electron plasma frequency. The set of equations (2.52–2.57) allows us to update the magnetic field and the transversal and longitudinal electric field. Equation (2.57) plays the same role as the Poisson equation in plasma models not requiring quasineutrality. A numerical method for solving (2.52–2.57) will be discussed in Sect. 5.6. A detail description of the numerical methods for conventional hybrid codes will be given in Chaps. 3–7. The conventional hybrid models have a wide application for simulation of space, astrophysical and laboratory plasma, in particular, the processes of ion heating and acceleration at collisionless shocks, magnetic reconnection and Kelvin–Helmholtz instability inside the current sheets. The simulation of these processes will be considered in Part 2.

2.2.4 Particle-Ion–Fluid-Ion–Fluid-Electron Model

In multidimensional, multiscale simulations, a sufficient approximation of the Vlasov–Maxwell equations sometimes may be reached only by a further simplification of the simulation model. We can combine the conventional hybrid codes and MHD approximation for background ions [543]. The basic equations for this combined model are: for active ions – the Vlasov equation for the ion distribution function (2.40), and the equation of ion motion (2.21); for background ions – the MHD equations of continuity and momentum.

The equation of continuity for background ions is

$$\frac{\partial n_i^*}{\partial t} + \nabla(n_i^* \mathbf{u}_i^*) = 0. \qquad (2.58)$$

The momentum equation for background ions is

$$M_i n_i^* \left(\frac{\partial \mathbf{u}_i^*}{\partial t} + (\mathbf{u}_i^* \cdot \nabla)\mathbf{u}_i^*\right) = Z_i e n_i^* \left(\mathbf{E} + \frac{1}{c}\mathbf{u}_i^* \times \mathbf{B}\right). \qquad (2.59)$$

The other equations are the generalized Ohm's law (2.22), Faraday's induction equation (2.35), and the equation for the electron pressure (2.40) or (2.46).

The average electron and ion bulk velocities U_e and U_i and the density and current of electrons are calculated from (2.42), (2.45) and (2.46) taking into account the density and velocity of the background ion component. The combined model describes the kinetic processes at a location of active ions (beams, current sheets, and shocks) well, and at the same time this model describes all MHD effects well, including the Hall effect, and the electron inertia effect in a background plasma, where the kinetic effects are not very important.

2.2.5 Particle-Ion-Element–Fluid-Electron Model

Further simplification of the above hybrid models may be done by using the so-called ion fluid equation for particle ion elements as of [508]:

$$\frac{d\boldsymbol{v}_i}{dt} = \frac{1}{4\pi\varrho}(\nabla \times \boldsymbol{B}) \times \boldsymbol{B} - \frac{1}{\varrho}\nabla \cdot p, \tag{2.60}$$

where ϱ is the ion mass density and p is the total plasma pressure. The difference between (2.60) and (2.21) arises from the degree of ion trajectory resolution in the equation of motion. Equation (2.21) resolves the gyromotion, whereas (2.60) averages the gyration effects. On the other hand, (2.60) and (2.22) are used in order to compute $\boldsymbol{E}$. Thus, the code which exploits (2.60) does not carry ion gyromotion anymore and corresponds to even lower frequencies than the conventional hybrid codes.

One of the important effects introduced by this model is the Hall term. It has been shown [508] that the Hall term leads to the bifurcation of the Alfvén wave into two branches, i.e., the Alfvén-ion cyclotron wave and the whistler wave. This split of frequencies $\Delta\omega$ is negligible for small wavenumbers, where the ideal MHD should be valid,

$$\Delta\omega = k_\parallel v_A (k_\parallel v_A / \Omega_i),$$

but becomes pronounced for wavenumbers of the order

$$k_\parallel \geq \Omega_i / v_A.$$

The Hall term is often neglected in the MHD study and may give rise to an important effect such as the possible stabilization of the tilt mode of the magnetic field reverse configuration [69, 155, 449].

2.2.6 Gyrokinetic-Ion–Fluid-Ion–Fluid-Electron Models. δF Method

In this section we shall consider low-frequency (compared to the ion cyclotron frequency) perturbations in a plasma consisting of low-temperature isotropic and hot anisotropic components with density $n_h \ll n_b$ and temperature

$T_h \gg T_b$ [39, 96]. Here h denotes the hot ions and b (bulk) is used for the rest of the plasma (cold electrons and cold ions). In order to include perpendicular wavelengths with $k_\perp \varrho_h \propto 1$, the gyrocenter equations of motion are used to advance energetic ions, while the dynamics of the bulk plasma is described by nonlinear, compressional, one-fluid MHD equations.

2.2.6.1 Fluid-Ion Background. Three-dimensional MHD-gyrokinetic simulations were first used in [415] to study nonlinear energetic particle effects in tokamaks. Two sets of equations, the pressure coupling and current coupling schemes, were derived. In the pressure coupling scheme the off-diagonal elements of the hot ion pressure tensor are usually neglected, whereas the current coupling scheme avoids this assumption. In addition, the calculation of the first velocity moment has an advantage over the pressure moment calculation in terms of the numerical noise level.

For these reasons the current coupling scheme is used in [39], and the hot ions are coupled to the fluid equations through their current, which appears in the bulk plasma momentum equation [415],

$$\varrho_b \frac{d\boldsymbol{U}_b}{dt} = -\nabla p_b + (\boldsymbol{j} - \boldsymbol{j}_h) \times \boldsymbol{B}/c - en_h \boldsymbol{E}, \tag{2.61}$$

where $\varrho_b, \boldsymbol{U}_b$, and p_b are the bulk plasma density, velocity, and pressure; $\boldsymbol{j}_h$ is the hot ion current density; $\boldsymbol{j}$ is total current density; and $\boldsymbol{B}$ and $\boldsymbol{E}$ are magnetic and electric fields. Note that the second term on the right side of (2.61) is Ampere's force acting on the bulk component and the third term represents the electric force acting on the excess electrons in the bulk plasma (quasineutrality is assumed).

We can write other equations for the bulk plasma:

$$\boldsymbol{E} = -\boldsymbol{U}_b \times \boldsymbol{B}/c, \tag{2.62}$$

$$\boldsymbol{B} = \boldsymbol{B}_0 + \nabla \times \boldsymbol{A}, \tag{2.63}$$

$$\partial \boldsymbol{A}/\partial t = -c\boldsymbol{E}, \tag{2.64}$$

$$\boldsymbol{j} = \frac{c}{4\pi} \nabla \times \boldsymbol{B}, \tag{2.65}$$

$$\partial p_b^{1/\gamma}/\partial t + \nabla \cdot (\boldsymbol{U}_b p_b^{1/\gamma}) = 0, \tag{2.66}$$

$$\partial \varrho_b/\partial t + \nabla \cdot (\boldsymbol{U}_b \varrho_b) = 0. \tag{2.67}$$

Here $\boldsymbol{B}_0$ is the equilibrium magnetic field, $\boldsymbol{A}$ is a modified vector potential (different from the usual vector potential by a term involving the gradient of the scalar potential) and an adiabatic equation of state is used. We also assume that the hydromagnetic approach for disturbances is justified by the presence of the dense cold population, which allows the parallel electric fields to be neglected.

2.2.6.2 Gyrokinetic Hot Ions.

To include the hot ion kinetic effects such as finite Larmor radius and wave–particle resonances, the low-frequency gyrokinetic equations are employed to describe the hot particle dynamics. Here we assume that

$$\frac{\omega}{\Omega_i} \propto \frac{\varrho_{\text{ch}}}{L} \propto \frac{k_\parallel}{k_\perp} \propto \frac{e\varphi}{T} \propto \frac{\delta B}{B} = O(\epsilon), \quad k_\perp \varrho_{\text{ch}} = O(1), \quad (2.68)$$

is adopted in this paper. Here ϱ_{ch} is the gyroradius, L is the equilibrium scale length, $k_\parallel$ and $k_\perp$ are the parallel and perpendicular wave numbers, δB and φ are the perturbation of magnetic field and electrostatic potential, and $\epsilon \ll 1$ is the smallness parameter.

Let us first apply Catto's gyrokinetic change of variables from $\boldsymbol{x}, \boldsymbol{v}$ to $\boldsymbol{R}$, $\mu, v_\parallel, \theta$ to the Vlasov equation in general geometry [93, 294],

$$\frac{\partial F}{\partial t} + \left(\boldsymbol{v}_\parallel + \frac{q_k}{M_k}\frac{\boldsymbol{E}\times\hat{\boldsymbol{b}}}{\Omega_k}\right)\cdot\frac{\partial F}{\partial \boldsymbol{R}} - \Omega_k\frac{\partial F}{\partial\theta}$$
$$+\boldsymbol{v}\cdot\left[\left(\frac{\partial\boldsymbol{\varrho}}{\partial\boldsymbol{x}}\right)\cdot\frac{\partial F}{\partial \boldsymbol{R}} + \frac{\partial\mu}{\partial\boldsymbol{x}}\frac{\partial F}{\partial\mu} + \frac{\partial v_\parallel}{\partial\boldsymbol{x}}\frac{\partial F}{\partial v_\parallel} + \frac{\partial\theta}{\partial\boldsymbol{x}}\frac{\partial F}{\partial\theta}\right]$$
$$+\frac{q_k}{M_k}\boldsymbol{E}\cdot\left(\frac{\boldsymbol{v}_\perp}{B}\frac{\partial F}{\partial\mu} + \hat{\boldsymbol{b}}\frac{\partial F}{\partial v_\parallel} + \frac{\hat{\boldsymbol{b}}\times\boldsymbol{v}_\perp}{v_\perp^2}\frac{\partial F}{\partial\theta}\right) = 0, \quad (2.69)$$

where $F(\boldsymbol{R},\mu,v_\parallel,\theta,t)$ is the distribution function, $\boldsymbol{E}(\boldsymbol{x})$ is the perturbed electric field, $B(\boldsymbol{x}) = |\boldsymbol{B}|$, and $\hat{\boldsymbol{b}}(\boldsymbol{x}) = \boldsymbol{B}/B$, where $\boldsymbol{B}$ is the external magnetic field, q_k and M_k are charge and mass of particle of sort k, $\mu = v_\perp^2/2B$, $\boldsymbol{\varrho} = \boldsymbol{v}_\perp\times\hat{\boldsymbol{b}}/\Omega_k$, $\Omega_k = q_k B/M_k c$, $\boldsymbol{v}_\perp = v_\perp(\cos\theta\hat{\boldsymbol{e}}_1 + \sin\theta\hat{\boldsymbol{e}}_2)$, $\boldsymbol{v}_\parallel = v_\parallel\hat{\boldsymbol{b}}$, $\hat{\boldsymbol{b}} = \hat{\boldsymbol{e}}_1\times\hat{\boldsymbol{e}}_2$, $\boldsymbol{R} = \boldsymbol{x} + \boldsymbol{\varrho}$, and

$$\frac{\partial\mu}{\partial\boldsymbol{x}} = -\frac{\mu}{B}\frac{\partial B}{\partial\boldsymbol{x}} - \frac{v_\parallel}{B}\left(\frac{\partial\hat{\boldsymbol{b}}}{\partial\boldsymbol{x}}\right)\cdot\boldsymbol{v}_\perp,$$

$$\frac{\partial v_\parallel}{\partial\boldsymbol{x}} = \left(\frac{\partial\hat{\boldsymbol{b}}}{\partial\boldsymbol{x}}\right)\cdot\boldsymbol{v}_\perp,$$

$$\frac{\partial\theta}{\partial\boldsymbol{x}} = \frac{v_\parallel}{v_\perp^2}\left(\frac{\partial\hat{\boldsymbol{b}}}{\partial\boldsymbol{x}}\right)\cdot(\boldsymbol{v}_\perp\times\hat{\boldsymbol{b}}) + \left(\frac{\partial\hat{\boldsymbol{e}}_2}{\partial\boldsymbol{x}}\right)\cdot\hat{\boldsymbol{e}}_1. \quad (2.70)$$

The electric field is given by

$$\boldsymbol{E} = -\frac{\partial\varphi}{\partial\boldsymbol{x}}, \quad (2.71)$$

where $\varphi(\boldsymbol{x})$ is the electrostatic potential. Invoking the gyrokinetic ordering of $\omega/\Omega_k \sim \epsilon$, $\varrho/L \sim \epsilon$, and $L \sim L_\parallel$, where ϵ is a smallness parameter, ϱ is the gyroradius, L is the equilibrium scale length and $L_\parallel$ is the perturbed parallel scale length, one can write the lowest order equation of (2.70) as

$$\Omega_k \frac{\partial F}{\partial \theta} = 0. \tag{2.72}$$

Here, the perturbed fields are also considered to be $O(\epsilon)$. In addition,

$$F = f + \epsilon g(\theta), \tag{2.73}$$

where f is the solution of (2.72) and is independent of phase, θ. Equation (2.70) to the next order reduces to

$$\frac{\partial f}{\partial t} + \left(\boldsymbol{v}_\parallel + \frac{q_k}{M_k} \frac{\boldsymbol{E} \times \hat{\boldsymbol{b}}}{\Omega_k} \right) \cdot \frac{\partial f}{\partial \boldsymbol{R}} - \Omega_k \frac{\partial}{\partial \theta} \left(g - \frac{q_k}{M_k} \frac{\varphi}{B} \frac{\partial f}{\partial \mu} \right)$$

$$+ \boldsymbol{v} \cdot \left[\left(\frac{\partial \boldsymbol{\varrho}}{\partial \boldsymbol{x}} \right) \cdot \frac{\partial f}{\partial \boldsymbol{R}} + \frac{\partial \mu}{\partial \boldsymbol{x}} \frac{\partial f}{\partial \mu} + \frac{\partial v_\parallel}{\partial \boldsymbol{x}} \frac{\partial f}{\partial v_\parallel} \right] + \frac{q_k}{M_k} \boldsymbol{E} \cdot \hat{\boldsymbol{b}} \frac{\partial f}{\partial v_\parallel} = 0. \tag{2.74}$$

Note the relations

$$\frac{\partial \varphi}{\partial \boldsymbol{x}} \approx \frac{\partial \varphi}{\partial \boldsymbol{R}}, \tag{2.75}$$

which is correct to order ϵ, and

$$\Omega_k \frac{\partial \varphi}{\partial \theta} = -\boldsymbol{v}_\perp \cdot \frac{\partial \varphi}{\partial \boldsymbol{R}} \tag{2.76}$$

have been used in order to arrive at (2.74). By taking the gyrophase average of (2.74), we then recover the usual drift-kinetic equation

$$\frac{\partial f}{\partial t} + \left(\boldsymbol{v}_\parallel + \boldsymbol{v}_d + \frac{q_k}{M_k} \frac{\langle \boldsymbol{E} \rangle \times \hat{\boldsymbol{b}}}{\Omega_k} \right) \cdot \frac{\partial f}{\partial \boldsymbol{R}}$$

$$+ \left(a_\parallel + \frac{q_k}{M_k} \langle \boldsymbol{E} \rangle \cdot \hat{\boldsymbol{b}} \right) \frac{\partial f}{\partial v_\parallel} = 0, \tag{2.77}$$

where $\langle \boldsymbol{E} \rangle = (2\pi)^{-1} \oint \boldsymbol{E} d\theta$,

$$\boldsymbol{v}_d = \frac{1}{2\pi} \oint \boldsymbol{v} \cdot \left(\frac{\partial \boldsymbol{\varrho}}{\partial \boldsymbol{x}} \right) d\theta = \hat{\boldsymbol{b}} \times \left[\frac{v_\parallel^2}{\Omega_k} \left(\hat{\boldsymbol{b}} \cdot \frac{\partial}{\partial \boldsymbol{R}} \right) \hat{\boldsymbol{b}} + \frac{v_\perp^2}{2\Omega_k} \frac{\partial \ln B}{\partial \boldsymbol{R}} \right] \tag{2.78}$$

and

$$a_\parallel = \frac{1}{2\pi} \oint \boldsymbol{v} \cdot \frac{\partial v_\parallel}{\partial \boldsymbol{x}} d\theta = -\frac{v_\perp^2}{2} \left(\hat{\boldsymbol{b}} \cdot \frac{\partial \ln B}{\partial \boldsymbol{R}} \right) \tag{2.79}$$

are the curvature, ∇B drifts and parallel acceleration, respectively. The parallel drift is ignored in (2.77) because it is of higher order in ϵ (see [93]) and $\oint \boldsymbol{v} \cdot (\partial \mu / \partial \boldsymbol{x}) d\theta = 0$. Together with (2.71) and

$$\mu = v_\perp^2/2B - \text{const.},\tag{2.80}$$

(2.77) can be used for particle pushing for electrons in an electrostatic plasma in the general geometry with $\langle E \rangle \approx E$. In general, g can be determined uniquely through the integration in θ of the difference between (2.74) and (2.77). However, if we ignore the term $O(k_\perp^4 \varrho^4)$ in g, it can be shown that

$$g = \frac{q_k}{M_k B} \frac{\partial f}{\partial \mu}(\varphi - \langle \varphi \rangle),\tag{2.81}$$

where the gyrophase-averaged potential $\langle \varphi \rangle$ is introduced here to insure that g contains only the phase-dependent part of the distribution. Equation (2.73) then gives

$$F = f + \frac{q_k}{M_k B} \frac{\partial f}{\partial \mu}(\varphi - \langle \varphi \rangle).\tag{2.82}$$

Substituting (2.82) into (2.69), and again utilizing (2.75–2.76) together with

$$\frac{\partial \varphi}{\partial v_\parallel} = 0 \quad \text{and} \quad \frac{\partial \varphi}{\partial \mu} = -\frac{B}{v_\perp^2} \varrho \cdot \frac{\partial \varphi}{\partial \mathbf{R}},\tag{2.83}$$

we can write the gyrokinetic equation as

$$\frac{\partial F}{\partial t} + \left(v_\parallel + \frac{q_k}{M_k} \frac{\mathbf{E} \times \hat{\mathbf{b}}}{\Omega_k}\right) \cdot \frac{\partial F}{\partial \mathbf{R}} + \frac{q}{M_k} \mathbf{E} \cdot \hat{\mathbf{b}} \frac{\partial F}{\partial v_\parallel}$$
$$+ \mathbf{v} \cdot \left[\left(\frac{\partial \varrho}{\partial \mathbf{x}}\right) \cdot \frac{\partial F}{\partial \mathbf{R}} + \frac{\partial \mu}{\partial \mathbf{x}} \frac{\partial F}{\partial \mu} + \frac{\partial v_\parallel}{\partial \mathbf{x}} \frac{\partial F}{\partial v_\parallel} + \frac{\partial \theta}{\partial \mathbf{x}} \frac{\partial F}{\partial \theta}\right]$$
$$+ \frac{\Omega_k}{B^2}\left(\frac{q_k}{M_k}\right)^2 \left(\frac{1}{2} \frac{\partial^2 f}{\partial \mu^2} \frac{\partial \varphi^2}{\partial \theta} - \frac{\partial f}{\partial \mu} \frac{\partial \varphi}{\partial \theta} \frac{\partial \langle \varphi \rangle}{\partial \mu}\right) = 0.\tag{2.84}$$

The last terms in the above equation can be eliminated through gyrophase averaging. Thus, it is suitable for the development of gyrokinetic particle simulation schemes for electrostatic plasmas.

In the previous paragraph we took into account the external magnetic field only. In the case of a nonstationary magnetic field, one can use the gyrokinetic equations derived from a systematic Hamiltonian theory [67, 93, 138, 212].

The dynamical equations, which were derived using the action-variational method, are as follows [39]:

$$\dot{\mathbf{R}} = \frac{1}{B_\parallel^{**}}[\mathbf{B}^{**} U - \hat{\mathbf{b}} \times (\langle \mathbf{E} \rangle - \mu \nabla B_0 + \nabla \langle \mathbf{v}_\perp \cdot \mathbf{A} \rangle)],\tag{2.85}$$

$$\dot{U} = \frac{\mathbf{B}^{**}}{B_\parallel^{**}} \cdot (\langle \mathbf{E} \rangle - \mu \nabla B_0 + \nabla \langle \mathbf{v}_\perp \cdot \mathbf{A} \rangle),\tag{2.86}$$

$$\dot{\mu} = 0,\tag{2.87}$$

where $\bm{B}^{**} = \bm{B}_0 + \langle \delta \bm{B} \rangle$, $B_\parallel^{**} = B_0 + \langle \delta B_\parallel \rangle$, $\delta \bm{B} = \nabla \times \bm{A}$; $\hat{\bm{b}}$ is the unit vector along the equilibrium magnetic field; $\bm{v}_\perp$ is particle perpendicular velocity; $\langle \rangle = \oint \mathrm{d}\theta/2\pi$ and as units in which $e = m = c = 1$ is used. In (2.85–2.86) $B_0 = B_0(\bm{R})$ is the value of the equilibrium magnetic field at the gyrocenter position, while all the perturbed fields are taken at the particle position $\bm{x} = \bm{R} + \bm{\varrho}_\mathrm{h}$, where $\bm{\rho}_\mathrm{h}$ is gyroradius vector. The particle magnetic moment μ is an invariant of motion and may be treated as a constant parameter in the simulation.

The gyrocenter particles are described in terms of the gyrocenter distribution function $F(\bm{R}, U, \mu)$, so that transformation of the zero and the first velocity momentums is required to calculate the hot ion density and current density in physical space:

$$n_\mathrm{h}(\bm{x}) = \int \delta(\bm{R} + \bm{\varrho}_\mathrm{h} - \bm{x}) F(\bm{R}, U, \mu) \mathrm{d}^3 \bm{R} \mathrm{d}U \mathrm{d}\mu \mathrm{d}\theta,$$

$$\bm{j}_\mathrm{h}(\bm{x}) = \int (\dot{\bm{R}} + \bm{v}_\perp) \delta(\bm{R} + \bm{\varrho}_\mathrm{h} - \bm{x}) F(\bm{R}, U, \mu) \mathrm{d}^3 \bm{R} \mathrm{d}U \mathrm{d}\mu \mathrm{d}\theta. \qquad (2.88)$$

In the code described here [39] this is done by distributing each gyrokinetic particle as four or eight subparticles uniformly around the gyro-orbit with the center at $\bm{R}$ and radius $\bm{\varrho}_\mathrm{h}$, and summing up the contributions from all subparticles when calculating current and density. The resulting n_h and $\bm{j}_\mathrm{h}$ are then substituted into the bulk momentum equation (2.61). Spatial gyrophase averaging in (2.85–2.86) and the transformation to the physical space in (2.87–2.88) are done in the same way.

Equations (2.61–2.67) and (2.85–2.88) constitute a hybrid MHD-gyrokinetic model for describing low-frequency MHD-type phenomena in the case where the parallel electric field effects can be neglected.

2.2.6.3 δF Method. Recently, improvements of the nonlinear particle simulation approach have been proposed in the form of the so-called δF methods [277, 509]. Additional development of the δF algorithm may be found in [34, 82, 132]. In these algorithms, the full distribution f is separated into a known or "background" part f_0 and a perturbed part δf:

$$f = f_0 + \delta f. \qquad (2.89)$$

The perturbed part of the distribution function is calculated with a set of characteristics of (2.85–2.86) by assigning a weight $w \sim \delta f$ to each simulation particle. The particle weights are then used to calculate the perturbed hot ion density δn_h and current density $\delta \bm{j}_\mathrm{h}$. Since the particle weight is no longer a constant of the motion (as in the conventional particle simulation), the evolution equation for the weight has to be added to the equations of motion (2.85–2.86). We can write the gyrocenter distribution function $F(\bm{R}, U, \mu, t)$ assuming that F includes the Jacobian of the transformation from physical-space coordinates to gyrocenter coordinates as follows [352]:

2.2 Hybrid Quasineutral Models

$$F = B_\parallel^{**} f(\mathbf{R}, U, \mu, t), \tag{2.90}$$

where $J = B_\parallel^{**}$ is the Jacobian, $B_\parallel^{**} = B_0 + \langle \delta B_\parallel \rangle$, and $f(\mathbf{R}, U, \mu, t)$ is the particle distribution function expressed in gyrocenter coordinates. The number of particles in a space volume element $d\Gamma = B_\parallel^{**} d\mathbf{Z}$ is equal to $dN = F d\mathbf{Z} = f d\Gamma$, where $\mathbf{Z} = (\mathbf{R}, U, \mu)$. The conservation of particles implies [39]:

$$\frac{\partial}{\partial t} F + \frac{\partial}{\partial \mathbf{R}} \cdot (\dot{\mathbf{R}} F) + \frac{\partial}{\partial U} (\dot{U} F) = 0. \tag{2.91}$$

Since the equations of motion (2.85–2.87) preserve the phase-space volume $d\Gamma$, the Liouville theorem holds,

$$\frac{\partial}{\partial t} B_\parallel^{**} + \frac{\partial}{\partial \mathbf{X}} \cdot (\dot{\mathbf{X}} B_\parallel^{**}) + \frac{\partial}{\partial U} (\dot{U} B_\parallel^{**}) = 0; \tag{2.92}$$

(2.91–2.92) become the Vlasov equation in gyrocenter coordinates:

$$\frac{df}{dt} = \frac{\partial f}{\partial t} + \dot{\mathbf{R}} \cdot \frac{\partial f}{\partial \mathbf{R}} + \dot{U} \frac{\partial f}{\partial U} = 0. \tag{2.93}$$

The moments of F in the simulation are calculated using a limited number of points in phase space at which the value of F is known. The positions of these points correspond to the positions of the simulation particles which are regarded as Lagrangian markers [19, 129, 242]. We shall distinguish between the physical particle described by the distribution function F and the simulation (marker) particles, which are described by a discrete distribution function $\hat{P}(\mathbf{Z}, t)$. The marker distribution can be written in the Klimontovich representation as

$$\hat{P}(\mathbf{Z}, t) = \sum_{m=1}^{M} \delta(\mathbf{R} - \mathbf{R}_m) \delta(U - U_m) \delta(\mu - \mu_m), \tag{2.94}$$

where m is the marker index and M is the total number of marker particles in the simulation. The particle pushing is equivalent to solving the equation

$$\frac{\partial}{\partial t} \hat{P} + \frac{\partial}{\partial \mathbf{Z}} \cdot (\dot{\mathbf{Z}} \hat{P}) = 0, \tag{2.95}$$

which has the same form as (2.91) for F.

We define the smooth marker distribution function $P(\mathbf{Z}, t)$ as an ensemble average of $\hat{P}$, and then define particle weight as

$$w = \frac{B_\parallel^{**} \delta f}{P}, \tag{2.96}$$

where $\delta f = f - f_0$. If P is chosen such that $P(\mathbf{Z}, t) = F(\mathbf{Z}, t)$, then from (2.90) we have

$$B_\parallel^{**} f_0 = (1-w)P, \tag{2.97}$$

and the following time evolution equation for w can be obtained using (2.91) and (2.93):

$$\dot{w} = -(1-w)\frac{1}{f_0}\frac{df_0}{dt}. \tag{2.98}$$

Since $\delta F = [w + (1-w)\langle \delta B_\parallel \rangle/B_\parallel^{**}]P$, in the simulation δF can be approximated as the weighted Klimontovich distribution [39]:

$$\delta F \approx \sum_{m=1}^{M} d_m \delta(\boldsymbol{R}-\boldsymbol{R}_m)\delta(U-U_m)\delta(\mu-\mu_m), \tag{2.99}$$

where we have defined

$$d_m = w_m + (1-w_m)\frac{\langle \delta B_\parallel \rangle}{B_\parallel^{**}}\bigg|_{\boldsymbol{z}=\boldsymbol{z}_m}. \tag{2.100}$$

Let us consider the calculation of a phase-space integral of the general form

$$I(A) = \int A(\boldsymbol{Z},t) F(\boldsymbol{Z},t) d\boldsymbol{Z}. \tag{2.101}$$

Separating the zero-order part and the perturbation, we can rewrite (2.101) as

$$I(A) = \int A_0 F_0 d\boldsymbol{Z} + \int (A_0 \delta F + \delta A F) d\boldsymbol{Z}, \tag{2.102}$$

where $A = A_0 + \delta A$ and $F = F_0 + \delta F$. We let $F \approx \hat{P}$ and use (2.99) to obtain the following from (2.102):

$$I(A) = \int A_0 F_0 d\boldsymbol{Z} + \sum_{m=1}^{M}(A_0 d_m + \delta A)|_{\boldsymbol{z}=\boldsymbol{z}_m}. \tag{2.103}$$

Let $A = A_0 = \langle \delta(\boldsymbol{R}+\boldsymbol{\varrho}_0-\boldsymbol{x})\rangle$, so that the particle density is

$$n_\mathrm{h}(\boldsymbol{x}) = n_{\mathrm{h}0}(\boldsymbol{x}) + \sum_{m=1}^{M} d_m \langle \delta(\boldsymbol{R}_m + \boldsymbol{\varrho}_m - \boldsymbol{x})\rangle. \tag{2.104}$$

Choosing $A_0 = \langle (\dot{\boldsymbol{R}}_0 + \boldsymbol{V}_\perp)\delta(\boldsymbol{R}+\boldsymbol{\varrho}_\mathrm{h}-\boldsymbol{x})\rangle$ and $\delta A = \delta\dot{\boldsymbol{R}}\langle \delta(\boldsymbol{R}+\boldsymbol{\varrho}_\mathrm{h}-\boldsymbol{x})\rangle$ in (2.103) gives

$$\boldsymbol{j}_\mathrm{h}(\boldsymbol{x}) = \boldsymbol{j}_{\mathrm{h}0}(\boldsymbol{x}) + \sum_{m=1}^{M}\langle [(\dot{\boldsymbol{R}}_{0m}+\boldsymbol{V}_{\perp m})d_m + \delta\dot{\boldsymbol{R}}_m]\delta(\boldsymbol{R}_m+\boldsymbol{\varrho}_m-\boldsymbol{x})\rangle, \tag{2.105}$$

where $\dot{\boldsymbol{R}}_0$ is the zero-order particle drift and $\delta\dot{\boldsymbol{R}} = \dot{\boldsymbol{R}} - \dot{\boldsymbol{R}}_0$. Assuming that gyrophase averages are calculated by distribution of the particle charge at N_g points on the ring $|\boldsymbol{x}-\boldsymbol{R}_m| = \varrho_m$, we can rewrite (2.104) and (2.105) as

$$n_\mathrm{h}(\boldsymbol{x}) = n_\mathrm{h0}(\boldsymbol{x}) + \sum_{m=1}^{M} \frac{1}{N_\mathrm{g}} \sum_{s=1}^{N_\mathrm{g}} d_m \langle \delta(\boldsymbol{R}_m + \boldsymbol{\varrho}_{m,s} - \boldsymbol{x}) \rangle, \quad (2.106)$$

$$\boldsymbol{j}_\mathrm{h}(\boldsymbol{x}) = \boldsymbol{j}_\mathrm{h0}(\boldsymbol{x}) + \sum_{m=1}^{M} \frac{1}{N_\mathrm{g}} \sum_{s=1}^{N_\mathrm{g}} [(\dot{\boldsymbol{R}}_{0m} + \boldsymbol{V}_{\perp m,s}) d_m + \delta \dot{\boldsymbol{R}}_m] \delta(\boldsymbol{R}_m + \boldsymbol{\varrho}_{m,s} - \boldsymbol{x}),$$
$$(2.107)$$

where index s denotes the sth point on the ring. Equations (2.106) and (2.107) were used in this code for calculation of the perturbed hot ion density and current. Here the definition of the weight is different from that of [242], where the general case of compressible particle dynamics was considered.

Despite superficial resemblances between the δF algorithm and the Vlasov hybrid simulation (VHS) algorithm (Sect. 1.4.3, see also [397]), the two are quite different. In the δF algorithm particles are distributed evenly in the phase fluid, and the phase volume associated with each particle is presumed to be preserved during the simulation. Consequently no velocity grid is required, and $\{\delta F \boldsymbol{v} d\boldsymbol{x} d\boldsymbol{v}\}$ is distributed to the nearest grid points as in particle in cell (PIC) codes. By contrast, with VHS, particle density in the phase fluid is freely variable above a certain minimum and F (or δF) must be truly interpolated onto a velocity-space grid.

The δF method has been applied to the simulation of the one-dimensional Vlasov equation, to a purely magnetic electron drift-kinetic Ampere's law and to other problems. The Vlasov code was found to considerably outperform corresponding PIC codes. In [277], it was pointed out that for many fusion problems $\delta f \ll f$, and in these cases PIC codes are highly noisy and very inefficient. The hybrid MHD–gyrokinetic model considered above is suitable for self-consistent study of the interaction of energetic particles with low-frequency MHD waves in high-beta space plasmas. The gyrokinetic model is also suitable for studying linear and nonlinear low-frequency micro-instabilities and associated anomalous transport in magnetically confined plasmas. The gyrokinetic description enables one to remove the restrictions on the particle time step dictated by the gyromotion, while the δF algorithm strongly reduces the numerical noise level in the simulation plasma. Therefore, considerably larger time steps and a smaller number of particles can be used in the simulations as compared to conventional methods.

2.2.7 Guiding-Center-Ion–Fluid-Electron Model

In the simulation of the low-frequency electromagnetic phenomena in kinetic plasma, such as kinetic Alfvén waves and magnetic field reconnection, there is a need to develop codes which include the substantial wave–particle interactions in its longitudinal (non-MHD) electric field [219]. The full particle methods provide a kinetic simulation of the low-frequency electromagnetic phenomena while eliminating the high-frequency electron plasma oscillations.

However, one difficulty with these methods was the small time step necessary to reproduce the diamagnetic and magnetization effects of the electrons in a strong magnetic field and high-beta plasmas. The approach developed below allows us to simulate these phenomena using a much larger time step ($\Omega_{i,e}\Delta t \gg 1$). In a strong magnetic field, the gyroradius of the particles may be smaller than the space scale $\varrho_{ci}, \varrho_{ce} \ll L$, and the gyrofrequency may be higher than the time scale $\Omega_{i,e} \gg \partial \ln(B)/\partial t$. So for this plasma system it is promising to use the drift approximation [396, 490].

The Vlasov equation for the guiding center distribution function f may be written as following [see Sect. 2.2.6 and (2.77–2.80)]

$$\frac{\partial}{\partial t}f + (\boldsymbol{U}_\perp + \hat{\boldsymbol{b}}v_\|)\nabla f + \dot{v}_\| \frac{\partial}{\partial v_\|} f = 0, \tag{2.108}$$

where $f = f(t, \boldsymbol{x}, v_\|, \mu)$ and $v_\|$ and $\boldsymbol{U}_\perp$ are particle velocities along and perpendicular to the magnetic field. It is comfortable to consider the guiding center distribution function instead of the particle distribution function:

$$\mathrm{d}n = f(\boldsymbol{x}, \boldsymbol{v}, t)\mathrm{d}^3\boldsymbol{x}\mathrm{d}^3\boldsymbol{v} = F(\boldsymbol{R}_{\mathrm{gc}}, \boldsymbol{V}, t)\mathrm{d}^3\boldsymbol{R}_{\mathrm{gc}}\mathrm{d}^3\boldsymbol{V}, \tag{2.109}$$

where F is a number of guiding centers in the volume $\mathrm{d}^3\boldsymbol{R}_{\mathrm{gc}}$. Finally we have

$$f(\boldsymbol{x}, \boldsymbol{v}, t)\mathrm{d}^3\boldsymbol{v} = F(\boldsymbol{R}_{\mathrm{gc}}, \boldsymbol{V}, t) J\left(\frac{\boldsymbol{R}_{\mathrm{gc}}}{\boldsymbol{x}}\right)\mathrm{d}^3\boldsymbol{V},$$

$$\boldsymbol{R}_{\mathrm{gc}} = \boldsymbol{x} + \boldsymbol{V} \times \hat{\boldsymbol{b}}/\omega + O(\varepsilon^2), \tag{2.110}$$

where $\boldsymbol{R}_{\mathrm{gc}}$ and $\boldsymbol{V}$ are the radius vector of the position and the velocity of a guiding center.

The equation of guiding center motion may be written as

$$\frac{M}{Ze}\frac{\mathrm{d}\dot{\boldsymbol{R}}_{\mathrm{gc}}}{\mathrm{d}t} = \boldsymbol{E} + \frac{\dot{\boldsymbol{R}}_{\mathrm{gc}} \times \boldsymbol{B}}{c} - \frac{\mu \nabla B}{Ze} + O(\varepsilon^2). \tag{2.111}$$

For transverse and longitudinal particle motion, we have

$$\boldsymbol{U}_\perp = \left(c\boldsymbol{E} - \frac{\mu c \nabla B}{Ze} - \frac{Mc\ddot{\boldsymbol{R}}_{\mathrm{gc}}}{Ze}\right) \times \frac{\hat{\boldsymbol{b}}}{B}, \tag{2.112}$$

$$\frac{M\ddot{\boldsymbol{R}}_{\mathrm{gc}} \cdot \hat{\boldsymbol{b}}}{Ze} = \boldsymbol{E} \cdot \boldsymbol{b} - \frac{\mu}{Ze}\frac{\partial B}{\partial s} + O(\varepsilon^2), \tag{2.113}$$

$$\frac{\mathrm{d}\boldsymbol{x}}{\mathrm{d}t} = \boldsymbol{U}_\perp + v_\|\hat{\boldsymbol{b}}. \tag{2.114}$$

Equations (2.112) and (2.113) may be written in terms of $v_\|$ and $E_\|$:

$$\boldsymbol{U}_\perp = \left(c\boldsymbol{E} - \frac{\mu c \nabla B}{Ze} - \frac{Mc}{Ze}v_\|^2\frac{\partial \hat{\boldsymbol{b}}}{\partial x_\|}\right) \times \frac{\hat{\boldsymbol{b}}}{B}, \tag{2.115}$$

2.2 Hybrid Quasineutral Models

$$\frac{d}{dt}v_{\|} = \frac{ZeE_{\|}}{M} - \frac{\mu}{M}\frac{\partial B}{\partial x_{\|}}. \tag{2.116}$$

In the considered model we assume that magnetic moment of the particles μ is adiabatically invariant:

$$\mu_{i,e} = \frac{p_{i,e\perp}}{nB} = \frac{\beta_{i,e}B_{\infty}}{8\pi n_{\infty}} = \text{const.}, \quad \varepsilon = \frac{\varrho_{ci}}{L} \quad \text{and} \quad \varrho_{ci} = \frac{v_{\perp\infty}}{\Omega_i}, \tag{2.117}$$

where $\hat{\boldsymbol{b}} = \boldsymbol{B}/|B|$.

In some cases, for example, in solar wind, there is a relation between the velocities, $v_{Ti} < v_{ph} \ll v_{Te}$, and hence the electron distribution function, f_e, may be considered to be the Maxwellian [9, 92]:

$$f_e \propto \exp\left[-\frac{m(v_{\|} - U_{\|})^2}{2kT_e} - \frac{\mu_e B}{kT_e} - \frac{e}{kT_e}\int_{\infty}^{r} d\boldsymbol{x}\left(\boldsymbol{E} + \frac{\boldsymbol{v}_e \times \boldsymbol{B}}{c}\right)\right]. \tag{2.118}$$

The condition of quasineutrality, $n_e \approx n_i$, allows us to consider the ion dynamics only. Ampere's law, taking the drift current into account, has the following form:

$$\nabla \times \boldsymbol{B} = \frac{4\pi}{B^2}\left[-p_{\perp}\left(\nabla B \times \hat{\boldsymbol{b}}\right) + p_{\|}B\boldsymbol{k} \times \hat{\boldsymbol{b}} - \varrho B\dot{\boldsymbol{v}} \times \hat{\boldsymbol{b}}\right] - 4\pi\nabla \times \left(\frac{p_{\perp}\hat{\boldsymbol{b}}}{B}\right), \tag{2.119}$$

where

$$p_{\perp} = B\int d^3v\mu F_i \quad \text{and} \quad p_{\|} = M_i\int d^3v(v_{\|} - U_{\|})^2 F_i. \tag{2.120}$$

Here $p_{\perp}$ and $p_{\|}$ are the total pressures in the transverse and longitudinal directions, $\boldsymbol{k}$ is the curvature of the magnetic field line, and R_{cur} is the radius of curvature of the magnetic field line, $|\boldsymbol{k}| = R_{cur}^{-1}$. Curvature $\boldsymbol{k}$ and parallel pressure of plasma $p_{\|}$ are

$$\boldsymbol{k} = \hat{\boldsymbol{b}} \times (\nabla \times \hat{\boldsymbol{b}}) \quad \text{and} \quad p_{\|} \propto \varrho^3/B^2. \tag{2.121}$$

The calculation of the moments of the electron distribution function (2.131) gives the following equation (Ohm's law) for the electric field:

$$\boldsymbol{E} = -\frac{\boldsymbol{v}_e \times \boldsymbol{B}}{c} - \frac{\nabla(nkT_e)}{en}. \tag{2.122}$$

The total current is a sum of the current through the ion guiding center, the diamagnetic ion current and the electron current:

$$\boldsymbol{J} = Zen\langle\dot{\boldsymbol{R}}_{gc}\rangle - c\nabla \times \left(\frac{p_{\perp i}\boldsymbol{B}}{B^2}\right) - en\boldsymbol{v}_e. \tag{2.123}$$

The rest equations for the magnetic field are

$$\nabla \times \boldsymbol{B} = \frac{4\pi}{c}\boldsymbol{J}, \quad \nabla \cdot \boldsymbol{B} = 0, \tag{2.124}$$

$$\nabla \times \boldsymbol{E} = -\frac{1}{c}\frac{\partial}{\partial t}\boldsymbol{B}. \tag{2.125}$$

The above model was used successfully in the solar-wind–nonmagnetic-planet interaction problems (see, e.g., [315]). Application for the solar-wind–Moon interaction will be considered in Sect. 12.5.

2.2.8 Particle-Electron–Fluid-Ion Model

Let us consider, as an example, an electrostatic drift wave destabilized by Landau damping of the electrons [509]. The frequency of the wave is given [257] by

$$\omega \approx \omega^* = k_\perp \varrho_s (\rho_s/L_n)\Omega_i,$$

where ϱ_s is the ion gyroradius at the electron temperature defined by $\varrho_s = (T_e/M)^{1/2}/\Omega_i$ and L_n is the density scale length of the plasma perpendicular to an external magnetic field. For $k_\parallel \varrho_s \leq 1$, which are the most important wavenumbers for drift instability, it is clear that $\omega \ll \Omega_i$, since the condition $\varrho_s/L_n \ll 1$ is usually satisfied. The drift waves can become unstable in the presence of electron Landau damping, and the resonance condition is given by

$$\omega \propto \omega^* \approx k_\parallel v_e.$$

It is thus found that

$$\frac{k_\parallel}{k_\perp} = \left(\frac{\varrho_s}{L_n}\right)\left(\frac{c_s}{v_e}\right) = \left(\frac{m}{M}\right)^{1/2}\left(\frac{\varrho_s}{L_n}\right) \ll \left(\frac{m}{M}\right)^{1/2},$$

as expected.

Let us consider an electrostatic drift wave in a uniform magnetic field, $\boldsymbol{B}_0$. Since quasineutrality is satisfied for a low-frequency wave, one has $\nabla \cdot \boldsymbol{J} = 0$. From this condition we find

$$\nabla_\parallel J_\parallel = -\nabla_\perp \cdot \boldsymbol{J}_\perp = -\nabla_\perp \cdot (\boldsymbol{J}_p + \boldsymbol{J}_{E\times B}), \tag{2.126}$$

where the $\boldsymbol{E} \times \boldsymbol{B}$ drift current, $\boldsymbol{J}_{E\times B}$, and the polarization current are given in [155] as follows:

$$\boldsymbol{J}_p = \frac{Mn}{B_0^2}\frac{d\boldsymbol{E}_\perp}{dt} = \frac{Mn_0}{B_0^2}\left[\frac{\partial}{\partial t} + \left(\frac{c\boldsymbol{E}\times\boldsymbol{B}_0}{B_0^2}\right)\cdot\nabla_\perp\right]\boldsymbol{E}_\perp, \tag{2.127}$$

$$\boldsymbol{J}_{E\times B} = en(1+\varrho_i^2\nabla^2)\frac{c\boldsymbol{E}\times\boldsymbol{B}_0}{B_0^2} - en\frac{c\boldsymbol{E}\times\boldsymbol{B}_0}{B_0^2}$$

$$= (en_0\varrho_i^2\nabla^2)\frac{c\boldsymbol{E}\times\boldsymbol{B}_0}{B_0^2}. \tag{2.128}$$

Since we are considering a plasma in a nearly uniform magnetic field, the currents due to the gradient and curvature of the magnetic field can be neglected in the case of the electrostatic waves.

Substituting $\boldsymbol{J}_\mathrm{p}$ and $\boldsymbol{J}_{\boldsymbol{E}\times\boldsymbol{B}}$ into $\nabla\cdot\boldsymbol{J}=0$ gives

$$\frac{\partial}{\partial t}\nabla_\perp^2\phi + \nabla_\perp\cdot\left(en_0\varrho_\mathrm{i}^2\nabla_\perp^2\frac{c\boldsymbol{E}\times\boldsymbol{B}_0}{B_0^2}\right) + \nabla_\perp\cdot\left[\left(\frac{c\boldsymbol{E}\times\boldsymbol{B}_0}{B_0^2}\cdot\nabla_\perp\right)\nabla_\perp\phi\right]$$

$$= \frac{B_0^2}{Mn_0}\nabla_\parallel J_\parallel^\mathrm{e}, \qquad (2.129)$$

where we neglect the field-aligned current associated with the ions, $\boldsymbol{J}_\parallel = \boldsymbol{J}_\parallel^\mathrm{e}$. As mentioned early, $\boldsymbol{J}_\parallel^\mathrm{e}$ is determined from the electrons simulated by the guiding-center drift approximations. It is clear from (2.129) that the ions are now represented by a fluid [401] moving under the influence of an electric field perpendicular to the magnetic field, with the finite gyroradius effects given by the second term on the left-hand side of (2.129). In addition to (2.129), which determines the potential ϕ, we use the following equations for electron dynamics:

$$\boldsymbol{v}_\perp = c\frac{\boldsymbol{E}\times\boldsymbol{B}}{B^2}, \qquad \frac{\mathrm{d}\boldsymbol{x}_\perp}{\mathrm{d}t} = \boldsymbol{v}_\perp, \qquad (2.130)$$

$$\frac{\mathrm{d}v_\parallel}{\mathrm{d}t} = \frac{q}{m}E_\parallel, \qquad \frac{\mathrm{d}\boldsymbol{x}_\parallel}{\mathrm{d}t} = \boldsymbol{v}_\parallel. \qquad (2.131)$$

Since in our model the parallel currents associated with the ions are neglected, ion-acoustic waves ($\omega^2 = k_\parallel^2 c_s^2$) are not retained in the model. In the presence of ∇B and curvature drifts, it is necessary to determine the currents perpendicular to the magnetic field associated with these drifts. Since these drifts include the particle energy parallel and perpendicular to the magnetic fields, equations which determine the ion pressure $p_\perp^\mathrm{i}$ and $p_\parallel^\mathrm{i}$, must be supplemented [220], while the electron pressure may be calculated from the moment of the electron distribution function. The further modification of this model, which includes fluid approximation for ions, may be found in [256, 356].

2.3 Particle Nonneutral Models

The section is devoted to nonneutral full particle models which are appropriated for the simulation of the multiscale electromagnetic phenomena ranging between electron plasma and ion cyclotron frequencies. At first we consider briefly the different numerical full particle models. The second section will address modified particle (marker) models, namely, particle–guiding-center, and guiding-center–guiding-center models.

2.3.1 Full Particle Models

In this section we shall consider the full particle models in which the ion and electron are described in a kinetic approach. In some models we use the nonradiative Maxwell's equations and in other models we take into account the electromagnetic radiation. Why do we have to know how the full particle codes operate? First of all, in some regimes we have to compare the results of hybrid simulation with results from full particle simulation. Full particle models can also take into account the charge separation effects. Second, many numerical ideas of implicit time integration of a particle trajectory or the electromagnetic field equation (the moment method, the direct implicit method) may be successfully used for multiscale hybrid code design.

The fundamental equations of full particle models are as follows: The Vlasov equation for ion distribution function $f_s(t, \boldsymbol{x}, \boldsymbol{v})$ is of the form

$$\frac{\partial}{\partial t} f_s + \boldsymbol{v} \frac{\partial}{\partial \boldsymbol{x}} f_s + \frac{\boldsymbol{F}}{M_s} \frac{\partial}{\partial \boldsymbol{v}} f_s = 0. \tag{2.132}$$

The equations of particle motion are of the form

$$\frac{\mathrm{d}\boldsymbol{x}_{sl}}{\mathrm{d}t} = \boldsymbol{v}_{sl} \quad \text{and} \quad \frac{\mathrm{d}\boldsymbol{v}_{sl}}{\mathrm{d}t} = \frac{Z_s e}{M_s} \left(\boldsymbol{E} + \frac{\boldsymbol{v}_{sl} \times \boldsymbol{B}}{c} \right), \tag{2.133}$$

where s denotes the particle species and l the particle index. $f_s(\boldsymbol{x}, \boldsymbol{v})$ is the distribution function for particles of species s in $(\boldsymbol{x}, \boldsymbol{v})$. The net charge density and current are given by the moments of the distribution,

$$\varrho = \sum_s \rho_s = \sum_s q_s \int f_s(\boldsymbol{x}, \boldsymbol{v}) \mathrm{d}\boldsymbol{v} \tag{2.134}$$

and

$$\boldsymbol{J} = \sum_s \boldsymbol{J}_s = \sum_s q_s \int \mathrm{d}\boldsymbol{v} \boldsymbol{v} f_s(\boldsymbol{x}, \boldsymbol{v}). \tag{2.135}$$

The Maxwell's equations are

$$\frac{1}{c} \frac{\partial \boldsymbol{B}}{\partial t} + \nabla \times \boldsymbol{E} = 0, \quad \nabla \cdot \boldsymbol{B} = 0,$$

$$\frac{1}{c} \frac{\partial \boldsymbol{E}}{\partial t} - \nabla \times \boldsymbol{B} = -\frac{4\pi}{c} \boldsymbol{J} \quad \text{and} \quad \nabla \cdot \boldsymbol{E} = 4\pi\varrho, \tag{2.136}$$

where $\boldsymbol{E}$ is the electric field, $\boldsymbol{B}$ is the magnetic induction, ϱ is the net charge density and $\boldsymbol{J}$ is the net current density. The numerical approach for the above Vlasov–Maxwell model will be considered briefly in the following paragraphs.

2.3.1.1 Darwin (Nonradiative) Models.
As is well known, explicit formulations of the Vlasov–Maxwell equations are only stable for values of the time step Δt and mesh interval Δx that resolve all time and space scales [292, 382]. In electromagnetic plasma simulation, for example, one is required to use time steps which resolve light waves and space steps which resolve the Debye length, even when neither radiation nor charge separation effects are important. In many cases, the time and length intervals of interest are very large compared with the values of Δx and Δt that satisfy the stability conditions, and then many time and space steps are required to integrate over them. This prevents the application of explicit plasma simulation methods to many problems [559].

A number of algorithms have been developed to remove these restrictions. In some, the equations are split, as in one developed in [393], in which the field advancement algorithm is not subject to the Courant condition [393]. In others, reduced equations are formulated in which the fast time scales are eliminated. Two level of reduction in two-dimensional electromagnetic plasma simulation algorithms are the Darwin model, in which Maxwell's equations are solved in the nonrelativistic limit [392], and the Darwin model with fluid electrons, in which the electrons are modeled by a collisional, sometimes massless fluid [80, 83, 227, 231]. In the Darwin formulation, light waves do not propagate, and the corresponding limit $kc\Delta t < 2$ is replaced by the less restrictive condition, $\omega_{\mathrm{pe}}\Delta t < 2$.

The proper small v/c approximation of the equation of the Maxwell–Lorentz system was first given in [122], where it was used to generalize the Bohr–Sommerfeld atom. Readable accounts of this derivation are given in [250, 287]. Additional features of the model are deduced in [266]. In the present section we shall describe the Darwin approximation in accordance with [392] and [228]. As was indicated in Sect. 2.2, in the nonradiative models we neglect only the inductive or solenoidal part of the displacement current. It is also convenient to split the electric field into longitudinal and transverse parts, denoted by $\boldsymbol{E}_\mathrm{l}$ and $\boldsymbol{E}_\mathrm{t}$, respectively, defined by the requirements $\nabla \times \boldsymbol{E}_\mathrm{l} = 0$ and $\nabla \cdot \boldsymbol{E}_\mathrm{t} = 0$ (see Sect. 2.2.3). In time integration of the full Maxwell equations, the time derivatives may be used to advance the field, while the divergence equations serve as initial conditions which are preserved in time because of the charge continuity equation. However, any attempt to use the time derivatives in the Darwin field equations in a similar way will lead to violent numerical instabilities. These instabilities [392] reflect the fact that the equations are now elliptic and represent instantaneous action-at-a-distance. Any attempt to propagate information across the mesh at a finite velocity is inconsistent with the nonretardation approximation. Instead, in [392], the Maxwell's equations are formulated via the elliptic-type equations. The latter were solved by means of successive overrelaxation [114, 131]. The reader must refer, for example, to [228, 392] for details.

2.3.1.2 Explicit Models.

In an explicit plasma simulation method, the equations are integrated in time with a time step Δt. To advance the solution from a moment of time $n\Delta t$ to $(n+1)\Delta t$, the charge density and current are calculated by particle weighting [(2.134) and (2.135)], and finally the particle update uses (2.133).

A convergence of the numerical solution to an exact solution is determined by the accuracy and numerical stability of the finite-difference schemes used, some of which are described in [74, 292]. In general, though, all explicit methods are conditionally stable; that is, the time step must satisfy the inequality

$$\max_s \omega_{\mathrm{ps}} \Delta t < 2,$$

where $\omega_{\mathrm{ps}} = (4\pi n q_s^2 / m_s)$ is the plasma frequency. One of the best example of a three-dimensional explicit relativistic electromagnetic particle code is TRISTAN, which was developed in [74]. It removes this constraint on Δt, which we are led to consider as an implicit formulation of the equations.

2.3.1.3 Implicit Moment Models.

The possibility of using implicit field computations for particle simulation was considered in [289], and it was concluded that a direct inversion of the implicit partial difference equations was impractical. In [366], it was shown that including only the cold fluid contribution in the time-advanced plasma response was sufficient for stability. An alternative to this moment-implicit method [128, 366] is the direct implicit method [106, 163, 228, 290, 291]. The moment method has been applied to the two-dimensional electromagnetic plasma simulation [59, 61] and the three-dimensional [293] electromagnetic plasma simulation, and the direct method has been applied to the two-dimensional electrostatic plasma simulation [33] and to the two-dimensional electromagnetic plasma simulation [228]. In an implicit formulation, the solution is advanced from a moment of time $n\Delta t$ to $(n+1)\Delta t$ by marching the equations backward from a moment of time $(n+1)\Delta t$ to $n\Delta t$. The essential idea is that what would be numerically unstable solution with growth rate λ for $\Delta t > 0$ is also a numerically unstable solution with growth rate λ for $\Delta t < 0$. Since the solution is known at time $n\Delta t$, however, the effect of solving the equations backwards to match this solution is to reduce the mode at time $(n+1)\Delta t$ by the factor $\exp(-\lambda \Delta t)$ from its value at time $n\Delta t$. Thus, the same difference equations which are unstable when solved forward can be stable when solved backward in time [59].

The analysis of the numerical stability shows that the lower bound on the time step imposed by the need to suppress the finite grid instability and the upper bound imposed by the convergence of the moment expansion restrict time step Δt to the interval

$$O(0.1) < v_T \Delta t / \Delta x = (\lambda_D / \Delta x)(\omega_{\mathrm{ps}} \Delta t) < O(1.0), \tag{2.137}$$

where ω_{ps} is a plasma frequency for particles of species s. For more details the reader may refer to [59, 61].

2.3.1.4 Direct Implicit Models. The basic idea of the direct approach is to use an intermediate time level for both particle positions and velocities, which are obtained by using all of the particle dynamical equations except that due to the fields at the advanced time levels [228, 229, 290]. By exploiting the definitions of plasma charge and current densities as sums over the particle shape factor, the advanced plasma source terms can be expressed in pieces. One part is obtained by summing over the intermediate level, and a second contribution is expressed as an operator on the advanced field. These expressions can now be used directly in the field solution. Once the advanced fields are obtained, the last piece of the particle advance from the intermediate level to the new positions and velocities can be completed. As with the moment method, an inconsistency can exist, depending on the care taken in the preparation of the operators multiplying the advanced field, between these new fields and the source terms that now could be obtained from the new positions and velocities. An iteration could be employed over this final correction of particle coordinates but this has not proven to be necessary for this method either.

The important advantage of this scheme over the moment method is that at no point is it necessary to introduce any auxiliary equations; we always work directly with the particle or field equations. As a consequence this method avoids any question of momentum–moment closure. This method exploits the D1 difference scheme for updating the particle velocities and positions (see, e.g., [228]).

Tests of several variants of the direct implicit algorithms demonstrated that a nonlinear instability occurs when the parameter

$$4\chi_1 = \frac{q^2 \Delta t^2}{2m|\Delta x|} = \frac{(\omega_{\text{pe}} \Delta t)^2}{2N_{\text{c}}}$$

exceeds unity, where $N_{\text{c}} = n|\Delta x|$ is the number of particles per cell. For more details the reader may refer to [228, 229, 290].

2.3.2 Particle-Ion–Guiding-Center-Electron Model

It may be promising to use the guiding center approach for electrons while keeping the kinetic description for ions to study the large space scale and low-frequency electromagnetic phenomena occurring in inhomogeneous plasma. In this model the electron dynamics is described by the guiding center approximation [2.77–2.79, 2.108–2.117] whereas the ions follow the particle model [see (2.133)]. The electromagnetic fields are described by the full Maxwell's equation and the finite-difference schemes may be written as follows

$$\frac{1}{c}\left(\frac{\partial \boldsymbol{E}}{\partial t}\right)^{n+\frac{1}{2}} = \nabla \times \boldsymbol{B}^{n+\alpha} - \frac{4\pi}{c}\boldsymbol{J}^{n+\alpha}, \qquad (2.138)$$

$$\frac{1}{c}\left(\frac{\partial \boldsymbol{B}}{\partial t}\right)^{n+\frac{1}{2}} = -\nabla \times \boldsymbol{E}^{n+\alpha}, \tag{2.139}$$

$$\nabla \cdot \boldsymbol{E}^{n+1} = 4\pi \varrho^{n+1}, \tag{2.140}$$

$$\nabla \cdot \boldsymbol{B}^{n+1} = 0. \tag{2.141}$$

Here, $\boldsymbol{E}$ and $\boldsymbol{B}$ are the electric and magnetic fields, respectively, and α is a decentering (implicitness) parameter.

The equations of motion for the ions are the standard Newton–Lorentz equations, except those for the time level of the electromagnetic field, which are given by

$$\left(\frac{d\boldsymbol{v}_l}{dt}\right)^{n+\frac{1}{2}} = \frac{Z_s e}{M_s}\left(\boldsymbol{E}^{n+\alpha}(\boldsymbol{x}_l) + \frac{\boldsymbol{v}_l^{n+1/2} \times \boldsymbol{B}^{n+\alpha}}{c}\right), \tag{2.142}$$

$$\left(\frac{d\boldsymbol{x}_l}{dt}\right)^{n+\frac{1}{2}} = \boldsymbol{v}_l^{n+1/2}. \tag{2.143}$$

The equations of motion for the electrons are considered in the guiding center approximation in order to eliminate the electron cyclotron time scale, Ω_e^{-1}. The equations of motion are decomposed into the parallel and perpendicular components with respect to the local magnetic field, which are given by (see also Sect. 2.2.7)

$$\left(\frac{dv_{\|l}}{dt}\right)^{n+1/2} = \left(\frac{-e}{m}\right)E_\|^{n+\alpha} - \left(\frac{\mu_l}{m}\right)\frac{\partial}{\partial x_\|}B^{n+\alpha}; \tag{2.144}$$

$$\boldsymbol{v}_{\perp l}^{n+\alpha} = c\left(\frac{\boldsymbol{E} \times \boldsymbol{B}}{B^2}\right)^{n+\alpha} - \left[\left(\frac{mc}{eB}\right)\hat{\boldsymbol{b}} \times \left(\frac{\mu}{m}\nabla B + v_{\|l}^2 \frac{\partial \hat{\boldsymbol{b}}}{\partial x_\|}\right)\right]^{n+\alpha}, \tag{2.145}$$

$$\left(\frac{d\boldsymbol{x}_l}{dt}\right)^{n+1/2} = \left(\boldsymbol{v}_{\|l}^{n+1/2} + \boldsymbol{v}_{\perp l}^{n+\alpha}\right). \tag{2.146}$$

In (2.144) $v_{\|l}$ is a scalar velocity along the magnetic field, $\mu_l = mv_{T,l}^2/2B(\boldsymbol{x})$ is the magnetic moment of the jth electron, and the unit vector $\hat{\boldsymbol{b}} = \boldsymbol{B}/B$. The three terms in (2.145) represent the $\boldsymbol{E} \times \boldsymbol{B}$, gradient B and curvature drifts, respectively. Here it is important to note that the time indices of each term in (2.144–2.146) must be consistent with their counterparts in the Newton–Lorentz equation. For example, the time level of the perpendicular velocity in (2.146) should be $t = t^{n+\alpha}$. Otherwise, the electrons and ions would show different responses ($\boldsymbol{E} \times \boldsymbol{B}$ drift etc.) to the low-frequency component of the electromagnetic field. The parallel velocity in (2.146) is defined by

$$\boldsymbol{v}_{\|l}^{n+1/2} = v_{\|l}^{n+1/2}\hat{\boldsymbol{b}}^{n+\alpha}(\boldsymbol{x}_l). \tag{2.147}$$

Equations (2.138–2.146) are a basic equations of the three-dimensional electromagnetic implicit code HIDENEK, which was developed in [512]. The details of the numerical algorithm may be found in [513]. A similar idea was used in the two-dimensional and three-dimensional models which were developed in [185]. These models allow us to simulate multiscale plasmas with sufficiently large time and space steps: $\omega_{pe}\Delta t \gg 1$, $\Omega_e \Delta t \gg 1$, and $\Delta x \geq c/\omega_{pe}$. All the following effects can be simulated: (a) various kinetic (particle) effects such as the Landau, cyclotron, and bounce resonances with low-frequency waves and those due to finite Larmor radius and complicated particle trajectories; (b) a space-charge electric field and a finite-speed plasma relaxation arising from nonzero electron inertia, and (c) nonlinear plasma processes under the nonmicroscopic time and space scales, i.e., duration time $\gg \omega_{pe}^{-1}, \Omega_e^{-1}$ and spatial scale $\gg \lambda_e$, ϱ_{ce}. This model was successfully used to study the electromagnetic ion-beam–plasma instability, the Alfvén-ion-cyclotron instability and the kink instability [512].

2.3.3 Guiding-Center-Ion–Guiding-Center-Electron (Drift-Kinetic) Model

In this section we present the basic equations that primarily govern the plasma dynamics in the case of a uniform magnetic field $B = (B_x, 0, B_z)$ in the x–z plane. We assume that the electrostatic type of instability takes place in a frequency regime lower than the ion cyclotron frequency, and we use the guiding center drift approximation [189]. The electron and ion velocity can be written explicitly in the following form (see Sects. 2.2.6 and 2.2.7):

$$\boldsymbol{v}_{e,i} = \boldsymbol{v}_{\|e,i} + \boldsymbol{v}_\perp$$

and

$$\boldsymbol{v}_\perp = \frac{\boldsymbol{E} \times \boldsymbol{B}}{B^2}.$$

The drift-kinetic Vlasov equation is written as follows (see Sect. 2.2.6):

$$\frac{\partial f_{i,e}}{\partial t} + \boldsymbol{v}_\| \cdot \frac{\partial f_{i,e}}{\partial \boldsymbol{r}_\|} + \frac{\boldsymbol{E} \times \boldsymbol{B}}{B^2} \cdot \frac{\partial f_{i,e}}{\partial \boldsymbol{r}_\perp} + \frac{e}{m_{i,e}} \boldsymbol{E}_\| \cdot \frac{\partial f_{i,e}}{\partial v_\|} = 0. \quad (2.148)$$

The electron and ion distribution functions are reduced to a three-dimensional phase space function $f_e(x, y, v_\|)$ and $f_i(x, y, v_\|)$, where $v_\|$ is the velocity variable along the magnetic field.

The electric fields are then given by

$$E_x = -\frac{\partial}{\partial x}\varphi \quad (2.149)$$

and

$$E_y = -\frac{\partial}{\partial y}\varphi, \quad (2.150)$$

where the electrostatic potential φ obeys Poisson's equation,

$$\frac{\partial^2 \varphi}{\partial x^2} + \frac{\partial^2 \varphi}{\partial y^2} = 4\pi e(n_e - n_i). \qquad (2.151)$$

The drift-kinetic Vlasov code was used to study $\boldsymbol{E} \times \boldsymbol{B}$ instabilities, for example, the Kelvin–Helmholtz instability and the ion temperature gradient instability [189]. In particular, the code provides excellent spatial resolution which allows a detailed examination of the mutual interaction of the arms rotating around the central structure of vortices. This code may be used for study the electrostatic instabilities inside the lunar wave, where the finite gyroradius effects may be not very important (see Sect. 12.5).

2.3.4 Particle-Electron–Immobile-Ion Model

A wide spectrum of plasma problem may be simulated by the model in which the active particle (electrons) are described as particles, whereas the passive particle (ions) constitute a stationary background. This model allow us to focus on the dynamics of whistlers and the electron dissipation region without problems arising from inadequate separation of scales due to unrealistic ion-to-electron mass ratios [483]. The electric and magnetic fields are stepped forward using the full Maxwell equations. At the end of each time step, the electric field is modified by adding an electrostatic component to ensure that $\nabla \cdot \boldsymbol{E} = 4\pi\varrho$. This model was used for the magnetic field reconnection problems in [483, 485].

2.4 Photo-ionization and Charge Exchange Processes

In many types of interaction of the solar wind with planets, comets, and the local interstellar medium (LISM) we have to include in simulation the dynamics of neutral components, taking into account the charge exchange process, photo-ionization and gravitation. The global interaction of the LISM with the solar wind is a fundamental problem of heliospheric physics. It requires the solution of a highly nonlinear coupled set of integro-MHD-Boltzmann equations which describes the dynamics of the ionized solar wind and LISM together with interstellar and heliospheric neutral atoms. At the same time the processes in the cometary tails, at the interplanetary shocks and the termination shock may demand the use of hybrid Boltzmann–kinetic models.

2.4.1 Hybrid Particle-Neutral-Component–Fluid-Plasma Models

Let us consider now the hybrid Boltzmann–MHD model, which was developed at first for simulation of the interaction of the LISM with the solar wind. In this model the dynamics of the ionized solar wind and the LISM are

2.4 Photo-ionization and Charge Exchange Processes

described in a gas dynamical approximation, while the neutral component is studied using a kinetic approach. The gas dynamical model for the ionized component is described in [420, 591]. For the neutral component, different approaches to solving the Boltzmann equation have been used, including a trajectory splitting Monte Carlo method [29, 358], a direct solution by the method of characteristics [148, 409, 443], an integral method [213, 453], and a moment equations method [553]. For nonstationary simulations with the Knudsen number $Kn = \lambda/L$ varying over a wide range, it can be useful to use direct finite-difference methods (see, e.g., the $\Delta - \varepsilon$ method of [511]), or modifications of direct particle simulation methods such as using discrete coordinates [274], direct Monte Carlo simulations [50, 387]; low-discrepancy methods [20], or statistical particle methods [37, 444]. The particle-mesh method has been developed extensively for fluid and plasma simulations over the last 30 years (see, e.g., [236]). Since the neutral particles interact with the plasma fluid by means of charge-exchange collisions in this model, it is natural to use the mesh for calculating the "collision processes". This allows us to use standard particle-mesh methods for the kinetic simulation of the interstellar neutral H component [347].

The Boltzmann equation for the neutral component velocity distribution is

$$\frac{\partial}{\partial t}f + \boldsymbol{v} \cdot \nabla f + \frac{\boldsymbol{F}}{M} \cdot \nabla_v f = P - L, \tag{2.152}$$

where P denotes the production rate of neutral particles and L the loss rate of particles at $(\boldsymbol{x}, \boldsymbol{v}, t)$. P and L are generally functions of the neutral and plasma distribution functions f and f_p. The loss term has the form

$$L = f(\boldsymbol{x}, \boldsymbol{v}, t)\beta(\boldsymbol{x}, \boldsymbol{v}, t), \tag{2.153}$$

and

$$\beta(\boldsymbol{x}, \boldsymbol{v}, t) = \int f_\mathrm{p}(\boldsymbol{x}, \boldsymbol{v}_\mathrm{p}, t) V_{\mathrm{rel,p}} \sigma_{\mathrm{ex}}(V_{\mathrm{rel,p}}) \mathrm{d}^3 \boldsymbol{v}_\mathrm{p} + \beta_{\mathrm{ph}}, \tag{2.154}$$

where β is the total loss rate in s^{-1}, $V_{\mathrm{rel,p}} = |\boldsymbol{v} - \boldsymbol{v}_\mathrm{p}|$ and β_{ph} is the particle loss due to photo-ionization.

The production term for neutral hydrogen experiencing charge exchange is given by

$$P_{\mathrm{ex}}(\boldsymbol{x}, \boldsymbol{v}, t) = f_\mathrm{p}(\boldsymbol{x}, \boldsymbol{v}, t) \int f(\boldsymbol{x}, \boldsymbol{v}_\mathrm{H}, t) V_{\mathrm{rel,H}} \sigma_{\mathrm{ex}}(V_{\mathrm{rel,H}}) \mathrm{d}^3 \boldsymbol{v}_\mathrm{H}, \tag{2.155}$$

where $V_{\mathrm{rel,H}} = |\boldsymbol{v} - \boldsymbol{v}_\mathrm{H}|$. Let us consider a group of atoms which have approximately the same velocities and coordinates as one macroparticle [236]. The dynamics of this macroparticle is governed by the same Newtonian equations of motion as a real atom. We assume that the distribution function for centers of the neutral macroparticles may be represented by a set of δ functions,

$$f = \sum_{i=1}^{N_\mathrm{p}} \alpha_i \delta(\boldsymbol{x} - \boldsymbol{x}_i) \delta(\boldsymbol{v} - \boldsymbol{v}_i), \tag{2.156}$$

where $\alpha_i = \alpha_i(t)$ denotes the weight of an individual macroparticle with number i and $\boldsymbol{x}_i = \boldsymbol{x}_i(t)$, $\boldsymbol{v}_i = \boldsymbol{v}_i(t)$ are the coordinates of macroparticles in phase space. Summation is over all macroparticles N_p. The neutral macroparticle is described by 8 (in the 2.5-dimensional model by 7) scalar components, i.e., the weight of a macroparticle α_i, the position $\boldsymbol{x}_i$, the velocity $\boldsymbol{v}_i$, and the survival probability w of the ith macroparticle against photo-ionization and charge exchange. The equation of motion for the neutral particles is, of course,

$$\frac{\mathrm{d}\boldsymbol{v}_i}{\mathrm{d}t} = \boldsymbol{F}_i/M_\mathrm{n} \quad \text{and} \quad \frac{\mathrm{d}\boldsymbol{x}_i}{\mathrm{d}t} = \boldsymbol{v}_i, \qquad (2.157)$$

where the force acting on the neutral particle $\boldsymbol{F}_i = \boldsymbol{F}_\mathrm{g} + \boldsymbol{F}_\mathrm{r}$ results from a balance of the solar gravitational force $\boldsymbol{F}_\mathrm{g}$ and radiation pressure $\boldsymbol{F}_\mathrm{r}$. Here M_n is the mass of the atom. Neutral atoms experience charge exchange with the ionized component of the LISM and solar wind, and may also be affected photoionization. The charge exchange cross-section σ_ex adopted here is

$$\sigma_\mathrm{ex} = \sigma(v). \qquad (2.158)$$

If we suppose that the ionized component has a Maxwellian distribution, then β_ex may be approximated as [443]

$$\beta_\mathrm{ex}(\boldsymbol{x}_i, \boldsymbol{v}_i, t) \simeq n_\mathrm{p}(\boldsymbol{x}_i, t) V_\mathrm{rel,p} \sigma_\mathrm{ex}(V_\mathrm{rel,p}). \qquad (2.159)$$

Here, $V_\mathrm{rel,p}$ is the average velocity of protons relative to an H atom with velocity $\boldsymbol{v}_i$ [443].

Let the time interval t^* with respect to the change exchange and photo-ionization processes be a random variable with the following distribution function:

$$w_i(t^*) = \exp\left(-\int_{t_0}^{t^*} (\beta_{\mathrm{ex},i} + \beta_{\mathrm{ph},i})\mathrm{d}t\right). \qquad (2.160)$$

Since charge exchange and photo-ionization are independent processes, we may split the total survival probability into the product of a survival probability against the charge exchange event w_ex and a survival probability against photo-ionization w_ph, thus

$$w_i(t) = w_{\mathrm{ex},i}(t) w_{\mathrm{ph},i}(t), \qquad (2.161)$$

where

$$w_{\mathrm{ex},i}(t) = \exp\left(-\int_{t_0}^{t} \beta_{\mathrm{ex},i}\mathrm{d}t\right) \quad \text{and} \quad w_{\mathrm{ph},i}(t) = \exp\left(-\int_{t_0}^{t} \beta_{\mathrm{ph},i}\mathrm{d}t\right).$$

The index i indicates that the integration is over the ith particle trajectory. Note here that to investigate, for example, the entry of neutral helium, oxygen, and other heavy species into the heliosphere, we have to consider the charge exchange and photoionization processes together {(2.12a) and (3.4)

2.4 Photo-ionization and Charge Exchange Processes

from [358]}. Photo-ionization results in the loss of neutral particles only; thus, it is convenient to include this probability into the particle weight time evolution, so that now

$$\alpha_i(t) = \alpha_i(t_0) \exp\left(-\int_{t_0}^{t} \beta_{\text{ph},i} dt\right). \tag{2.162}$$

At the time of creation (either at the boundary of the calculation domain or at the moment of charge exchange), a neutral macroparticle has initial coordinates $\boldsymbol{x}_i(t_0) = \boldsymbol{x}_{i,0}$ and $\boldsymbol{v}_i(t_0) = \boldsymbol{v}_{i,0}$, a weight $\alpha_i(t_0) = \alpha_{i,0}$, and a survival probability $w_{\text{ex},i}(t_0) = 1$. For each new neutral particle i, we have to determine the critical probability, $w^*_{\text{ex},i}$, when charge exchange will occur, and this is done using the relation

$$w^*_{\text{ex},i} = \xi, \tag{2.163}$$

where ξ is a random number drawn from the interval $[0, 1]$. During the calculation we have to identify those particles for which the probability of survival satisfies the condition

$$w_{\text{ex},i} \leq w^*_{\text{ex},i}. \tag{2.164}$$

If the particle satisfies condition (2.164), then we have to exchange the velocity of this neutral macroparticle with the velocity of a proton from the ionized component (in Sect. 12.6, the ions of the LISM or the solar wind). This is accomplished using a random number generator for the probability (or the frequency) of charge exchange of the atom with velocity $\boldsymbol{v}$ and the proton with velocity $\boldsymbol{v}_p$ (see [358]),

$$\nu(\boldsymbol{v}, \boldsymbol{v}_p) \sim |\boldsymbol{v} - \boldsymbol{v}_p| \sigma_{\text{ex}} \exp\left(-\frac{(\boldsymbol{v}_p - \boldsymbol{U}_p)^2}{2v_{T,p}^2}\right). \tag{2.165}$$

In the present simulations, we do not take the cross-section σ into account (as was done in [358] in (2.165) because of the weak dependence on $|\boldsymbol{v} - \boldsymbol{v}_p|$. If charge exchange occurs, then a new neutral macroparticle begins its motion with $w_{\text{ex},i} = 1$.

If we know the neutral particle distribution function at time $t = (m+1)\Delta t$, we can in principle update the plasma distribution by means of the following hydrodynamical equations:

$$\frac{\partial \varrho_p}{\partial t} + \nabla \cdot (\varrho_p \boldsymbol{U}_p) = Q_\varrho, \tag{2.166}$$

$$\frac{\partial (\varrho_p \boldsymbol{U}_p)}{\partial t} + \nabla \cdot (\varrho_p \boldsymbol{U}_p \boldsymbol{U}_p + p\mathbf{I}) = \boldsymbol{Q}_m, \tag{2.167}$$

$$\frac{\partial}{\partial t}\left(\frac{1}{2}\varrho_p U_p^2 + \frac{p}{\gamma - 1}\right) + \nabla \cdot \left(\frac{1}{2}\rho_p U_p^2 \boldsymbol{U}_p + \frac{\gamma}{\gamma - 1}\boldsymbol{U}_p p\right) = Q_e, \tag{2.168}$$

2. Particle-Mesh Models

where $\varrho_p = n_p M$, $\boldsymbol{U}_p$, p, and $Q_{(\varrho,m,e)}$ denote respectively the plasma density, bulk velocity, pressure, and source terms for the density, momentum, and energy equations. I denotes the unit tensor and γ $(= 5/3)$ is the adiabatic index. The source terms in (2.166–2.168) are due to photo-ionization and charge exchange:

$$Q_\varrho = Q_{\varrho,\text{ph}} + Q_{\varrho,\text{ex}}, \quad \boldsymbol{Q}_m = \boldsymbol{Q}_{m,\text{ph}} + \boldsymbol{Q}_{m,\text{ex}}, \quad Q_e = Q_{e,\text{ph}} + Q_{e,\text{ex}}.$$

The source terms may be evaluated from

$$Q_{\varrho,\text{ph}} = M_n \int \beta_{\text{ph}} f(\boldsymbol{v}) \mathrm{d}\boldsymbol{v}, \tag{2.169}$$

$$Q_{\varrho,\text{ex}} = \frac{M_n - M}{n_p} \int \int \beta_{\text{ex}} f(\boldsymbol{v}) f_p(\boldsymbol{v}_p) \mathrm{d}\boldsymbol{v} \mathrm{d}\boldsymbol{v}_p, \tag{2.170}$$

$$\boldsymbol{Q}_{m,\text{ph}} = M_n \int \beta_{\text{ph}} \boldsymbol{v} f(\boldsymbol{v}) \mathrm{d}\boldsymbol{v}, \tag{2.171}$$

$$\boldsymbol{Q}_{m,\text{ex}} = \frac{1}{n_p} \int \int \beta_{\text{ex}} (M_n \boldsymbol{v} - M \boldsymbol{v}_p) f(\boldsymbol{v}) f_p(\boldsymbol{v}_p) \mathrm{d}\boldsymbol{v} \mathrm{d}\boldsymbol{v}_p, \tag{2.172}$$

$$Q_{e,\text{ph}} = M_n \int \beta_{\text{ph}} \frac{v^2}{2} f(\boldsymbol{v}) \mathrm{d}\boldsymbol{v}, \tag{2.173}$$

$$Q_{e,\text{ex}} = \frac{1}{n_p} \int \int \beta_{\text{ex}} \left(\frac{M_n v^2 - M v_p^2}{2} \right) f(\boldsymbol{v}) f_p(\boldsymbol{v}_p) \mathrm{d}\boldsymbol{v} \mathrm{d}\boldsymbol{v}_p. \tag{2.174}$$

For the sake of generality, we have assumed that $M_n - M$ is not necessarily zero (M_n the neutral mass). Here we neglect the average photo-electron energy transferred to the plasma during each photo-ionization event.

The substitution of (2.156) into the source term relations (2.169–2.174) gives

$$Q_{\varrho,\text{ph}} = M_n \sum_{i=1}^{N_p} W(\boldsymbol{x} - \boldsymbol{x}_i) \alpha_i \beta_{\text{ph},i} \approx n M_n \langle \beta_{\text{ph}} \rangle, \tag{2.175}$$

$$Q_{\varrho,\text{ex}} = \frac{M_n - M}{n_p} \sum_{i=1}^{N_p} W(\boldsymbol{x} - \boldsymbol{x}_i) \alpha_i \int \beta_{\text{ex},i} f_p(\boldsymbol{v}_p) \mathrm{d}\boldsymbol{v}_p$$
$$\approx n(M_n - M)\langle \beta_{\text{ex}} \rangle, \tag{2.176}$$

$$\boldsymbol{Q}_{m,\text{ph}} = M_n \sum_{i=1}^{N_p} W(\boldsymbol{x} - \boldsymbol{x}_i) \alpha_i \beta_{\text{ph},i} \boldsymbol{v}_i$$
$$\approx n M_n \boldsymbol{u} \langle \beta_{\text{ph}} \rangle, \tag{2.177}$$

2.4 Photo-ionization and Charge Exchange Processes

$$Q_{m,ex} = \frac{1}{n_p} \sum_{i=1}^{N_p} W(\boldsymbol{x} - \boldsymbol{x}_i)\alpha_i \int \beta_{ex,i}(M_n\boldsymbol{v}_i - M\boldsymbol{v}_p)f_p(\boldsymbol{v}_p)d\boldsymbol{v}_p$$
$$\approx n(M_n\boldsymbol{u} - M\boldsymbol{U}_p)\langle \beta_{ex}\rangle, \tag{2.178}$$

$$Q_{e,ph} = M_n \sum_{i=1}^{N_p} W(\boldsymbol{x} - \boldsymbol{x}_i)\alpha_i \beta_{ph,i} \frac{v_i^2}{2}$$
$$\approx n M_n \langle \frac{v^2}{2}\rangle \langle \beta_{ph}\rangle, \tag{2.179}$$

$$Q_{e,ex} = \frac{1}{n_p} \sum_{i=1}^{N_p} W(\boldsymbol{x} - \boldsymbol{x}_i)\alpha_i \int \beta_{ex,i}\left(\frac{M_n v_i^2 - M v_p^2}{2}\right) f_p(\boldsymbol{v}_p)d\boldsymbol{v}_p$$
$$\approx n\langle \frac{M_n v^2 - M v_p^2}{2}\rangle \langle \beta_{ex}\rangle. \tag{2.180}$$

Here we have used the mesh value of the macroscopic parameter at the point $\boldsymbol{x}$ and the independence of β_{ph} from the particle velocity so that

$$Q_\psi(\boldsymbol{x},t) = \int\int_{\Delta\boldsymbol{x}} \psi(\boldsymbol{v})f d\boldsymbol{x}d\boldsymbol{v} = \sum_{i=1}^{N_p} W(\boldsymbol{x} - \boldsymbol{x}_i)\psi(\boldsymbol{v}_i)\alpha_i,$$

where $W(\boldsymbol{x} - \boldsymbol{x}_i)$ is the charge-assignment-force-interpolation function [236]. In our problem, the terms with φ_e are very small compared to other terms, justifying their neglection. Since a neutral particle interacts with only one random ion during a charge-exchange event, we can approximate the integrals in (2.175–2.180) by means of a Monte-Carlo technique.

To calculate the total loss rate β at the position of the neutral macroparticle, we need to interpolate the mesh values of the density $n_{p,m}$, bulk velocity $\boldsymbol{u}_p$ and thermal velocities $v_{T,p}$ of the ionized component to the position of the neutral macroparticle. In the particle-mesh method, the relation between these parameters (ϕ and ϕ_m) is given by

$$\phi(\boldsymbol{x},t) = \sum_{k=1}^{N_g} W(\boldsymbol{x} - \boldsymbol{x}_k)\phi_m(\boldsymbol{x}_k,t), \tag{2.181}$$

where ϕ can be $n_p(\boldsymbol{x},t)$, $\boldsymbol{u}_p(\boldsymbol{x},t)$, or $v_{T,p}(\boldsymbol{x},t)$, and N_g is the number of mesh nodes.

To update the macroparticle coordinate and velocity, a general leapfrog scheme [236] was used. The probability of survival $w_{ex,i}$ (2.164) at time $t = (m+1)\Delta t$ was approximated by

$$\ln w_{ex,i}^{m+1} = \ln w_{ex,i}^m - \Delta t \beta_{ex,i}^m, \tag{2.182}$$

where $\Delta t \beta_{\text{ex},i}^k$ denotes the probability of charge exchange during the time step Δt.

During the time step Δt, we have to calculate the source terms for the mass, momentum and energy equations of the plasma. The photo-ionization source terms were estimated explicitly from equations (2.175), (2.177) and (2.179). In principle, since we know the distribution function for the neutral macroparticles, we could estimate the source terms by performing the six-dimensional phase space integral. However, since only a small number of neutral macroparticles experience charge exchange during the interval $(t, t+\Delta t)$, we may estimate the source terms using

$$\Delta t \cdot Q_{l,\text{ex}}(\boldsymbol{x},t) = \sum_{i=1}^{N_{\text{p,ex}}} W(\boldsymbol{x}-\boldsymbol{x}_i)\alpha_i(t)\psi_l; \quad l = \varrho, \text{m}, \text{e}, \tag{2.183}$$

where

$$\psi_\varrho = (M_\text{n} - M), \quad \psi_\text{m} = (M_\text{n}\boldsymbol{v}_i - M\boldsymbol{v}_\text{p}), \quad \psi_\text{e} = \left(\frac{M_\text{n} v_i^2 - M v_\text{p}^2}{2}\right).$$

Here, $\boldsymbol{v}_i$ and $\boldsymbol{v}_\text{p}$ are the neutral macroparticle velocity and the proton macroparticle velocity at the moment of charge exchange, and W is the same weight function as used in the interpolation procedure.

Summation is over those particles which experienced charge exchange during the time interval $(t, t+\Delta t)$. The application of this model to simulation of the solar wind–heliosphere interaction will be considered in Sect. 12.6.

2.4.2 Hybrid Particle-Neutral-Component–Kinetic-Plasma Models

The Boltzmann particle-mesh method may be emploed for studying nonstationary problems which might include instabilities and the generation of turbulence. Furthermore, particle-mesh or particle-in-cell–Boltzmann code may be coupled to kinetic (full particle or hybrid) codes. Note that both codes have the same structure and use the same scheme for particle coordinate and velocity advancing and the same particle and force (field) weighting procedures and diagnostics procedures. This would allow the investigation of the kinetic processes in a partially ionized gas, e.g., the dynamics of particles and electromagnetic fields at collisionless shock transition layers, the excitation of turbulence in regions of mass loading in the solar wind by pickup ions, and magnetic field reconnection and particle acceleration in current layers. A version of such a hybrid Boltzmann–Vlasov approach is used, for example, to study the extremely ultraviolet (EUV)/X-ray sources near comets and non-magnetic planets, and the formation of a ray-like tail structure [339]. The use of a "moment description" for ions and atoms to describe charge-exchange

2.4 Photo-ionization and Charge Exchange Processes

processes (K + H$^+$ ⇒ K$^+$ + H, K$^+$ + H ⇒ K + H$^+$) rather than a kinetic approach may not always be appropriate [346].

This direct simulation model allows us to study self-consistently the processes of ionization within a kinetic framework. Let us consider the charge-exchange process between an H atom and protons described in a kinetic approximation.

The dynamics of this macroparticle is governed by the same Newtonian equations of motion as a real atom and ion. We assume that the distribution function for the centers of the macro-atoms and macro-ions may be represented by a set of δ functions,

$$f = \sum_{i=1}^{N_{p1}} \alpha_i \delta(\boldsymbol{x} - \boldsymbol{x}_i)\delta(\boldsymbol{v} - \boldsymbol{v}_i), \tag{2.184}$$

$$f_{p} = \sum_{i=1}^{N_{p2}} \gamma_i \delta(\boldsymbol{x} - \boldsymbol{x}_i)\delta(\boldsymbol{v} - \boldsymbol{v}_i), \tag{2.185}$$

where $\alpha_i = \alpha_i(t)$ and $\gamma_i = \beta_i(t)$ denote the weight of an individual macroparticle with number i, and $\boldsymbol{x}_i = \boldsymbol{x}_i(t)$ and $\boldsymbol{v}_i = \boldsymbol{v}_i(t)$ are the coordinates of macroparticles in phase space. Summation is over all macro-atoms N_{p1} and macro-ions N_{p2}. The neutral macroparticle is described by 8 (in the 2.5-dimensional model by 7) scalar components, i.e., the weight of a macroparticle α_i, the position $\boldsymbol{x}_i$, the velocity $\boldsymbol{v}_i$, and the survival probability w of the ith macroparticle against photo-ionization and charge exchange.

Substitution of (2.185) into (2.154) gives the following relation for the loss rate β_{ex} due to charge exchange:

$$\beta_{\text{ex}}(\boldsymbol{x}, \boldsymbol{v}, t) = \int \sum_{j=1}^{N_{p2}} \gamma_j \delta(\boldsymbol{x} - \boldsymbol{x}_j)\delta(\boldsymbol{v}_p - \boldsymbol{v}_j)|\boldsymbol{v} - \boldsymbol{v}_p|\sigma(|\boldsymbol{v} - \boldsymbol{v}_p|)d\boldsymbol{v}_p^3$$

$$= \sum_{j=1}^{N_{p2}} \gamma_j \delta(\boldsymbol{x} - \boldsymbol{x}_j)|\boldsymbol{v} - \boldsymbol{v}_j|\sigma(|\boldsymbol{v} - \boldsymbol{v}_j|), \tag{2.186}$$

where β_{ex} is the loss rate in s^{-1}. To calculate the grid value of the loss rate, we have to use the assignment function W (Sect. 4.2):

$$\beta_{\text{ex}}(\boldsymbol{x}_k, \boldsymbol{v}, t) = \sum_{j=1}^{N_{p2}} \gamma_j W(\boldsymbol{x}_k - \boldsymbol{x}_j)|\boldsymbol{v} - \boldsymbol{v}_j|\sigma(|\boldsymbol{v} - \boldsymbol{v}_j|). \tag{2.187}$$

Here, the $\boldsymbol{v}$ is an H atom velocity, and a summation is done over the particles which have nonzero weight function W at grid point $\boldsymbol{x}_k$. During the calculation we have to identify those particles for which the probability of survival satisfies the condition (2.164).

If the particle satisfies the condition (2.164), then we have to exchange the velocity of this neutral macroparticle with the velocity of a proton from the ionized component (in Sect. 2.4.1, the ions of the LISM or the solar wind). This is accomplished using a random number generator for the probability (or the frequency) of charge exchange of the atom with velocity v and an individual macroproton with number j located in the vicinity of grid point x_k,

$$\nu(x_k, v, j) \sim \gamma_j(x_k - x_j)|v - v_j|. \tag{2.188}$$

Here, we do not take the cross-section σ into account (as was done in Sect. 2.4.1) because of the weak dependence on $|v - v_p|$. If charge exchange occurs, then a new neutral macroparticle begins its motion with $w_{\text{ex},i} = 1$. The cumulative distribution function has a form

$$D(j) = \frac{\sum_{l=Npk1}^{j} \gamma_l W(x_k - x_l)|v - v_l|}{\sum_{l=Npk1}^{Npk2} \gamma_l W(x_k - x_l)|v - v_l|}. \tag{2.189}$$

Equating $D(j_s)$ to a uniform distribution of numbers χ_s ($0 < \chi_s < 1$) will produce the j_s corresponding to the distribution $\nu(j_s)$. Here the summation is done over the macroprotons located in the vicinity of the cell with index k, $|x_k^m - x_l^m| < K(W)h^m$, and $m = 1, 2, 3$ denotes the coordinates x, y, z. The function K depends on a type assignment function. In the case of the NGP assignment scheme, (2.189) may be reduce to

$$D(j) = \frac{\sum_{l=Npk1}^{j} \gamma_l |v - v_l|}{\sum_{l=Npk1}^{Npk2} \gamma_l |v - v_l|}, \tag{2.190}$$

where summation is done over the set of macroprotons which belong the cell with index k. The practical charge exchange procedure will be considered in Exercises.

Summary

In this chapter we have tried to give a survey of the issues involved in selecting a method (model) to describe different species of particles and a brief account of the main features of the most important models. We hope this will be adequate to indicate the type of method likely to be the most useful in any particular case and lead to the avoidance of grossly inefficient methods. The optimal choice of an efficient computational model is dictated by the physical processes of the considered phenomena. The computational model must be as simple as possible and at the same time it must provide a good approximation to the main processes. There is a wide range of possible particle and fluid models from which to choose.

Exercise

2.1 Derive the algorithm for the charge-exchange procedure between the macro-atom with index i and the macro-ion with index j (2.189–2.190).

3. Time Integration of the Particle Motion Equations

3.1 Introduction

In this chapter we shall consider the different methods for time integration of the particle equations of motion. We shall describe the explicit (Sect. 3.2) and implicit (Sect. 3.3) schemes. In some applications it is useful to exploit the splitting of the general equation of particle motion into a set of equations with a more simple structure (Sect. 3.4). The basic analysis of the numerical stability of the time integration scheme is presented in Sect. 3.5. Examples of C1 and D1 implicit time integration schemes are given in Sect. 3.6. The Runge–Kutta schemes, which are more accurate in some cases than leap-frog schemes, are presented in Sect. 3.7. And, finally, the scheme for relativistic particle motion is considered in Sect. 3.8.

Let us consider the particle motion equations in general [142], Cartesian, cylindrical and spherical coordinates. Let us introduce the tensor formulation and vector identities in accordance with [495]. Let the reference Cartesian coordinates have components (x^1, x^2, x^3) and unit vectors $(\hat{x}_1, \hat{x}_2, \hat{x}_3)$, so that a vector, x, may be written $x = x^i \hat{x}_i$, where summation over the repeated index i $(= 1, 2, 3)$ is implied. If (x^1, x^2, x^3) are expressed as functions of the curvilinear coordinate components $(\bar{x}^1, \bar{x}^2, \bar{x}^3)$, then $x^1 = x^1(\bar{x}^1, \bar{x}^2, \bar{x}^3)$ etc., or more concisely $x = x(\bar{x})$.

The basis vectors e_i and the reciprocal basis vectors e^i may be written in the following form

$$e_i = \frac{\partial x}{\partial \bar{x}^i} \quad \text{and} \quad e^i = \nabla \bar{x}^i. \tag{3.1}$$

These vectors are orthogonal, $e_i \cdot e^j = \delta_i^j$, where δ_j^i is the Kronecker delta $(= 1$ if $i = j, 0$ otherwise). $e^i = e_j \times e_k / J$ and vice versa $e_i = (e^j \times e^k) J$, where the Jacobian $J = \sqrt{g} = |e_1 \cdot e_2 \times e_k|$ is the square root of the determinant g of the metric tensor.

The (covariant) metric tensor is defined as $g_{ij} = e_i \cdot e_j$, and its reciprocal tensor, the contravariant metric tensor, is $g^{ij} = e^i \cdot e^j$. In Cartesian coordinates, the covariant, contravariant and physical components are identical and the metric tensor reduces to $g_{ij} = \delta_{ij}$. For orthogonal systems, $g_{ij} = 0$ for $i \neq j$. In general, g_{ij} is symmetric with six distinct elements, $g_{ij} = g_{ji}$.

3. Time Integration of the Particle Motion Equations

A vector $\boldsymbol{A}$ may be expressed in terms of its contravariant component A^i, covariant component A_i or physical components $A(i)$:

$$\boldsymbol{A} = A^i \boldsymbol{e}_i = A_i \boldsymbol{e}^i = A(i)\hat{\boldsymbol{e}}_i, \tag{3.2}$$

where

$$\hat{\boldsymbol{e}}_i = \frac{\boldsymbol{e}_i}{|\boldsymbol{e}_i|}. \tag{3.3}$$

The permutation symbols e^{ijk} and e_{ijk} are 1 for cyclic indices, -1 for anti-cyclic indices, and 0 otherwise; they are related to the permutation tensors by

$$\epsilon^{ijk} = e^{ijk}/\sqrt{g}, \quad \epsilon_{ijk} = \sqrt{g}\, e_{ijk}. \tag{3.4}$$

The permutation tensors satisfy the identity

$$\epsilon^{ijk}\epsilon_{klm} = \delta^i_l \delta^j_m - \delta^i_m \delta^j_l. \tag{3.5}$$

Vector dot and cross products become:

$$\boldsymbol{a} \cdot \boldsymbol{b} = a^i b_i = a_i b^i,$$
$$(\boldsymbol{a} \times \boldsymbol{b})^i = \epsilon^{ijk} a_j b_k,$$
$$(\boldsymbol{a} \times \boldsymbol{b})_i = \epsilon_{ijk} a^j b^k. \tag{3.6}$$

The common vector operations are:

$$(\nabla \varrho)_i = \frac{\partial \varrho}{\partial \bar{x}^i}, \tag{3.7}$$

$$\nabla \cdot \boldsymbol{a} = \frac{1}{\sqrt{g}} \frac{\partial}{\partial \bar{x}^i} \sqrt{g}\, a^i, \tag{3.8}$$

$$(\nabla \times \boldsymbol{a})^i = \epsilon^{ijk} \frac{\partial a_k}{\partial \bar{x}^j}, \tag{3.9}$$

$$\boldsymbol{a} \cdot \nabla \varrho = a^i \frac{\partial}{\partial \bar{x}^i} \varrho, \tag{3.10}$$

$$\nabla^2 \varrho = \frac{1}{\sqrt{g}} \frac{\partial}{\partial \bar{x}^i} \left(g^{ij} \sqrt{g}\, \frac{\partial \varrho}{\partial \bar{x}^j} \right). \tag{3.11}$$

The basic equations of a relativistic particle motion may be written in the following form:

$$\frac{d\boldsymbol{x}}{dt} = \boldsymbol{v} \quad \text{and} \quad \frac{d\boldsymbol{p}}{dt} = \frac{Z_s e}{M_s}\left(\boldsymbol{E} + \frac{\boldsymbol{v} \times \boldsymbol{B}}{c}\right), \tag{3.12}$$

where

$$\boldsymbol{p} = \gamma m_0 \boldsymbol{v},$$

3.1 Introduction

$$\gamma^2 = 1 + \frac{|\boldsymbol{p}|^2}{(m_0 c)^2} = 1 / \left(1 - \frac{|\boldsymbol{v}|^2}{c^2}\right). \tag{3.13}$$

In general coordinates $(\bar{x}^1, \bar{x}^2, \bar{x}^3)$, these equations become [142]

$$\frac{\mathrm{d}\bar{x}^k}{\mathrm{d}t} = \bar{v}^k \tag{3.14}$$

and

$$\frac{\mathrm{d}\bar{p}_m}{\mathrm{d}t} - \Gamma^r_{ml} \bar{p}_r \bar{v}^l = Z_s e (\bar{E}_m + e_{mlr} \bar{v}^l b^r), \tag{3.15}$$

where the Christoffel symbol of this kind is given by

$$\Gamma^i_{jk} = \boldsymbol{e}^i \cdot \frac{\partial \boldsymbol{e}_k}{\partial \bar{x}^j}. \tag{3.16}$$

In the following sections, except Sect. 3.8, we shall consider nonrelativistic ion motion.

The basic equations of a nonrelativistic particle motion may be written in the following form:

$$\frac{\mathrm{d}\boldsymbol{x}_l}{\mathrm{d}t} = \boldsymbol{v}_l \quad \text{and} \quad \frac{\mathrm{d}\boldsymbol{v}_l}{\mathrm{d}t} = \frac{Z_s e}{M_s}\left(\boldsymbol{E} + \frac{\boldsymbol{v}_l \times \boldsymbol{B}}{c}\right). \tag{3.17}$$

These equations may be written in Cartesian coordinates,

$$\frac{\mathrm{d}}{\mathrm{d}t}\begin{pmatrix} v_x \\ v_y \\ v_z \end{pmatrix} = \frac{Z_s e}{M_s} \begin{pmatrix} E_x + \dfrac{v_y B_z - v_z B_y}{c} \\ E_y + \dfrac{v_z B_x - v_x B_z}{c} \\ E_z + \dfrac{v_x B_y - v_y B_x}{c} \end{pmatrix} \quad \text{and} \quad \frac{\mathrm{d}}{\mathrm{d}t}\begin{pmatrix} x \\ y \\ z \end{pmatrix} = \begin{pmatrix} v_x \\ v_y \\ v_z \end{pmatrix}; \tag{3.18}$$

in cylindrical coordinates,

$$v_\varrho = \dot{\varrho}, \quad v_\phi = \varrho \dot{\phi}, \quad v_z = \dot{z}, \tag{3.19}$$

$$\frac{\mathrm{d}}{\mathrm{d}t} v_\varrho = w_\varrho = \ddot{\varrho} - \varrho \dot{\phi}^2 = \frac{Z_s e}{M_s}\left(E_\varrho + \frac{v_\phi B_z - v_z B_\phi}{c}\right), \tag{3.20}$$

$$\frac{\mathrm{d}}{\mathrm{d}t} v_\phi = w_\phi = \frac{1}{\varrho}\frac{\mathrm{d}}{\mathrm{d}t}(\varrho^2 \dot{\phi}) = \frac{Z_s e}{M_s}\left(E_\phi + \frac{v_z B_\varrho - v_\varrho B_z}{c}\right), \tag{3.21}$$

$$\frac{\mathrm{d}}{\mathrm{d}t} v_z = w_z = \ddot{z} = \frac{Z_s e}{M_s}\left(E_z + \frac{v_\varrho B_{phi} - v_\phi B_\varrho}{c}\right), \tag{3.22}$$

where w_ϱ, w_ϕ and w_z denote the components of the particle acceleration; and in spherical coordinates,

$$v_r = \dot{r}, \quad v_\theta = r\dot{\theta}, \quad v_\phi = r\dot{\phi}\sin\theta, \tag{3.23}$$

$$\frac{d}{dt}v_r = w_r = \ddot{r} - r\dot{\theta}^2 - r\dot{\phi}^2\sin^2\theta = \frac{Z_s e}{M_s}\left(E_r + \frac{v_\theta B_\phi - v_\phi B_\theta}{c}\right), \quad (3.24)$$

$$\frac{d}{dt}v_\theta = w_\theta = \frac{1}{r}\left(\frac{d}{dt}(r^2\dot{\theta}) - r^2\dot{\phi}^2\sin\theta\cos\theta\right) = \frac{Z_s e}{M_s}\left(E_\theta + \frac{v_\phi B_r - v_r B_\phi}{c}\right), \quad (3.25)$$

$$\frac{d}{dt}v_\phi = w_\phi = \frac{1}{r\sin\theta}\frac{d}{dt}(r^2\dot{\phi}\sin^2\theta) = \frac{Z_s e}{M_s}\left(E_\phi + \frac{v_r B_\theta - v_\theta B_r}{c}\right), \quad (3.26)$$

where w_r, w_θ and w_ϕ denote the components of the particle acceleration.

In general, a multistep scheme has the form

$$\sum_{i=0}^{k} a_{k-i}\boldsymbol{x}^{n+k-i} = \frac{e\Delta t}{M}\sum_{i=0}^{k} b_{k-i}\boldsymbol{F}^{n+k-i}. \quad (3.27)$$

If b_k is zero, the scheme becomes explicit and the unknown vector $\boldsymbol{x}^{n+k}$ can be solved directly in terms of known quantities. If b_k is nonzero the scheme becomes implicit, and, unless $\boldsymbol{F}$ is a simple function of $\boldsymbol{x}$, the new particle position $\boldsymbol{x}^{n+k}$ must be found iteratively at every time step. We can solve these nonlinear equations using a Newton–Raphson technique. Although explicit schemes are simpler to implement than implicit schemes, they are more severely limited by stability restrictions. For an optimal algorithm we have to choose schemes from (3.27) which satisfy certain criteria. The differential equations (3.17) are time-reversible, i.e., if a particle is integrated forwards in time in a given force field, and then time (and velocity) is reversed, the particle will retrace its path and return to its starting point. Time-reversible schemes are obtained by defining time-centered derivatives. However, it is not always practicable to exploit property time-centered schemes, since they generally lead to implicit equations in variables at the new time level. If the magnetic field is not a simple algebraic function of $\boldsymbol{x}$, one has to use iterations in order to update the position and velocity of particles. Consequently, the cost of solving the implicit equations is acceptable for small numbers of particles but not for the large ensembles used in many particle-mesh models. For this reason, explicit schemes which are not property time-centered are sometimes employed. The convergence of the numerical solution to the analytical one is determined by the accuracy of the finite-difference approximation and the numerical stability of finite-difference schemes. Accuracy is concerned with the round-off errors resulting from the finite word length of numbers within a computer and the truncation errors caused by the finite-difference approximation of the particle equation of motion. Efficiency is very important in particle-mesh simulations because of the large number of macroparticles used. Most important of all, storage limitation leads to the use of schemes with as few time levels as possible in the time integration of the particle equations of motion. Time limitation points to schemes with a small numbers of operations per particle per time step.

3.2 Explicit Leapfrog Method

We shall first consider the explicit method. In this method the pair of variables x and v are advanced in time alternately. This algorithm was used in [42, 174, 333, 349, 351, 417]. The equations of motion (3.17) are advanced as follows:

$$\frac{v^{n+1/2} - v^{n-1/2}}{\Delta t} = \frac{q_s}{M_s}\left(E(x^n) + \frac{v^{n+1/2} + v^{n-1/2}}{2c} \times B(x^n)\right), \quad (3.28)$$

$$\frac{x^{n+1} - x^n}{\Delta t} = v^{n+1/2}. \quad (3.29)$$

The superscript denotes the time level of the variable, for example, $x^{n+1} = x((n+1)\Delta t)$, and $v^{n+1/2} = v((n+1/2)\Delta t)$. Equation (3.28) can be solved for $v^{n+1/2}$, by taking its dot and cross product with B and substituting back into (3.28) [159, 567]:

$$v^{n+1/2} = \frac{1}{1 + \frac{\Omega_s^2 \Delta t^2}{4}}\left[v^{n-1/2}\left(1 - \frac{\Omega_s^2 \Delta t^2}{4}\right) + \frac{\Delta t q_s}{M_s}\left(E + \frac{v^{n-1/2} \times B}{c}\right)\right.$$

$$\left. + \frac{\Delta t^2 q_s^2}{2M_s^2 c} E \times B + \frac{\Delta t^2 q_s^2}{2M_s^2 c^2}\left(v^{n-1/2} \cdot B\right)B + \frac{\Delta t^3 q_s^3}{4M_s^3 c^2}(E \cdot B)B\right]. \quad (3.30)$$

To order $(\Delta t)^2$, (3.30) becomes:

$$v^{n+1/2} = v^{n-1/2}\left(1 - \frac{\Omega_s^2 \Delta t^2}{2}\right) + \frac{\Delta t q_s}{M_s}\left(E + \frac{v^{n-1/2} \times B}{c}\right)$$

$$+ \frac{\Delta t^2 q_s^2}{2M_s^2 c} E \times B + \frac{\Delta t^2 q_s^2}{2M_s^2 c^2}\left(v^{n-1/2} \cdot B\right)B, \quad (3.31)$$

where $\Omega_s = q_s B^n / M_s c = \Omega_s(x^n)$. Given $v^{n-1/2}$ and x^n, one solves (3.30) for $v^{n+1/2}$, and (3.29) is used subsequently to calculate x^{n+1}.

3.3 Implicit Method

In this method the pair of variables x and v are defined at the same time levels and are therefore advanced together [42, 59, 318, 319, 334], instead of being advanced alternately as in the leapfrog method. Equations (3.28–3.29), are advanced as

$$\frac{v^{n+1} - v^n}{\Delta t} = \frac{q_s}{M_s}\left(E^{n+\theta} + \frac{v^{n+1} + v^n}{2c} \times B^{n+\theta}\right), \quad (3.32)$$

72   3. Time Integration of the Particle Motion Equations

$$\frac{x^{n+1} - x^n}{\Delta t} = \frac{v^{n+1} + v^n}{2}, \tag{3.33}$$

where
$$E^{n+\theta} = E\left(x^{n+\theta}\right), \quad B^{n+\theta} = B\left(x^{n+\theta}\right)$$

and
$$x^{n+\theta} = \theta x^{n+1} + (1-\theta)x^n. \tag{3.34}$$

As before, the superscript denotes the time level of the variable. Given x^n and v^n, (3.32–3.33) is used to calculate x^{n+1} and v^{n+1}. It is important to recognize that, since the electric and magnetic field must be evaluated at the mid-orbit position ($\theta \geq 1/2$), one must iterate to solve (3.32–3.33).

Direct Solution for Updated Velocity

Equations (3.32–3.34) include implicit terms on their right-hand side. Moving the implicit terms to the left gives a matrix equation of the form

$$A^{n+1/2} v^{n+1} = C^{n+1/2}. \tag{3.35}$$

Fortunately, $A^{n+1/2}$ always has an inverse matrix (see [238]) given by

$$(A^{n+1/2})^{-1} = \frac{1}{\delta} \begin{pmatrix} 1+\alpha^2 & \alpha\beta+\gamma & \alpha\gamma-\beta \\ \alpha\beta-\gamma & 1+\beta^2 & \beta\gamma+\alpha \\ \alpha\gamma+\beta & \beta\gamma-\alpha & 1+\gamma^2 \end{pmatrix}, \tag{3.36}$$

where
$$\alpha = \frac{q_s \Delta t}{2 M_s c} B_x^{n+1/2},$$

$$\beta = \frac{q_s \Delta t}{2 M_s c} B_y^{n+1/2},$$

$$\gamma = \frac{q_s \Delta t}{2 M_s c} B_z^{n+1/2}$$

and
$$\delta = 1 + \alpha^2 + \beta^2 + \gamma^2.$$

Thus, the vector $v^{n+1/2}$ may be found directly from

$$v^{n+1} = (A^{n+1/2})^{-1} C^{n+1/2}, \tag{3.37}$$

which is now completely explicit except that the electromagnetic field E, B contains the implicit terms x^{n+1}.

3.4 Operator Splitting Method

The direct time integration of particle equations of motion is not always effective. To improve the efficiency of particle movers, one has to use a different way of splitting the particle equations of motion. The equation of motion may be split by taking into account the different physical processes, for example, the separate motion in the electric and magnetic fields, or the separate motion along and perpendicular to the magnetic field. In this section we shall consider the most popular numerical schemes based on the operator splitting method.

3.4.1 Splitting of the Particle Motion Equations. Boris's Scheme

Let us consider the schemes in which the particle motion equations are split into a set of equations with a more simple structure. In the first method [73], the drift velocity $\boldsymbol{E} \times \boldsymbol{B}/B^2$ is subtracted from the total velocity $\boldsymbol{v}$, as

$$\boldsymbol{v}_1 = \boldsymbol{v}^n - c\boldsymbol{E} \times \boldsymbol{B}/B^2. \tag{3.38}$$

Then we rotate the transversal velocity $\boldsymbol{v}_\perp$ and allow free acceleration of the longitudinal velocity $\boldsymbol{v}_\parallel$,

$$\frac{\boldsymbol{v}_2 - \boldsymbol{v}_1}{\Delta t} = \frac{q_s}{M_s}\boldsymbol{E}_\parallel + (\boldsymbol{v}_1 + \boldsymbol{v}_2) \times \boldsymbol{\Omega}/2, \tag{3.39}$$

and finally we find a new velocity by adding a drift velocity

$$\boldsymbol{v}^{n+1} = \boldsymbol{v}_2 + c\boldsymbol{E} \times \boldsymbol{B}/B^2, \tag{3.40}$$

where $\boldsymbol{\Omega}$ is a vector of gyrofrequency in the magnetic field $\boldsymbol{B}$. To advance the particle coordinates, we may use the following scheme:

$$\frac{\boldsymbol{x}^{n+1} - \boldsymbol{x}^n}{\Delta t} = \frac{\boldsymbol{v}^{n+1} + \boldsymbol{v}^n}{2}. \tag{3.41}$$

In another method we split the particle motion into three separate steps, namely, the motion due to the electric field, the motion due to the magnetic field, and again the motion due to the electric field [57]:

$$\frac{\boldsymbol{v}_1 - \boldsymbol{v}^{n-1/2}}{\Delta t/2} = \frac{q_s}{M_s}\boldsymbol{E}^n, \tag{3.42}$$

$$\frac{\boldsymbol{v}_2 - \boldsymbol{v}_1}{\Delta t} = (\boldsymbol{v}_1 + \boldsymbol{v}_2) \times \frac{\boldsymbol{\Omega}_s^n}{2}, \tag{3.43}$$

$$\frac{\boldsymbol{v}^{n+1/2} - \boldsymbol{v}_2}{\Delta t/2} = \frac{q_s}{M_s}\boldsymbol{E}^n, \tag{3.44}$$

74   3. Time Integration of the Particle Motion Equations

$$\frac{x^{n+1} - x^n}{\Delta t} = v^{n+1/2}. \tag{3.45}$$

In this scheme the motion in the magnetic field centered at the moment of time $t = n\Delta t$. In the general case, when the electromagnetic field is oriented arbitrarily to the coordinate axis, it is convenient to use the following realization for the second step of the above schemes, (3.39) and (3.43), suggested in [57] (see also [51, 236]):

$$v_3 = v_1 + a_1 v_1 \times B \quad \text{and} \quad v_2 = v_1 + a_2 v_3 \times B, \tag{3.46}$$

where $a_1 = B^{-1}\tan(q_s \Delta t B / 2 M_s c)$, $a_2 = 2a_1/(1 + a_1^2 B^2)$ and $B = |B|$. According to [57], the algorithm (3.46) takes about 35% of the number of operations less than a realization without the splitting of the second step in (3.39) and (3.43).

The finite-difference approximation to (3.43) gives velocities lying on a circle of radius $|v|$ in the velocity space and positions lying on a circle of radius R' in coordinate space. The finite time step cause the frequency to be higher than the correct frequency Ω and the radius R' to differ from the cyclotron radius $R = |v|/\Omega$. The frequency ω or the angle of particle rotation for one time step $\theta = \omega \Delta t$ may be estimated from geometry of the vector rotation:

$$\left|\tan \frac{\theta}{2}\right| = \frac{|v^{n+1} - v^n|}{|v^{n+1} + v^n|} = \frac{\Omega \Delta t}{2}. \tag{3.47}$$

Hence our finite-difference scheme produces a rotation through the angle

$$\theta = 2\arctan\left(\frac{q_s B}{M_s}\frac{\Delta t}{2}\right) = \Omega \Delta t \left\{1 - \frac{(\Omega \Delta t)^2}{12} + ...\right\},$$

which has an error of about 1% for $\Omega \Delta t < 0.35$. The radius R' may be found from (3.43) and (3.47) and with the result that $|v| = \text{const.}$:

$$R' = \frac{|v|}{\Omega} \sec\left(\frac{\omega \Delta t}{2}\right). \tag{3.48}$$

As in the case of the leapfrog scheme for an harmonic oscillator (Sect. 3.5), the error in the frequency can be eliminated by adjusting the frequency that appears in the difference equations.

3.4.2 Analytical Time Integration. Buneman's Scheme

Let us consider the schemes which are based on analytical integration of the particle motion equations. The general form of these schemes [73, 583] may be written as

$$v^{n+1/2} = \frac{c}{B} E^n \times b + \left\{\left(v^{n-1/2} - \frac{c}{B} E^n \times b\right)\left(1 - \frac{\Omega_s^2 \Delta t^2}{4}\right)\right.$$
$$\left. + \left(v^{n-1/2} - \frac{c}{B} E^n \times b\right) \times \Omega_s \Delta t\right\} \left(1 + \frac{\Omega_s^2 \Delta t^2}{4}\right)^{-1}, \tag{3.49}$$

where Ω_s is a gyrofrequency in the magnetic field $\boldsymbol{B}^n$ and $\boldsymbol{b} = \boldsymbol{B}^n/B^n$. This scheme provides an approximation accurate to the second order $[O(\Delta t^2)]$ to the particle motion equations. In the limit of a large time step integration, $\Omega_s \Delta t \to \infty$, we have

$$(\boldsymbol{v}^{n-1/2} + \boldsymbol{v}^{n+1/2})/2 = \frac{c}{(B^n)^2} \boldsymbol{E}^n \times \boldsymbol{B}^n + O(\Delta t^{-1}). \qquad (3.50)$$

This solution tends towards the drift of the particles in the electromagnetic field.

3.4.3 Time Integration with $\Omega \Delta t \gg 1$

In some multiscale and multisort ion simulations, it is very important to use a large time step, $\Omega \Delta t \gg 1$. The schemes considered above may result in an increasing numerical gyroradius (explicit leapfrog scheme) or in a decreasing numerical drift velocity (implicit Boris algorithm). To improve these schemes, inclusion of the effective force $-\mu \nabla B$ into implicit numerical scheme (3.32–3.33) was suggested [544]:

$$\frac{\boldsymbol{v}^{n+1} - \boldsymbol{v}^n}{\Delta t} = \frac{q_s}{M_s} \left(\boldsymbol{E}^{n+1/2} + \frac{\boldsymbol{v}^{n+1} + \boldsymbol{v}^n}{2c} \times \boldsymbol{B}^{n+1/2} \right) - \mu \nabla B^{n+1/2},$$

$$\frac{\boldsymbol{x}^{n+1} - \boldsymbol{x}^n}{\Delta t} = \frac{\boldsymbol{v}^{n+1} + \boldsymbol{v}^n}{2}, \qquad (3.51)$$

where

$$\mu = \frac{[(\boldsymbol{v}^{n+1} - \boldsymbol{v}^n) - (\boldsymbol{v}^{n+1} - \boldsymbol{v}^n) \cdot \boldsymbol{B}^{n+1/2} \boldsymbol{B}^{n+1/2}/(B^{n+1/2})^2]^2}{8 B^{n+1/2}}. \qquad (3.52)$$

One can show that the above expression for the magnetic moment μ can be recast in a form different from that given in [61], so that μ does not explicitly depend on the time-advanced velocity $\boldsymbol{v}^{n+1}$. This is achieved by solving for $\boldsymbol{v}^{n+1}$ in (3.51) and substituting this expression for $\boldsymbol{v}^{n+1}$ into (3.52). The resulting equation is a quadratic equation for μ:

$$A^* \mu^2 - B^* \mu + C^* = 0,$$

$$A^* = \frac{\boldsymbol{\xi}_\perp \cdot \boldsymbol{\xi}_\perp \Delta t^2}{8 B^{n+1/2}},$$

$$B^* = 1 + \frac{\Delta t \boldsymbol{\xi}_\perp \cdot (\hat{\boldsymbol{v}}_\perp - \boldsymbol{v}_\perp^n)}{2 B^{n+1/2}},$$

$$C^* = \frac{(\hat{\boldsymbol{v}}_\perp - \boldsymbol{v}_\perp^n) \cdot (\hat{\boldsymbol{v}}_\perp - \boldsymbol{v}_\perp^n)}{2 B^{n+1/2}}, \qquad (3.53)$$

where

$$\left\{\begin{matrix}\xi\\\hat{v}\end{matrix}\right\} = \frac{\mathsf{I} - \mathsf{I} \times \left(\frac{\Delta t}{2}\Omega^{n+1/2}\right) + \left(\frac{\Delta t}{2}\Omega^{n+1/2}\right)\left(\frac{\Delta t}{2}\Omega^{n+1/2}\right)}{1 + \left(\frac{\Delta t}{2}\Omega^{n+1/2}\right)^2}$$

$$\cdot \left\{\begin{matrix}\nabla B^{n+1/2}\\ v^n + \left(\frac{q_s \Delta t}{2 M_s} E^{n+1/2}\right)\end{matrix}\right\}, \quad (3.54)$$

$$\left\{\begin{matrix}\xi_\perp\\\hat{v}_\perp\\v^n_\perp\end{matrix}\right\} = \left(\mathsf{I} - \frac{B^{n+1/2}B^{n+1/2}}{(B^{n+1/2})^2}\right)\cdot\left\{\begin{matrix}\xi\\\hat{v}\\v^n\end{matrix}\right\}, \quad (3.55)$$

where I is the unit dyad and $\Omega^{n+1/2} = q_s B^{n+1/2}/M_s c$. The quadratic equation for μ yields two roots. One can show that in order to obtain the correct asymptotic behavior for μ in the limits of small and large time steps, the following root for μ must be used:

$$\mu = \frac{1}{2A^*}\{B^* - [(B^*)^2 - 4A^*C^*]^{1/2}\}. \quad (3.56)$$

Given x^n and v^n, one uses (3.51) and (3.53–3.55) to calculate x^{n+1} and v^{n+1}. Note that one needs to have an iteration procedure, just as in the implicit method, to solve for x^{n+1} and v^{n+1} from (3.51) and (3.53–3.55). For simplicity, we solve the equations approximately using a predictor–corrector procedure.

3.5 Stability and Accuracy of the Leapfrog Schemes

The analysis of the stability and the convergence of the finite-difference schemes for particle motion is a very complicated problem because of the nonlinear nature of the equations considered. In some simple cases, for example, the leapfrog scheme for an harmonic oscillator, we can compare the finite-difference solution with the exact one (see [51, 236]). Let us consider the second-order equation for a harmonic oscillator,

$$\frac{d^2 x}{dt^2} = -\omega_0^2 x. \quad (3.57)$$

This equation has solutions

$$x(t, t_0) = A(t_0)\cos\omega_0 t + B(t_0)\sin\omega_0 t. \quad (3.58)$$

The substitution of the finite-difference approximation of derivatives into a homogeneous equation of motion gives

$$x^{n+1} - 2x^n + x^{n-1} = -\omega_0^2 \Delta t^2 x^n, \quad (3.59)$$

where the time is $t - n\Delta t$. This is a standard finite-difference equation, which can be readily solved by assuming solutions of the form

$$x^n = Ae^{-i\omega n \Delta t}, \quad (3.60)$$

where A is an initial value and ω is the unknown. Substituting this value (and $x^{n-1} = Ae^{-i\omega(n-1)\Delta t}$ etc.) into (3.59), we obtain

$$\sin\left(\omega \frac{\Delta t}{2}\right) = \pm \omega_0 \frac{\Delta t}{2}. \quad (3.61)$$

The solution (3.61) of (3.59) is plotted in Fig. 3.1. We see that for $\omega_0 \Delta t/2 \ll 1$, $\omega \approx \omega_0$, as desired. However, we see that as $\omega_0 \Delta t$ increases beyond 2, the (initially wholly) real solution for ω becomes complex, with growing and decaying roots, which indicates numerical instability.

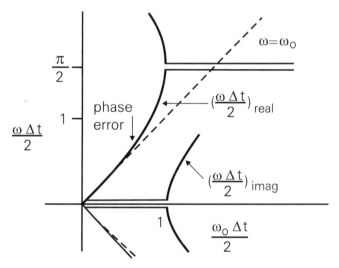

Fig. 3.1. The solution for ω in terms of ω_0 for simple harmonic motion, from the leap-frog finite-difference equation. Frequency ω agrees with ω_0 for small $\omega_0 \Delta t$, but it is larger than ω_0 as $\omega_0 \Delta t$ increases. For $\omega_0 \Delta t/2 > 1$, the solution becomes complex, with growing and decaying roots for ω (this is numerical instability). Note the double root at $\omega = 0, \pi/2$. (From [51])

The amplitude error is zero for $\omega_0 \Delta t < 2$. The phase error for $\omega_0 \Delta t \ll 1$ may be estimated as

$$\omega \frac{\Delta t}{2} \approx \omega_0 \frac{\Delta t}{2} \left[1 + \frac{1}{6}\left(\omega_0 \frac{\Delta t}{2}\right)^2 + \ldots \right], \quad (3.62)$$

showing a quadratic error term, as desired. The cumulative phase after N steps is

$$\text{phase} - \text{error} = \omega N \Delta t \approx \omega_0 N \frac{\Delta t}{6} \left(\omega_0 \frac{\Delta t}{2} \right)^2 = \frac{N}{24} (\omega_0 \Delta t)^3. \quad (3.63)$$

It is important to know that restricting the allowables error to small values limits the number of steps (and cycles), and increasing the step increases the error as the cube of the step size. Analysis of another scheme will be given later in this chapter. The other approaches to analyze the finite-difference scheme (the Amplification Matrix Method and the Root Locus Method) can be found in [236].

3.6 Implicit Time Integration. C1 and D1 Class Schemes

The problem of implicit time integration of the particle motion equations has been discussed in detail in the literature [3, 32, 106, 107, 128, 291]. An implicit time scheme may be thought of as a low pass filter: one wishes to have an accurate solution for frequencies lower than some value ω_0 and suppress as efficiently as possible modes with frequencies higher than ω_0, because they cannot be accurately described with the time step chosen. The ideal implicit scheme should be like an ideal low pass filter, i.e., a step function. In this section we shall see that by keeping information on enough time levels it is possible to design a third-order damping implicit scheme.

3.6.1 C1 Class Scheme

The first class of algorithms we shall consider is called the class C1 scheme [104, 107]. The finite-difference scheme may be written as follows:

$$\frac{v^{n+1/2} - v^{n-1/2}}{\Delta t} = a^n,$$

$$\frac{x^{n+1} - x^n}{\Delta t} = v^{n+1/2} + C_0 \Delta t (a^{n+1} - a^n) + C_1 \Delta t^2 (a^n - a^{n-1}) \ldots \quad (3.64)$$

Insight into the behavior of this scheme can be gained by appling it to the problem of the harmonic oscillator. The roots of the dispersion relation for small values of $\omega_0 \Delta t$ are given by

$$\text{Re} \frac{\omega}{\omega_0} = \pm \left[1 + \frac{1}{2} (\omega_0 \Delta t)^2 \left(\frac{1}{12} - C_0 - C_1 \ldots \right) + O(\Delta t^3) \right],$$

$$\text{Im} \frac{\omega}{\omega_0} = -\frac{1}{2} (\omega_0 \Delta t)^3 (C_1 + 2C_2 + \ldots) + O(\Delta t^4). \quad (3.65)$$

At this stage the constants $C_0, C_1, C_2, \ldots$ have to be chosen to give optimal properties to the scheme. It is obvious from (3.64) that for $C_0 = C_1 = C_2 =$

... − 0 one recovers the standard explicit leapfrog scheme. The algorithm used by Denavit [128] for advancing particles is the following:

$$v^{n+1} - v^n = \left(\frac{3}{4}a^{n+1} + \frac{1}{4}a^{n-1}\right)\Delta t,$$

$$x^{n+1} - x^n = \left(\frac{3}{4}v^{n+1} + \frac{1}{4}v^{n-1}\right)\Delta t, \quad (3.66)$$

which can be cast in the form of (3.64) with $C_0 = 9/16$, $C_1 = 1/8$ and $C_2 = 1/16$. Then from (3.65) one has

$$\text{Re}\frac{\omega}{\omega_0} = \pm\left(1 - \frac{(\omega_0\Delta t)^2}{3} + ...\right) \quad \text{and} \quad \text{Im}\frac{\omega}{\omega_0} = -\frac{(\omega_0\Delta t)^3}{8}. \quad (3.67)$$

Another Denavit's algorithm may be written as follows:

$$\frac{v^{n+1} - v^n}{\Delta t} = \frac{1}{16}(9a^{n+1} + 6a^n + a^{n-1}),$$

$$\frac{x^{n+1} - x^n}{\Delta t} = \frac{1}{16}(9v^{n+1} + 6v^n + v^{n-1}). \quad (3.68)$$

The corresponding coefficients are $C_0 = 81/256$, $C_1 = 14/256$ and $C_2 = 1/256$, and (3.65) gives

$$\text{Re}\frac{\omega}{\omega_0} = \pm\left(1 - \frac{7}{48}(\omega_0\Delta t)^2\right) \quad \text{and} \quad \text{Im}\frac{\omega}{\omega_0} = -\left(\frac{\omega_0\Delta t}{32}\right)^3. \quad (3.69)$$

Let us compare the phase properties and damping properties of (3.67) and (3.69). In the second scheme the error on the phase (i.e., the real part of ω) has been reduced by almost a factor 2, but simultaneously the damping of unwanted frequencies has been reduced by four.

3.6.2 D1 Class Scheme

Let us consider the so-called D1 class scheme for time integration. According to [106, 184], this scheme may be written as follows:

$$x^{n+1} - x^n = v^{n+1/2}\Delta t,$$

$$v^{n+1/2} - v^{n-1/2} = \frac{1}{2}a^{n+1}\Delta t + \frac{1}{2}(v^{n-1/2} - v^{n-3/2}). \quad (3.70)$$

For a harmonic oscillation one can obtain the following dispersion relation:

$$\text{Re}\left(\frac{\omega}{\omega_0}\right) = \pm\left(1 - \frac{11}{24}(\omega_0\Delta t)^2 + ...\right) \quad \text{and} \quad \text{Im}\left(\frac{\omega}{\omega_0}\right) = -\frac{1}{2}(\omega_0\Delta t)^3 + ... \quad (3.71)$$

A comparison of (3.71) and (3.67) shows that the damping rate of high frequencies ($\omega_0 \Delta t \gg 1$) in this scheme is 50 times larger than for the optimized C1 scheme [128].

However, the phase error for $\omega_0 \Delta t \ll 1$ is also multiplied by 4, and it becomes much larger than in the case of the standard explicit leapfrog scheme. This means that, if one limits the phase error to some maximum value for a given frequency of interest, a time step smaller by a factor of 2 has to be used. This in turn reduces the damping by a factor of 8. Even under this constraint the D1 scheme remains superior to the C1 scheme; however, it does become twice as expensive.

3.7 Runge–Kutta Schemes

The particle motion equations may be considered in the following general form:
$$\frac{d\boldsymbol{x}}{dt} = \boldsymbol{v} \quad \text{and} \quad \frac{d\boldsymbol{v}}{dt} = \frac{\boldsymbol{F}}{M}(\boldsymbol{x}, \boldsymbol{v}). \tag{3.72}$$

We concentrate on two specific examples from the Runge–Kutta family of multistage schemes (see, e.g., [81]). First, the 2-stage Runge–Kutta algorithm often called the Heun method is written

$$\boldsymbol{a}_x = \Delta t \boldsymbol{v}^n, \quad \boldsymbol{a}_v = \Delta t \boldsymbol{F}(\boldsymbol{v}^n, \boldsymbol{x}^n),$$

$$\boldsymbol{b}_x = \Delta t (\boldsymbol{v}^n + \boldsymbol{a}_v), \quad \boldsymbol{b}_v = \Delta t \boldsymbol{F}(\boldsymbol{v}^n + \boldsymbol{a}_v, \boldsymbol{x}^n + \boldsymbol{a}_x), \tag{3.73}$$

$$\boldsymbol{x}^{n+1} = \boldsymbol{x}^n + \frac{1}{2}(\boldsymbol{a}_x + \boldsymbol{b}_x),$$

$$\boldsymbol{v}^{n+1} = \boldsymbol{v}^n + \frac{1}{2}(\boldsymbol{a}_v + \boldsymbol{b}_v).$$

The second scheme is the classic 4-stage Runge–Kutta algorithm [44, 284, 454] and can be written

$$\boldsymbol{a}_x = \Delta t \boldsymbol{v}^n, \quad \boldsymbol{a}_v = \Delta t \boldsymbol{F}(\boldsymbol{v}^n, \boldsymbol{x}^n),$$

$$\boldsymbol{b}_x = \Delta t \left(\boldsymbol{v}^n + \frac{1}{2}\boldsymbol{a}_v \right), \quad \boldsymbol{b}_v = \Delta t \boldsymbol{F}\left(\boldsymbol{v}^n + \frac{1}{2}\boldsymbol{a}_v, \boldsymbol{x}^n + \frac{1}{2}\boldsymbol{a}_x \right),$$

$$\boldsymbol{c}_x = \Delta t \left(\boldsymbol{v}^n + \frac{1}{2}\boldsymbol{b}_v \right), \quad \boldsymbol{c}_v = \Delta t \boldsymbol{F}\left(\boldsymbol{v}^n + \frac{1}{2}\boldsymbol{b}_v, \boldsymbol{x}^n + \frac{1}{2}\boldsymbol{b}_x \right), \tag{3.74}$$

$$\boldsymbol{d}_x = \Delta t (\boldsymbol{v}^n + \boldsymbol{c}_v), \quad \boldsymbol{d}_v = \Delta t \boldsymbol{F}(\boldsymbol{v}^n + \boldsymbol{c}_v, \boldsymbol{x}^n + \boldsymbol{c}_x),$$

$$\boldsymbol{x}^{n+1} = \boldsymbol{x}^n + \frac{1}{6}(\boldsymbol{a}_x + 2\boldsymbol{b}_x + 2\boldsymbol{c}_x + \boldsymbol{d}_x),$$

$$\boldsymbol{v}^{n+1} = \boldsymbol{v}^n + \frac{1}{6}(\boldsymbol{a}_v + 2\boldsymbol{b}_v + 2\boldsymbol{c}_v + \boldsymbol{d}_v).$$

A difficulty with multistage schemes is that they frequently require velocity data at intermediate times between t^n and t^{n+1}. For example, RK4 requires velocity data at the midpoints. Since velocity data is only available at t^n, the velocity at intermediate times must be interpolated from the previous or the current velocity fields. The RK4 method is also used in some hybrid codes for solving Maxwell equations [572].

3.8 Relativistic Particle Motion Equations

Relativistic generalization of the above numerical schemes for the particle motion equations may be done without any difficulties. The finite-difference equations of particle motion in the relativistic limit may be written as

$$\frac{\boldsymbol{P}^{n+1} - \boldsymbol{P}^n}{\Delta t} = q\boldsymbol{E}^{n+1/2} + (\boldsymbol{P}^{n+1} + \boldsymbol{P}^n) \times \frac{\boldsymbol{\omega}'}{2}, \quad (3.75)$$

where $\boldsymbol{P} = M\boldsymbol{v}(1 - v^2/c^2)^{-1/2}$ is the particle momentum and

$$\boldsymbol{\omega}' = \boldsymbol{\Omega}^{n+1/2} \cdot (1 + P^2/M^2c^2)^{-1/2}.$$

Let us consider the next splitted scheme for this equation:

$$\frac{\boldsymbol{P}_1 - \boldsymbol{P}^n}{\Delta t/2} = q\boldsymbol{E}, \quad \frac{\boldsymbol{P}_2 - \boldsymbol{P}_1}{\Delta t} = (\boldsymbol{P}_1 + \boldsymbol{P}_2) \times \frac{\boldsymbol{\omega}'}{2},$$

$$\frac{\boldsymbol{P}^{n+1} - \boldsymbol{P}_2}{\Delta t/2} = q\boldsymbol{E}. \quad (3.76)$$

In the second step of this scheme, we rotate the particle momentum vector only, without changing its value. Hence $\boldsymbol{P}_2^2 = \boldsymbol{P}_1^2$, and the value $(1-v^2/c^2)^{1/2}$ is a constant, and we can take it into account when we calculate the value of $\boldsymbol{\Omega}$. Equation (3.76) may be written as follows:

$$\boldsymbol{P}_3 = \boldsymbol{P}_1 + a_1 \boldsymbol{P}_1 \times \boldsymbol{B}, \quad \boldsymbol{P}_2 = \boldsymbol{P}_1 + a_2 \boldsymbol{P}_3 \times \boldsymbol{B}, \quad (3.77)$$

where $a_1 = B^{-1} \tan(q\Delta tB/2Mc)$, $a_2 = 2a_1/(1 + a_1^2 B^2)$ and $B = |\boldsymbol{B}|$.

The scheme predictor-corrector for relativistic particle motion was considered in [243, 428]. The system of particle motion equations may be written as follows:

$$\frac{d\boldsymbol{v}}{dt} = \boldsymbol{a}, \quad \frac{d\boldsymbol{x}}{dt} = \boldsymbol{v} \quad \text{and} \quad \boldsymbol{a} = \frac{q\sqrt{1-v^2/c^2}}{M}\left(\boldsymbol{E} - \boldsymbol{v}\frac{(\boldsymbol{v}\cdot\boldsymbol{E})}{c^2} + \frac{\boldsymbol{v}\times\boldsymbol{B}}{c}\right). \quad (3.78)$$

The previous values of the velocities and coordinates are determined by the following equations

$$\boldsymbol{v}_1 = \boldsymbol{v}^n + \boldsymbol{a}^n \Delta t \quad \text{and} \quad \boldsymbol{x}_1 = \boldsymbol{x}^n + (\boldsymbol{v}^n + \boldsymbol{v}_1)\Delta t/2. \quad (3.79)$$

Using the values of x_1 and v_1, we find the previous value of a_1. Then we execute the corrector step according to the following formulae:

$$v^{n+1} = v^n + (a^n + a_1)\Delta t/2,$$

$$x^{n+1} = x^n + (v^n + v^{n+1})\Delta t/2. \tag{3.80}$$

Equations (3.79) and (3.80) represent a scheme accurate in time to the second order. It is very simple, but it is not reversible in time.

As a conclusion we can state that if the time step for the particle motion equations is smaller than the time step for the field equations, it may result in the additional smoothing of the grid values of the plasma density and current [450].

Summary

Efficiency is the main characteristic of the time integration schemes for the particle motion equations. The compromise between accuracy and efficiency can be altered in two ways – either by using a higher-order scheme and a larger time step or by using a low-order scheme and a smaller time step. The first approach suffers because (a) the time step is limited by the natural frequency, (b) higher-order schemes often have more restrictive stability limits on the time step, and (c) higher-order schemes need force values at several time levels. For typical particle-mesh simulation the best compromise between accuracy, stability and efficiency is found by using simple schemes accurate to the second order (such as leapfrog) and adjusting the time step accordingly. However, in the case of particle motion in the vicinity of discontinuities or other thin structures (e.g., acceleration by shock surfing) the use of higher-order schemes such as the Runge–Kutta scheme may be reasonable.

Exercises

3.1 Show that the numerical scheme (3.28–3.29) is reversible in time.

3.2 Show that the numerical scheme (3.32–3.34) is reversible in time.

3.3 Show that the numerical scheme (3.38–3.41) is reversible in time.

3.4 Show that the numerical scheme (3.42–3.45) is reversible in time.

3.5 Derive the algorithm for the implicit scheme (3.32–3.34) in coordinate form.

4. Density and Current Assignment. Force Interpolation. Conservation Laws

4.1 Introduction

In this chapter we introduce the methods for particle and force weighting in Cartesian as well as in cylindrical and spherical coordinates (Sects. 4.2 and 4.3). We also analyze the application of mass, momentum and energy conservation laws to the hybrid models (Sect. 4.4).

4.2 Cloud and Assignment Function Shapes

Let us suppose that the plasma consists of N_p particles of a given sort in volume V. The each particle has a charge q_l and mass M_l; its position in phase space is characterized by the vector $x_l(t)$ and the velocity $v_l(t)$. The distribution function for these point-type particles may be written as follows:

$$f(\boldsymbol{x}, \boldsymbol{v}, t) = \sum_{l=1}^{N_p} \delta\left(\boldsymbol{x} - \boldsymbol{x}_l(t)\right) \delta\left(\boldsymbol{v} - \boldsymbol{v}_l(t)\right). \tag{4.1}$$

Direct use of such a type of distribution function in the two- and three-dimensional problems may result in a strong "shot" noise, and the singularities of the electric field may occur at small scales, because of the limited number of particles involved in simulation. In order to decrease the level of fluctuation, a different modification of the finite size particles is proposed. Such a type of distribution may result in smoothing of the small-scale forces, but large-scale collective forces are described well. The finite size particle distribution function may be written as [42, 51]

$$f_c(\boldsymbol{x}, \boldsymbol{v}, t) = \int f(\boldsymbol{x}', \boldsymbol{v}, t) S(\boldsymbol{x}, \boldsymbol{x}') \mathrm{d}\boldsymbol{x}' = \sum_{l=1}^{N_p} S\left(\boldsymbol{x}, \boldsymbol{x}_l(t)\right) \delta\left(\boldsymbol{v} - \boldsymbol{v}_l(t)\right). \tag{4.2}$$

The value of the density, charge and current at the point $\boldsymbol{x}$ are assigned in accordance with the following relations:

$$n_c(\boldsymbol{x}, t) = \sum_{l=1}^{N_p} S\left(\boldsymbol{x}, \boldsymbol{x}_l(t)\right), \quad \varrho_c(\boldsymbol{x}, t) = \sum_{l=1}^{N_p} q_l S\left(\boldsymbol{x}, \boldsymbol{x}_l(t)\right),$$

4. Density and Current Assignment. Force Interpolation

$$\boldsymbol{j}_c(\boldsymbol{x},t) = \sum_{l=1}^{N_p} q_l \boldsymbol{v}_l(t) S(\boldsymbol{x}, \boldsymbol{x}_l(t)). \tag{4.3}$$

Here the shape of macroparticles S satisfies the normalization condition $\int_V S(\boldsymbol{x},\boldsymbol{x}')\mathrm{d}\boldsymbol{x} = 1$. Integration of (4.3) over the volume V gives the total charge of the system:

$$Q = \int_V \varrho_c(\boldsymbol{x},t)\mathrm{d}\boldsymbol{x} = \sum_{l=1}^{N_p} q_l \int_V S(\boldsymbol{x},\boldsymbol{x}_l)\mathrm{d}\boldsymbol{x} \tag{4.4}$$

$$= \sum_{l=1}^{N_p} q_l = \int_V \varrho(\boldsymbol{x},t)\mathrm{d}\boldsymbol{x}.$$

In the three-dimensional case the charge corresponding to the cell with indices i,j,k is

$$\varrho_{ijk} = \frac{1}{V_{ijk}} \int_{V_{ijk}} \varrho_c(\boldsymbol{x})\mathrm{d}\boldsymbol{x}, \tag{4.5}$$

where V_{ijk} is the volume of the cell with indices i,j,k; $V_{ijk} = h_1 h_2 h_3$. In the three-dimensional case we have

$$\varrho_{ijk} = \frac{1}{V_{ijk}} \int_{x_i-\frac{h_1}{2}}^{x_i+\frac{h_1}{2}} \int_{y_j-\frac{h_2}{2}}^{y_j+\frac{h_2}{2}} \int_{z_k-\frac{h_3}{2}}^{z_k+\frac{h_3}{2}} \varrho_c(\boldsymbol{x})\mathrm{d}\boldsymbol{x}, \tag{4.6}$$

where

$$\varrho_c(x,y,z) = \sum_{l=1}^{N_p} q_l S(x - x_l, y - y_l, z - z_l). \tag{4.7}$$

Substitution of (4.7) into (4.6) gives

$$\varrho_{ijk} = \sum_{l=1}^{N_p} q_l W(x_i - x_l, y_j - y_l, z_k - z_l), \tag{4.8}$$

where W is the assignment function.

The cloud shape $S(x')$ of a particle with unit charge is its charge density, where x' measures the distances from the center of the particle. The fraction of the charge assigned from a particle of shape S at position x' to mesh point p at x_p is given by the overlap of the cloud with cell p, i.e.,

$$W(x_p - x) = W_p(x) = \int_{x_p-h/2}^{x_p+h/2} S(x' - x)\mathrm{d}x'. \tag{4.9}$$

Using (4.9) and the top-hat function Π from (4.14), the relationship between the cloud shape S and the assignment function W can be obtained:

$$W(x) = \int \Pi\left(\frac{x'}{h}\right) S(x'-x) \mathrm{d}x'. \tag{4.10}$$

If we restrict S to even functions, then (4.10) may be more compactly expressed as

$$W(x) = \Pi\left(\frac{x'}{h}\right) * S(x), \tag{4.11}$$

where $*$ denotes a convolution operator.

The relation between S and W in three-dimensional case is the following:

$$W(x_i - x, y_j - y, z_k - z)$$
$$= \frac{1}{V_{ijk}} \int_{x_i - \frac{h_1}{2}}^{x_i + \frac{h_1}{2}} \int_{y_j - \frac{h_2}{2}}^{y_j + \frac{h_2}{2}} \int_{z_k - \frac{h_3}{2}}^{z_k + \frac{h_3}{2}} S(x - x', y - y', z - z') \mathrm{d}x' \mathrm{d}y' \mathrm{d}z'. \tag{4.12}$$

4.3 NGP, CIC and TSC Weighting

4.3.1 Cloud (S) and Assignment (W) Function Hierarchy

Cloud shapes S for the hierarchy of schemes may be obtained by using an approach similar to that used for the assignment function [42, 236]. Using the definitions of the functions Π and Λ given below, we can write the cloud and charge assignment functions for the NGP, CIC, TSC, and other one-dimensional schemes as follows:

(1) Zero-order particle weighting or nearest grid point interpolation (NGP) with a stepwise distribution of force:

$$S(x, x') = \delta(x - x'),$$
$$W(x) = \Pi\left(\frac{x}{h}\right) = \frac{1}{h}\Pi\left(\frac{x}{h}\right) * \delta\left(\frac{x}{h}\right), \tag{4.13}$$

where $*$ denotes a convolution operator.

(2) First-order particle weighting or cloud-in-cell model (CIC) with a continuous piecewise linear distribution of force:

$$S(x, x') = \frac{1}{h}\Pi\left(\frac{x}{h}\right) = \begin{cases} (2h)^{-1} & \text{for } |x - x'| \leq h/2, \\ 0 & \text{for } |x - x'| > h/2, \end{cases}$$

$$W(x) = \Lambda\left(\frac{x}{h}\right) = \frac{1}{h}\Pi\left(\frac{x}{h}\right) * \Pi\left(\frac{x}{h}\right). \tag{4.14}$$

(3) Triangular-shaped density cloud scheme (TSC) with a continuous value and the first derivative of the force distribution:

$$S(x, x') = \frac{1}{h}\Lambda\left(\frac{x}{h}\right) = \begin{cases} h^{-1}(1 - |x - x'|/h) & \text{for } |x - x'| \leq h, \\ 0 & \text{for } |x - x'| > h, \end{cases}$$

$$W(x) = \frac{1}{h} * \Lambda\left(\frac{x}{h}\right) * \Pi\left(\frac{x}{h}\right) = \frac{1}{h^2} \Pi\left(\frac{x}{h}\right) * \Pi\left(\frac{x}{h}\right) * \Pi\left(\frac{x}{h}\right). \quad (4.15)$$

(4) Quadratic-spline interpolation scheme (PQS) with a continuous value and the first and second derivatives of the force distribution:

$$S(x, x') = \Lambda\left(\frac{x}{h}\right) * \Pi\left(\frac{x}{h}\right),$$

$$W(x) = \frac{1}{h^3} \Pi\left(\frac{x}{h}\right) * \Pi\left(\frac{x}{h}\right) * \Pi\left(\frac{x}{h}\right) * \Pi\left(\frac{x}{h}\right). \quad (4.16)$$

(5) Cosine scheme with continuous derivatives of the force distribution:

$$S(x, x') = \begin{cases} (2h)^{-1}(1 + \cos \pi |x - x'|/h) & \text{for } |x - x'| \leq h, \\ 0 & \text{for } |x - x'| > h, \end{cases}$$

$$W(x) = \begin{cases} 0 & \text{for } |x| \geq \frac{3h}{2}, \\ \frac{1}{2h^2}\left(x + \frac{3h}{2} + \frac{h}{\pi} \sin \frac{\pi(x + h/2)}{h}\right) & \text{for } -\frac{3h}{2} \leq x \leq -\frac{h}{2}, \\ \frac{1}{2h^2}\left(h + \frac{2h}{\pi} \cos \frac{\pi x}{h}\right) & \text{for } -\frac{h}{2} \leq x \leq \frac{h}{2}, \\ \frac{1}{2h^2}\left(-x + \frac{3h}{2} + \frac{h}{\pi} \sin \frac{\pi(h/2 - x)}{h}\right) & \text{for } \frac{h}{2} \leq x \leq \frac{3h}{2}. \end{cases}$$
$$(4.17)$$

(6) Gaussian particle shape function with continuous derivatives of the force distribution:

$$S(x, x') = (2\pi)^{-\frac{1}{2}} h^{-1} \exp\left[-(x - x')^2/2h^2\right],$$

$$W(x) = \frac{\sqrt{\pi}}{2h} \left[\operatorname{erf}\left(\frac{x + h/2}{\sqrt{2}h}\right) - \operatorname{erf}\left(\frac{x - h/2}{\sqrt{2}h}\right)\right]. \quad (4.18)$$

The cloud and assignment functions must satisfy the charge conservation condition:

$$\sum_{i=1}^{N_g} W(x_i - x) = 1 \quad \text{and} \quad \int S(x' - x) dx' = 1. \quad (4.19)$$

Figure 4.1 shows the cloud shape interpretation. The particle carries with it the cloud shape appropriate to the assignment scheme. The area of overlap of the cloud shapes with a cell determines the fraction of the charge assigned to the mesh point in that cell. Figure 4.2 shows the corresponding situation for the assignment function interpretation of charge assignment. In this case, the particles carry the assignment function, and the value of the function at a mesh point gives the fraction of the charge assigned to that point.

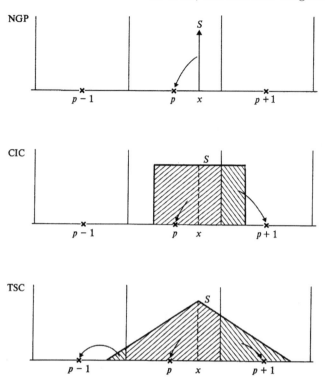

Fig. 4.1. The cloud shape interpretation of charge assignment. The fraction of charge from a particle at position x to a given mesh point is equal to the area of overlap of the cloud shape S with the cell containing that point [236]

4.3.2 Weighting in Two- and Three-Dimensional Space

In the two- and three-dimensional case for Cartesian coordinates, we can use a decomposition of the weighting function. To retain the smoothness properties of the one-dimensional charge-assignment-force-interpolation functions, the assignment functions for two-dimensional schemes must take the product form [236]

$$W(\boldsymbol{x}) = W(x,y) = W(x)W(y), \qquad (4.20)$$

where both of the separate functions in the product on the right side of (4.20) are the one-dimensional assignment functions with the required order of smoothness. Similarly, in three dimensions, the assignment functions must be of the form

$$W(\boldsymbol{x}) = W(x,y,z) = W(x)W(y)W(z) \qquad (4.21)$$

to obtain continuity of value (first order), continuity of value and the first derivative (second order), etc., everywhere. These product forms have the additional advantage that the factorization enables them to be computed with

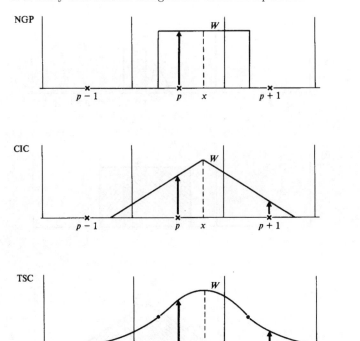

Fig. 4.2. The assignment function shape interpretation of charge assignment. The fraction of charge assigned from a particle at position x to a given mesh point is equal to the value of the assignment function W at that point [236]

a relatively small number of arithmetic operations. Figure 4.3 demonstrates the two-dimensional CIC or area-weighting scheme. The fraction of charge assigned to the four neighboring mesh points from a particle at position x is given by the area of overlap of its cloud shape with the cells containing those neighboring mesh points.

Let us consider the weighting in cylindrical and spherical coordinates. At first we consider cylindrical coordinates with a homogeneous mesh in all directions. We can weight the particle charge q_i *bilinearly* (see also [51]) in r–ϕ to the nearest four grid points, similar to the operation performed in rectangular coordinates (see CIC scheme). Another method is bilinear in (r^2, ϕ), or *area weighting*, as shown in Fig. 4.4. Let the particle be located at (r_i, ϕ_i). The fraction of the charge q_i assigned to point $A(r_j, \phi_k)$ is (area a)/(areas $a+b+c+d$)=$f_{j,k}$, leading to

$$Q_A = Q_{j,k} = q_i \frac{(r_{j+1}^2 - r_i^2)(\phi_{k+1} - \phi_i)}{(r_{j+1}^2 - r_j^2)(\phi_{k+1} - \phi_k)} = q_i f_{j,k}. \qquad (4.22)$$

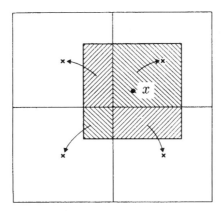

Fig. 4.3. Particle weighting to x–y grid point (from [236])

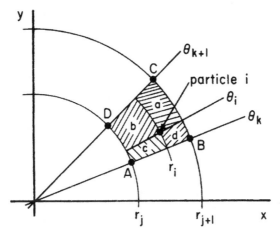

Fig. 4.4. Particle weighting to an r–θ grid point using area ratio $a/(a+b+c+d)$ for assignment to A etc., interpreted as area or r^2 weighting (from [51])

Note that these weightings are valid for $r_i < r_1$, where grid points A and D are at the origin. In case of full cylindrical coordinates, r–ϕ–z, we can use the following weighting function:

$$W(r, \phi, z) = W(r^2)W(\phi)W(z). \tag{4.23}$$

Let us consider now spherical coordinates with a homogeneous mesh in all directions. In the first method we can weight the particle charge q_i *bilinearly*, as in the cylindrical case, in r–θ–ϕ to the nearest eight grid points, similar to the operation performed in rectangular coordinates (see CIC scheme),

$$W(r, \theta, \phi) = W(r)W(\theta)W(\phi). \tag{4.24}$$

Another method is bilinear in (r^3, θ), or area weighting. Let the particle be located at (r_i, θ_i, ϕ_i). The fraction of the charge q_i assigned to point $A(r_j, \theta_m, \phi_k)$ may be expressed as follows:

$$Q_A = Q_{j,m,k} = q_i \frac{(r_{j+1}^3 - r_i^3)(\theta_{m+1} - \theta_i)(\phi_{k+1} - \phi_i)}{(r_{j+1}^3 - r_j^3)(\theta_{m+1} - \theta_m)(\phi_{k+1} - \phi_k)}. \tag{4.25}$$

Although it is taken for granted that the total charge in the system must be conserved in the process of conventional particle-to-grid weighting, conservation of charge density is not guaranteed for arbitrary weighting schemes, particularly for cylindrical and spherical coordinates [457]. For these latter two systems, it was shown in [457] that the much-used PIC and CIC weighting schemes do not result in uniform charge densities on the grid, even for a uniform distribution of particles. This effect is most pronounced at the boundaries of the grid, but is, actually, intrinsic to the definition of the shape factor for the whole grid. Thus, one cannot apply a correction factor only at the boundaries to remedy the situation.

It was shown in [457], quite generally, how the nonconservation of charge density can be overcome; one can derive, specifically, a number of alternative shape factors that conserve charge as well as charge density. These are alternative PIC and CIC weightings in cylindrical coordinates [(4.2) and (4.3), [457]] and an alternative PIC weighting in spherical coordinates [(4.4), [457]].

In a second approach, no modification to a given shape factor is required, but, instead, the location of cell boundaries is modified. Thus, different normalization volumes are obtained by which charges on the grid can be converted to charge densities on the grid. Since, however, cell boundaries are normally chosen as the basis for a finite difference formulation of the electrostatic or electromagnetic interactions between the plasma particles, the conceptual difficulty of having two different sets of cell boundaries in the same calculation arises with this alternative approach. The reader is referred to [457] for a detailed consideration of this subject.

4.4 Force Interpolation

Let us consider a set of equations that describes the dynamics of moments of the distribution function

$$\frac{\partial}{\partial t} \int \boldsymbol{v}^N f d\boldsymbol{v} + \frac{\partial}{\partial x_l} \int \boldsymbol{v}^N v_l f d\boldsymbol{v} - \frac{1}{M} \int \left(\frac{\partial (F_l \boldsymbol{v}^N)}{\partial v_l} \right) f d\boldsymbol{v} = 0, \tag{4.26}$$

where $N = 0, 1, \ldots$ and a summation is made over the index l. The substitution of the distribution function of finite size particle f_c (4.2) into (4.26) gives the following set of equations for coordinates $\boldsymbol{x}_j(t)$ and velocities $\boldsymbol{v}_j(t)$ of particles [488]:

$$\sum_{j=1}^{N_\mathrm{p}} \left(\dot{\boldsymbol{x}}_j \frac{\partial S(\boldsymbol{x},\boldsymbol{x}_j)}{\partial \boldsymbol{x}_j} + \boldsymbol{v}_j \frac{\partial S(\boldsymbol{x},\boldsymbol{x}_j)}{\partial \boldsymbol{x}} \right) v_j^N + \frac{1}{M} \sum_{j=1}^{N_\mathrm{p}} (M\dot{\boldsymbol{v}}_{jl} - F_l) S(\boldsymbol{x},\boldsymbol{x}_j) \frac{\partial v_j^N}{\partial v_{jl}} = 0, \tag{4.27}$$

where N_p is a total number of the particles. Let us integrate (4.27) over the volume occupied by the system considered:

$$\sum_{j=1}^{N_\mathrm{p}} (\dot{\boldsymbol{x}}_j - \boldsymbol{v}_j) v_j^N \int_V \frac{\partial S(\boldsymbol{x},\boldsymbol{x}_j)}{\partial \boldsymbol{x}_j} \mathrm{d}\boldsymbol{x}$$

$$+ \sum_{j=1}^{N_\mathrm{p}} \left(\dot{v}_{jl} \int_V S(\boldsymbol{x},\boldsymbol{x}_j) \mathrm{d}\boldsymbol{x} - \int_V \frac{F_l}{M} S(\boldsymbol{x},\boldsymbol{x}_j) \mathrm{d}\boldsymbol{x} \right) \frac{\partial v_j^N}{\partial v_{jl}} = 0. \tag{4.28}$$

This system of equations will be satisfied in the case of symmetrical shapes of a cloud $S(\boldsymbol{x},\boldsymbol{x}_j) = S(\boldsymbol{x} - \boldsymbol{x}_j)$ only for the following conditions:

$$\dot{\boldsymbol{x}}_j(t) = \boldsymbol{v}_j(t), \tag{4.29}$$

$$M\dot{\boldsymbol{v}}_j(t) = \int \boldsymbol{F}(\boldsymbol{x},t) S(\boldsymbol{x} - \boldsymbol{x}_j) \mathrm{d}\boldsymbol{x} = \boldsymbol{F}_j(t), \quad j = 1,...,N_\mathrm{p}. \tag{4.30}$$

The integral in (4.30) denotes a force which acts on the particle with number j and with shape S. In order to calculate the force at the particle location, one has to interpolate the force from its values at the mesh points. The use of different interpolation formulae allows us to obtain different approximations for the force at the particle location.

Let us express the force $\boldsymbol{F}(\boldsymbol{x},t)$ through its value $\boldsymbol{F}(\boldsymbol{x}_k,t) \equiv \boldsymbol{F}_k(t)$ at the mesh points in the following way:

$$\boldsymbol{F}(\boldsymbol{x},t) = \sum_{k=1}^{N_\mathrm{g}} \boldsymbol{F}_k(t) R(\boldsymbol{x} - \boldsymbol{x}_k), \tag{4.31}$$

where R denotes an interpolation function which satisfies the normalization condition

$$\sum_{k=1}^{N_\mathrm{g}} R(\boldsymbol{x} - \boldsymbol{x}_k) = 1$$

for all values of $\boldsymbol{x}$, and N_g is the total number of mesh points. The shape of R determines the interpolation of the force. The force at the particle location may be written as follows:

$$\boldsymbol{F}_j = \int_V \sum_{k=1}^{N_\mathrm{g}} \boldsymbol{F}_k R(\boldsymbol{x} - \boldsymbol{x}_k) S(\boldsymbol{x} - \boldsymbol{x}_j) \mathrm{d}\boldsymbol{x}. \tag{4.32}$$

In a momentum-conserving scheme, as will be shown in Sect. (4.5), we have to use the same function for the force interpolation as for the particle weighting (density/current assignment), $R(\boldsymbol{x}) = W(\boldsymbol{x})$. Finally, the force acting on the particle at the location $\boldsymbol{x}$ may be estimated as

$$\boldsymbol{F}(\boldsymbol{x}_j, t) = \sum_{k=1}^{N_g} \boldsymbol{F}_k(t) W(\boldsymbol{x}_j - \boldsymbol{x}_k). \qquad (4.33)$$

4.5 Mass, Momentum and Energy Conservation

The equations for macroscopic parameters, the so-called transport equations, may be obtained from the kinetic equation. The integration of the kinetic equation with factors 1, $M_k \boldsymbol{v}$ and $M_k v^2/2$ over the velocity space gives the divergence, momentum and energy transport equations, respectively. In order to study the conservation laws for the numerical models one may use the semi-discretization (exact time advancement) form of the Vlasov–Maxwell equations.

4.5.1 Mass Conservation

4.5.1.1 Plasma System. The divergence equation may be written as follows:

$$\frac{\partial}{\partial t} \left(\sum_{k=1}^{N_s} \int M_k f_k \mathrm{d}\boldsymbol{v} + m n_e \right) + \frac{\partial}{\partial x_\beta} \left(\sum_{k=1}^{N_s} \int M_k v_\beta f_k \mathrm{d}\boldsymbol{v} + m n_e U_{e\beta} \right) = 0. \qquad (4.34)$$

Since we do not consider the charge exchange processes here, we can write the mass conservation law for each component of ions as follows:

$$\frac{\partial}{\partial t} \left(\int M_k f_k \mathrm{d}\boldsymbol{v} \right) + \frac{\partial}{\partial x_\beta} \left(\int M_k v_\beta f_k \mathrm{d}\boldsymbol{v} \right) = 0, \quad k = 1, ..., N_s. \qquad (4.35)$$

The integration of (4.35) over the volume V and the time interval $t_2 - t_1$ gives the integral mass conservation law for the volume considered V:

$$M_k \left[N_k(t_2) - N_k(t_1) \right] + \int_{t_1}^{t_2} \mathrm{d}t \oint \mathrm{d}\boldsymbol{s} M_k \boldsymbol{v}_k n_k = 0, \quad k = 1, ..., N_s. \qquad (4.36)$$

4.5.1.2 Numerical Model. Let us write the distribution function for the finite-size particles (4.2) as follows:

$$f_c(\boldsymbol{x}, \boldsymbol{v}, t) = \sum_{k=1}^{N_s} \sum_{l=1}^{N_{pk}} S(\boldsymbol{x} - \boldsymbol{x}_{kl}) \delta(\boldsymbol{v} - \boldsymbol{v}_{kl}), \qquad (4.37)$$

where N_{pk} denotes the total number of the particles of species k.

4.5 Mass, Momentum and Energy Conservation

The integration of (2.20) with a factor $M_k \boldsymbol{v}^N$ and the distribution function (4.37) over the computational phase space gives

$$\frac{d}{dt}\sum_{k=1}^{N_s}\sum_{l=1}^{N_{pk}} M_k \boldsymbol{v}_{kl}^N \int d\boldsymbol{x} S(\boldsymbol{x}-\boldsymbol{x}_{kl}) + \oint d\boldsymbol{s} \sum_{k=1}^{N_s}\sum_{l=1}^{N_{pk}} M_k \boldsymbol{v}_{kl}^{N+1} S(\boldsymbol{x}-\boldsymbol{x}_{kl})$$

$$= N \sum_{k=1}^{N_s}\sum_{l=1}^{N_{pk}} \boldsymbol{v}_{kl}^{N-1} \int d\boldsymbol{x} \boldsymbol{F}(\boldsymbol{x},\boldsymbol{v}_{kl}) S(\boldsymbol{x}-\boldsymbol{x}_{kl}), \qquad (4.38)$$

where

$$\int d\boldsymbol{x} S(\boldsymbol{x}-\boldsymbol{x}_{kl}) = 1,$$

and summation is made over the particles which are inside the computational domain, $\boldsymbol{x}_{kl} \in V$.

We used the following relation for the right-hand side of (4.38)

$$\int_V \int d\boldsymbol{x} d\boldsymbol{v} \boldsymbol{v}^N \frac{\partial}{\partial \boldsymbol{v}}(\boldsymbol{F}f) = \boldsymbol{v}^N \boldsymbol{F}f|_{-\infty}^{\infty} - N \int_v \int d\boldsymbol{x} d\boldsymbol{v} \boldsymbol{F} f \boldsymbol{v}^{N-1},$$

where the first term on the right-hand side is equal to zero.

Substitution of $N = 0$ into (4.38) gives the following mass conservation law for the numerical plasma system inside a finite computational domain

$$\frac{d}{dt}\sum_{k=1}^{N_s}\sum_{l=1}^{N_{pk}} M_k + \oint d\boldsymbol{s} \sum_{k=1}^{N_s}\sum_{l=1}^{N_{pk}} M_k \boldsymbol{v}_{kl} S(\boldsymbol{x}-\boldsymbol{x}_{kl}) = 0.$$

Let us consider the isolated systems or systems with periodical boundary conditions. The mass conservation law for these systems may be expressed as:

$$\frac{d}{dt}\sum_{k=1}^{N_s}\sum_{l=1}^{N_{pk}} M_k = 0.$$

The local mass conservation law (4.35) may not be satisfied because of microscopic inconsistencies between $n_k U_k$ and n_k due to the use of the mesh and weights. Let us consider the semi-discretization form of the one-dimensional divergence equation for ions at grid point k

$$\frac{\partial n_k}{\partial t} + \frac{(nU)_{k+\frac{1}{2}} - (nU)_{k-\frac{1}{2}}}{\Delta x} = 0. \qquad (4.39)$$

In accordance with definition the first term may be expressed as

$$\frac{\partial n_k}{\partial t} = \sum_{l=1}^{N_p} W'(x_l - x_k)\frac{dx_l}{dt},$$

and the second term is

$$\frac{(nU)_{k+\frac{1}{2}} - (nU)_{k-\frac{1}{2}}}{\Delta x} = \sum_{l=1}^{N_p} u_l \frac{W(x_l - x_{k+1}) - W(x_l - x_{k-1})}{2\Delta x},$$

where the prime denotes the analytical derivative at a point $|x_l - x_k|$. Summation of the two expressions above gives

$$\sum_{l=1}^{N_p} u_l \left\{ W'(x_l - x_k) + \frac{W(x_l - x_{k+1}) - W(x_l - x_{k-1})}{2\Delta x} \right\} = 0. \quad (4.40)$$

It is clear that the local finite-difference equation for an individual particle,

$$\left\{ W'(x_l - x_k) + \frac{W(x_l - x_{k+1}) - W(x_l - x_{k-1})}{2\Delta x} \right\} = 0,$$

cannot be satisfied for an arbitrary weight function W, but the error of approximation will be smaller for macroparticles whose width is much larger than the size of the cell. However, if the number of particles per cell is sufficiently large, one can expect a good approximation to the local divergency equation for any arbitrary function W. Note that the particle-mesh method does not satisfy the divergence equation in an individual cell. However, the integral mass conservation law is satisfied for the whole computational domain. There is another way to satisfy the local divergence equation. One can calculate the charge at the center of a cell, whereas the current must be calculated as a flux across the faces of the cell [541]. This finite-difference approximation is suitable for full particle code with special grid location for computation of the electric and magnetic field components. In the conventional hybrid codes the conservation of mass for the whole system is satisfied; this is also the case in any general particle-mesh algorithm.

4.5.2 Momentum Conservation

4.5.2.1 Plasma System. Combining (2.22) and (2.23), one obtains the momentum transport equation for a conventional hybrid model:

$$\frac{\partial}{\partial t} \left(\sum_{k=1}^{N_s} \int M_k \boldsymbol{v} f_k \mathrm{d}\boldsymbol{v} + mn_e \boldsymbol{U}_e \right) + \frac{\partial}{\partial x_\beta} \left(\sum_{k=1}^{N_s} \int M_k \boldsymbol{v} v_\beta f_k \mathrm{d}\boldsymbol{v} + mn_e \langle \boldsymbol{v}_e v_{e\beta} \rangle \right)$$

$$= \frac{\boldsymbol{J} \times \boldsymbol{B}}{c} \quad (4.41)$$

or in the equivalent form

$$\frac{\partial}{\partial t}\boldsymbol{p} + \frac{\partial}{\partial x_\beta} \mathsf{K}_{\alpha\beta} = \frac{\boldsymbol{J} \times \boldsymbol{B}}{c}, \quad (4.42)$$

where $\boldsymbol{p}$ is the momentum density of particles,

$$\boldsymbol{p} = \sum_{k=1}^{N_s} M_k n_k \boldsymbol{U}_k + mn_e \boldsymbol{U}_e, \qquad (4.43)$$

and K is the kinetic energy tensor

$$\mathsf{K} = \sum_{k=1}^{N_s} M_k n_k \langle v_k v_{k\beta} \rangle + mn_e U_e U_{e\beta} + p_e + \pi^e_{\alpha\beta}. \qquad (4.44)$$

$\boldsymbol{U}_k$ and $\boldsymbol{U}_e$ denote the ion and electron bulk velocities, p_e and $\pi^e_{\alpha\beta}$ denote the electron pressure and viscosity stress tensor, N_s is the number of ion species.

Integration of (4.42) over the volume V and the time interval $t_2 - t_1$ gives the dependence of the total momentum on time:

$$\boldsymbol{P}(t_2) - \boldsymbol{P}(t_1) + \int_{t_1}^{t_2} dt \oint_S \mathsf{K}\hat{\boldsymbol{n}} ds = \int_{t_1}^{t_2} dt \int_V \frac{\boldsymbol{J} \times \boldsymbol{B}}{c} d\boldsymbol{x}, \qquad (4.45)$$

where

$$\boldsymbol{P}(t) = \int_V \boldsymbol{p} d\boldsymbol{x} \qquad (4.46)$$

and $\hat{\boldsymbol{n}}$ denotes a unit vector normal to the surface. The third term on left side of (4.45) represents the flux of the momentum across the boundary surface S. The right side of (4.45) may be also written as

$$\frac{4\pi}{c} \int_V (\boldsymbol{J} \times \boldsymbol{B}) d\boldsymbol{x} = -\frac{1}{2} \oint_S B^2 \hat{\boldsymbol{n}} ds + \int_V (\boldsymbol{B} \cdot \nabla) \boldsymbol{B} d\boldsymbol{x}, \qquad (4.47)$$

where the second term is determined by the curvature of the magnetic field.

4.5.2.2 Numerical Model. Let us consider the conservation of the momentum in the numerical model. Substitution of $N = 1$ into (4.38) gives the following momentum conservation law for the numerical plasma system inside the finite computational domain

$$\frac{d}{dt} \sum_{k=1}^{N_s} \sum_{l=1}^{N_{pk}} M_k \boldsymbol{v}_{kl} + \oint d\boldsymbol{s} \sum_{k=1}^{N_s} \sum_{l=1}^{N_{pk}} M_k v_{kl}^2 S(\boldsymbol{x} - \boldsymbol{x}_{kl})$$

$$= \sum_{k=1}^{N_s} \sum_{l=1}^{N_{pk}} \int d\boldsymbol{x} \boldsymbol{F}(\boldsymbol{x}, \boldsymbol{v}_{kl}) S(\boldsymbol{x} - \boldsymbol{x}_{kl}). \qquad (4.48)$$

For the isolated systems or systems with periodical boundary conditions one can write

$$\frac{d}{dt} \sum_{k=1}^{N_s} \sum_{l=1}^{N_{pk}} M_k \boldsymbol{v}_{kl} = \sum_{k=1}^{N_s} \sum_{l=1}^{N_{pk}} \int d\boldsymbol{x} \boldsymbol{F}(\boldsymbol{x}, \boldsymbol{v}_{kl}) S(\boldsymbol{x} - \boldsymbol{x}_{kl}),$$

or
$$\frac{\mathrm{d}}{\mathrm{d}t}\sum_{k=1}^{N_\mathrm{s}} \boldsymbol{P}_k = \sum_{k=1}^{N_\mathrm{s}}\sum_{l=1}^{N_\mathrm{pk}} \int \mathrm{d}\boldsymbol{x}\, \boldsymbol{F}_{kl},$$

where
$$\boldsymbol{P}_k = \sum_{l=1}^{N_\mathrm{pk}} M_k \boldsymbol{v}_{kl}.$$

The total momentum of all the ion species may be expressed as

$$\sum_{k=1}^{N_\mathrm{s}} \boldsymbol{P}_k = \sum_{l=1}^{N_\mathrm{p}} M_l \boldsymbol{v}_l, \qquad (4.49)$$

which results from the relation

$$\int_V \int_V \int \sum_{l=1}^{N_\mathrm{p}} M_l \boldsymbol{v}^N S(\boldsymbol{x} - \boldsymbol{x}')\delta(\boldsymbol{x}' - \boldsymbol{x}_l)\delta(\boldsymbol{v} - \boldsymbol{v}_l)\mathrm{d}\boldsymbol{x}'\mathrm{d}\boldsymbol{x}\mathrm{d}\boldsymbol{v}$$

$$= \sum_{l=1}^{N_\mathrm{p}} M_l \boldsymbol{v}_l^N \int_V S(\boldsymbol{x} - \boldsymbol{x}_l)\mathrm{d}\boldsymbol{x} = \sum_{l=1}^{N_\mathrm{p}} M_l \boldsymbol{v}_l^N, \qquad (4.50)$$

where V denotes the total volume of the computational domain. The summation is made over all particles. The time derivation of the total ion momentum may be estimated as

$$\frac{\mathrm{d}}{\mathrm{d}t}\sum_{k=1}^{N_\mathrm{s}} \boldsymbol{P}_k = \sum_{l=1}^{N_\mathrm{p}} \boldsymbol{F}_l. \qquad (4.51)$$

Let us assume that the density (charge) assignment function has the same form as a force interpolation function, namely, $W_F = W_d = W$, in our computational model. In this case the force interpolation and the density and current assignment relations have the following form:

$$\boldsymbol{F}(\boldsymbol{x}_l) = q_l \sum_{j=1}^{N_\mathrm{g}} \left(\boldsymbol{E}^*{}_j + \frac{\boldsymbol{v}_l \times \boldsymbol{B}_j}{c}\right) W(\boldsymbol{x}_j - \boldsymbol{x}_l), \qquad (4.52)$$

$$\varrho_j = \frac{1}{V_\mathrm{c}} \sum_{l=1}^{N_\mathrm{p}} q_l W(\boldsymbol{x}_j - \boldsymbol{x}_l), \qquad (4.53)$$

$$\boldsymbol{J}_j = \frac{1}{V_\mathrm{c}} \sum_{l=1}^{N_\mathrm{p}} q_l \boldsymbol{v}_l W(\boldsymbol{x}_j - \boldsymbol{x}_l), \qquad (4.54)$$

where V_c denotes the cell volume, $V_\mathrm{c} = h_1 h_2 h_3$. The summation is made over all mesh points and all particles. Here we use $\boldsymbol{E}^*$ rather than $\boldsymbol{E}$ in order to satisfy momentum conservation.

4.5 Mass, Momentum and Energy Conservation

Substitution of the finite-difference approximation for the generalized Ohm's law (2.22) into (4.51) and using (4.52), (4.53) and (4.54) give

$$\frac{d}{dt}\sum_{k=1}^{N_s} P_k = -\sum_{l=1}^{N_p}\frac{q_l}{c}\sum_{j=1}^{N_g} W(\boldsymbol{x}_j - \boldsymbol{x}_l)\left(\boldsymbol{U}_i \times \boldsymbol{B} - \boldsymbol{v}_l \times \boldsymbol{B}\right)_j$$
$$+\sum_{l=1}^{N_p}\frac{q_l}{e}\sum_{j=1}^{N_g} W(\boldsymbol{x}_j - \boldsymbol{x}_l)\left[\frac{\boldsymbol{J}\times\boldsymbol{B}}{cn_e} - \frac{(\nabla_h p_e + \partial\pi^e_{\alpha\beta}/\partial x_\beta)}{n_e}\right]_j$$
$$-\sum_{l=1}^{N_p}\frac{q_l}{e}\sum_{j=1}^{N_g} W(\boldsymbol{x}_j - \boldsymbol{x}_l)\frac{m}{n_{ej}}\left[\frac{\partial n_e\boldsymbol{U}_e}{\partial t} + (\nabla_h \boldsymbol{U}_e)n_e\boldsymbol{U}_e\right]_j. \quad (4.55)$$

∇_h denotes the finite-difference approximation to the operator ∇. Let us express the first term on the right-hand side of (4.55) as

$$-\sum_{l=1}^{N_p} q_l \sum_{j=1}^{N_g} W(\boldsymbol{x}_j - \boldsymbol{x}_l)(\boldsymbol{U}_i \times \boldsymbol{B})_j = -\sum_{j=1}^{N_g}\left(\sum_{l=1}^{N_p} q_l W(\boldsymbol{x}_j - \boldsymbol{x}_l)\right)(\boldsymbol{U}_i \times \boldsymbol{B})_j$$
$$= -\sum_{j=1}^{N_g} \varrho_j \boldsymbol{U}_{ij} \times \boldsymbol{B}_j.$$

In accordance with the relation

$$\boldsymbol{U}_{ij} = \frac{\sum_{l=1}^{N_p} q_l \boldsymbol{v}_l W(\boldsymbol{x}_j - \boldsymbol{x}_l)}{\sum_{l=1}^{N_p} q_l W(\boldsymbol{x}_j - \boldsymbol{x}_l)}$$

the second term gives

$$\sum_{j=1}^{N_g}\left(\sum_{l=1}^{N_p} q_l W(\boldsymbol{x}_j - \boldsymbol{x}_l)\boldsymbol{v}_l\right)\times\boldsymbol{B}_j,$$

and, hence, the first and the second terms are mutually excluded. The Hall term may be expressed as

$$\sum_{j=1}^{N_g}\left(\sum_{l=1}^{N_p} q_l W(\boldsymbol{x}_j - \boldsymbol{x}_l)\right)\frac{(\boldsymbol{J}\times\boldsymbol{B})_j}{en_{ej}c} = \sum_{j=1}^{N_g}\frac{(\boldsymbol{J}\times\boldsymbol{B})_j}{c},$$

and the term with the electron pressure becomes

$$-\sum_{j=1}^{N_g}\frac{\left[\sum_{l=1}^{N_p} q_l W(\boldsymbol{x}_j - \boldsymbol{x}_l)\right]}{en_{ej}}(\nabla_h p_e + \partial\pi^e_{\alpha\beta}/\partial x_\beta)_j = -\sum_{j=1}^{N_g}(\nabla_h p_e + \partial\pi^e_{\alpha\beta}/\partial x_\beta)_j.$$

The summation of this term is taken over the computational domain (in the case of a symmetrical finite-difference approximation and a rectangular domain), and it is equal to the difference in the electron pressure between the two opposite sides of the surface.

Finally, the last term of (4.55) (the electron momentum) becomes

$$-\sum_{l=1}^{N_p} q_l \sum_{j=1}^{N_g} W(\boldsymbol{x}_j - \boldsymbol{x}_l) \left[\frac{m}{en_{ej}} \left(\frac{\partial n_e \boldsymbol{U}_e}{\partial t} \right)_j + \frac{m}{en_{ej}} [(\nabla_h \boldsymbol{U}_e) n_e \boldsymbol{U}_e]_j \right]$$

$$= -\sum_{j=1}^{N_g} \left(\sum_{l=1}^{N_p} q_l W(\boldsymbol{x}_j - \boldsymbol{x}_l) \right) \left[\frac{m}{en_{ej}} \left(\frac{\partial n_e \boldsymbol{U}_e}{\partial t} \right)_j + \frac{m}{en_{ej}} [(\nabla_h \boldsymbol{U}_e) n_e \boldsymbol{U}_e]_j \right]$$

$$= -\sum_{j=1}^{N_g} \left(\frac{\partial m n_e \boldsymbol{U}_e}{\partial t} \right)_j - \sum_{j=1}^{N_g} [(\nabla_h \boldsymbol{U}_e) m n_e \boldsymbol{U}_e]_j.$$

The summation of the second term is taken over the computational domain (in the case of a symmetrical finite difference approximation and a rectangular domain), and it gives the difference in the electron momentum flux at the opposite side of the surface. Permutation of the first term of the above relation onto the left side of (4.55) gives, finally,

$$\frac{d}{dt}\boldsymbol{P} = \frac{d}{dt}\sum_{s=1}^{N_s} \boldsymbol{P}_s + \frac{d}{dt}\sum_{j=1}^{N_g} (mn_e\boldsymbol{U}_e)_j = \sum_{j=1}^{N_g} \left\{ \frac{\boldsymbol{J} \times \boldsymbol{B}}{c} \right\}_j - \sum_{j=1}^{N'_g} (\mathsf{K}^e \hat{\boldsymbol{n}})_j, \quad (4.56)$$

where $\mathsf{K}^e \hat{\boldsymbol{n}}$ denotes the electron flux of momentum at the surface of the computational domain (4.44) and a summation was taken over the grid points N'_g, which are located at the boundary surface. The electron momentum flux on the right-hand side of (4.56) must be dropped for isolated plasma systems or systems with periodical boundary conditions.

In the general case (nonperiodical boundary conditions or particle injection at the boundaries) one has to add the ion momentum flux to the right-hand side of (4.56). The resulting momentum conservation law becomes

$$\frac{d}{dt}\boldsymbol{P} + \sum_{j=1}^{N'_g} \{\mathsf{K}\hat{\boldsymbol{n}}\}_j = \sum_{j=1}^{N_g} \left\{ \frac{\boldsymbol{J} \times \boldsymbol{B}}{c} \right\}_j,$$

where the second term denotes the total (ion and electron) momentum flux and $\hat{\boldsymbol{n}}$ is a unit normal to the boundary of the computational domain.

4.5.3 Energy Conservation

4.5.3.1 Plasma System.
Combining (2.29) for electrons and ions, one obtains the energy transport equation:

4.5 Mass, Momentum and Energy Conservation

$$\frac{\partial}{\partial t}\left(\frac{1}{2}\sum_{k=1}^{N_s}\int M_k v^2 f_k d\boldsymbol{v} + \frac{mn_e\langle v_e^2\rangle}{2}\right)$$

$$+\nabla\left(\frac{1}{2}\sum_{k=1}^{N_s}\int M_k v^2 \boldsymbol{v} f_k d\boldsymbol{v} + \frac{mn_e\langle v_e^2 \boldsymbol{v}_e\rangle}{2}\right) = \boldsymbol{E}\cdot\boldsymbol{J}. \quad (4.57)$$

According to the vector relation

$$\nabla\cdot(\boldsymbol{E}\times\boldsymbol{B}) = \boldsymbol{B}\cdot(\nabla\times\boldsymbol{E}) - \boldsymbol{E}\cdot(\nabla\times\boldsymbol{B}), \quad (4.58)$$

the Maxwell equations provide an energy relation for the field (the Poynting theorem in the approximation $E << B$):

$$\frac{1}{8\pi}\frac{\partial B^2}{\partial t} = \frac{c}{4\pi}\nabla\cdot(\boldsymbol{B}\times\boldsymbol{E}) - \boldsymbol{E}\cdot\boldsymbol{J}. \quad (4.59)$$

Combining (4.57), (4.58) and (4.59), one obtains a conservation law for the total energy:

$$\frac{\partial}{\partial t}\epsilon + \nabla\boldsymbol{G} = 0, \quad (4.60)$$

where the total energy ϵ is

$$\epsilon = \frac{1}{2}\left(\sum_{k=1}^{N_s} M_k n_k\langle v_k^2\rangle + mn_e U_e^2\right) + \frac{3p_e}{2} + \frac{B^2}{8\pi} \quad (4.61)$$

and the flux of energy $\boldsymbol{G}$ is

$$\boldsymbol{G} = \frac{1}{2}\left(\sum_{l=1}^{N_p} M_l n_l \langle v_l v_l^2\rangle + mn_e \boldsymbol{U}_e U_e^2\right) + \frac{5p_e \boldsymbol{U}_e}{2} + \frac{c\boldsymbol{E}\times\boldsymbol{B}}{4\pi} + \boldsymbol{q} + (\pi_{\alpha\beta}^e \cdot U_{e\alpha}). \quad (4.62)$$

Here

$$\boldsymbol{q}_e = n_e m\langle \frac{v'^2}{2}\boldsymbol{v}'\rangle, \quad (4.63)$$

where $\boldsymbol{q}$ denotes the electron heat density flux. The first term of (4.60) represents the variation of the total energy of the particles and the field. The second term is the full flux of the energy, which contains the transport of the full energy with bulk velocity $\boldsymbol{U}$, the microtransport of the energy, and the work of the pressure and other forces, and also the heat release.

The integration of (4.60) over the volume V and the time interval $t_2 - t_1$ gives

$$E_W(t_2) - E_W(t_1) + \int_{t_1}^{t_2} dt \oint_S \boldsymbol{G}\hat{\boldsymbol{n}} ds = 0, \quad (4.64)$$

where the total energy of the ions and electrons is

$$E_W = \int_V \epsilon d\boldsymbol{x} \quad (4.65)$$

and the second term of (4.64) represents the flux of the energy across the boundary surface.

4.5.3.2 Numerical Model.

Let us consider the conservation of energy in the conventional hybrid models. Substitution of $N = 2$ into (4.38) gives the energy conservation law for numerical plasma systems inside the computational domain

$$\frac{d}{dt}\sum_{k=1}^{N_s}\sum_{l=1}^{N_{pk}}\frac{M_k v_{kl}^2}{2} + \oint d\mathbf{s}\sum_{k=1}^{N_s}\sum_{l=1}^{N_{pk}}\frac{M_k v_{kl}^2 \mathbf{v}_{kl}}{2}S(\mathbf{x}-\mathbf{x}_{kl})$$
$$= \sum_{k=1}^{N_s}\sum_{l=1}^{N_{pk}} Z_k e \mathbf{v}_{kl} \int d\mathbf{x}\, \mathbf{E}^*(\mathbf{x}) S(\mathbf{x}-\mathbf{x}_{kl}). \tag{4.66}$$

The term with $\mathbf{v}_{kl} \cdot \{\mathbf{v}_{kl} \times \int d\mathbf{x}\, \mathbf{B}(\mathbf{x}) S(\mathbf{x}-\mathbf{x}_{kl})\}$ on the right-hand side of (4.66) was omitted.

Let us express the right-hand side of (4.66) as

$$\sum_{k=1}^{N_s}\sum_{l=1}^{N_{pk}} Z_k e \mathbf{v}_{kl} \int d\mathbf{x}\, \mathbf{E}^*(\mathbf{x}) S(\mathbf{x}-\mathbf{x}_{kl}) = \sum_{k=1}^{N_s}\sum_{l=1}^{N_{pk}} Z_k e \mathbf{v}_{kl} \sum_{j=1}^{N_g} \mathbf{E}_j^* W(\mathbf{x}_j - \mathbf{x}_{kl})$$
$$= \sum_{j=1}^{N_g}\left\{\sum_{k=1}^{N_s} Z_k e \sum_{l=1}^{N_{pk}} \mathbf{v}_{kl} W(\mathbf{x}-\mathbf{x}_{kl})\right\} \mathbf{E}_j^* = \sum_{j=1}^{N_g}\left\{\sum_{k=1}^{N_s} Z_k e \mathbf{E}^* n_k \mathbf{U}_k\right\}_j. \tag{4.67}$$

Then the total kinetic energy of all ion species is governed by the work of all the forces which act on the particles:

$$\frac{d}{dt}\sum_{l=1}^{N_p}\frac{M_l v_l^2}{2} = \sum_{j=1}^{N_g}\left\{\sum_{k=1}^{N_s} Z_k e \mathbf{E}^* n_k \mathbf{U}_k\right\}_j. \tag{4.68}$$

Summation of the products of the electron bulk velocity and the finite-difference approximation to the electron momentum equation (2.25) over all grid nodes of the computational domain gives

$$\sum_{j=1}^{N_g}\left\{m\mathbf{U}_e\left(\frac{\partial n_e \mathbf{U}_e}{\partial t} + (\nabla_h \mathbf{U}_e) n_e \mathbf{U}_e\right)\right\}_j + \sum_{j=1}^{N_g}\{\mathbf{U}_e(\nabla_h p_e + \partial \pi_{\alpha\beta}^e/\partial x_\beta)\}_j$$
$$= -e\sum_{j=1}^{N_g}\{\mathbf{E}^* n_e \mathbf{U}_e\}_j, \tag{4.69}$$

where $\mathbf{E}^*$ has been taken from (2.21). Summation of the finite-difference approximation to the equation for the electron pressure (2.40) over the computational domain results in

4.5 Mass, Momentum and Energy Conservation

$$\sum_{j=1}^{N_g} \left\{ \frac{3}{2}\frac{\partial p_e}{\partial t} + \frac{3}{2}\nabla_h(p_e \cdot \boldsymbol{U}_e) + p_e \nabla_h \cdot \boldsymbol{U}_e \right\}_j + \sum_{j=1}^{N_g} \left\{ \pi_{\alpha\beta}^e \frac{\partial U_{e\alpha}}{\partial x_\beta} \right\}_j$$

$$= \sum_{j=1}^{N_g} \{-\nabla_h \cdot \boldsymbol{q}_e + Q_e\}_j. \qquad (4.70)$$

We assume that the central finite differences satisfy the following relation:

$$\nabla_h(ab) = a\nabla_h b + b\nabla_h a.$$

Summation of the finite-difference approximation to (4.59) gives

$$\sum_{j=1}^{N_g} \frac{1}{8\pi}\left\{\frac{\partial B^2}{\partial t}\right\}_j - \sum_{j=1}^{N_g} \frac{c}{4\pi}\{\nabla_h \cdot (\boldsymbol{B} \times \boldsymbol{E})\}_j = -\sum_{j=1}^{N_g}\{\boldsymbol{E} \cdot \boldsymbol{J}\}_j. \qquad (4.71)$$

We also have to use the following finite-difference relation for electrons:

$$\frac{1}{2}\sum_{j=1}^{N_g}\left\{\frac{\partial}{\partial t}(mn_e U_\alpha U_\alpha) + \frac{\partial}{\partial x_\beta}(mn_e U_\alpha U_\alpha U_\beta)\right\}_j$$

$$= \frac{1}{2}\sum_{j=1}^{N_g}\left\{2U_\alpha\left(\frac{\partial}{\partial t}mn_e U_\alpha + \frac{\partial}{\partial x_\beta}mn_e U_\alpha U_\beta\right)\right\}_j - \delta_D, \qquad (4.72)$$

where

$$\delta_D = \frac{1}{2}\sum_{j=1}^{N_g}\left\{U_\alpha U_\alpha \left(\frac{\partial}{\partial t}mn_e + \frac{\partial}{\partial x_\beta}mn_e U_\beta\right)\right\}_j,$$

which results from the divergence equation for the electron component (4.34). Remember that in accordance with (2.21) one has

$$\sum_{j=1}^{N_g}\{\boldsymbol{E}^* \cdot \boldsymbol{J}\}_j = \sum_{j=1}^{N_g}\{\boldsymbol{E} \cdot \boldsymbol{J}\}_j - \sum_{j=1}^{N_g}\left\{\frac{J^2}{\sigma_e}\right\}_j. \qquad (4.73)$$

Finally, summation of (4.68–4.73) gives a conservation law for the total energy in the computational domain:

$$\frac{d}{dt}E_W + \sum_{j=1}^{N'_g}(\boldsymbol{G}^e \cdot \hat{\boldsymbol{n}})_j = \delta_D, \qquad (4.74)$$

where, in accordance with (4.50),

$$E_W = \sum_{l=1}^{N_p}\frac{M_l v_l^2}{2} + \sum_{j=1}^{N_g}\left\{\frac{mn_e U_e^2}{2} + \frac{3p_e}{2} + \frac{B^2}{8\pi}\right\}_j$$

and $\boldsymbol{G}_j^{\mathrm{e}}$ denotes the projection of flux $\boldsymbol{G}^{\mathrm{e}}$ (4.62) (without the ion energy flux) on the grid space

$$\boldsymbol{G}_j^{\mathrm{e}} = \frac{1}{2}\left(mn_{\mathrm{e}}U_{\mathrm{e}}^2 \boldsymbol{U}_{\mathrm{e}}\right)_j + \frac{5}{2}\left(p_{\mathrm{e}}\boldsymbol{U}_{\mathrm{e}}\right)_j + \frac{c}{4\pi}\left(\boldsymbol{E}\times\boldsymbol{B}\right)_j + \boldsymbol{q}_j + \left(\pi_{\alpha\beta}^{\mathrm{e}}U_{\mathrm{e}\alpha}\right)_j. \quad (4.75)$$

Here δ_D is the scheme's sources of energy due to the finite-difference approximation to the Vlasov–Maxwell equations. The energy flux $\boldsymbol{G}_j^{\mathrm{e}}$ on the left-hand side of (4.74) must be dropped for isolated plasma systems or systems with periodical boundary conditions.

In the general case (nonperiodical boundary conditions or particle injection at the boundaries) one has to add the ion energy flux to the left-hand side of (4.74). The resulting energy conservation law becomes

$$\frac{\mathrm{d}}{\mathrm{d}t}E_W + \sum_{j=1}^{N_g'}\{\boldsymbol{G}\hat{\boldsymbol{n}}\}_j = \delta_{\mathrm{D}},$$

where the second term denotes the total (ion, electron and electromagnetic) energy flux and $\hat{\boldsymbol{n}}$ is a unit normal to the boundary of the computational domain.

4.6 Periodic Systems. Multipole Expansion Method

The multipole method is a field algorithm in which the particle is considered to be a finite-size cloud and the field is represented by a truncated Fourier series [123].

The charge density is represented by

$$\varrho(x) = e\sum_{j}^{N_p} S(x - x_j), \quad (4.76)$$

where S is an arbitrary smoothing factor. Let the computational domain have homogeneous cells with a size h and x_l be the center of a cell. Let us suppose that S has an infinite number of derivatives. Then, Taylor expansion transforms (4.76) to

$$\varrho(x,t) = e\sum_l \sum_{j\in l} W(x_j(t) - x)$$

$$= e\sum_l \sum_{j\in l}\left(S(x_l - x) + \Delta x_j S'(x_l - x) + \frac{(\Delta x_j)^2}{2!}S''(x_l - x) + ...\right)$$

$$= e\sum_l \left(S(x_l - x)\varrho_0(l,t) + S'(x_l - x)\varrho_1(l,t) + \frac{1}{2!}S''(x_l - x)\varrho_2(l,t) + ...\right), \quad (4.77)$$

where

$$\varrho_m(l,t) = \sum_{j \in l}[x_j(t) - x_l]^m \tag{4.78}$$

is a multiple moment with index m which is calculated over all the particles in the cell with index l; $\Delta x_j = x_j(t) - x_l$ is the distance between the jth particle and the node of cell x_l. A truncated Fourier series [123] gives

$$\varrho(k,t) = eS(k)\sum_{m=n}^{\infty}\frac{(-ik)^m}{m!}\sum_{l}\varrho_m(l,t)e^{-ikx_l}, \tag{4.79}$$

where $|kh| < \pi$. Thus, having the multiple moments (4.78), we can find the amplitude of the Fourier harmonics of the density (4.79). Then the Poisson equation gives

$$E(k,t) = \frac{4\pi i}{k}\varrho(k,t). \tag{4.80}$$

In accordance with (4.30), the force which acts on the particle at point $x_j(t)$ is

$$F(x_j,t) = e\int E(x,t)R(x - x_j)\mathrm{d}x = e\int E(x + x_j,t)R(x)\mathrm{d}x. \tag{4.81}$$

Taylor expansion gives

$$F(x_j,t) = F_0(x_l,t) + \Delta x_j F_1(x_l,t) + \frac{(\Delta x_j)^2}{2!}F_2(x_l,t) + ..., \tag{4.82}$$

where the moment of force with index m is

$$F_m(x_l,t) = e\int \frac{\partial^m E(x + x_l,t)}{\partial x^m}R(x)\mathrm{d}x.$$

In [282], it was suggested that the storage required by the multipole method could be reduced by using the subtracted multipole scheme. In this method, the derivatives of the force at the grid points are formed by using a finite-difference operator on the grid. Hence, one need only to calculate and store the force at each grid point. The force may be obtained by finite-difference operation on the potential. The multipole densities are combined into a charge density by difference operators symmetric to those used for the force. Fast Fourier transforms are retained for Poisson solution and smoothing, and perhaps also in differentiating the potential [51].

Summary

One of the main steps of simulation is to compare the results of the simulation with the analytical solution in linear and nonlinear regimes. The

4. Density and Current Assignment. Force Interpolation

simulation results must also be compared with experimental data. There are several possibilities by which to receive incorrect results from the computer simulation. The finite time step, the finite grid spacing, the finite size of the particle and the "shot noise" may result in artificial effects, the heating and acceleration of particles, artificial wave generation and nonlinear numerical instabilities. These effects have been studied well for full particle codes (see, e.g., [51, 236]). The effects of finite time and space steps for hybrid codes still have to be studied. However, some of them, e.g., finite-grid instabilities, have the same nature as instabilities in full particle codes (see [62, 441]), and one can use the same procedure to suppress numerical instability, e.g., randomly jiggled grid spacing [62].

The computational model must satisfy the conservation laws. Momentum conservation may be satisfied by appropriately choosing the weighting function, whereas energy conservation is determined by the time integration scheme, and the difference approximation of the spatial derivatives.

5. Time Integration of the Field and Electron Pressure Equations

5.1 Introduction

The integration in time of the electromagnetic field equations is an important part of the hybrid simulation. The induction equation contains four main terms which describe the following processes. First of all a convection of the magnetic field is conditioned by a macroscopic flux of plasma. The so-called Hall term may generate the low-frequency waves, namely, Alfvén waves, helicons, and a whistler, and the formation of so-called magnetic barriers, the magnetopause, overshoots at the shock fronts etc. The term with the anomalous resistivity may cause dissipation effects: heating of electrons at the shocks and widening of the shock fronts. The anomalous resistivity may also play a crucial role in the magnetic field reconnection. The electron pressure gradient term plays an important role at the plasma discontinuities. Our discrete model must describe all of the processes as well as the initial differential induction equation. So we have to make a good finite-difference approximation for the main terms: the convection term, the dispersion (Hall) term, the resistance term, and the electron pressure gradient term. The hybrid model also includes the electron inertia term, so we have to make a good approximation for this term as well. The approximation of the convective term, the diffusion term, and the electron pressure gradient has been discussed widely in the papers and books on computational hydrodynamics. For the study of whistler generation problems, we have to provide a good (explicit or implicit) approximation for the Hall term. Several recent developments in numerical methods have allowed the MHD (magnetohydrodynamic) and Maxwell equations to be solved and modified. These include numerical algorithms such as implicit–explicit schemes, the predictor–corrector method, the modified leapfrog scheme, the two-step Lax–Wendroff scheme, the Runge–Kutta method, the Crank–Nicolson scheme and other improved schemes. Some of these schemes will be considered in this chapter.

The nature of the considered problems very often demands the use of an adequate coordinate system. Sometimes it is desirable for the boundaries of a computational domain to be the coordinate surfaces. Our mesh must provide a good approximation to the considered problem assuming that the number of grid points is limited. The satisfactory approximation of the initial differential equations and the numerical stability of the difference schemes lead to the convergence of the numerical solution to the that of the differential equations. The numerical schemes must have damping and dispersion that are as small as possible over the wide range of wavelengths and frequencies which determine the physical phenomena. The numerical schemes must be monotonous to avoid nonphysical numerical oscillations, and they must provide the correct propagation of information from one grid point to another (transportive property). All of these considerations will be dealt with in this chapter.

First of all we have to present the field equation (Ohm's law, the induction equation, Faraday's law, Ampere's law) in general, Cartesian, cylindrical and spherical coordinate systems.

General Curvilinear Coordinates

Let us use the tensor formulation and vector identities from Sect. 2.1 to write Maxwell equations in tensor form:
Ohm's law is

$$E_i = -\epsilon_{ijk}\frac{U_e^j B^k}{c} - \frac{1}{en_e}\frac{\partial p_e}{\partial \bar{x}^i} + \frac{J_i}{\sigma_{\text{eff}}} - \frac{m}{e}\frac{d}{dt}U_{ei}. \quad (5.1)$$

Faraday's law is

$$\frac{\partial B^i}{\partial t} = -c\epsilon^{ijk}\frac{\partial E_k}{\partial \bar{x}^j}. \quad (5.2)$$

Ampere's law is

$$J^i = \frac{c}{4\pi}\epsilon^{ijk}\frac{\partial B_k}{\partial \bar{x}^j}. \quad (5.3)$$

It is convenient to introduce extensive current and charge variables,

$$I^i = \sqrt{g}J^i \quad \text{and} \quad Q = \sqrt{g}\varrho, \quad (5.4)$$

and volume-scaled flux quantities,

$$b^i = \sqrt{g}B^i, \quad (5.5)$$

where g denotes the determinant of the metric tensor (see Chap. 3). If in addition we write the permutation tensor in terms of the permutation symbol, Maxwell equations become

$$\frac{\partial b^i}{\partial t} = -ce^{ijk}\frac{\partial E_k}{\partial \bar{x}^j}, \quad (5.6)$$

$$I^i = \frac{c}{4\pi}e^{ijk}\frac{\partial B_k}{\partial \bar{x}^j}. \quad (5.7)$$

The time evolution of the electron pressure is described by the following equation when the viscosity stress tensor π^e and the heat flux q_e are neglected:

$$\frac{3}{2}\frac{\partial p_e}{\partial t} + \frac{3}{2\sqrt{g}}\frac{\partial}{\partial \tilde{x}^i}\sqrt{g}U_e^i p_e + p_e \frac{1}{\sqrt{g}}\frac{\partial}{\partial \tilde{x}^i}\sqrt{g}U_e^i = \frac{c^2}{16\pi^2 \sigma_{\text{eff}}}\left(\epsilon^{ijk}\frac{\partial B_k}{\partial \tilde{x}^j}\right)^2, \quad (5.8)$$

or

$$\frac{\partial p_e}{\partial t} + U_e^i \frac{\partial}{\partial \tilde{x}^i} p_e + \gamma p_e \frac{1}{\sqrt{g}}\frac{\partial}{\partial \tilde{x}^i}\sqrt{g}U_e^i = \frac{(\gamma-1)c^2}{16\pi^2 \sigma_{\text{eff}}}\left(\epsilon^{ijk}\frac{\partial B_k}{\partial \tilde{x}^j}\right)^2. \quad (5.9)$$

The Cartesian, cylindrical and spherical coordinate forms for the equations above are presented in Appendices, Sects. 13.1.1–13.1.3.

The structure of the field equation is very complicated, and it is difficult to develop a direct solution for the three-dimensional case. Later in this chapter we shall consider a different approximation of the initial equation in which the complicated system (in space and in physical processes) of equations is split into a set of equations (usually one-dimensional) with a more simple structure.

5.2 Predictor–Corrector Methods

We shall follow the variant of the predictor–corrector methods which was developed to solve the gas dynamics problems [110, 279, 462]. The predictor–corrector method allows us to improve the time approximation of the Maxwell equations and to satisfy to the conservation laws.

Let us suppose that the Maxwell equations for conventional hybrid model have the following form:

$$\frac{\partial \boldsymbol{B}}{\partial t} + c\nabla \times \boldsymbol{E} = 0, \quad \boldsymbol{E} = \boldsymbol{E}\{n_e, \boldsymbol{J}_e, \boldsymbol{B}\}, \quad (5.10)$$

where $\boldsymbol{B}$ and $\boldsymbol{E}$ denote the magnetic and electric fields.

The predictor–corrector method is realized in two time steps: the predictor step and the corrector step.

(1) Predictor step: We estimate the value of the magnetic field at time level $(n+1/2)$ by any explicit or implicit numerical scheme:

$$\frac{\boldsymbol{B}_h^{n+\frac{1}{2}} - \boldsymbol{B}_h^n}{\Delta t/2} + c\nabla_h^k \times \boldsymbol{E}_h^{n+\theta} = 0. \quad (5.11)$$

If $\theta = 0$, the scheme is explicit; otherwise, it is implicit.

(2) Corrector step: First of all we have to estimate the electric field at time level $(n+1/2)$,

$$\boldsymbol{E}_h^{n+\frac{1}{2}} = \boldsymbol{E}^{n+\frac{1}{2}}\left(n_{eh}^{n+\frac{1}{2}}, \boldsymbol{J}_{eh}^{n+\frac{1}{2}}, \boldsymbol{B}_h^{n+\frac{1}{2}}\right), \quad (5.12)$$

and then update the magnetic field at time level $(n+1)$,

$$\frac{\boldsymbol{B}_h^{n+1} - \boldsymbol{B}_h^n}{\Delta t} + c\nabla_h^k \times \boldsymbol{E}_h^{n+\frac{1}{2}} = 0. \qquad (5.13)$$

Equation (5.13) approximates the system of (5.10) with an $O(\Delta t^2 + h^k)$ order accurate in space and time. This scheme is fully conservative. Here ∇_h^k is the finite-difference approximation to the operator ∇ with the error of order $O(h^k)$ in space.

5.2.1 The Upwind Method

In this section we present a class of second-order conservative finite-difference algorithms for solving numerically time-dependent problems for hyperbolic conservation laws in several space variables [110]. These methods are upwind and multidimensional, in that the numerical fluxes are obtained by solving the characteristic form of the full multidimensional equations at the zone edge, and in that all fluxes are evaluated and differenced at the same time; in particular, operator splitting is not used.

To demonstrate this method we consider the scalar advection equation for one-dimensional variables. The starting point for the discussion of upwind schemes is the method of [119]:

$$\frac{u_i^{n+1} - u_i^n}{\Delta t} + c\frac{\Delta_i u_i^n}{\Delta x} = 0, \quad c > 0, \qquad (5.14)$$

where the backward difference operator $\Delta_i U_i = U_i - U_{i-1}$. The analysis by Saltzman [462] shows that this scheme is both stable and consistent, i.e., it is convergent. Conservation is easily seen because mass leaving one cell goes into the next. The first-order rate of convergence is ascertained from the first truncation error of the scheme.

The method of Godunov is a generalization of the method of Courant, Issacson and Rees. The difference equation for a hyperbolic system in one dimension is

$$\frac{U_i^{n+1} - U_i^n}{\Delta t} + \frac{\Delta_i F_1\left(U_{i+\frac{1}{2}}^{n+\frac{1}{2}}\right)}{\Delta x} = 0, \qquad (5.15)$$

where

$$U_{i+\frac{1}{2}}^{n+\frac{1}{2}} = R_1(U_i^n, U_{i+1}^n). \qquad (5.16)$$

The function $R_1(U_i^n, U_{i+1}^n)$ is a solution of a Riemann problem specified by the two states U_i^n and U_{i+1}^n. The method of Courant, Issacson and Rees can be derived from Godunov's method by substituting a linear flux function ($F(u) = cu$) and noting that the solution of the Riemann problem is the upstream state. The upstream state for a linear flux function with positive coefficient ($c > 0$) is the left state. Fourier analysis of the linearized system yields CFL numbers constrained not exceed unity. The Courant–Fridrich–Levy (CFL) numbers would use the characteristic speeds of the gradient of the flux function instead of an advection velocity [462].

Let us consider now a second-order scheme. An algebraic description of van Leer's algorithm for advection equation is [462]

$$\frac{u_i^{n+1} - u_i^n}{\Delta t} + c\frac{\Delta_i u_{i+\frac{1}{2}}^{n+\frac{1}{2}}}{\Delta x} = 0, \tag{5.17}$$

where

$$u_{i+\frac{1}{2}}^{n+\frac{1}{2}} = u_i^n + \frac{1}{2}\left(1 - \frac{c\Delta t}{\Delta x}\right)\bar{\Delta}_i u_i^n$$

and

$$\bar{\Delta}_i u_i^n = \begin{cases} \delta\,\mathrm{sgn}(u_{i+1} - u_{i-1}), & \text{if } \Delta_i u_{i+1} > 0, \\ 0, & \text{otherwise,} \end{cases}$$

such that

$$\delta = \min\left(2|\Delta_i u_{i+1}|, 2|\Delta_i u_i|, \frac{1}{2}|u_{i+1} - u_{i-1}|\right).$$

We can write the corresponding generalization of van Leer's scheme again using the solution of a Riemann problem [462]:

$$\frac{U_i^{n+1} - U_i^n}{\Delta t} + \frac{\Delta_i F_{i+\frac{1}{2}}^{n+\frac{1}{2}}}{\Delta x} = 0, \tag{5.18}$$

$$U_{i+\frac{1}{2}}^{n+\frac{1}{2}} = R_1\left({}_{(-)}U_{i+\frac{1}{2}}^{n+\frac{1}{2}}, {}_{(+)}U_{i+\frac{1}{2}}^{n+\frac{1}{2}}\right), \tag{5.19}$$

$${}_{(-)}U_{i+\frac{1}{2}}^{n+\frac{1}{2}} = U_i^n + \frac{1}{2}\left(I - \frac{\Delta t}{\Delta x}\frac{\partial F_1}{\partial U}\right)\bar{\Delta}_i U_i^n, \tag{5.20}$$

$${}_{(+)}U_{i+\frac{1}{2}}^{n+\frac{1}{2}} = T_i\left[U_i^n - \frac{1}{2}\left(I + \frac{\Delta t}{\Delta x}\frac{\partial F_1}{\partial U}\right)\bar{\Delta}_i U_i^n\right], \tag{5.21}$$

where T denotes the translation operator $T_i U_i = U_{i+1}$.

The corresponding unsplit upwind finite-difference scheme for the three-dimensional advection equation and the appropriate generalization for the system may be found in [462]. Different modifications of the above scheme have been used in MHD simulations over the past 20 years (see, e.g., [395, 503, 576] for a list of references). It uses the artificial viscosity of von Neumann and Richtmyer to smear shocks and the second-order van Leer upstream-weighted interpolation algorithms.

5.2.2 The Leapfrog Scheme

One of the ways to solve the Maxwell equations and the electron pressure equation is to apply the leapfrog scheme to hyperbolic equations (see, e.g., [399, 550] and references therein). These equations are of the following form:

$$\frac{\partial \phi}{\partial t} + \nabla \cdot (\boldsymbol{F}) - D\nabla^2 \phi + F(\xi) = 0. \tag{5.22}$$

ξ and ϕ are two scalar functions; $\boldsymbol{F} = \boldsymbol{U}\phi$, where $\boldsymbol{U} = U_z \hat{\boldsymbol{z}} - \nabla \phi \times \hat{\boldsymbol{z}}$. The explicit time-stepping scheme is the trapezoidal leapfrog algorithm [207, 585]

$$\hat{\phi}^{n+\frac{1}{2}} = \frac{1}{2}\phi^{n-1} + \frac{1}{2}\phi^n + \Delta t[-\nabla \cdot (\boldsymbol{F})^n + D\nabla^2 \phi^{n-1} - F(\xi^n)], \tag{5.23}$$

$$\phi^{n+1} = \phi^n + \Delta t \left(-\nabla \cdot (\hat{\boldsymbol{F}})^{n+\frac{1}{2}} + D\nabla^2 \phi^n - F(\hat{\xi}^{n+\frac{1}{2}})\right). \tag{5.24}$$

The dominant nonlinearities in the equations arise from the convective terms, which are written in conservative form [586]. A common fix is to introduce upwind differencing. However, this approach effectively introduces a stabilizing velocity-dependent diffusion, $U_x \Delta x/2$, which is often excessive. Instead, in [207], a velocity-dependent hyperviscosity is introduced. In one dimension the convective part of (5.22) becomes

$$\frac{\partial \phi}{\partial t} + \frac{\partial F_x}{\partial x} + \mu \Delta x^3 \frac{\partial}{\partial x} |U_x| \frac{\partial^3}{\partial x^3} \phi = 0, \tag{5.25}$$

with $\mu \sim 1$ is an arbitrary coefficient and Δx is the grid spacing. In practical simulations, the gradients of the flux are evaluated to the fourth order in Δx, and $\mu = 1/12$ is chosen to generalize the conventional upwind differencing or donor cell differencing to a higher order. Let us follow [207] to illustrate the finite-difference approximation to the term $\partial F_x/\partial x$:

$$\frac{\partial F_x}{\partial x} = \frac{2}{3\Delta x}\left(F^x_{i+1,j,k} - F^x_{i-1,j,k}\right) - \frac{1}{12\Delta x}\left(F^x_{i+2,j,k} - F^x_{i-2,j,k}\right) + H^x, \tag{5.26}$$

where H^x is the hyperviscosity,

$$H^x = \frac{1}{24\Delta x}\{|U^x_{i+1,i,j} + U^x_{i,j,k}|\left[(\phi_{i+2,j,k} - \phi_{i-1,j,k}) - 3(\phi_{i+1,j,k} - \phi_{i,j,k})\right]$$
$$-|U^x_{i,j,k} + U_{i-1,j,k}|\left[(\phi_{i+1,j,k} - \phi_{i-2,j,k}) - 3(\phi_{i,j,k} - \phi_{i-1,j,k})\right]\}. \tag{5.27}$$

The same approximation must be used for derivatives along other coordinates.

The above algorithm was used in [207] for a three-dimensional study of the turbulence and sheared flow generated by the drift-resistive ballooning modes in tokamak edge plasmas. The extension of this high-precision algorithm to hybrid/gyrokinetic models was made in [39, 484, 485]. The high-precision schemes allow us to reduce essentially the number of cells and macroparticles in multiscale multi-dimensional simulations. Despite the inclusion of the scheme's viscosity to suppress spurious oscillations, one has yet to estimate the effective Reynolds number. The use of an effective scheme viscosity (a small parameter at the highest order derivative) may be a crucial point in the magnetic field reconnection simulation. The other high-precision schemes, in particular, compact difference schemes, will be considered in Sect. 5.5.

5.2.3 Lax Wendroff Scheme. Explicit Calculation of the Electric Field

In this section we apply the modified two-step Lax–Wendroff scheme to a set of Maxwell equations and the generalized Ohm's law. The electron inertia effects can also be included in this model. Ampere's law in the Darwin limit (2.35), the definition of the total current (2.36) and the electron momentum transport equation (2.22) give the equation for the electric field:

$$\boldsymbol{E} = \frac{4\pi e}{\omega_{pe}^2 m} \left\{ \frac{1}{c} \boldsymbol{J}_e \times \boldsymbol{B} - \nabla p_e \right\}, \tag{5.28}$$

where

$$\boldsymbol{J}_e = \frac{c}{4\pi} \nabla \times \boldsymbol{B} - \sum_{k=1}^{N_s} Z_k e n_k \boldsymbol{U}_k - \boldsymbol{J}_{ext}. \tag{5.29}$$

Here for simplicity we have dropped the terms with effective anomalous resistivity and electron inertia.

Let us introduce the two meshes: the first mesh with location at grid points i, j, k and the second one with location at grid points $i \pm 1/2, j \pm 1/2, k \pm 1/2$. Let us suppose that all plasma parameters n_i, $\boldsymbol{U}_i$, p_e are known at time level $n+1/2$ and located at grid points $i \pm 1/2, j \pm 1/2, k \pm 1/2$. The electromagnetic field is known at time level n and located at grid points i, j, k.

In the predictor step we first have to calculate the magnetic field $\boldsymbol{B}$ at time level $n + 1/2$ and located at points $i \pm 1/2, j \pm 1/2, k \pm 1/2$. The second Maxwell equation gives

$$\boldsymbol{B}^{n+\frac{1}{2}}_{i\pm\frac{1}{2},j\pm\frac{1}{2},k\pm\frac{1}{2}} = \boldsymbol{B}^{n}_{i\pm\frac{1}{2},j\pm\frac{1}{2},k\pm\frac{1}{2}} - \frac{c\Delta t}{2} \nabla \times \boldsymbol{E}^{n}_{i,j,k}. \tag{5.30}$$

The interpolation from the eight nearest grid points gives the value of the magnetic field in the second mesh:

$$\boldsymbol{B}^{n}_{i+\frac{1}{2},j+\frac{1}{2},k+\frac{1}{2}} = \frac{1}{8} \left(\boldsymbol{B}^{n}_{i,j,k} + \boldsymbol{B}^{n}_{i+1,j,k} + \boldsymbol{B}^{n}_{i,j+1,k} + \boldsymbol{B}^{n}_{i+1,j+1,k} + \boldsymbol{B}^{n}_{i,j,k+1} \right.$$
$$\left. + \boldsymbol{B}^{n}_{i+1,j,k+1} + \boldsymbol{B}^{n}_{i,j+1,k+1} + \boldsymbol{B}^{n}_{i+1,j+1,k+1} \right). \tag{5.31}$$

Then we can evaluate the magnetic field at time level $n + 1/2$ at step (1):

(1) Predictor step

$$B^{x,n+\frac{1}{2}}_{i+\frac{1}{2},j+\frac{1}{2},k+\frac{1}{2}} = B^{x,n}_{i+\frac{1}{2},j+\frac{1}{2},k+\frac{1}{2}} - \frac{c\Delta t}{8\Delta y}\delta_y E^z + \frac{c\Delta t}{8\Delta z}\delta_z E^y, \tag{5.32}$$

where

$$\delta_y E^z = E^{z,n}_{i+1,j+1,k+1} + E^{z,n}_{i+1,j+1,k} + E^{z,n}_{i,j+1,k+1} + E^{z,n}_{i,j+1,k}$$
$$- E^{z,n}_{i+1,j,k+1} - E^{z,n}_{i+1,j,k} - E^{z,n}_{i,j,k+1} - E^{z,n}_{i,j,k}$$

and
$$\delta_z E^y = E^{y,n}_{i+1,j+1,k+1} + E^{y,n}_{i+1,j,k+1} + E^{y,n}_{i,j+1,k+1} + E^{y,n}_{i,j,k+1}$$
$$- E^{y,n}_{i+1,j+1,k} - E^{y,n}_{i+1,j,k} - E^{y,n}_{i,j+1,k} - E^{y,n}_{i,j,k},$$

$$B^{y,n+\frac{1}{2}}_{i+\frac{1}{2},j+\frac{1}{2},k+\frac{1}{2}} = B^{y,n}_{i+\frac{1}{2},j+\frac{1}{2},k+\frac{1}{2}} - \frac{c\Delta t}{8\Delta z}\delta_z E^x + \frac{c\Delta t}{8\Delta x}\delta_x E^z, \quad (5.33)$$

where
$$\delta_z E^x = E^{x,n}_{i+1,j+1,k+1} + E^{x,n}_{i+1,j,k+1} + E^{x,n}_{i,j+1,k+1} + E^{x,n}_{i,j,k+1}$$
$$- E^{x,n}_{i+1,j+1,k} - E^{x,n}_{i+1,j,k} - E^{x,n}_{i,j+1,k} - E^{x,n}_{i,j,k}$$

and
$$\delta_x E^z = E^{z,n}_{i+1,j+1,k+1} + E^{z,n}_{i+1,j+1,k} + E^{z,n}_{i+1,j,k+1} + E^{z,n}_{i+1,j,k}$$
$$- E^{z,n}_{i,j+1,k+1} - E^{z,n}_{i,j+1,k} - E^{z,n}_{i,j,k+1} - E^{z,n}_{i,j,k},$$

$$B^{z,n+\frac{1}{2}}_{i+\frac{1}{2},j+\frac{1}{2},k+\frac{1}{2}} = B^{z,n}_{i+\frac{1}{2},j+\frac{1}{2},k+\frac{1}{2}} - \frac{c\Delta t}{8\Delta x}\delta_x E^y + \frac{c\Delta t}{8\Delta y}\delta_y E^x, \quad (5.34)$$

where
$$\delta_x E^y = E^{y,n}_{i+1,j+1,k+1} + E^{y,n}_{i+1,j+1,k} + E^{y,n}_{i+1,j,k+1} + E^{y,n}_{i+1,j,k}$$
$$- E^{y,n}_{i,j+1,k+1} - E^{y,n}_{i,j+1,k} - E^{y,n}_{i,j,k+1} - E^{y,n}_{i,j,k}$$

and
$$\delta_y E^x = E^{x,n}_{i+1,j+1,k+1} + E^{x,n}_{i+1,j+1,k} + E^{x,n}_{i,j+1,k+1} + E^{x,n}_{i,j+1,k}$$
$$- E^{x,n}_{i+1,j,k+1} - E^{x,n}_{i+1,j,k} - E^{x,n}_{i,j,k+1} - E^{x,n}_{i,j,k}.$$

Provided that we know all the values at grid points $i\pm 1/2, j\pm 1/2, k\pm 1/2$, we can use the corrector step (2) to update the magnetic field:

$$\boldsymbol{B}^{n+1} = \boldsymbol{B}^n - c\Delta t \nabla \times \boldsymbol{E}^{n+\frac{1}{2}}, \quad (5.35)$$

where the electric field at time level $n+1/2$ was estimated by

$$\boldsymbol{E}^{n+\frac{1}{2}}_{i+\frac{1}{2},j+\frac{1}{2},k+\frac{1}{2}} = \boldsymbol{E}^{n+\frac{1}{2}}\{n_\mathrm{i}, \boldsymbol{U}_\mathrm{i}, \boldsymbol{B}, p_\mathrm{e}\}_{i+\frac{1}{2},j+\frac{1}{2},k+\frac{1}{2}}. \quad (5.36)$$

(2) Corrector step

$$B^{x,n+1}_{i,j,k} = B^{x,n}_{i,j,k} - \frac{c\Delta t}{4\Delta y}\delta_y E^z + \frac{c\Delta t}{4\Delta z}\delta_z E^y, \quad (5.37)$$

where

$$\delta_y E^z = E^{z,n+\frac{1}{2}}_{i+\frac{1}{2},j+\frac{1}{2},k+\frac{1}{2}} + E^{z,n+\frac{1}{2}}_{i+\frac{1}{2},j+\frac{1}{2},k-\frac{1}{2}} + E^{z,n+\frac{1}{2}}_{i-\frac{1}{2},j+\frac{1}{2},k+\frac{1}{2}} + E^{z,n+\frac{1}{2}}_{i-\frac{1}{2},j+\frac{1}{2},k-\frac{1}{2}}$$

$$-E^{z,n+\frac{1}{2}}_{i+\frac{1}{2},j-\frac{1}{2},k+\frac{1}{2}} - E^{z,n+\frac{1}{2}}_{i+\frac{1}{2},j-\frac{1}{2},k-\frac{1}{2}} - E^{z,n+\frac{1}{2}}_{i-\frac{1}{2},j-\frac{1}{2},k+\frac{1}{2}} - E^{z,n+\frac{1}{2}}_{i-\frac{1}{2},j-\frac{1}{2},k-\frac{1}{2}}$$

and

$$\delta_z E^y = E^{y,n+\frac{1}{2}}_{i+\frac{1}{2},j+\frac{1}{2},k+\frac{1}{2}} + E^{y,n+\frac{1}{2}}_{i+\frac{1}{2},j-\frac{1}{2},k+\frac{1}{2}} + E^{y,n+\frac{1}{2}}_{i-\frac{1}{2},j+\frac{1}{2},k+\frac{1}{2}} + E^{y,n+\frac{1}{2}}_{i-\frac{1}{2},j-\frac{1}{2},k+\frac{1}{2}}$$

$$-E^{y,n+\frac{1}{2}}_{i+\frac{1}{2},j+\frac{1}{2},k-\frac{1}{2}} - E^{y,n+\frac{1}{2}}_{i+\frac{1}{2},j-\frac{1}{2},k-\frac{1}{2}} - E^{y,n+\frac{1}{2}}_{i-\frac{1}{2},j+\frac{1}{2},k-\frac{1}{2}} - E^{y,n+\frac{1}{2}}_{i-\frac{1}{2},j-\frac{1}{2},k-\frac{1}{2}},$$

$$B^{y,n+1}_{i,j,k} = B^{y,n}_{i,j,k} - \frac{c\Delta t}{4\Delta z}\delta_z E^x + \frac{c\Delta t}{4\Delta x}\delta_x E^z, \quad (5.38)$$

where

$$\delta_z E^x = E^{x,n+\frac{1}{2}}_{i+\frac{1}{2},j+\frac{1}{2},k+\frac{1}{2}} + E^{x,n+\frac{1}{2}}_{i+\frac{1}{2},j-\frac{1}{2},k+\frac{1}{2}} + E^{x,n+\frac{1}{2}}_{i-\frac{1}{2},j+\frac{1}{2},k+\frac{1}{2}} + E^{x,n+\frac{1}{2}}_{i-\frac{1}{2},j-\frac{1}{2},k+\frac{1}{2}}$$

$$-E^{x,n+\frac{1}{2}}_{i+\frac{1}{2},j+\frac{1}{2},k-\frac{1}{2}} - E^{x,n+\frac{1}{2}}_{i+\frac{1}{2},j-\frac{1}{2},k-\frac{1}{2}} - E^{x,n+\frac{1}{2}}_{i-\frac{1}{2},j+\frac{1}{2},k-\frac{1}{2}} - E^{x,n+\frac{1}{2}}_{i-\frac{1}{2},j-\frac{1}{2},k-\frac{1}{2}}$$

and

$$\delta_x E^z = E^{z,n+\frac{1}{2}}_{i+\frac{1}{2},j+\frac{1}{2},k+\frac{1}{2}} + E^{z,n+\frac{1}{2}}_{i+\frac{1}{2},j+\frac{1}{2},k-\frac{1}{2}} + E^{z,n+\frac{1}{2}}_{i+\frac{1}{2},j-\frac{1}{2},k+\frac{1}{2}} + E^{z,n+\frac{1}{2}}_{i+\frac{1}{2},j-\frac{1}{2},k-\frac{1}{2}}$$

$$-E^{z,n+\frac{1}{2}}_{i-\frac{1}{2},j+\frac{1}{2},k+\frac{1}{2}} - E^{z,n+\frac{1}{2}}_{i-\frac{1}{2},j+\frac{1}{2},k-\frac{1}{2}} - E^{z,n+\frac{1}{2}}_{i-\frac{1}{2},j-\frac{1}{2},k+\frac{1}{2}} - E^{z,n+\frac{1}{2}}_{i-\frac{1}{2},j-\frac{1}{2},k-\frac{1}{2}},$$

$$B^{z,n+1}_{i,j,k} = B^{z,n}_{i,j,k} - \frac{c\Delta t}{4\Delta x}\delta_x E^y + \frac{c\Delta t}{4\Delta y}\delta_y E^x, \quad (5.39)$$

where

$$\delta_x E^y = E^{y,n+\frac{1}{2}}_{i+\frac{1}{2},j+\frac{1}{2},k+\frac{1}{2}} + E^{y,n+\frac{1}{2}}_{i+\frac{1}{2},j+\frac{1}{2},k-\frac{1}{2}} + E^{y,n+\frac{1}{2}}_{i+\frac{1}{2},j-\frac{1}{2},k+\frac{1}{2}} + E^{y,n+\frac{1}{2}}_{i+\frac{1}{2},j-\frac{1}{2},k-\frac{1}{2}}$$

$$-E^{y,n+\frac{1}{2}}_{i-\frac{1}{2},j+\frac{1}{2},k+\frac{1}{2}} - E^{y,n+\frac{1}{2}}_{i-\frac{1}{2},j+\frac{1}{2},k-\frac{1}{2}} - E^{y,n+\frac{1}{2}}_{i-\frac{1}{2},j-\frac{1}{2},k+\frac{1}{2}} - E^{y,n+\frac{1}{2}}_{i-\frac{1}{2},j-\frac{1}{2},k-\frac{1}{2}}$$

and

$$\delta_y E^x = E^{x,n+\frac{1}{2}}_{i+\frac{1}{2},j+\frac{1}{2},k+\frac{1}{2}} + E^{x,n+\frac{1}{2}}_{i+\frac{1}{2},j+\frac{1}{2},k-\frac{1}{2}} + E^{x,n+\frac{1}{2}}_{i-\frac{1}{2},j+\frac{1}{2},k+\frac{1}{2}} + E^{x,n+\frac{1}{2}}_{i-\frac{1}{2},j+\frac{1}{2},k-\frac{1}{2}}$$

$$-E^{x,n+\frac{1}{2}}_{i+\frac{1}{2},j-\frac{1}{2},k+\frac{1}{2}} - E^{x,n+\frac{1}{2}}_{i+\frac{1}{2},j-\frac{1}{2},k-\frac{1}{2}} - E^{x,n+\frac{1}{2}}_{i-\frac{1}{2},j-\frac{1}{2},k+\frac{1}{2}} - E^{x,n+\frac{1}{2}}_{i-\frac{1}{2},j-\frac{1}{2},k-\frac{1}{2}}.$$

The total final electric field $\boldsymbol{E}$ at the time level $n+1$ may be estimated by extrapolation from values at time levels $n+1/2$ and n:

$$\boldsymbol{E}^{n+1} = 2\boldsymbol{E}^{n+\frac{1}{2}} - \boldsymbol{E}^n. \quad (5.40)$$

This method uses a second-order approximation in coordinate space and in time. For the solution of the magnetic field equation the error will be $O(\Delta x^2 + \Delta y^2 + \Delta z^2 + \Delta t^2)$. The numerical stability is provided if all the possible speeds satisfy the CFL number condition CFL < 1.

The anomalous resistivity may be included directly in step (2) by explicit or implicit ways. In the latter case the diffusive term must be approximated using the Crank–Nicolson scheme. The electron inertia term may be included in same way as it was in Sect. 5.7.

5.2.4 Implicit Calculation of the Electric Field

Use of the implicit calculation of the electric field at the predictor step results in much better numerical stability when updating the electromagnetic field. In addition, some of the electron inertia effects can be included without too much difficulty [334, 486].

Let us write the general equation for the electric field in the following form (2.38):

$$c^2 \nabla \times (\nabla \times \boldsymbol{E}) + \omega_{\text{pe}}^2 \left(1 + \sum_{k=1}^{N_s} \frac{n_k Z_k^2 m}{n_e M_k}\right) \boldsymbol{E}$$

$$+ \frac{e}{m}\left[\frac{4\pi}{c}\sum_{k=1}^{N_s}\left(1 + \frac{Z_k m}{M_k}\right)\boldsymbol{J}_k - \nabla \times \boldsymbol{B} + \frac{4\pi}{c}\boldsymbol{J}_{\text{ext}}\right] \times \boldsymbol{B}$$

$$= -\frac{4\pi e}{m}\left(\nabla p_{\text{e}} + \partial/x_\beta \pi_{\alpha\beta}^{\text{e}}\right) + 4\pi \left(\sum_{k=1}^{N_s} Z_k e \nabla \cdot \boldsymbol{\mathsf{K}}_k + (\nabla \boldsymbol{U}_{\text{e}})\boldsymbol{J}_{\text{e}}\right)$$

$$+ \frac{4\pi \partial \boldsymbol{J}_{\text{ext}}}{\partial t} - \frac{4\pi e}{m}\left(\boldsymbol{R}_{\text{e}} - \sum_{k=1}^{N_s}\frac{Z_k m}{M_k}\boldsymbol{R}_k\right), \tag{5.41}$$

where $\boldsymbol{J}_{\text{ext}}$ is the external current and $\boldsymbol{R}_{\text{e}}$ and $\boldsymbol{R}_k$ are determined by (2.42) and (2.43). In the case of a collisionless plasma, one can neglect these terms; in the case of weak collisions, one can consider explicitly an external source in (5.41). However, in some cases it is important to include the effective anomalous resistivity. For regimes where the dissipation length is comparable to the size of the grid, one has to approximate the resistive term implicitly.

This equation is evaluated at a time level between the n and $n+1$ levels. The electric and magnetic fields at this time level, $n+\theta$, are given by

$$\boldsymbol{E}^{n+\theta} = \theta \boldsymbol{E}^{n+1} + (1-\theta)\boldsymbol{E}^n, \tag{5.42}$$

$$\boldsymbol{B}^{n+\theta} = \theta \boldsymbol{B}^{n+1} + (1-\theta)\boldsymbol{B}^n. \tag{5.43}$$

In the predictor step one needs to calculate $\boldsymbol{E}$ and $\boldsymbol{B}$ at time level $n+\theta$, and for this purpose we used second Maxwell equation, which gives

$$\boldsymbol{B}^{n+\theta} = \boldsymbol{B}^n - \theta c \Delta t \nabla \times \boldsymbol{E}^{n+\theta}. \tag{5.44}$$

The evaluation of (5.41) at time level $n+\theta$ in combination with (5.44) results in an equation for $\boldsymbol{E}^{n+\theta}$, i.e., the electric field at time level $n+\theta$

$$d\nabla \times (\nabla \times \boldsymbol{E}^{n+\theta}) + A\boldsymbol{E}^{n+\theta} + (\nabla \times \boldsymbol{E}^{n+\theta}) \times \boldsymbol{I}$$

$$+ g\left\{[\nabla \times (\nabla \times \boldsymbol{E}^{n+\theta})] \times \boldsymbol{B}^n\right\} = \boldsymbol{Q}, \tag{5.45}$$

where

$$d = c^2, \quad A = \omega_{\text{pe}}^2 \left(1 + \sum_{k=1}^{N_s} \frac{n_k Z_k^2 m}{n_e M_k}\right), \quad g = \frac{\theta e c \Delta t}{m}, \tag{5.46}$$

$$\boldsymbol{I} = \frac{\theta e c \Delta t}{m} \left[\frac{4\pi}{c} \sum_{k=1}^{N_s} \left(1 + \frac{Z_k m}{M_k}\right) \boldsymbol{J}_k^{n+\frac{1}{2}} - \nabla \times \boldsymbol{B}^n + \frac{4\pi}{c} \boldsymbol{J}_{\text{ext}}^{n+\frac{1}{2}}\right], \tag{5.47}$$

$$\boldsymbol{Q} = -\frac{e}{m} \left[\frac{4\pi}{c} \sum_{k=1}^{N_s} \left(1 + \frac{Z_k m}{M_k}\right) \boldsymbol{J}_k^{n+\frac{1}{2}} - \nabla \times \boldsymbol{B}^n + \frac{4\pi}{c} \boldsymbol{J}_{\text{ext}}^{n+\frac{1}{2}}\right] \times \boldsymbol{B}^n$$

$$-\frac{4\pi e}{m} \left(\nabla p_{\text{e}} + \partial/\partial x_\beta \pi_{\alpha\beta}^{\text{e}}\right) + 4\pi \left(\sum_{k=1}^{N_s} Z_k e \nabla \cdot \mathbf{K}_k + (\nabla \boldsymbol{U}_{\text{e}}) \boldsymbol{J}_{\text{e}}\right)$$

$$-\frac{4\pi e}{m} \left(\boldsymbol{R}_{\text{e}} - \sum_{k=1}^{N_s} \frac{Z_k m}{M_k} \boldsymbol{R}_k\right) + \frac{4\pi \partial \boldsymbol{J}_{\text{ext}}}{\partial t}. \tag{5.48}$$

In some interesting applications one can cancel the small parameters such as m/M_k in the same term. In this approximation the resistive terms may be replaced by the following simple expression:

$$-\frac{4\pi e}{m} \boldsymbol{R}_{\text{e}} = c\nu_{\text{ei}} \nabla \times \boldsymbol{B}. \tag{5.49}$$

The magnetic field must be estimated at time level $n + \theta$ using the (5.44). One can also omit some terms that depend on $\boldsymbol{J}_{\text{e}}^2$ and $\boldsymbol{J}_k^2$. In [486], the resistivity and electron pressure were also omitted. However, to study space and astrophysical plasma, one has to keep these terms. We also assume that $p_{\text{e}} \geq p_k$ and that $\pi_{\alpha\beta}^{\text{e}(s)} \ll p_{\text{e}(k)}$. In this very important case we have

$$d = c^2 + \theta c^2 \nu_{\text{ei}} \Delta t, \quad A = \omega_{\text{pe}}^2, \quad g = \frac{\theta e c \Delta t}{m}, \tag{5.50}$$

$$\boldsymbol{I} = \frac{\theta e c \Delta t}{m} \left(\frac{4\pi}{c} \sum_{k=1}^{N_s} \boldsymbol{J}_k^{n+\frac{1}{2}} - \nabla \times \boldsymbol{B}^n + \frac{4\pi}{c} \boldsymbol{J}_{\text{ext}}^{n+\frac{1}{2}}\right), \tag{5.51}$$

$$\boldsymbol{Q} = -\frac{e}{m} \left(\frac{4\pi}{c} \sum_{k=1}^{N_s} \boldsymbol{J}_k^{n+\frac{1}{2}} - \nabla \times \boldsymbol{B}^n + \frac{4\pi}{c} \boldsymbol{J}_{\text{ext}}^{n+\frac{1}{2}}\right) \times \boldsymbol{B}^n - \frac{4\pi e}{m} \nabla p_{\text{e}}$$

$$+ c\nu_{\text{ei}} \nabla \times \boldsymbol{B}^n + \frac{4\pi \partial \boldsymbol{J}_{\text{ext}}}{\partial t}. \tag{5.52}$$

However, one can keep the terms with $\boldsymbol{J}_{\text{e}}^2$ and $\boldsymbol{J}_k^2$ and consider them as external terms. In the case with adiabatic electrons one can split the total electric field into the sum of inductive ($\boldsymbol{E}_1$) and electrostatic ($\boldsymbol{E}_2$) fields:

$$\boldsymbol{E} = \boldsymbol{E}_1 + \boldsymbol{E}_2, \tag{5.53}$$

where E_2 satisfies the condition

$$\nabla \times E_2 = 0. \tag{5.54}$$

Then we can solve (5.45) for component E_1 neglecting ∇p_e. The electrostatic electric field E_2 can be calculated from p_e [334] because

$$E_2 = -\frac{1}{en_e}\nabla p_e. \tag{5.55}$$

Let us consider two meshes. The first mesh contains the nodes i,j,k, which are located in the center of a cell. It is used for the computation of the density, current, bulk velocity, electron pressure and inductive electric field. The second mesh contains the nodes $i \pm 1/2, j \pm 1/2, k \pm 1/2$, which are located at the corners of a cell. This mesh is used for the computation of the magnetic field components, electrostatic field and final electric field. Let us assume that $J_i^{n+1/2}$, ω_{pe} and n_i are known at time level $n + 1/2$. Then, expressing $\nabla \times E_1^{n+\theta}$ and $\nabla \times (\nabla \times E_1^{n+\theta})$ via central finite differences at each cell center, one can obtain the following 3×3 matrix equation:

$$L \cdot E_{i,j,k}^{n+\theta} = F\left(E_{i\pm 1,j\pm 1,k\pm 1}^{n+\theta}\right). \tag{5.56}$$

Equation (5.56) may be solved by iteration. In each iteration the electric field on the right side is given. The iteration continues until some convergence criterion is satisfied. At the same time the electrostatic field E_2 (5.55) is calculated at points $i \pm 1/2, j \pm 1/2, k \pm 1/2$ at each cell corner. The use of the multigrid technique (see, for example, [154, 595] and references therein) for solving (5.56) may significantly reduce the number of iterations.

In the corrector step, when E_2 is obtain, one can update the magnetic field, using again the second Maxwell equation:

$$B^{n+1} = B^n - \frac{c\Delta t}{2}\nabla \times (E^{n+1} + E^n). \tag{5.57}$$

The total final electric field E is calculated at each cell corner – $i \pm 1/2, j \pm 1/2, k \pm 1/2$ – by means of interpolation of E_1 from points i, j, k and adding it to the electrostatic field E_2. The electric field at time level $n+1$ may be estimated by extrapolation from the values at time levels $n+\theta$ and n:

$$E^{n+1} = \frac{E^{n+\theta}}{\theta} - \frac{(1-\theta)E^n}{\theta}. \tag{5.58}$$

This method is accurate to the second order in space and in time. In order to solve for the magnetic field, the equation is approximated with the truncation error $O(\Delta x^2 + \Delta y^2 + \Delta z^2 + \Delta t^2)$. The scheme is numerically stable for all values of Δt provided $\theta \geq 1/2$.

The other implicit algorithm for the calculation of the electric field was proposed in [362]. Let us evaluate the electron current at the $n+\theta$ time level:

$$\boldsymbol{J}_{\mathrm{e}}^{n+\theta} = \theta \boldsymbol{J}_{\mathrm{e}}^{n+1} + (1-\theta)\boldsymbol{J}_{\mathrm{e}}^{n}. \qquad (5.59)$$

Thus (5.44) can be inserted into Ampere's law at the $n+\theta$ time level to give

$$\nabla \times \nabla \times \boldsymbol{E}^{n+\theta} = -\frac{4\pi}{c^2 \theta \Delta t}(\boldsymbol{J}_{\mathrm{e}}^{n+\theta} + \boldsymbol{J}_{\mathrm{i}}^{n+\frac{1}{2}}) + \frac{1}{c\theta\Delta t}\nabla \times \boldsymbol{B}^n. \qquad (5.60)$$

To determine $\boldsymbol{J}_{\mathrm{e}}^{n+\theta}$, we consider the momentum balance equation (2.25) at time level $n+\theta$:

$$\frac{\boldsymbol{J}_{\mathrm{e}}^{n+\theta} - \boldsymbol{J}_{\mathrm{e}}^{n}}{\theta \Delta t} = \frac{\omega_{\mathrm{pe}}^2}{4\pi}\boldsymbol{E}^{n+\theta} - \nu_{\mathrm{e}}\boldsymbol{J}_{\mathrm{e}}^{n+\theta} + \boldsymbol{J}_{\mathrm{e}}^{n+\theta} \times \boldsymbol{\Omega}_{\mathrm{e}}^{n+\theta} + \boldsymbol{j}_{\mathrm{e}}^{n+\theta}, \qquad (5.61)$$

where

$$\nu_{\mathrm{e}} = \sum_k \nu_{\mathrm{S}}^{\mathrm{ek}}$$

and

$$\boldsymbol{j}_{\mathrm{e}} = \frac{e}{m}\nabla p_{\mathrm{e}} - en\sum_k \nu_{\mathrm{S}}^{\mathrm{ek}}\boldsymbol{U}_k.$$

In the above equation, $\boldsymbol{U}_k$ represents the local mean velocity of ion species k. Now (5.61) can be solved formally for $\boldsymbol{J}_{\mathrm{e}}^{n+\theta}$ [362]:

$$\boldsymbol{J}_{\mathrm{e}}^{n+\theta} = \boldsymbol{O}^{n+\theta}\left(\frac{\omega_{\mathrm{pe}}^2 \theta \Delta t}{4\pi}\boldsymbol{E}^{n+\theta} + \boldsymbol{J}_{\mathrm{e}}^{n} + \theta \Delta t \boldsymbol{j}_{\mathrm{e}}^{n+\theta}\right), \qquad (5.62)$$

where the operator

$$\boldsymbol{O}^{n+\theta}(\boldsymbol{A}) = [(1+\nu\theta\Delta t)^2 + (\Omega_{\mathrm{e}}\theta\Delta t)^2]^{-1}$$

$$\cdot \left((1+\nu\theta\Delta t)\boldsymbol{A} - \theta\Delta t \boldsymbol{\Omega}_{\mathrm{e}} \times \boldsymbol{A} + \frac{(\theta\Delta t)^2}{(1+\nu\theta\Delta t)}\boldsymbol{\Omega}_{\mathrm{e}}(\boldsymbol{\Omega}_{\mathrm{e}} \cdot \boldsymbol{A})\right). \qquad (5.63)$$

In (5.63), $\boldsymbol{A}$ is an arbitrary vector and all quantities on the right hand side are evaluated at the $n+\theta$ time level. Substituting (5.62) into (5.60), one obtains an equation for $\boldsymbol{E}^{n+\theta}$:

$$\nabla \times \nabla \times \boldsymbol{E}^{n+\theta} + \frac{\omega_{\mathrm{pe}}^2}{c^2}\boldsymbol{O}^{n+\theta}\left(\boldsymbol{E}^{n+\theta}\right)$$

$$= \frac{1}{c\theta\Delta t}\left(\nabla \times \boldsymbol{B}^n - \frac{4\pi}{c}\boldsymbol{J}_{\mathrm{i}}^{n+\frac{1}{2}}\right) - \frac{4\pi}{c^2}\boldsymbol{O}^{n+\theta}\left(\frac{1}{\theta\Delta t}\boldsymbol{J}_{\mathrm{e}}^{n} + \boldsymbol{j}_{\mathrm{e}}\right), \qquad (5.64)$$

where we can see that the operator $\boldsymbol{O}$ plays the role of a dielectric tensor. Assuming (5.64) has been solved, one determines the values of $\boldsymbol{E}^{n+\theta}$, $\boldsymbol{B}^{n+\theta}$ and $\boldsymbol{J}_{\mathrm{e}}^{n+\theta}$ from Faraday's law and the interpolation formulae (5.42) and (5.59). The main problem remaining is to find a suitable way of solving (5.64). The original CIDER algorithm achieved time centering via an integration on an explicit expression for the electric field, similar to (5.61) [362]. This iteration was subject to von Neumann convergence condition of the form

$$\Omega_i \Delta t \leq \left(\frac{\omega_{pi} \Delta x_{min}}{c}\right)^2, \tag{5.65}$$

where Ω_i and ω_{pi} are the background ion cyclotron and plasma frequencies, respectively. As discussed in [361], violation of (5.65) to any significant degree causes the calculation to fail due to divergent values of the electric and magnetic fields. Although (5.65) is expressed in terms of ion quantities, it is in fact a condition in which the time step must be shorter than the period of a whistler wave with the minimum allowed wavelength. This condition can be viewed as a result of including the double curl operator in (5.64) only iteratively when solving for $\boldsymbol{E}^{n+\theta}$. This is corrected in the following scheme, where $\boldsymbol{E}^{n+\theta} = \lim_{l\to\infty} \boldsymbol{E}_l$ and $\boldsymbol{E}_l$ satisfies the recursion relation [362]:

$$k_*^2 \boldsymbol{E}_{l+1} + \frac{\omega_{pe}^2}{c^2} \boldsymbol{O}_l(\boldsymbol{E}_{l+1}) = k_*^2 \boldsymbol{E}_l + \frac{\omega_{pe}^2}{c^2} \boldsymbol{O}_l(\boldsymbol{E}_l)$$

$$+ \alpha \left(\boldsymbol{S} - \nabla \times \nabla \times \boldsymbol{E}_l - \frac{\omega_{pe}^2}{c^2} \boldsymbol{O}_l(\boldsymbol{E}_l) \right), \tag{5.66}$$

or

$$\boldsymbol{E}_{l+1} = \boldsymbol{E}_l + \alpha \boldsymbol{L}_l \left[\boldsymbol{S} - \left(\nabla \times \nabla \times + \frac{\omega_{pe}^2}{c^2} \boldsymbol{O}_l \right) \boldsymbol{E}_l \right],$$

where

$$\boldsymbol{L}_l = \left(k_*^2 + \frac{\omega_{pe}^2}{c^2} \boldsymbol{O}_l \right)^{-1}.$$

Here k_*^2 and α are two parameters which are adjusted to give optimum convergence. Typically, one chooses k_*^2 to be the largest eigenvalue of the double curl operator, and $\alpha < 1$. The quantity $\boldsymbol{S}$ is given by

$$\boldsymbol{S} = \frac{1}{c\theta \Delta t} \left(\nabla \times \boldsymbol{B}^n - \frac{4\pi}{c} \boldsymbol{J}_i^{n+\frac{1}{2}} \right) - \frac{4\pi}{c^2} \boldsymbol{O}_l \left(\frac{1}{\theta \Delta t} \boldsymbol{J}_e^n + \boldsymbol{j}_e \right). \tag{5.67}$$

The stability analysis indicates a good convergence of implicit algorithms for $\theta \geq 1/2$. The advantage to the implicit algorithm is its ability to treat low-density and vacuum regions. The explicit expression for the electric field contains terms that are inversely proportional to density, necessitating the imposition of a density minimum during a calculation; in effect, this density minimum places an upper limit on the Alfvén speed, thereby enforcing a Courant condition on Alfvén waves [362]. Equations (5.45) and (5.64), however, are well behaved for arbitrary values of density, and in fact reduce to the proper vacuum field equation, $\nabla \times \nabla \times \boldsymbol{E}^{n+\theta} = 0$, in regions where the plasma density vanishes.

5.3 Operator Splitting Methods

5.3.1 Splitting Schemes

To study nonstationary problems, sometimes it is sufficient to use the equation in a nondivergent form with a high finite-difference approximation in time and space. Such a numerical scheme may possess good numerical stability, and at the same time the conservation laws may be violated only at some points of mesh.

Let us suppose that the Maxwell equations for conventional hybrid models are described by (5.10), where B and E are vectors with dimensionality p. Suppose that vector f has the same dimensionality and that the following dependencies exist: $B = B(f)$, $E = E(f)$. Suppose also that the nonsingular matrices $A = \partial B/\partial f$, $B_k = \partial E_k/\partial f$ and A^{-1} exist. In case of $U = f$ we have $A = I$, where I is the identity matrix. Then the system of (5.10) may be written in the following nondivergent form (see, e.g., [41, 70, 227, 318, 321]):

$$\frac{\partial f}{\partial t} + \sum_{k=1}^{N} C_k \frac{\partial f}{\partial x_k} = Q, \tag{5.68}$$

where $C_k = A^{-1} B_k$, $A = \partial B/\partial f$ and $B_k = \partial E_k/\partial f$. Let us suppose that the matrix C_k may be represented as a sum of matrices q_k which have a more simple structure (see, for example, [279, 581]):

$$C_k = \sum_{i=1}^{q_k} C_{ki}. \tag{5.69}$$

For the simplicity we can rewrite (5.68) as

$$\frac{\partial f}{\partial t} + \sum_{p=1}^{M} L_p f = 0, \tag{5.70}$$

where

$$M = \sum_{k=1}^{N} q_k$$

and

$$\sum_{p=1}^{M} L_p f = \sum_{k=1}^{N} \sum_{i=1}^{q_k} C_{ki} \partial f/\partial x_k.$$

Let us suppose that the split system of

$$\frac{1}{M} \frac{\partial f}{\partial t} + L_p f = \frac{Q}{M} \tag{5.71}$$

may be solved by using the economic finite-difference scheme. Then, the scheme

$$\frac{f_h^{n+\frac{1}{M}} - f_h^n}{\Delta t} + L_{1h}^n f_h^{n+\frac{1}{M}} = \frac{Q}{M},$$

$$\cdots\cdots\cdots\cdots\cdots$$

$$\frac{f_h^{n+1} - f_h^{n+\frac{M-1}{M}}}{\Delta t} + L_{Mh}^n f_h^{n+1} = \frac{Q}{M}, \tag{5.72}$$

$$B_h^{n+1} = B\left(f_h^{n+1}\right),$$

$$E_h^{n+\frac{1}{2}} = E^{n+\frac{1}{2}}\left\{n_{eh}^{n+\frac{1}{2}}, J_{eh}^{n+\frac{1}{2}}, B_h^{n+\frac{1}{2}}\right\}$$

approximates the system of (5.68) with order $O(\Delta t^2 + h^k)$, and it needs the iteration because the coefficients of operators L_{ih}^n approximate L_i at time level n. The other equivalent methods are (a) factorization of operator method and (b) alternative direction implicit method. Thus, the multidimensional problem may be reduced to a set of simple one-dimensional problems (5.72).

The inductive equation for the magnetic field in the case of the isotropic electron pressure may be written in nondivergent form, (13.4), (13.9) and (13.14), which may be used in (5.72).

5.3.2 Predictor–Corrector/Operator Splitting Scheme

The splitting of the operator method may be used together with the idea of the predictor–corrector method. At the predictor step the finite-difference scheme with a high accumulation of numerical stability is developed, and at the corrector step the scheme conservation is restored and the order of approximation in time is also increased. So, the predictor–corrector schemes may be considered as the operator split method. At the predictor step the splitting over physical processes and space directions is used to approximate the equations in a nondivergent form. At the corrector step, the equations are approximated in a divergent form. The linear analysis shows that these schemes are unconditionally stable.

Then, the following two-step scheme approximates the system of (5.10) with an approximation error of about $O(\Delta t^2 + h^k)$:

(1) Splitting on the predictor step

$$\frac{f_h^{n+\frac{1}{2M}} - f_h^n}{\Delta t/2} + L_{1h}^n f_h^{n+\frac{1}{2M}} = \frac{Q}{M}, \tag{5.73}$$

$$\cdots\cdots\cdots\cdots\cdots$$

$$\frac{f_h^{n+\frac{1}{2}} - f_h^{n+\frac{M-1}{2M}}}{\Delta t/2} + L_{Mh}^n f_h^{n+\frac{1}{2}} = \frac{Q}{M}, \tag{5.74}$$

(2) Corrector step

$$B_h^{n+\frac{1}{2}} = B\left(f_h^{n+\frac{1}{2}}\right), \tag{5.75}$$

$$E_h^{n+\frac{1}{2}} = E^{n+\frac{1}{2}}\left\{n_{eh}^{n+\frac{1}{2}}, J_{eh}^{n+\frac{1}{2}}, B_h^{n+\frac{1}{2}}\right\}, \tag{5.76}$$

$$\frac{B_h^{n+1} - B_h^n}{\Delta t} + c\nabla^k \times E_h^{n+\frac{1}{2}} = 0. \tag{5.77}$$

It does not need iteration because the coefficients of operators L_{ih}^n approximate L_i at time level n, and this scheme is fully conservative. Here ∇_h^k presents the finite-difference approximation to the operator ∇ with the error of order $O(h^k)$. Examples of the application of the splitting of the operator method will be considered in the exercises.

5.4 The Transportive Property

We shall say that a finite difference of an induction equation possesses the transportive property (see analysis of fluid equations [445]) if the effect of perturbation in a transport property is advected only in the direction of the velocity. Innocuous and obvious as this definition may read, the fact is that the most frequently used methods do not possess this property. All methods which use centered-space derivatives for the advection terms do not possess this property.

The emphasis is on the word "advected". A physical perturbation in the magnetic field will spread in all directions due to diffusion. But it should be carried along only in the direction of the velocity. Let us consider the one-dimensional ideal model equation of motion/magnetic field transport:

$$\frac{\partial b}{\partial t} = -\frac{\partial U_e b}{\partial x}. \tag{5.78}$$

The finite-difference scheme of (5.78) with a central approximation of space derivative is

$$\frac{b_i^{n+1} - b_i^n}{\Delta t} = -\frac{ub_{i+1}^n - ub_{i-1}^n}{2\Delta x}. \tag{5.79}$$

Let us consider a local perturbation at mesh point m, $\varepsilon_m = \delta$, with $\varepsilon = 0$ for all $i \neq m$; we suppose that the electron bulk velocity is constant and $u > 0$. Then at point $(m+1)$, downstream of the perturbation, we have

$$\frac{b_{m+1}^{n+1} - b_{m+1}^n}{\Delta t} = -\frac{(0 - u\delta)}{2\Delta x} = +\frac{u\delta}{2\Delta x}, \tag{5.80}$$

which satisfies the transport property. But, at point m, where the perturbation of the magnetic field occurs,

$$\frac{b_m^{n+1} - b_m^n}{\Delta t} = -\frac{(0 - 0)}{2\Delta x} = 0, \tag{5.81}$$

which is not reasonable. More importantly that at point $i = m - 1$ upstream of the disturbance

$$\frac{b_{m-1}^{n+1} - b_{m-1}^n}{\Delta t} = -\frac{(u\delta - 0)}{2\Delta x} = -\frac{u\delta}{2\Delta x}. \tag{5.82}$$

Thus, a perturbation effect appears upstream of the perturbation and the transportive property is violated. At the next time step, a positive disturbance will appear at point $m - 2$, and so forth.

We can compare the property of this solution with the upwind scheme. Let us suppose again that $u > 0$. Then we have

$$\frac{b_i^{n+1} - b_i^n}{\Delta t} = -\frac{ub_i^n - ub_{i-1}^n}{\Delta x}. \tag{5.83}$$

Then for $\varepsilon = \delta$ as before, at the point $m+1$, downstream of the disturbance, we have

$$\frac{b_{m+1}^{n+1} - b_{m+1}^n}{\Delta t} = -\frac{(0 - u\delta)}{\Delta x} = +\frac{u\delta}{\Delta x}, \tag{5.84}$$

which is reasonable. At point m we have

$$\frac{b_m^{n+1} - b_m^n}{\Delta t} = -\frac{(u\delta - 0)}{\Delta x} = -\frac{u\delta}{\Delta x}, \tag{5.85}$$

which means that the perturbation is being transported out of the affected region, as it should. Finally, at the point $m - 1$, upstream of the disturbance, we have

$$\frac{b_{m-1}^{n+1} - b_{m-1}^n}{\Delta t} = \frac{(0 - 0)}{\Delta x} = 0, \tag{5.86}$$

which indicates that no perturbation effect is carried upstream. Thus, this method possesses the transportive property. It maintains the "unidirectional flow of information" [162].

"Upstream differencing" is not equivalent to the transportive property. Let us consider the simple example of upstream-differencing scheme which was used for the two-dimensional flow problem [446] to illustrate the violation of the transportive property. We suppose that the fluxes are defined by space averages in both directions. For clarity, we assume that the electron velocity components are constant

$$\frac{\partial b}{\partial t} = -u\frac{\partial b}{\partial x} - v\frac{\partial b}{\partial y}, \tag{5.87}$$

$$\frac{b_{i,j}^{n+1} - b_{i,j}^n}{\Delta t} = -u\frac{\hat{b}_R - \hat{b}_L}{\Delta x} - v\frac{\hat{b}_T - \hat{b}_B}{\Delta y}, \tag{5.88}$$

where u and v are the electron bulk velocity components. The upwind differencing has been written for positive velocity components; the average magnetic field components are defined as follows:

$$b_{\rm R} = \frac{b_{i,j+1} + 2b_{i,j} + b_{i,j-1}}{4},$$

$$b_{\rm L} = \frac{b_{i-1,j+1} + 2b_{i-1,j} + b_{i-1,j-1}}{4},$$

$$b_{\rm T} = \frac{b_{i-1,j} + 2b_{i,j} + b_{i+1,j}}{4}, \tag{5.89}$$

$$b_{\rm B} = \frac{b_{i-1,j-1} + 2b_{i,j-1} + b_{i+1,j-1}}{4}.$$

In this method, the upwind cell property $b_{\rm R}$ is the parabolic weighted average over the column from which b is transported. Now we consider a steady-state solution with all $v = 0$, and introduce a magnetic field perturbation $\varepsilon_{a,b} = \delta$; all other $\varepsilon = 0$. Then, at point $(a, b-1)$,

$$\frac{b_{a,b-1}^{n+1} - b_{a,b-1}^{n}}{\Delta t} = -u \frac{\delta/4 - 0}{\Delta x} - 0 = -u\delta/4. \tag{5.90}$$

Thus, the perturbation effect at point (a, b) is transported in the v-direction to $(a, b-1)$, even though the v-velocity is equal to zero. The method violates the transportive property, although it is based on a kind of upstream-differencing scheme. Space-centered differences are more accurate than the upwind one-sided differences insofar as the formal Taylor series expansion indicates. As discussed in Sect. III-A-3 of [445] with regard to the conservative property, it is also possible to more accurately represent a derivative by using a nonconservative method, but the whole system is not more accurate if one's criterion for accuracy includes the conservative property.

The transportive property appears to be as physically significant as the conservative property. Upstream-differencing schemes which possess the transportive property are more accurate in this sense, although not in the sense of order of the truncation error, than schemes with space-centered first derivatives. Accuracy in the finite-difference representation of the differentials is not equivalent to accuracy in representation of the differential equations. The transportive property should be kept in mind when one attempts to design a finite-difference scheme with advection terms. One way to increase the approximation to advection terms upto the third order is to use the non-central upwind compact difference scheme (see, e.g., [528]). The conservation property is usually violated at the sonic point, and it may cause serious error. However, there are ways to construct the scheme which provide second-order accuracy near sonic points and third-order accuracy elsewhere. The reduction in the approximation order near sonic points does not significantly influence the accuracy in the L_2-norm if the number of such points is not large in comparison with the total amount of nodes [528]. Here, we can point out the class of essentially nonoscillatory (ENO) schemes (see, e.g., [1, 218]).

We can also point out some schemes which are conservative as well as transportive: (a) Explicit schemes – one-step, two-time-level upstream-differencing method [446]; "donor cell" method [186]; one-step, three-time-level, forward-time scheme [309]; one-step, two-level, second-order-accurate method [298]. (b) Implicit schemes [247, 502]. (c) Alternating direction implicit (ADI) and Alternating direction explicit (ADE) methods (see references in [445]).

5.5 High-Order Schemes

Many physical phenomena possess a range of space and time scales. Turbulent processes are a common example of such phenomena. Direct numerical simulation of these processes requires all the relevant scales to be properly represented in the numerical model. These requirements have led to the development of spectral methods (see, e.g., [88, 203]). The use of spectral methods is, however, limited to problems in simple domains and with simple boundary conditions. These difficulties may be overcome by employing alternative numerical representations, in particular, high-order upwind-biased finite-difference schemes [440] and spectral element methods [187, 263, 419].

The high precision of schemes with high-order approximation, which is achieved in smooth solutions, stimulates the development of schemes with an order of more than two. Such schemes may be classified as follows: multipoint schemes, schemes which use the differential consequences of the initial equations and compact approximation schemes.

5.5.1 Multipoint Stencil Schemes

The straightforward way to design a high-order finite-difference scheme is to include more grid points in its stencil. However, this method is not effective in the general case. For implicit schemes, the inversion of the matrices with an increased number of nonzero diagonals may be too costly. The use of an explicit scheme is disadvantageous due to its poor stability. In both cases, a considerable number of spurious solutions may cause a high manifestation of nonphysical oscillations.

However, the reasonable use of high-order difference formulae allows one to design successfully third- and fourth-order schemes for hyperbolic equations, and in particular, Euler equations [23, 79, 455, 528, 549]. All of them can be combined in a many-parameter family of schemes based on the Runge–Kutta method. It is also worth mentioning that the third-order scheme [23] coincides with the maximum-order scheme for a four-point stencil, in the linear case [504]. At the boundaries one has to use the nonstandard schemes near-boundary node, which may results in additional difficulties. The efficient use of high-order approximations defined on a many-point stencil was demonstrated in the excellent work in [218], where high-resolution schemes for hyperbolic conservation laws were proposed. The idea of an adaptive stencil incorporated into these schemes permits solutions of a very high quality to be obtained.

5.5.2 Differential Consequences from the Governing Equations

As a starting point for constructing high-order schemes, one may take advantage of the fact that the exact solution of a difference equation rather than an arbitrary function is used in the definition of the approximation. This means that some terms in the truncation error may be found to be zero if the functions involved in these terms satisfy the exact problem. A two-level, three-point weighted scheme which approximates the heat conduction equation to the fourth order for the special choice of the weight factor is a typical example.

Another approach is to make use of additional equations obtained via differentiation of the equations which are to be solved [532].

Consider, for example, the equation

$$\frac{\partial u}{\partial t} = \frac{\partial F(u)}{\partial x}. \tag{5.91}$$

Differentiating it with respect to x gives the prolongated system [451]

$$\frac{\partial u_x^{(k)}}{\partial t} = \frac{\partial^k F(u)}{\partial x^k}, \quad k = 0, 1, ..., n, \tag{5.92}$$

where $u_x^{(k)}$ denotes $\partial^k u / \partial x^k$. The derivative $\partial^k F(u)/\partial x^k$ in (5.92) is the known function of $u = u_x^{(0)}, u_x^{(1)}, ..., u_x^{(k)}$. Therefore, it is possible to approximate each equation of system (5.92) by some suitable difference scheme [e.g., by the same scheme as (5.91)] and obtain $u_x^{(k)}$, $k = 1, ..., n$ at every time level. These high-order derivatives may then be utilized to annihilate the low-order truncation error for (5.91). In general, this approach permits the construction of difference schemes of an arbitrary order. However, this technique becomes excessively tedious for large n.

The similar ideas were proposed in [206], the slight difference being due to the way $u_x^{(k)}$ was used. The idea of using differentiation of the governing equations to obtain higher-order schemes was also exploited in [98], when constructing fourth- and sixth-order methods for the convection–diffusion equation.

5.5.3 Compact Schemes with Spectral-Like Resolution

In this subsection we shall consider finite-difference schemes for use on problems with a range of spatial scales. Compared to the traditional finite-difference approximations, the schemes presented here provide a better representation of the shorter length scales. This feature brings them closer to the spectral methods, while the freedom to choose the mesh geometry and the boundary conditions is maintained.

5.5.3.1 Approximation of the First Derivative.

For simplicity consider a uniformly spaced mesh where the nodes are indexed by i (see, e.g., [99, 299, 357, 528]). Examples of compact schemes which use an inhomogeneous mesh may be found in [196]. The independent variables at the nodes are $x_i = h(i-1)$ for $1 \leq i \leq N$, and the function values at the nodes $f_i = f(x_i)$ are given. The finite-difference approximation f'_i to the first derivative $(\mathrm{d}f/\mathrm{d}x)(x_i)$ at the node i depends on the function values at the nodes near i. For second- and fourth-order central differences, the approximation f'_i depends on the sets (f_{i-1}, f_{i+1}) and $(f_{i-2}, f_{i-1}, f_{i+1}, f_{i+2})$, respectively. In the spectral methods, however, the value of f'_i depends on all the nodal values. The Padé or compact finite-difference schemes [4, 235, 281, 299, 363, 452, 528] mimic this global dependence.

These generalizations are derived by writing an approximation of the following form:

$$\beta f'_{i-2} + \alpha f'_{i-1} + f'_i + \alpha f'_{i+1} + \beta f'_{i+2}$$
$$= c\frac{f_{i+3} - f_{i-3}}{6h} + b\frac{f_{i+2} - f_{i-2}}{4h} + a\frac{f_{i+1} - f_{i-1}}{2h}. \quad (5.93)$$

The relations between the coefficients a, b, c and α, β are derived by matching the Taylor series coefficients of various orders. The first unmatched coefficient determines the formal truncation error of the approximation (5.93). These constraints are

$$a + b + c = 1 + 2\alpha + 2\beta \quad \text{(second order)}, \quad (5.94)$$

$$a + 2^2 b + 3^2 c = 2\frac{3!}{2!}(\alpha + 2^2 \beta) \quad \text{(fourth order)}, \quad (5.95)$$

$$a + 2^4 b + 3^4 c = 2\frac{5!}{4!}(\alpha + 2^4 \beta) \quad \text{(sixth order)}, \quad (5.96)$$

$$a + 2^6 b + 3^6 c = 2\frac{7!}{6!}(\alpha + 2^6 \beta) \quad \text{(eighth order)}, \quad (5.97)$$

$$a + 2^8 b + 3^8 c = 2\frac{9!}{8!}(\alpha + 2^8 \beta) \quad \text{(tenth order)}. \quad (5.98)$$

If the dependent variables are periodic in x, then the system of relations (5.93–5.98) written for each node can be solved together as a linear system of equations for the unknown derivative values. This linear system is a cycle pentadiagonal (tridiagonal) when β is nonzero (zero). The general nonperiodic case requires additional relations appropriate for the linear boundary nodes. The above relations (5.93–5.98), along with a mathematically defined mapping between a nonuniform physical mesh and a uniform computational mesh, provides derivatives on a nonuniform mesh. It is also possible to derive relations analogous to (5.93–5.98) for a nonuniform mesh directly.

An analysis of the dispersive errors of schemes (5.93) [299] shows the improved representation of the shorter length scales (i.e., spectral-like resolution) of the schemes presented here. The analysis also leads to schemes with

5.5 High-Order Schemes

very small dispersive errors (almost spectral). Let us consider the tridiagonal schemes ($\beta = 0$). If a choice of $c = 0$ is made, a one-parameter (α) family of fourth-order tridiagonal schemes is obtained. For these schemes we have

$$\beta = 0, \quad a = \frac{2}{3}(\alpha + 2), \quad b = \frac{1}{3}(4\alpha - 1), \quad c = 0. \tag{5.99}$$

As $\alpha \leftarrow 0$, this family merges into the well-known fourth-order central difference scheme. Similarly, for $\alpha = 1/4$, the classical Padé scheme is recovered:

$$f'_{i-1} + 4f'_i + f'_{i+1} = 6\frac{f_{i+1} - f_{i-1}}{2h}. \tag{5.100}$$

Furthermore, for $\alpha = 1/3$, the leading-order truncation error coefficient vanishes, and the scheme is formally accurate to the sixth-order. The specific tridiagonal schemes obtained for $\alpha = 1/4$ and $\alpha = 1/3$ have been given in [113].

Let us consider now the finite-difference approximation to the first derivative at the boundary. The first derivative at the boundary $i = 1$ may be obtained from a relation of the form

$$f'_1 + \alpha f'_2 = \frac{1}{h}(af_1 + bf_2 + cf_3 + df_4), \tag{5.101}$$

coupled to (5.93) written for the interior nodes. Requiring (5.101) to be at least accurate to the second order constrains the coefficients to

$$a = -\frac{3 + \alpha + 2d}{2}, \quad b = 2 + 3d, \quad c = -\frac{1 - \alpha + 6d}{2}. \tag{5.102}$$

If higher-order formal accuracy is desired, schemes with an accuracy to the third and fourth orders may be derived. These are given in [299]:

$$a = -\frac{11 + 2\alpha}{6}, \quad b = \frac{6 - \alpha}{2},$$

$$c = \frac{2\alpha - 3}{2}, \quad d = \frac{2 - \alpha}{6} \quad \text{(third order)}, \tag{5.103}$$

$$\alpha = 3, \quad a = -\frac{17}{6}, \quad b = \frac{3}{2},$$

$$c = \frac{3}{2}, \quad d = -\frac{1}{6} \quad \text{(fourth order)}. \tag{5.104}$$

The leading-order truncation error [on the right hand side of (5.101)] for these boundary approximations is given by $[(2 - \alpha - 6d)/3!]h^2 f_i^{(3)}$ for the second-order schemes, by $[2(\alpha - 3)/4!]h^3 f_i^{(4)}$ for the third-order schemes and by $(6/5!)h^4 f_i^{(5)}$ for fourth-order schemes. It may be noted that for the even-order schemes the leading-order truncation error is of dispersive type, while for the third-order schemes it is dissipative. A detailed analysis of higher-order approximations may be found in [299].

5.5.3.2 Approximation of the Second Derivative.

The derivation of compact approximations for the second derivative proceeds exactly analogously to the first derivative. Once again, the point is a relation of the form

$$\beta f''_{i-2} + \alpha f''_{i-1} + f''_i + \alpha f''_{i+1} + \beta f''_{i+2}$$
$$= c\frac{f_{i+3} - 2f_i + f_{i-3}}{9h^2} + b\frac{f_{i+2} - 2f_i + f_{i-2}}{4h^2}$$
$$+ a\frac{f_{i+1} - 2f_i + f_{i-1}}{h^2}, \quad (5.105)$$

where f''_i represents the finite-difference approximation to the second derivative at node i. The relations between the coefficients a, b, c and α, β are derived by matching the Taylor series coefficients of various orders. The first unmatched coefficient determines the formal truncation error of the approximation (5.105). These constraints are

$$a + b + c = 1 + 2\alpha + 2\beta \quad \text{(second order)}, \quad (5.106)$$

$$a + 2^2 b + 3^2 c = \frac{4!}{2!}(\alpha + 2^2\beta) \quad \text{(fourth order)}, \quad (5.107)$$

$$a + 2^4 b + 3^4 c = \frac{6!}{4!}(\alpha + 2^4\beta) \quad \text{(sixth order)}, \quad (5.108)$$

$$a + 2^6 b + 3^6 c = \frac{8!}{6!}(\alpha + 2^6\beta) \quad \text{(eighth order)}, \quad (5.109)$$

$$a + 2^8 b + 3^8 c = \frac{10!}{8!}(\alpha + 2^8\beta) \quad \text{(tenth order)}. \quad (5.110)$$

The form of these constraints is very close to those derived for the first derivative approximations, but the multiplying factors on the right hand side are different. By choosing $\beta = 0$ and $c = 0$, a one-parameter family of fourth-order schemes is generated:

$$\beta = 0, \quad c = 0, \quad a = \frac{4}{3}(1 - \alpha), \quad b = \frac{1}{3}(-1 + 10\alpha). \quad (5.111)$$

As $\alpha \leftarrow 0$, this family coincides with the well-known fourth-order central difference scheme. For $\alpha = 1/10$, the classical Padé scheme is recovered:

$$f''_{i-1} + 10 f''_i + f''_{i+1} = 12\frac{f_{i+1} - 2f_i + f_{i-1}}{h^2}. \quad (5.112)$$

For $\alpha = 2/11$, a sixth-order tridiagonal scheme is obtained. The specific tridiagonal schemes obtained for $\alpha = 1/10$ and $\alpha = 2/11$ were given in [113]. The detail analysis of higher-order approximations may be found in [299].

5.5 High-Order Schemes

5.5.3.3 Compact Schemes on a Cell-Centered Mesh. This section presents the compact finite-difference schemes for the first and second derivatives on a cell-centered mesh. The nodes, on which the derivatives are evaluated below, are staggered by a half-cell ($h/2$) from the nodes on which the function values are prescribed. Such grid configurations arise naturally from a finite-volume discretization of conservation equations.

Let us first consider a first derivative. Starting from an approximation of the form

$$\beta f'_{i-2} + \alpha f'_{i-1} + f'_i + \alpha f'_{i+1} + \beta f'_{i+2}$$
$$= c\frac{f_{i+5/2} - f_{i-5/2}}{5h} + b\frac{f_{i+3/2} - f_{i-3/2}}{3h} + a\frac{f_{i+1/2} - f_{i-1/2}}{h}, \quad (5.113)$$

the constraints on the coefficients are derived by matching the Taylor series coefficients (at least up to fourth order). Tridiagonal schemes analogous to the standard Padé scheme are obtained with $\beta = 0$, $c = 0$. These fourth-order schemes are defined in [299]

$$\beta = 0, \quad a = \frac{3}{8}(3 - 2\alpha), \quad b = \frac{1}{8}(22\alpha - 1). \quad (5.114)$$

The truncation error (on the r.h.s. of (5.113)) is $((9 - 62\alpha)/1920)h^4 f^{(5)}$.

For $\alpha = 1/22$ the coefficient b vanishes, generating the most compact scheme. For $\alpha = 9/62$ a sixth-order tridiagonal scheme is generated. As expected the schemes on the cell-centered mesh have considerably lower differencing errors compared to the unstaggered schemes (5.93). For example, the sixth-order tridiagonal scheme ($\alpha = 9/62$ with (5.114)) is quite close to exact differentiation [299].

To obtain compact approximations for the second derivatives we start from the equation

$$\beta f''_{i-2} + \alpha f''_{i-1} + f''_i + \alpha f''_{i+1} + \beta f''_{i+2}$$
$$= 4c\frac{f_{i+5/2} - 2f_i + f_{i-5/2}}{25h^2} + 4b\frac{f_{i+3/2} - 2f_i + f_{i-3/2}}{9h^2}$$
$$+ 4a\frac{f_{i+1/2} - 2f_i + f_{i-1/2}}{h^2}, \quad (5.115)$$

where the constraints on the coefficients are derived by matching the Taylor series coefficients (at least up to fourth order). Tridiagonal schemes analogous to the standard Padé scheme are obtained with $\beta = 0$, $c = 0$. These fourth-order schemes are defined by [299]

$$\beta = 0, \quad a = \frac{3}{8}(3 - 10\alpha), \quad b = \frac{1}{8}(46\alpha - 1). \quad (5.116)$$

The truncation error (on the r.h.s. of (5.115)) is $1/640(1 + 2\alpha)h^4 f^{(6)}$. For $\alpha = 1/46$ the coefficient b vanishes generating the standard Padé scheme.

For $\alpha = -1/2$ a sixth-order tridiagonal scheme is generated. Thus on a cell-centered mesh compact schemes can provide almost exact second derivatives. Note that the use of schemes on the cell-centered meshes for solution of the electromagnetic equations may provide much better resolution than the use of unstaggered schemes.

5.5.3.4 Fourier Analysis of Differencing Errors. For the purposes of Fourier analysis the dependent variables are assumed to be periodic over the domain $[0, L]$ of the independent variable, i.e. $f_1 = f_{N+1}$ and $h = L/N$ [299]. The dependent variables may be decomposed into Fourier coefficients

$$f(x) = \sum_{k=-N/2}^{k=N/2} \hat{f}_k \exp\left(\frac{2\pi ikx}{L}\right), \qquad (5.117)$$

where $i = \sqrt{-1}$. Since the dependent variables are real-valued, the Fourier coefficients satisfy $\hat{f}_k = \hat{f}^*_{-k}$ for $1 \le k \le N/2$ and $\hat{f}_0 = \hat{f}^*_0$, where $*$ denotes the complex conjugate.

It is convenient to introduce a scaled wavenumber $w = 2\pi kh/L = 2\pi k/N$ and a scaled coordinate $s = x/h$. The Fourier modes in terms of these are simply $\exp(iws)$. The domain of the scaled wavenumber w is $[0, \pi]$. The exact first derivative of (5.117) (with respect to s) generates a function with Fourier coefficients $\hat{f}'_k = iw\hat{f}_k$. The differencing error of the first derivative scheme may be assessed by comparing the Fourier coefficients of the derivative obtained from the differencing scheme $(\hat{f}'_k)$ with the exact Fourier coefficients $\hat{f}'_k$. For the central difference schemes it may be shown that $(\hat{f}'_k)_{fd} = iw'\hat{f}_k$, where the modified wavenumber w' is real-valued [299]. Each finite-difference scheme corresponds to a particular function $w'(w)$. Exact differentiation corresponds to the straight line $w' = w$. Spectral methods provide $w' = w$ for $w \ne \pi$ (and $w' = 0$ for $w = \pi$). The range of wavenumbers $[2\pi/N, w_f]$ over which the modified wavenumber $w'(w)$ approximates the exact differentiation $w'(w) = w$ within a spectral error tolerance defines the set of well-resolved waves. It should also be noted that w_f depends only on the scheme and not on the number of points N used in the discretization. The modified wavenumber w' can alternatively determine the dispersive error in terms of the error in the phase speed of waves of different wavenumbers.

The finite-difference approximations of the first derivative (5.93–5.98) correspond to

$$w'(w) = \frac{a\sin(w) + (b/2)\sin(2w) + (c/3)\sin(3w)}{1 + 2\alpha\cos(w) + 2\beta\cos(2w)}. \qquad (5.118)$$

The finite-difference approximations of the first derivative on a cell-centered mesh (5.113) correspond to

$$w'(w) = 2\frac{a\sin(w/2) + (b/3)\sin(3/2w) + (c/5)\sin(5/2w)}{1 + 2\alpha\cos(w) + 2\beta\cos(2w)}. \qquad (5.119)$$

As expected the schemes on the cell-centered mesh have considerably lower differencing errors compared to the unstaggered schemes (5.93). For example, the sixth-order tridiagonal scheme ($\alpha = 9/62$ with (5.114)) is quite close to exact differentiation (see Figs. 1, 2 and 16 from [299]). The further analysis demonstrates that compact schemes on a cell-centered mesh can provide almost exact second derivatives (see Figs. 5 and 17 from [299]).

5.5.4 Advection and Diffusion Equations

Now we consider the time step limits which need to be maintained for a numerically stable time advancement (Runge–Kutta scheme). In order to make the analysis simple, the stability bounds are presented only for model problems with periodic boundary conditions. The analysis is restricted to the case of pure advection and pure diffusion operators.

First we consider the pure advection case (in a periodic domain):

$$\frac{\partial f}{\partial t} + c\frac{\partial f}{\partial x} = 0. \tag{5.120}$$

The stability limit in this case is given in [299, 540]:

$$\frac{c\Delta t}{\Delta x} \leq \frac{\sigma_i}{w'_m}, \tag{5.121}$$

where $[-i\sigma_i, i\sigma_i]$ is the segment of the imaginary axis in the stable region for the time-advancement scheme and w'_m is the maximum value of the modified wavenumber for the first derivative operator (Fig. 1, in [299]). In particular, for the Padé scheme the value of w'_m is 1.732 (or $\sqrt{3}$). The values of σ_r for the second-, third- and fourth-order Runge–Kutta schemes are 0, $\sqrt{3}$ and 2.85, respectively.

The modified wavenumber w' can alternatively determine the dispersive error in terms of the error in the phase speed of waves of different wavenumbers. It may be shown by considering the semi-discrete (exact time advancement) form of the advection equation (5.120). The phase speed for a wave of wavenumber w is given by the finite-difference scheme as $(c_{ph})_{fd} = w'(w)/w$. The partial differential equation (5.120) has a phase speed of 1 for all wavenumbers, thus $(c_{ph})_{fd} - 1$ is the measure of the phase error [88]. Our analysis demonstrates the improved phase error of the compact schemes. The use of the third-order compact upwind differencing schemes will be considered in the exercises.

Next we consider now the pure diffusion case (in a periodic domain):

$$\frac{\partial f}{\partial t} = \nu \frac{\partial^2 f}{\partial x^2} = 0. \tag{5.122}$$

The stability limit in this case is given in [299, 540].

$$\frac{\nu \Delta t}{(\Delta x)^2} \leq \frac{\sigma_\text{r}}{w''_\text{m}}, \tag{5.123}$$

where $[-\sigma_\text{r}, 0]$ is the segment of the real axis in the stable region for the time-advancement scheme and w''_m is the maximum value of w'' for the second derivative operator (Fig. 5 in [299]). In particular, for the Padé scheme the value of w''_m is 6.0. The values of σ_r for the second-, third- and fourth-order Runge–Kutta schemes are 2, 2.5 and 2.9, respectively. The use of the third-order compact upwind differencing schemes will be considered in the exercises.

5.5.5 Maxwell's Equations

The relevant time-dependent Maxwell equations for the electromagnetic field in the radiative limit can be written in the following form:

$$\frac{1}{c}\frac{\partial \boldsymbol{B}}{\partial t} + \nabla \times \boldsymbol{E} = 0, \tag{5.124}$$

$$\frac{1}{c}\frac{\partial \boldsymbol{E}}{\partial t} - \nabla \times \boldsymbol{B} = -\frac{4\pi}{c}\boldsymbol{J}. \tag{5.125}$$

The system of equations expressed in vector form on a Cartesian frame is as follows:

$$\frac{\partial \boldsymbol{U}}{\partial t} + A\frac{\partial \boldsymbol{U}}{\partial x} + B\frac{\partial \boldsymbol{U}}{\partial y} + C\frac{\partial \boldsymbol{U}}{\partial z} = -4\pi \boldsymbol{J}, \tag{5.126}$$

where the coefficient matrices (Jacobian of flux vector) A, B and C are

$$A = \begin{bmatrix} 0 & 0 & 0 & 0 & 0 & 0 \\ 0 & 0 & 0 & 0 & 0 & c \\ 0 & 0 & 0 & 0 & -c & 0 \\ 0 & 0 & 0 & 0 & 0 & 0 \\ 0 & 0 & -c & 0 & 0 & 0 \\ 0 & c & 0 & 0 & 0 & 0 \end{bmatrix}, \tag{5.127}$$

$$B = \begin{bmatrix} 0 & 0 & 0 & 0 & 0 & -c \\ 0 & 0 & 0 & 0 & 0 & 0 \\ 0 & 0 & 0 & c & 0 & 0 \\ 0 & 0 & c & 0 & 0 & 0 \\ 0 & 0 & 0 & 0 & 0 & 0 \\ -c & 0 & 0 & 0 & 0 & 0 \end{bmatrix}, \tag{5.128}$$

$$C = \begin{bmatrix} 0 & 0 & 0 & 0 & c & 0 \\ 0 & 0 & 0 & -c & 0 & 0 \\ 0 & 0 & 0 & 0 & 0 & 0 \\ 0 & -c & 0 & 0 & 0 & 0 \\ c & 0 & 0 & 0 & 0 & 0 \\ 0 & 0 & 0 & 0 & 0 & 0 \end{bmatrix}, \tag{5.129}$$

$$\boldsymbol{U} = [E_x, E_y, E_z, B_x, B_y, B_z], \tag{5.130}$$

$$\boldsymbol{J} = [J_x, J_y, J_z, 0, 0, 0]^\mathrm{T}. \tag{5.131}$$

One may to use Runge–Kutta schemes for time integration. The adopted four-stage method can be expressed as

$$U^{n=1} = U^n + \frac{\Delta t}{6}(U'_1 + 2U'_2 + U'_3 + U'_4),$$

$$U'_1 = U'_1(t, U^n),$$
$$U'_2 = U'_2(t + \Delta t/2, U^n + \Delta t/2 \cdot U'_1),$$
$$U'_3 = U'_3(t + \Delta t/2, U^n + \Delta t/2 \cdot U'_2),$$
$$U'_4 = U'_4(t + \Delta t/2, U^n + \Delta t/2 \cdot U'_3). \tag{5.132}$$

Spatial derivatives are evaluated by compact finite-difference schemes [482]. The numerical procedure is implicit in the derivative evaluation and requires additional data at the boundaries.

Numerical stability is a critical concern for compact-difference methods, which must include a stable boundary scheme to preserve the global accuracy [91]. Usually the numerical boundary value is obtained by a one-sided difference approximation. For example, for a sixth-order or higher scheme, the stencil dimension involves five points or more and a boundary scheme becomes necessary. The basic idea of using the staged one-order-lower approximation for the boundary scheme is derived from the summation-by-parts energy-mode analysis [91, 209]. Therefore, a formally sixth-order interior scheme must be supported by a fifth-order boundary scheme. In most practical applications, this procedure is the source of numerical instability manifested in spurious high-frequency oscillations. Fourth-order and fifth-order explicit one-sided difference formulae are given as [482]

$$U'_i = (-25U_i + 48U_{i+1} - 36U_{i+2} + 12U_{i+3} - 3U_{i+4})/12h,$$

$$U'_i = (25U_i - 48U_{i-1} + 36U_{i-2} - 12U_{i-3} + 3U_{i-4})/12h, \tag{5.133}$$

$$U'_i = (-137U_i + 300U_{i+1} - 300U_{i+2} + 200U_{i+3} - 75U_{i+4} + 12U_{i+5})/12h,$$

$$U'_i = (137U_i - 300U_{i+1} + 300U_{i+2} - 200U_{i+3} + 75U_{i+4} - 12U_{i+5})/12h. \tag{5.134}$$

Another method for time integration is to use an approximation that is accurate to the second order in time. The standard leapfrog scheme was used in [424], whereas a fractional-step method was used for solving three-dimensional problems [481]. One of the main properties of numerical schemes is to produce an essentially nonoscillatory solution in the vicinity of shocks and discontinuities. In these algorithms we have to exploit the non-symmetrical Padé operators to approximate the convective terms. These schemes have a small numerical dissipation which damps the spurious oscillation upstream and downstream of the shocks and discontinuities (see, e.g., [442, 481, 529, 556]).

The relevant time-dependent Maxwell equations for the electromagnetic field in an hybrid approximation can be written in the following form:

$$\frac{1}{c}\frac{\partial \boldsymbol{B}}{\partial t} + \nabla \times \boldsymbol{E} = 0, \qquad (5.135)$$

$$\boldsymbol{E} = \boldsymbol{E}(\boldsymbol{B}, \boldsymbol{J}_\mathrm{i}, n_\mathrm{i}, \nabla p_\mathrm{e}). \qquad (5.136)$$

One can approximate the first derivatives in (5.135–5.136) by the formulae in Sect. 5.5.3. Another way is to substitute (5.136) into (5.135) and then to approximate the convective term, Hall's term and the resistive term using the compact finite-difference operators presented in Sect. 5.5.3.

5.5.6 Filtering of Spurious Oscillations

The existence of a smooth solution is the underlying assumption when constructing high-order centered-difference approximations. This assumption is obviously violated at shocks and contact discontinuities. Spurious oscillations can be expected. One way to overcome this problem is to introduce viscosity as a filter [209]. The same technique was used in fluid dynamics by using a so-called dissipator, which smooths out the strong gradients in solution while keeping away from the smooth parts (see, e.g., [425]). Let $\tilde{B}^{n+1}$ denote the output of (5.126) or (5.135). As a preliminary step in the derivation of the filter, we define the new time level as

$$B_i^{n+1} = \frac{1}{4}\left(\tilde{B}_{i-1}^{n+1} + 2\tilde{B}_i^{n+1} + \tilde{B}_{i+1}^{n+1}\right)$$

$$= \tilde{B}_i^{n+1} + \frac{1}{4}\Delta_+\Delta_-\tilde{B}_i^{n+1}, \quad i = 1, 2, ..., N-1. \qquad (5.137)$$

All points will be filtered as the scheme stands above. To avoid this effect, we introduce a switch r_i to turn off the filter outside the spurious region:

$$B_i^{n+1} = \tilde{B}_i^{n+1} + \frac{1}{4}\Delta_+\left(r_{i-1/2}\Delta_-\tilde{B}_i^{n+1}\right), \quad i = 1, 2, ..., N-1. \qquad (5.138)$$

One can use here a switch proposed by Jameson (private communication):

$$r_i = \left(\frac{|\Delta_+\tilde{B}_i - \Delta_-\tilde{B}_i|}{|\Delta_+\tilde{B}_i + \Delta_-\tilde{B}_i|}\right)^m, \quad i = 1, 2, ..., N-1, \qquad (5.139)$$

where we have omitted the time index $n+1$ in the right member for simplicity. If $\Delta_+\tilde{B}_i$ and $\Delta_-\tilde{B}_i$ do not have the same sign, which happens for high-frequency oscillations, then $r_i = 1$. For the remaining grid points, we obviously have $0 \leq r_i < 1$. At $m = \infty$, $r_i = 0$ away from the spurious regions. The complete numerical algorithm is thus defined by (5.137–5.139).

In the one-dimensional case the compact scheme reduces the number of grid points by a factor of two for the explicit fourth-order method and by a factor of three for the implicit operator. This is true for each space dimension. Thus, in three dimensions one would obtain a reduction by a factor of 8 or 27, respectively. Since the work grows linearly, it is natural to assume that high-order methods would be even more efficient for multidimensional problems [209]. On the other hand the use of large sizes of cell results in a reduction of the shot noise by factor $1/\sqrt{27}$ for the same number of particles in particle-mesh simulation. Finally, for the same accurate approximation and shot noise level the considered three-dimensional compact schemes give a reduction of grid points by a factor of 27 and the number of particles by a factor of 27 in particle-mesh simulations.

5.6 Time Integration of the Equations for Electromagnetic Potentials

In Sect. 2.2.2 we derived electromagnetic equations for a hybrid Darwin model in terms of the scalar and vector potentials ϕ and $\boldsymbol{A}$, respectively (as in [231]). This approach is very important when the plasma currents are connected to an external circuit. The resulting equations in the Coulomb gauge are

$$\boldsymbol{B} = \nabla \times \boldsymbol{A}, \tag{5.140}$$

where

$$\nabla^2 \boldsymbol{A} = -\frac{4\pi}{c} \boldsymbol{J}_t, \tag{5.141}$$

$$\nabla^2 \boldsymbol{E}_t = \frac{4\pi}{c^2} \dot{\boldsymbol{J}}_t, \tag{5.142}$$

and

$$\boldsymbol{E}_l = -\nabla \phi, \tag{5.143}$$

where

$$\nabla \cdot (\mu \nabla \phi) = \nabla \cdot (\boldsymbol{D} + \mu \boldsymbol{E}_t + \boldsymbol{\xi} \times \boldsymbol{B}), \tag{5.144}$$

where t and l denote the transversal and longitudinal components.

The total current and its derivability in time may be estimated by the moment equations [see (2.56), Sect. 2.2.3]:

$$(\dot{\boldsymbol{J}}_i + \dot{\boldsymbol{J}}_e) = \boldsymbol{D} + \mu(\boldsymbol{E}_l + \boldsymbol{E}_t) + \boldsymbol{\xi} \times \boldsymbol{B}, \tag{5.145}$$

where

$$\mu = \frac{\omega_{pe}^2 + \omega_{pi}^2}{4\pi},$$

$$\xi = \frac{e}{c}\left(\frac{Z_i \boldsymbol{J}_i}{M_i} - \frac{\boldsymbol{J}_e}{m}\right),$$

$$\boldsymbol{D} = -Z_i e \nabla \cdot \mathsf{K}_i + e \nabla \cdot \mathsf{K}_e,$$

and $\mathsf{K}_{i(e)}$ is the ion (electron) kinetic energy tensor.

The time advance of $\boldsymbol{J}_{et}$ is accomplished as follows. First, the temporary vector $\boldsymbol{J}_{es}$ is defined by

$$\boldsymbol{J}_{es} = \boldsymbol{J}_{el}^n + \boldsymbol{J}_{et}^{n-1/2}.$$

Then, the transverse current may be advanced by the two steps

$$\boldsymbol{J}_e^n = \boldsymbol{J}_{es} + 0.5\Delta t [\dot{\boldsymbol{J}}_e(\mathsf{K}_e^n, n_e^n, \boldsymbol{E}^n, \boldsymbol{J}_{es}, \boldsymbol{B}^n)]_t, \qquad (5.146)$$

$$\boldsymbol{J}_{et}^{n+1/2} = \boldsymbol{J}_{et}^{n-1/2} + \Delta t [\dot{\boldsymbol{J}}_e(\mathsf{K}_e^n, n_e^n, \boldsymbol{E}^n, \boldsymbol{J}_e^n, \boldsymbol{B}^n)]_t, \qquad (5.147)$$

in which each evaluation of $\dot{\boldsymbol{J}}_e$ is obtained from (2.22). This scheme is accurate in Δt to the second order. Equation (5.80) advances the transverse part of the electron current the additional one half time step required so that the result of (5.146) is the total electron current at time $n\Delta t$. Equation (5.147) produces second-order accuracy by virtue of time centering.

Equation (5.144) serves the same purpose for the quasineutral hybrid model as Poisson's equation serves for other models not requiring quasineutrality. Updating the magnetic field, accomplished by solving (5.141) and (5.140), does not present any principle difficulties. This is accomplished for the total current, as it is for the time derivative of the electron current, by subtraction of the longitudinal part, $\boldsymbol{J}_t = \boldsymbol{J} - \boldsymbol{J}_l$. The decomposition is obtained from

$$\boldsymbol{J}_t = \boldsymbol{J} + \nabla V, \qquad (5.148)$$

where

$$\nabla^2 V = -\nabla \cdot \boldsymbol{J}, \qquad (5.149)$$

in which the boundary conditions on V are determined by specifying the normal component of $\boldsymbol{J}_l$ at the simulation boundaries.

Calculation of $\boldsymbol{E}_t$ is somewhat more complicated than the preceding calculation of the magnetic field. It has been demonstrated that the calculation of $\boldsymbol{E}_t$ needs to be fully implicit in time [392], since this model exhibits instantaneous propagation. To avoid finite differencing in time, it is therefore necessary to obtain the time derivative of the total current by summing the ion and electron momentum (5.145). Obtaining $\dot{\boldsymbol{J}}_t$ from $\dot{\boldsymbol{J}}$ is accomplished using the same procedure used for the right-hand side of the vector potential calculation (5.141), namely, by subtracting the longitudinal part:

$$\dot{\boldsymbol{J}}_t = (\dot{\boldsymbol{J}} - \dot{\boldsymbol{J}}_l) = (\dot{\boldsymbol{J}} + \nabla V), \qquad (5.150)$$

for which V is generated by

5.6 Time Integration of the Equations for Electromagnetic Potentials

$$\nabla^2 V = -\nabla \cdot \dot{\boldsymbol{J}}, \tag{5.151}$$

with the boundary conditions on V determined by specifying the normal component of $\dot{\boldsymbol{J}}$ at the simulation boundaries. Since the right hand side requires an explicit expression of $\boldsymbol{E}_t$, a fully implicit calculation of $\boldsymbol{E}_t$ requires iteration. To this end, the rapidly converging iteration scheme described in [392] is implemented to provide the solution to (5.142). One additional complication is that the quasineutral Poisson's equation (5.144) requires $\boldsymbol{E}_t$ as a source term and the equation for $\boldsymbol{E}_t$ (5.142) with source given by (5.150) requires $\boldsymbol{E}_l$ as a source term. A fully implicit solution requires, therefore, iteration over both $\boldsymbol{E}_l$ and $\boldsymbol{E}_t$ calculations. In practice, the code runs stably with two or three iterations over both $\boldsymbol{E}_l$ and $\boldsymbol{E}_t$ calculations.

Let us consider the solution of the field equations. We can write the field equations in general form:

$$\nabla^2 \boldsymbol{A} - \eta \boldsymbol{A} = \boldsymbol{\xi} - \nabla \chi, \tag{5.152}$$

$$\nabla \cdot \boldsymbol{A} = 0. \tag{5.153}$$

The solution of (5.152–5.153) may be found by means of successive overrelaxation [114, 131] in the case of no coupling due to $\nabla \chi$ and $\nabla \cdot \boldsymbol{A} = 0$. With the coupling of the components of $\boldsymbol{A}$ by the $\nabla \chi$ term and the $\nabla \cdot \boldsymbol{A} = 0$ requirement, we must complicate the iteration algorithm considerably. The reader must refer to [228, 392, 565, 567] for details.

Let us consider another approach for the field induction equation. For vector potential $\boldsymbol{A}$ we have

$$\frac{\partial \boldsymbol{A}}{\partial t} = \frac{c}{en_e}\nabla p_e + \boldsymbol{U}_i \times \boldsymbol{B} - \frac{\nu c^2}{\omega_{pe}^2}\nabla \times \boldsymbol{B} + \frac{c}{4\pi e n_e}\nabla \times \boldsymbol{B} \times \boldsymbol{B}. \tag{5.154}$$

The explicit finite-difference scheme may be written as

$$\frac{\boldsymbol{A}^{n+1} - \boldsymbol{A}^n}{\Delta t} = \frac{c}{en_e}\nabla p_e + \boldsymbol{U}_i \times \boldsymbol{B}^n - \frac{\nu c^2}{\omega_{pe}^2}\nabla \times \boldsymbol{B}^n + \frac{c}{4\pi e n_e}\nabla \times \boldsymbol{B}^n \times \boldsymbol{B}^n. \tag{5.155}$$

The two- and three-dimensional simulations with the above scheme (5.155) (for an arbitrary finite-difference approximation in space) show numerical stability for $\Delta t < \nu \Delta x^2 / B_0^2$. For very small values of ν or $\nu = 0$, the above scheme becomes useless [596].

Consider the other scheme suggested in [596]:

$$\boldsymbol{B}^{(1)} \equiv \boldsymbol{B}^n = \nabla \boldsymbol{A}^{(1)}, \quad \boldsymbol{A}^{(1)} \equiv \boldsymbol{A}^n = (A_x^n, A_y^n, A_z^n),$$

$$\frac{A_x^{n+1} - A_x^n}{\Delta t} = \frac{c}{en_e}\frac{\partial p_e}{\partial x} + \left\{ \boldsymbol{U}_i \times \boldsymbol{B}^{(1)} - \frac{\nu c^2}{\omega_{pe}^2}\nabla \times \boldsymbol{B}^{(1)} \right.$$
$$\left. + \frac{c}{4\pi e n_e}\nabla \times \boldsymbol{B}^{(1)} \times \boldsymbol{B}^{(1)} \right\}_x, \tag{5.156}$$

$$B^{(2)} = \nabla \times A^{(2)}, \quad A^{(2)} = (A_x^{n+1}, A_y^n, A_z^n),$$

$$\frac{A_y^{n+1} - A_y^n}{\Delta t} = \frac{c}{en_e}\frac{\partial p_e}{\partial y} + \left\{ U_i \times B^{(2)} - \frac{\nu c^2}{\omega_{pe}^2}\nabla \times B^{(2)} \right.$$
$$\left. + \frac{c}{4\pi e n_e}\nabla \times B^{(2)} \times B^{(2)} \right\}_y, \quad (5.157)$$

$$B^{(3)} = \nabla \times A^{(3)}, \quad A^{(3)} = (A_x^{n+1}, A_y^{n+1}, A_z^n),$$

$$\frac{A_z^{n+1} - A_z^n}{\Delta t} = \frac{c}{en_e}\frac{\partial p_e}{\partial z} + \left\{ U_i \times B^{(3)} - \frac{\nu c^2}{\omega_{pe}^2}\nabla \times B^{(3)} \right.$$
$$\left. + \frac{c}{4\pi e n_e}\nabla \times B^{(3)} \times B^{(3)} \right\}_z. \quad (5.158)$$

The simulation with the scheme (5.156–5.158) shows that a time step must satisfy the condition $\Delta t < \Delta x^2/B_0$ for numerical stability. This scheme allows us to simulate the regimes with small value of ν and even $\nu = 0$.

5.7 Time Integration of the Generalized Field Equations

The conventional hybrid models may also be formulated via the generalized electromagnetic field. One may rewrite the generalized Ohm's law (2.22) as follows:

$$E = -\frac{1}{c}(U_e \times B) - \frac{\nabla \cdot P_e}{en_e} - \frac{m}{e}\left[\frac{\partial U_e}{\partial t} + (\nabla \cdot U_e)U_e\right]. \quad (5.159)$$

In the above equation the friction force is absent. Let us introduce the generalized magnetic and electric fields in the following form:

$$\hat{B} = B + \nabla \times (\delta_e^2 \nabla \times B), \quad (5.160)$$

where
$$\delta_e = c/\omega_{pe}$$

and
$$\hat{E} = E - \frac{\partial}{\partial t}\left(\frac{c}{\omega_{pe}^2}\nabla \times B\right). \quad (5.161)$$

Substitution of the generalized magnetic and electric fields (5.160) and (5.161) into the generalized Faraday's law

$$\frac{\partial \hat{B}}{\partial t} = -c\nabla \times \hat{E} \quad (5.162)$$

5.7 Time Integration of the Generalized Field Equations

exactly satisfies Faraday's law for a standard electromagnetic field (2.35). Combining (5.159) and (5.161) gives the final expression for the generalized electric field

$$\hat{E} = -\frac{1}{c}(U_e \times B) - \frac{\nabla \cdot P_e}{en_e} - \frac{m}{e}(\nabla \cdot U_e)U_e - \frac{m}{e}\frac{\partial U_i}{\partial t}. \quad (5.163)$$

The term $-m/e\partial U_i/\partial t$ may be estimated from (2.23) or directly from the particle distribution. The generalized electromagnetic fields were used first in one-dimensional hybrid simulations of the collisionless shocks [41, 174, 328, 330, 333, 345].

Further simplification of the above equations was used recently in two-dimensional simulations of the magnetic field reconnection [226, 484]. Let us first consider the algorithm of [226]. The generalized electric field $\hat{E}$ satisfies Ohm's law

$$\hat{E} = -\frac{1}{c}U_e \times B - \frac{\nabla \cdot P_e}{en_e} - \frac{m}{e}(U_e \cdot \nabla)U_e. \quad (5.164)$$

Here m denotes the electron mass and P_e is the electron pressure tensor. The resulting equations for a generalized magnetic field $\hat{B}$ take the following form:

$$\frac{\partial \hat{B}}{\partial t} = -c\nabla \times \hat{E}, \quad (5.165)$$

where

$$\hat{B} = B - \delta_e^2 \nabla^2 B. \quad (5.166)$$

Note that substitution of (5.161) and (5.166) into (5.165) does not satisfy Faraday's law (2.35) for a standard electromagnetic field. Here the ion motion and density and current variations on the electron spatial scales are neglected.

Another version of the generalized field equations was suggested in [484]. The magnetic field is described by the same equations as (5.165–5.166). However, Ohm's law has the form

$$\hat{E} = \frac{1}{en_e c}J \times \hat{B} - \frac{1}{en_e c}J_p \times B - \frac{\nabla p_e}{en_e}, \quad (5.167)$$

where p_e denotes the scalar electron pressure. Note that the electron bulk acceleration is dropped in this model and the standard electromagnetic fields (B and E) do not satisfy the Faraday's law (2.35). In this model we neglect the finite electron inertia correction to the electric field which is used to step the ion forward in time, because this correction only becomes important at spatial scales of c/ω_{pe}. Changes in the electric field over a distance of c/ω_{pe} have very little effect on the motion of the ions because of their large mass, at least in the magnetic field reconnection problem. However, the above model may be not appropriate for some problems, for example, ion acceleration by shock surfing, where the ion dynamics near fine structures becomes very important.

Note also that the above models do not allow calculations in the region with a small value of density, whereas the models with an implicit electric field (Sect. 5.2.4) work well in the computation domain with a strong variation in the plasma density.

The generalized fields $\hat{\boldsymbol{B}}$ and $\hat{\boldsymbol{E}}$ (5.164–5.166) are advanced in time by using the predictor–corrector scheme (Sect. 6.2.1, see also [562]). In contrast, (5.165–5.167) are advanced in time by using the trapezoidal leapfrog algorithm [484]. It is very important to note that the nonstationary electron inertia term was changed from Ohm's law to the generalized magnetic field (5.160) and (5.166), resulting in the elliptic type equations for the magnetic field. In contrast, the nonstationary electron inertia term included in Ohm's law (Sect. 2.2.3) results in the elliptic type equations for the electric field. Hence, one has to solve the elliptic type equations for the magnetic field (5.160) or (5.166), or the equation for the electric field (2.38) using the relevant numerical algorithms: the Fast Fourier Transform, the iterative methods or other effective numerical algorithms.

Further analysis of these models is done in the exercises.

5.8 Time Integration of the Electron Pressure Equation

The equation for electron pressure may be written as follows [see (2.47)]:

$$\frac{3}{2}\frac{\partial p_e}{\partial t} + \frac{3}{2}\nabla(p_e \cdot \boldsymbol{U}_e) + p_e \nabla \cdot \boldsymbol{U}_e = -\pi^e_{\alpha\beta}\frac{\partial U_{e\alpha}}{\partial x_\beta} - \nabla \cdot \boldsymbol{q}_e + Q_e.$$

When the viscosity stress tensor π^e and the heat flux $\boldsymbol{q}_e$ are neglected a more simple equation may be used:

$$\frac{\partial p_e}{\partial t} + \boldsymbol{U}_e \cdot \nabla p_e + \gamma p_e \nabla \cdot \boldsymbol{U}_e = (\gamma - 1)J^2/\sigma_{\text{eff}}.$$

Here, we do not consider the simplest case of the adiabatic electron gas $p_e \propto n_e^\gamma$, where the effective adiabatic index $\gamma = 5/3$. In the case of isotropic electron pressure one can solve the above equations using the predictor–corrector method, operator splitting schemes, leapfrog schemes etc. (see this chapter and Exercises 5.7 and 5.8). The solution of the equation for nongyrotropic pressure will be considered in Chap. 10.

Summary

In this chapter we have tried to give a survey of the issues involved in selecting a method for solving the electromagnetic field and electron pressure equations and a brief account of the main features of the most important methods. We hope this will be adequate to indicate the type of method likely to be the most useful in any particular case and to help avoid the choice of grossly inefficient methods. We mention again the growing interest in the high-order schemes and, in particular, compact schemes with spectral-like resolution. However, to suppress the shot noise effect, one has to use a smoothing procedure (a weak filtering or artificial viscosity). The other restriction on the choice of the field solver is dictated by the architecture of the computer used. The numerical algorithm must allow the vectorization or parallelization of the calculation depending on the type of computer used.

Exercises

5.1 Write the dimensionless form of (5.45).

5.2 Derive the equation for evaluating the implicit electric field ($m/M \neq 0$) in the case of implicit global time integration (Sect. 6.2.2).

5.3 Derive the finite-difference Crank–Nicolson scheme for the magnetic field induction equation in the one-dimensional case

$$\frac{\partial \boldsymbol{B}}{\partial t} + U_x \frac{\partial \boldsymbol{B}}{\partial x} - \frac{\partial}{\partial x} l_d^* \frac{\partial}{\partial x} \boldsymbol{B} = -\boldsymbol{\Phi}, \tag{5.168}$$

where $\boldsymbol{\Phi}$ is a source.

5.4 Derive the numerical scheme for the magnetic field equation in the two-dimensional case.

5.5 Derive the one-dimensional variant of the generalized field equations.

5.6 Analyze the approximation of (5.164) to the electron inertia term in the generalized Ohm's law:

$$\frac{m}{ne^2} \left\{ \frac{\partial}{\partial t} \boldsymbol{J}_e + (\nabla \cdot \boldsymbol{U}_e) \boldsymbol{J}_e \right\}. \tag{5.169}$$

5.7 Analyze the approximation of (5.167) to the electron inertia term in the generalized Ohm's law:

$$\frac{m}{ne^2} \left\{ \frac{\partial}{\partial t} \boldsymbol{J}_e + (\nabla \cdot \boldsymbol{U}_e) \boldsymbol{J}_e \right\}. \tag{5.170}$$

5.8 Derive a finite-difference scheme for the electron pressure equation in the one-dimensional case

$$\frac{\partial p}{\partial t} + U_x \cdot \frac{\partial p}{\partial x} + \gamma p \frac{\partial U_x}{\partial x} = 2(\gamma - 1)(J_y^2 + J_z^2)/\beta_e Re. \tag{5.171}$$

5.9 Derive the numerical scheme for the electron pressure equation in the two-dimensional case

$$\frac{\partial p}{\partial t} + U_x \cdot \frac{\partial p}{\partial x} + U_y \cdot \frac{\partial p}{\partial y} + \gamma p \left(\frac{\partial U_x}{\partial x} + \frac{\partial U_y}{\partial y} \right) = 2(\gamma - 1)(J_y^2 + J_z^2)/\beta_e Re. \tag{5.172}$$

5.10 Study the dispersion and dissipation properties of third-order schemes with compact upwind differencing (CUD-3) for the advection equation.

5.11 Derive the compact finite-difference schemes for the advection–diffusion equation

$$\frac{\partial u}{\partial t} + \frac{\partial \phi(u,x)}{\partial x} = \frac{\partial}{\partial x}\epsilon(u,x)\frac{\partial u}{\partial x} + f(u,x), \quad \epsilon > 0. \tag{5.173}$$

6. General Loops for Hybrid Codes. Multiscale Methods

6.1 Introduction

The hybrid simulation as well as all particle-mesh simulations includes the following principal steps:

- Density and current assignment to the mesh.
- Solution of the electromagnetic field equation on the appropriate mesh.
- Calculation the mesh-defined force field.
- Interpolation forces on particle position.
- Updating the particle velocity and position.
- Particle injection and absorption at the boundaries.

Since the general loops contain nonlinear implicit equations, to achieve the desirable accuracy, in some cases we have to iterate over some steps of this loop.

6.2 Examples of the Conventional Hybrid Simulation Loops

There is a wide spectrum of different representations of conventional hybrid codes stemming from simple first-order-accurate explicit schemes (see, e.g., [41, 97, 598]), implicit schemes [318], predictor–corrector schemes [66, 83, 217, 561] and iterative implicit schemes [238, 486]. In this chapter we shall consider the most popular and useful hybrid code loops, to demonstrate the possibility of simulating multiscale plasma systems.

6.2.1 General Predictor–Corrector Loop

The global time-advance predictor–corrector hybrid algorithm was suggested first in [83, 217]. Further implementation was made in [66, 361, 404, 561]. The basic time-stepping sequence is as follows (see, e.g., [561]): At a moment of time $n\Delta t$ (denoted by superscripts), we assume that $\boldsymbol{E}^n$ and $\boldsymbol{B}^n$ are known. The ions are moved by a standard leapfrog technique to obtain their positions

x_l^n and velocities $v_l^{n+1/2}$, collecting the charge density $n_i^{n+1/2}$ and current $J_i^{n+1/2}$ in the process.

(1) B^n is advanced to $B^{n+1/2}$ using Faraday's law:

$$B^{n+1/2} = B^n - \frac{c\Delta t}{2} \nabla \times E^n;$$

E^n is then advanced using Ohm's law:

$$E^{n+1/2} = F\left(B^{n+1/2}, n_i^{n+1/2}, J_i^{n+1/2}\right).$$

(2) Then, E and B are predicted at time level $n+1$:

$$E_p^{n+1} = 2E^{n+1/2} - E^n,$$

$$B_p^{n+1} = B^{n+1/2} - \frac{c\Delta t}{2} \nabla \times E_p^{n+1}.$$

(3) Using these predictor fields, the particles are advanced one time step in order to collect the predicted ion source terms $n_{ip}^{n+3/2}$, and $J_{ip}^{n+3/2}$.

(4) E_p and B_p are then computed at time level $n+3/2$, using the formulae from step 1:

$$B_p^{n+3/2} = B_p^{n+1} - \frac{c\Delta t}{2} \nabla \times E_p^{n+1},$$

$$E_p^{n+3/2} = F(B_p^{n+3/2}, n_{ip}^{n+3/2}, J_{ip}^{n+3/2}).$$

(5) Finally, E^{n+1} and B^{n+1} are determined:

$$E^{n+1} = \frac{1}{2}\left(E_p^{n+3/2} + E^{n+1/2}\right),$$

$$B^{n+1} = B^{n+1/2} - \frac{c\Delta t}{2} \nabla \times E^{n+1}.$$

(6) If needed, E^{n+1} and B^{n+1} computed in step 5 can be used as predictor fields and the process repeated by returning to step 3.

Note, that step 5 results in a smoothing of the electric field when one calculates the value of E at time level $n+1$ from time levels $n+3/2$ and $n+1/2$. The predictor–corrector method was successfully used in kinetic simulation of collisionless shocks, current sheets, beam dynamics and global simulation of the interaction of the solar wind with Venus, Mars and comets.

6.2.2 Implicit Time Integration of the Electromagnetic Equations

This method was designed in [238], and it uses the same time step for fields and particles. Let us suppose that the value of density is known at time level n, ion velocity at time level $n-1/2$, and the electromagnetic field at a moment of time $n\Delta t$. For particle advance we use the leapfrog scheme:

$$v^{n+1/2} = v^{n-1/2} + \frac{e\Delta t}{M}\left(E^n + \frac{(v^{n+1/2} + v^{n-1/2}) \times B^n}{2c}\right), \quad (6.1)$$

$$x^{n+1} = x^n + \Delta t v^{n+1/2}, \quad x^{n+1/2} = x^n + \frac{\Delta t}{2} v^{n+1/2}. \quad (6.2)$$

For the electromagnetic field we use

$$B_{k+1}^{n+1} = B^n - \frac{c\Delta t}{2} \nabla \times \left(E_k^{n+1} + E^n\right), \quad (6.3)$$

$$J_{e,k+1}^{n+1/2} = -J_i^{n+1/2} + \frac{c}{8\pi} \nabla \times \left(B_{k+1}^{n+1} + B^n\right), \quad (6.4)$$

$$E_{k+1}^{n+1} = -E^n + \frac{1}{en_i^{n+1/2} c} J_{e,k+1}^{n+1/2} \times \left(B_{k+1}^{n+1} + B^n\right). \quad (6.5)$$

Since the first of these equations requires knowledge of E at the new time step, we guess at its value and find a better guess from the last equation. Thus, this is an iteration scheme with the subscript k indicating the iteration.

The spatial derivatives are represented as central differences. To start the iteration we used $E_1^{n+1} = E^n$. Typically four to nine iterations give a relative error of 10^{-3}. In this scheme we have only one loop to advance the particle velocities and positions. The above implicit method was successfully used in kinetic simulation of the magnetic reconnection in current sheets.

6.2.3 The Moment Method

In [515], the field is substepped for each particle step, as in a cyclic leapfrog, and a second-order rational Runge–Kutta algorithm [548] is used to advance both fields and particles. In [572], a moment method is presented which was described in [438] as a variation of Terasawa's method. There are also subcycles, but with a fourth-order rational Runge–Kutta algorithm. We consider one complete time cycle of this scheme [572]. To start, assume that E and B are known at time level n, the density n_i is known at time level n, and the ion velocity U_i is known at time level $n - 1/2$. All quantities are defined at the cell center except B, which is defined at the cell corners. The ion macroparticle velocities v are known at the moment of time $(n - 1/2)\Delta t$, and the positions x are known at time level n. Quantities are advanced to the next time level by the following sequence of steps:

(1) Advance the macroparticle velocities using a standard leapfrog algorithm:

$$v^{n+1/2} = v^{n-1/2} + \frac{e\Delta t}{M}\left(E^n + \frac{v^n \times B^n}{c}\right). \quad (6.6)$$

(2) Advance the ion positions using leapfrog:

$$x^{n+1} = x^n + \Delta t v^{n+1/2}. \quad (6.7)$$

146   6. General Loops for Hybrid Codes. Multiscale Methods

(3) Calculate the bulk ion moments $n_i^{n+1/2}$, n_i^{n+1} and $\boldsymbol{U}_i^{n+1/2}$.

(4) Calculate $\boldsymbol{B}^{n+1}$ using a fourth-order Runge–Kutta algorithm and subcycling. Throughout this operation fix $\boldsymbol{U}_i$ and the density n_i at their values for time level $n + 1/2$. The subcycling is necessary because of the lack of convergence of the Runge–Kutta routine for large Δt.

Defining the time step for the field integration to be $\Delta t'$, and letting $\theta = \Delta t / \Delta t'$, where θ is a positive integer, the magnetic field at time level $n + \theta$ is

$$\boldsymbol{B}^{n+\theta} = \boldsymbol{B}^n + \frac{c\Delta t'}{6}(\boldsymbol{K}_1^n + 2\boldsymbol{K}_2^n + 2\boldsymbol{K}_3^n + \boldsymbol{K}_4^n), \tag{6.8}$$

where

$$\boldsymbol{K}_1^n = -\nabla \times \boldsymbol{F}(\boldsymbol{B}^n),$$

$$\boldsymbol{K}_2^n = -\nabla \times \boldsymbol{F}\left(\boldsymbol{B}^n + \frac{c\Delta t'}{2}\boldsymbol{K}_1^n\right),$$

$$\boldsymbol{K}_3^n = -\nabla \times \boldsymbol{F}\left(\boldsymbol{B}^n + \frac{c\Delta t'}{2}\boldsymbol{K}_2^n\right),$$

$$\boldsymbol{K}_4^n = -\nabla \times \boldsymbol{F}(\boldsymbol{B}^n + c\Delta t' \boldsymbol{K}_3^n)$$

and

$$\boldsymbol{F}(\boldsymbol{B}) = -\frac{\boldsymbol{U}_i^{n+\frac{1}{2}} \times \boldsymbol{B}}{c} - \frac{\nabla p_e^{n+\frac{1}{2}}}{en_i^{n+\frac{1}{2}}} - \frac{\boldsymbol{B} \times \nabla \times \boldsymbol{B}}{4\pi e n_i^{n+\frac{1}{2}}}.$$

$\boldsymbol{F}$ is the electric field, but calculated at mixed time levels. The magnetic field at time level $n + 1$ is obtained by repeating the above calculation a total of θ times.

(5) Having calculated $\boldsymbol{B}$ at time level $n+1$, we now need $\boldsymbol{E}$ at time level $n + 1$ to complete the time cycle. The only quantity missing at time level $n + 1$ necessary to calculate $\boldsymbol{E}$ is $\boldsymbol{U}_i$. This, in turn, can be obtained from

$$\boldsymbol{U}_i^{n+1} = \boldsymbol{U}_i^{n+\frac{1}{2}} - \frac{\Delta t}{2}(\boldsymbol{U}_i \cdot \nabla \boldsymbol{U}_i)^{n+\frac{1}{2}}$$

$$- \frac{\Delta t}{2 M_i n_i^{n+\frac{1}{2}}}\left[\nabla p_e + \nabla p_i^{n+\frac{1}{2}} - (\boldsymbol{J} \times \boldsymbol{B})^{n+1}\right]. \tag{6.9}$$

(6) Given $\boldsymbol{U}_i$ at time level $n + 1$, the electric field is

$$\boldsymbol{E} = -\frac{\boldsymbol{U}_i \times \boldsymbol{B}}{c} - \frac{\nabla p_e}{en_i} - \frac{\boldsymbol{B} \times \nabla \times \boldsymbol{B}}{4\pi e n_i}, \tag{6.10}$$

where all terms on the right-hand side of (6.10) are evaluated at time level $n + 1$.

The moment method was successfully used in the kinetic simulation of collisionless shocks, current sheets and beam dynamics.

6.2.4 The Richardson Extrapolation Method

It was observed in simulation that the magnetic field generates very short wavelength disturbances of the order of a few grid spacings, independent of how fine a grid was used. They were clearly unphysical and associated with the numerical method. The smoothing technique eliminated the presence of modes with a wavelength below a certain size (i.e., usually of the order of, or shorter than, the particle size). However, this technique results in very poor energy conservation. It was observed that by reducing the time step by a factor of two the energy conservation improved by a factor of two [265]. In [265], it was hypothesized that the B field smoothing introduced an error in the velocity computation which was to the lowest order quadratic in the time step. It was suggested that a time-integration method be used similar to the so-called Richardson extrapolation used for ordinary differential equations discussed in [120]. This method works as follows: if one somehow knows the dependence of a dependent variable (here v_i) on the independent variable step size (here Δt) in a difference scheme, then one can make two or more computations of that dependent variable using different step sizes and use those results to compare the zero-order dependence or the correct value of the dependent variable. In order to eliminate the error, one makes parallel runs with Δt and $\Delta t/2$, and at the end of each full time step, one corrects the velocity in the following manner: If v'_i indicates the velocity from the $\Delta t/2$ run, then upon going from time level $n + 1/2$ to $n + 3/2$, one introduces an error of $\epsilon_i (\Delta t/2)^2$ for each time step and v'_i is given by

$$v'^{n+\frac{3}{2}}_i = v^{n+\frac{3}{2}}_{ic} + 2\epsilon_i \left(\frac{\Delta t}{2}\right)^2,$$

where v_{ic} is the correct velocity at time level $n + 3/2$. If one pushes $v^{n+1/2}_i$ one time step Δt, one introduces an error of size $\epsilon_i \Delta t^2$ and v_i is

$$v^{n+\frac{3}{2}}_i = v^{n+\frac{3}{2}}_{ic} + \epsilon_i (\Delta t)^2.$$

Upon eliminating ϵ_i from the equations above, one obtains the following for the correct velocity:

$$v^{n+\frac{3}{2}}_{ic} = 2v'^{n+\frac{3}{2}}_i - v^{n+\frac{3}{2}}_i.$$

As a result, conservation of energy is improved by a factor of 400 for the same time step and time duration as without this correction. The improvement is adequate to increase the time step by a factor of 10. This code was successfully used for the study of the ion cyclotron emission in the Joint European Torus (JET) tokamak. Details of this algorithm may be found in [265].

6.3 Multiple-Time-Scale Methods

The conventional hybrid simulation loops use the same time-integration step for the field and particle motion equations to provide the resolution of the same time scales in the field and particle trajectory. To resolve the multiple time scales, we have to use the different time steps for the field and particle integrations.

6.3.1 Electromagnetic Field Subcycling. Current Advanced Methods and Cyclic Leapfrog Schemes

In many simulations the magnetic field requires a smaller time step than the particles to resolve its evolution, in particular, the effect of dispersion. Cyclic leapfrog enables it to be substepped through a cycle of smaller time steps for each particle update step [367].

6.3.1.1 Advancing the Ionic Current Density.
An equation of motion is now derived for advancing $\boldsymbol{J}_\mathrm{i}$ to the midpoint of the time step [367]. Multiply the "pre-push" equation of motion by the charge q_s, and sum the contributions of the terms at the grid point, using weights $W_{sj}^{1/2} = W(\boldsymbol{x}_s^{1/2}, \boldsymbol{x}_j)$ evaluated at particle positions $\boldsymbol{x}_s^{1/2}$:

$$\sum_s W_{sj}^{1/2} q_s \boldsymbol{v}_s^{1/2} = \sum_s W_{sj}^{1/2} q_s \boldsymbol{v}_s^0 + \frac{\Delta t}{2} \sum_s W_{sj}^{1/2} \frac{q_s^2}{M_s} \left(\boldsymbol{E}^* + \frac{\boldsymbol{v}_s^0 \times \boldsymbol{B}^{1/2}}{c} \right),$$

$$\boldsymbol{J}_\mathrm{i}^{1/2} = \boldsymbol{J}_\mathrm{i}^* + \frac{\Delta t}{2} (\Lambda \boldsymbol{E}^* + \boldsymbol{\Gamma} \times \boldsymbol{B}^{1/2}), \qquad (6.11)$$

where

$$\Lambda = \sum_s W_{sj}^{1/2} \frac{q_s^2}{M_s},$$

and

$$\boldsymbol{\Gamma} = \sum_s W_{sj}^{1/2} \frac{q_s^2}{M_s c} \boldsymbol{v}_s^0,$$

where summation is done over the ion species s.

6.3.1.2 Magnetic Field Substepping.
The electric field is evaluated as a function of the time-centered charge and current densities:

$$\boldsymbol{E}_p = \boldsymbol{E}(\varrho_\mathrm{c}^{1/2}, \boldsymbol{J}_\mathrm{i}^{1/2}, \boldsymbol{B}_p, T_\mathrm{e}),$$

where $\varrho_\mathrm{c} = \sum_s n_s q_s$, and $\boldsymbol{B}_0 = \boldsymbol{B}(t_0)$ is advanced from t_0 to $t_0 + \Delta t$ in a cycle of n substeps of size $\delta t = \Delta t/n$, so that

$$\boldsymbol{B}_p = \boldsymbol{B}(t_0 + p\delta t),$$

$$\boldsymbol{B}_1 = \boldsymbol{B}_0 - c\delta t \nabla \times \boldsymbol{E}_0,$$

$$\boldsymbol{B}_2 = \boldsymbol{B}_0 - 2c\delta t \nabla \times \boldsymbol{E}_1,$$

$$\vdots$$

$$\boldsymbol{B}_{p+1} = \boldsymbol{B}_{p-1} - 2c\delta t \nabla \times \boldsymbol{E}_p,$$

$$\vdots$$

where $p = 1, 2, ..., n-1$,

$$\boldsymbol{B}_n = \boldsymbol{B}_{n-2} - 2c\delta t \nabla \times \boldsymbol{E}_{n-1},$$

$$\boldsymbol{B}_n^* = \boldsymbol{B}_{n-1} - c\delta t \nabla \times \boldsymbol{E}_n,$$

and, finally,

$$\boldsymbol{B}(t_0 + \Delta t) = \frac{1}{2}(\boldsymbol{B}_n + \boldsymbol{B}_n^*). \tag{6.12}$$

Here we use two copies of $\boldsymbol{B}$, one for the *odd* solution (p odd, together with $\boldsymbol{B}_n^*$) and one for the *even* solution (p even), which leapfrog over each other. After n steps as described in (6.12), the two solutions may either be averaged or the cycle may be prolonged, starting with $\boldsymbol{B}_{n+1} = \boldsymbol{B}_n^* - c\delta t \nabla \times \boldsymbol{E}(\boldsymbol{B}_n^*)$. This may be prescribed by the algorithm, or, alternatively, the error between the two solutions $\boldsymbol{B}_n^*$ and $\boldsymbol{B}_n$ may be used as a criterion for averaging them and starting a new cycle, possibly with a different time step [367].

6.3.1.3 Time-Advance Algorithm. In general, given $\boldsymbol{x}^{1/2}, \boldsymbol{v}^0, \boldsymbol{B}^0, \varrho_c^0, \varrho_c^{1/2}, \boldsymbol{J}_i^0, \Lambda, \Gamma$ and $\boldsymbol{J}_i^+ \doteq \boldsymbol{J}_i^*(\boldsymbol{x}^{1/2}, \boldsymbol{v}^0)$ (with $T_e = \text{const}$), each step from t_0 to t_1 is made in a sequence:

(1) Advance $\boldsymbol{B}^0$ to $\boldsymbol{B}^{1/2}$, $\boldsymbol{J}_i^+$ to $\boldsymbol{J}_i^{1/2}$ and evaluate $\boldsymbol{E}^{1/2}$:

$$\boldsymbol{B}^{1/2} = \boldsymbol{B}^0 - c \int_0^{\Delta t/2} \nabla \times \boldsymbol{E}(\varrho_c^0, \boldsymbol{J}_i^0, \boldsymbol{B}(t), T_e) \mathrm{d}t,$$

$$\boldsymbol{E}^* = \boldsymbol{E}(\varrho_c^{1/2}, \boldsymbol{J}_i^0, \boldsymbol{B}^{1/2}, T_e),$$

$$\boldsymbol{J}_i^{1/2} = \boldsymbol{J}_i^+ + \frac{\Delta t}{2}(\Lambda \boldsymbol{E}^* + \Gamma \times \boldsymbol{B}^{1/2}),$$

$$\boldsymbol{E}^{1/2} = \boldsymbol{E}(\varrho_c^{1/2}, \boldsymbol{J}_i^{1/2}, \boldsymbol{B}^{1/2}, T_e).$$

(2a) Advance $\boldsymbol{v}^0$ to $\boldsymbol{v}^1$ and $\boldsymbol{x}^{1/2}$ to $\boldsymbol{x}^{3/2}$:

$$\boldsymbol{v}^{1/2} = \boldsymbol{v}^0 + \frac{\Delta t q}{2M}\left(\boldsymbol{E}^{1/2} + \frac{\boldsymbol{v}^0 \times \boldsymbol{B}^{1/2}}{c}\right),$$

$$\boldsymbol{v}^1 = \boldsymbol{v}^0 + \frac{\Delta t q}{M}\left(\boldsymbol{E}^{1/2} + \frac{\boldsymbol{v}^{1/2} \times \boldsymbol{B}^{1/2}}{c}\right),$$

$$\boldsymbol{x}^{3/2} = \boldsymbol{x}^{1/2} + \Delta t \boldsymbol{v}^1.$$

(2b) Collect the moments in the same loop through the particle:

$$\varrho_c^{3/2} = \varrho_c(\boldsymbol{x}^{3/2}),$$

$$J_i^- = J_i^*(x^{1/2}, v^1),$$
$$J_i^+ = J_i^*(x^{3/2}, v^1),$$
$$\Lambda = \Lambda(x^{3/2}, v^1),$$
$$\Gamma = \Gamma(x^{3/2}, v^1).$$

(3) ϱ_c^1 and J_i^1 are obtained as averages, and $B^{1/2}$ is advanced to B^1:

$$\varrho_c^1 = \frac{1}{2}(\varrho_c^{1/2} + \varrho_c^{3/2}),$$
$$J_i^1 = \frac{1}{2}(J_i^- + J_i^+),$$
$$B^1 = B^{1/2} - c \int_{\Delta t/2}^{\Delta t} \nabla \times E(\varrho_c^1, J_i^1, B(t), T_e) dt.$$

B is integrated in time by *cyclic leapfrog* (6.12). The whole loop is illustrated in Fig. 6.1.

At the beginning of a simulation x, v and B are known at time t_0, and at the end of a simulation, a data set with synchronous quantities is desirable. Therefore a first and last step are necessary:

First step – given x^0, v^0 and B^0,

$$\varrho_c^0 = \varrho_c(x^0),$$
$$J_i^0 = J_i(x^0, v^0),$$
$$x^{1/2} = x^0 + \frac{\Delta t}{2} v^0,$$
$$\varrho_c^{1/2} = \varrho_c(x^{1/2}),$$
$$J_i^+ = J_i^*(x^{1/2}, v^0),$$
$$\Lambda = \Lambda(x^{1/2}, v^0),$$
$$\Gamma = \Gamma(x^{1/2}, v^0).$$

Final step – retreat x a half step, collect ϱ_m^1 and u_i^1,

$$x^1 = x^{3/2} - \frac{\Delta t}{2} v^1,$$
$$\varrho_m^1 = \varrho_m(x^1),$$
$$u_i^1 = u_i(x^1, v^1).$$

The current advance method [367] has several advantages compared with the moment method: (1) Multiple ion species may be easy modeled. (2) The numerically awkward advective term is absent. The collection of moments at the time-centered positions a half-step ahead includes the transport of particle momenta. (3) No ionic pressure tensor is collected [for the same reason as in (2)].

The CAM-CL algorithm was successfully used in the kinetic simulation of collisionless shocks, current sheets and beam dynamics.

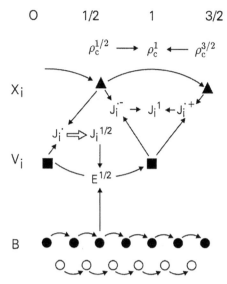

Fig. 6.1. Schematic of the time-advance scheme in CAM-CL. At the beginning of the step, x has been advanced to time level $1/2$ with v^0. Moments already collected are $\varrho_c^{1/2}$ and the "free-streaming" ionic current density $J_i^*(x^{1/2}, v^0)$, as well as ϱ_c^0 and J_i^0. Two solutions of B are advanced by substeps (*cyclic leapfrog*) to time level $1/2$, with $E(\varrho_c^0, J_i^0, B, T_e)$. The *current advance method* advances J_i^* to $J_i^{1/2}$, with the fields $B^{1/2}$ and $E(\varrho_c^{1/2}, J_i^0, B^{1/2}, T_e)$. The time-centred (for v) electric field is now evaluated at time level $1/2$: $E^{1/2} = E(\varrho_c^{1/2}, J_i^{1/2}, B^{1/2}, T_e)$. The particles are pushed, $v^0 \to v^1, x^{1/2} \to x^{3/2}$, and the moments are collected: $\varrho_c(x^{3/2})$, from which ϱ_c^1 is obtained as an average of $\varrho_c^{1/2}$ and $\varrho_c^{3/2}$; the backward and forward "free-streaming" currents $J_i^{*-}(x^{1/2}, v^1)$ and $J_i^{*+}(x^{3/2}, v^1)$, which are averaged to yield J_i^1. Finally $B^{1/2} \to B^1$. (From [367])

6.3.2 Light Ion (Electron) Subcycling

In many cases, when the mass ratio of the different species of particles (M/M_i or m/M) is very small, one may use different time steps for advancing ions, electrons and fields. In this section we follow [102]. An early example of subcycling is given in [57], where the field equations were multistepped in between successive particle advances. The integration was explicit, reversible, time-centered, and accurate in Δt to the second order. Boris's scheme was stable for sufficiently small time steps. The economy achieved here by subcycling derives from the longer time step used for the particle advance and, hence, the fewer particle pushes. This can be an appreciable saving because particle pushing accounts for the major fraction of the computational burden in a particle simulation.

In [292] subcycling was subsequently applied in work on two-dimensional electromagnetic simulation. In some applications a serious numerical instability was found when the light-wave frequency equalled the Nyquist frequency for the particle $\pi/\Delta t$. A similar instability occurs for subcycling in an electrostatic model [5].

In [194] the numerical Cherenkov instability in subcycled electromagnetic algorithms was studied. This instability involves a resonance between a spurious beaming mode and the electromagnetic normal modes. For particles advanced every N time steps, stability is ensured for

$$N\Delta t \leq \Delta x/v, \tag{6.13}$$

where v is the velocity of a single beam. Solution of Maxwell's equations using a Fourier transform technique [194] or the *Langdon–Dawson* advective algorithm [195] avoids the numerical Cherenkov instability. In recent work [5], subcycling has been introduced to the particle simulation of a multispecies plasma. The difference in ion and electron inertia was exploited to devise a scheme in which the cost of advancing the ions is negligible compared with that for the electrons. In a subcycled electrostatic algorithm, Poisson's equation is solved, and the electrons are advanced on the same time step Δt_e; ions are advanced after N steps with time step $\Delta t_i = N\Delta t_e$, $N \gg 1$. This scheme is explicit and centered. It can be made numerically stable with the addition of weak damping, but the stability of electron plasma waves requires $\omega_{\mathrm{pe}}\Delta t_e < O(1)$.

The method of electron subcycling may be demonstrated here by using the simplest explicit electrostatic algorithm [5, 102]. The electron equations of motion and Poisson's equation are solved in each time step Δt_e. The electric field felt by the ions is a filtered average of the electric fields seen by the electrons; a simple average works well [5].

The subcycling algorithm differs slightly, depending on whether N is even or odd (Fig. 6.2). In Fig. 6.2 ion densities are shown above the time line and electron densities below. The ion density used in conjunction with the electron density in Poisson's equation is shown below the electron densities. The super-scripts denote the time levels in units of the ion time step. The half-integer superscripts on the ion densities in Fig. 6.2 are simple averages of the ion densities at adjacent integer time levels. The ion and electron densities are never known at the same time in Fig. 6.2. The even-N schemes are less convenient than the odd-N schemes, and in [5] the latter are favored. Both odd-N and even-N schemes are time-centered, reversible and accurate in time to the second order.

The electron subcycling algorithm can be unstable when $\omega_{\mathrm{pe}}\Delta t_i \approx l\pi$, where l is an integer. This is a temporal aliasing phenomenon caused by the coupling of the two electron plasma normal modes $\omega = \pm\omega_{\mathrm{pe}}$ by the ions. Moreover, the manner in which the electric field seen by the ions is averaged and filtered can lead to a strong instability for $\omega_{\mathrm{pe}}\Delta t_i = 2\pi$. Damping in the particle equations can stabilize these instabilities at the expense of numerical cooling of the electrons [106].

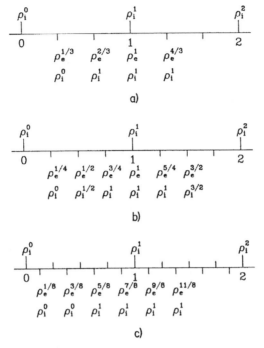

Fig. 6.2. Schematics of time levels for odd and even versions of the electrostatic subcycling algorithm. The leapfrog advance of the ion velocities and positions allows calculation of the ion density above the time line before it is needed below the line in the electric-field calculation. (a) Odd-N scheme; (b) even-N scheme; (c) even-N scheme. (From [102])

Numerical damping can be added to the electron equations of motion [106, 289], such that
$$v^{n+1/2} = v^{n-1/2} + a^n \Delta t,$$
$$x^{n+1} = x^n + v^{n+1/2}\Delta t + c_1 \Delta t^2 (a^n - a^{n-1}). \quad (6.14)$$

This introduces dissipation and a phase shift in the electron plasma waves; the complex frequency shift $\delta\omega$ is given by [102]

$$\text{Re}(\delta\omega/\omega_{\text{pe}}) = \frac{1 - 12c_1}{24}(\omega_{\text{pe}}\Delta t)^2 + O(\Delta t^3),$$

$$\text{Im}(\delta\omega/\omega_{\text{pe}}) = -\frac{c_1}{2}(\omega_{\text{pe}}\Delta t)^3 + ..., \quad (6.15)$$

where ω_{pe} is the plasma frequency. For $\omega_{\text{pe}}\Delta t_i \approx \pi$ and damping rates in (6.14) less than the maximum growth rate from the dispersion relation, the net growth rates in the simulation were reduced in agreement with (6.14). Stability was achieved when the damping rate in (6.14) exceeded the maximum growth rate. For $\omega_{\text{pe}}\Delta t_i \approx 2\pi$, (6.14) with $\omega_0 \sim \omega_{\text{pi}}$ overestimates the stabilizing influence of the dissipation and indicates that the electron dissipation has a much weaker influence on ion-acoustic waves than on electron plasma waves. The application of subcycling to electromagnetic hybrid codes was considered in Sect. 6.3.1.

6.3.3 Orbit Averaging

Another approach to save computer time is orbit averaging. Orbit averaging has been successfully applied to a magnetoinductive model of plasma [102]–[105, 108] and Ampere's model [349, 351].

In these models, charge separation effects are neglected. A particle species carries a current, which in turn supports a magnetic field. The time derivative of the magnetic field generates an inductive electric field that is determined by solving of Faraday's law. Strict charge neutrality is assumed, and the ambipolar electric field is neglected. This model is appropriate for a number of laboratory plasma experiments, including, for example, neutral-beam-injected mirror experiments [103, 104].

In the orbit-averaged magneto-inductive scheme, particles are advanced with a small time step Δt [102],

$$x^{n+1} = x^n + v^{n+1/2}\Delta t,$$

$$v^{n+1/2} = v^{n-1/2} + \frac{q\Delta t}{m}\sum_{j}^{N_g} S(x^n - x_j)\left(E^*_j + \frac{1}{2c}(v^{n+1/2} + v^{n-1/2}) \times B^*_j\right), \quad (6.16)$$

where $S(x^n - x_j)$ is the particle spline interpolation of the fields defined on the grid between positions x_j and x^n, where the particle is accelerated,

$$(E, B)^* = w_1(M, n')(E, B)^M + [1 - w_1(M, n')](E, B)^{M+1}, \quad (6.17)$$

and $w_1(M, n')$ is an interpolation function with $0 \leq n' \leq N-1, N = \Delta T/\Delta t$ and $n = MN + n'$. Maxwell's equations are used to calculate the fields:

$$\nabla \times [\alpha B_j^{M+1} + (1-\alpha)B_j^M] = \frac{4\pi}{c}\langle J_j\rangle^{M+1/2},$$

$$\nabla \times [\beta E_j^{M+1} + (1-\beta)E_j^M] = -\left(B_j^{M+1} - B_j^M\right)c\Delta T, \quad (6.18)$$

where

$$\langle J_j\rangle^{M+1/2} = \frac{1}{2}\sum_{n'=0}^{N-1} w_2(n')\sum_i q\left[S(x_i^n - x_j) + S(x_i^{n+1} - x_j)\right]v_i^{n+1/2}, \quad (6.19)$$

α and β are centering parameters controlling dissipation, with $0 \leq \alpha, \beta \leq 1$, and $w_2(n')$ is a digital filter with $\sum_{n'=0}^{N-1} w_2(n') = 1$. Figure 6.3 gives a

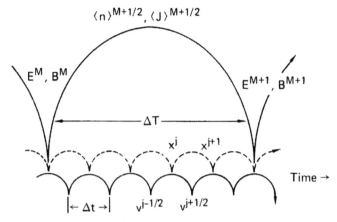

Fig. 6.3. Schematic of time levels in generic orbit averaging. The quantities $\langle n \rangle$ and $\langle J \rangle$ are averaged in time over the number and current densities accumulated at each small time step. (From [102])

schematic of orbit averaging. To approximately center $(\boldsymbol{E}, \boldsymbol{B})^*$, a predictor–corrector iteration is performed: $w_1(M, n') = 1$ and $(\boldsymbol{E}, \boldsymbol{B})^* = (\boldsymbol{E}, \boldsymbol{B})^M$ in the predictor step, and $w_1(M, n') = 1 - (n'/N)$ or $w_1(M, n') = 0$ in the corrector step with $(\boldsymbol{E}, \boldsymbol{B})^{M+1}$ used in $(\boldsymbol{E}, \boldsymbol{B})^*$ calculated earlier in the predictor step. Equations (6.16–6.17) are formally only accurate to the first order in ΔT, but with $(\alpha, \beta) = \frac{1}{2} + \varepsilon$, $\varepsilon \ll 1$, and $w_1 = 1 - (n'/N)$ on the corrector iteration, the algorithm is effectively accurate to the second order.

According to [104], as with subcycling, spatial grid effects and orbit averaging do not lead to any new unusual behavior. A quartic dispersion relation in ω was obtained for a one-dimensional slab model with wave propagation perpendicular to a uniform magnetic field. In the limit $\Omega_i^2 \Delta T^2 \ll 1$, two of the four roots are compressional Alfvén waves:

$$\omega^2 = \frac{k^2 v_A^2}{1 + \dfrac{k^2 c^2}{\omega_{pi}^2}}, \tag{6.20}$$

where v_A is the Alfvén speed $v_A = c\Omega_i/\omega_{pi}$, Ω_i the cyclotron frequency, ω_{pi} the ion plasma frequency, and k the wavenumber. The other two roots are unphysical.

With $w_1(M, n') = 0$ in the corrector iteration and $\beta = 1$, the Alfvén waves are neutrally stable for $\omega_{pi}^2/k^2 c^2 \ll 1$ and all values of α. The Alfvén waves acquire damping for finite $\omega_{pi}^2/k^2 c^2$ [104]. One of the unphysical modes is an odd–even (Nyquist) oscillation that is neutrally stable for $\alpha = 1/2$ and all values of $\omega_{pi}^2/k^2 c^2$, whereas the fourth normal mode purely grows for $\Omega_i^2 \Delta T^2 \ll 1$ and $\omega_{pi}^2/k^2 c^2 > 1$ and is heavily damped for $\omega_{pi}^2/k^2 c^2 \ll 1$. Numerical solution of the dispersion relation is shown in Fig. 5 of [102] for

$\alpha = \beta = 1$ and $\omega_{\text{pi}}^2/k^2c^2 = 3$; note that there is stability for $\Omega_i \Delta T > 2$ and strong instability for $\Omega_i \Delta T < 1$.

For $\Omega_i^2 \Delta T^2 \gg 1$ and $\alpha > 1/2$, all modes are stable. The Alfvén waves are weakly unstable if $\omega_{\text{pi}}^2/k^2c^2 > 1$ with $\Omega_i^2 \Delta T^2 \gg 1$ and $\alpha = 1/2$. The two unphysical modes are both damped Nyquist oscillations for $\alpha > 1/2, \Omega_i^2 \Delta T^2 \gg 1$ and all values of $\omega_{\text{pi}}^2/k^2c^2$. The Alfvén-wave frequencies approach $\pm O(\pi/\Delta T)$ for $\Omega_i^2 \Delta T^2 \gg 1$, and the modes acquire some damping for $\alpha > 1/2$. Thus, no instability is encountered even though $kv_A \Delta T \gg 1$. Simulation experience with orbit averaging has confirmed stability for $\Omega_i^2 \Delta T^2 > 1$ and instability at small time steps [102].

The inclusion of space-charge fields significantly alters the physical model and requires a different approach if stability is to be ensured with a large time step [102]. Langdon has given a proof of the necessity for implicit time differencing to achieve stability for $\omega_{\text{pi}}^2 \Delta T^2 \to \infty$ when solving Poisson's equation and the particle equation of motion. Cohen et al. [105, 106] combined orbit averaging with the implicit moment method [366] to produce a stable algorithm for $\omega_{\text{pi}}^2 \Delta T^2 \gg 1$ and operated with fewer particles than conventional explicit and implicit moment algorithms. Orbit-averaging explicit electrostatic algorithms were unstable for $\omega_{\text{pi}}^2 \Delta T^2 > O(1)$, which was consistent with Langdon's general considerations [102]. The readers who are interested in implicit simulation should consult [102, 108] for details.

Orbit averaging is also used in Darwin hybrid simulation, in which a MHD approximation is used for the background plasma and particle description for the beam. Such a model was used successfully to study the in situ acceleration of cosmic rays by supernova-remnant shock waves [584] and to study the magnetic field reconnection problems using Ampere's model [236, 349, 351, 514].

Further modification of orbit averaging for gyrokinetic particle and hybrid codes and their application in order to study low-frequency phenomena may be found in [39, 108].

6.4 Multiple-Space/Time-Scale Methods

Very often the plasma systems have a complex structure which contains discontinuities. Such discontinuities may play a very important role in energy transport from a background plasma to the particles by means of wave–particle interactions. The geometry and position of such discontinuities may change in time, so that to resolve these discontinuities one has to use an adaptive mesh. In two- and three-dimensional problems concerning the interaction of the solar wind with planets, comets and local interstellar medium, it may be sufficient to use homogeneous or inhomogeneous meshes in the cylindrical, spherical, elliptical, or toroidal coordinates. Examples of some meshes used in hybrid simulations will be given in Part II. However, in simulations with nonstationary structures one has to use adaptive meshes.

6.4.1 Variational Methods

One of the most popular methods for construction of the adaptive grids is based on the variational principle. In [278, 582], a general principle for the construction of grids was formulated, which may control the deviation from Lagrangian coordinates, deformation and node clustering. Let us consider the two-dimensional clustering transform $q_i = q_i(x_1, x_2, t)$, where x_i are the physical coordinates ($i = 1, 2$). Let us write the functional in the form

$$F[q_1, q_2] = \int\int_D \Phi I^\alpha \mathrm{d}x_1 \mathrm{d}x_2, \tag{6.21}$$

where D is the computational domain, and

$$I = (\partial x_1/\partial q_1)(\partial x_2/\partial q_2) - (\partial x_1/\partial q_2)(\partial x_2/\partial q_1).$$

Φ is some known function of the coordinates and unknown functions, which controls the coordinate line clustering. According to [278, 582], it is necessary to find the transform from q_i, which gives the minimum of (6.21). All q_i satisfy the boundary condition in D. Euler's equation for this problem may be written as

$$\frac{\partial}{\partial x_i}\left(\frac{1}{\Phi}\frac{\partial q_j}{\partial x_i}\right) = 0, \quad j = 1, 2, \tag{6.22}$$

or using new variables

$$g^{kn}\frac{\partial}{\partial q_k}\left(\Phi\frac{\partial x_l}{\partial q_n}\right) = 0, \quad l = 1, 2, \tag{6.23}$$

where g^{kn} are covariant components of the metrical tensor for the variable transform $x_i = x_i(q_1, q_2), i = 1, 2$. Let us choose the control function Φ in the following form

$$\Phi = (\varepsilon + |\nabla u|^\alpha)^\beta,$$

where $\varepsilon, \alpha, \beta$ are positive constants, and u is an unknown function. Let us consider the one-dimensional case. Let us expand the region where the derivative $\partial u/\partial x$ becomes large. One can rewrite (6.23) as follows:

$$\frac{1}{\Phi}\frac{\partial q_j}{\partial x} = \mathrm{const}, \quad \Phi = \left(\varepsilon + \left|\frac{\partial u}{\partial x}\right|^\alpha\right)^\beta. \tag{6.24}$$

From the equation above it is clear that $\partial q/\partial x$ is large where the value of $|\partial u/\partial x|$ becomes large, and it means the expansion of this region by transform $q(x)$.

Let us take $\alpha = \beta = 1$, then an integration of (6.24) gives

$$q = \int |\partial u/\partial x| \mathrm{d}x + \varepsilon + c.$$

In practice, automatic node expansion may be realized by the following evolution equation:

$$\frac{\partial x}{\partial t} = \frac{\partial}{\partial q}\left[\left(\left|\frac{\partial u}{\partial x}\right|^\alpha + \varepsilon\right)\frac{\partial x}{\partial q}\right]. \tag{6.25}$$

By solving (6.25) we can find the new coordinates x. Then we can determine $\partial x/\partial q$, and use it in transformed electromagnetic equations and in particle and force weighting. The algorithms for the generation of adaptive and unstructured grids in particle simulation were suggested in [142, 493].

In a region with a small grid size, the statistical fluctuations become very strong and one has to use the splitting of particle procedures to reduce shot noise (see, e.g., Sects. 2.4.2 and 6.4.2). In contrast, in a region of widely spaced grid points, one can join a group of small particles to a big particle provided that the energy, momentum and mass laws are satisfied.

6.4.2 Adaptive Mesh and Particle Refinement Methods

One way to handle multiple scales in a numerical model is through the use of adaptive mesh and particle refinement (AMPR). This approach represents an advancement of methodologies developed for neutral flows block-structured adaptive mesh refinement (AMR) to include the additional kinetic (particle) effect. AMR algorithms permit the underlying computational mesh to be modified in space and time to follow changing solution features. The application of AMR techniques to plasma simulation has been mainly limited to fusion plasma (e.g., [275] and references therein), MHD [109, 115, 130, 429, 500], high-density, low-pressure process plasmas [111] and Boltzmann simulations [179]. One of the first block-structured AMR algorithms was introduced in [45, 46] for hyperbolic conservation laws using finite-difference methods on an hierarchy of regular Cartesian grids. Early applications included problems arising in gas dynamics and shock physics [45, 222].

In this approach, each level of the grids hierarchy corresponds to a degree of spatial refinement, where the location and topology of the grids is determined by Richardson extrapolation estimates of the truncation error combined with cell tagging/clustering algorithms. The hyperbolic system is integrated on each refinement level separately using a time step appropriate to the Courant–Fridrich–Levy (CFL) stability requirement of the grid on that particular level. Communication among levels occurs through the use of temporally and spatially interpolated coarse grid data to define boundary conditions for the integrations on finer levels, as well as so-called "refluxing" operations performed to restore flux continuity and local conservation at coarse–fine boundaries.

The generalization of AMR for systems that are not purely hyperbolic, such as the plasma fluid model and different particle models, presents a number of additional challenges. Elliptic, parabolic and mixed systems lack the real characteristic structure and transit time isolation that make hyperbolic

AMR straightforward in comparison. The availability and high efficiency of multigrid algorithms on Cartesian grids is one of the primary motivations for using a block-structured AMR approach. In addition to the vast literature on multigrid methods, a good overview of multilevel adaptive methods can be found in [368].

6.4.2.1 Grid Description. Employing the composite grids described in [45], refined regions will be covered by a hierarchical sequence of nested, progressively finer levels $l = 1, 2, ..., l_{\text{finest}}$. Each level l is formed by a set of disjoint rectangular grids $G_{l,k}$, $k = 1, 2, ..., n_l$, that is,

$$G_l = \bigcup_k G_{l,k}, \tag{6.26}$$

with $G_{l,j} \cap G_{l,k} = \emptyset, j \neq k$ (two different grids in the same level do not overlap), which have the same spacing h_l and whose sides are aligned in the coordinate directions. As an example, {level 1} = $G_{1,1}$, where $G_{1,1}$ is a global uniform grid covering Ω, the rectangular domain used in the model problem considered.

We use a standard technique to create the grid hierarchy (see, e.g., [45, 448]):

(1) *Add the buffer zone.* A buffer zone of unflagged points is added around every grid. This ensures that discontinuities or other regions of high error do not propagate out from a fine grid into coarser regions before the next regridding time. This is possible because of the finite propagation speed of hyperbolic systems. The larger the buffer zone, the more expensive it is to integrate the solution on the fine grids, but the less often the error needs to be estimated on the coarse grid points that are sufficiently close to flagged points with high error estimates. A buffer zone of two cells in each direction is typical. By flagging neighboring points, instead of enlarging grids at a later step, the area of overlap between grids is reduced.

(2) *Flag every cell at level l corresponding to an interior cell in a level $l + 2$ grid.* This will maintain proper nesting, by ensuring that there will be a new level $l + 1$ grid containing every point in the level $l + 2$ grid, even if the level l grid error estimation did not report a high error. This procedure ensures that the fine grid error estimates are used instead of the coarse grid estimates at the same point. To ensure proper nesting, points within one cell of a nonphysical (interior) boundary of G_l are deleted from the list of flagged points.

(3) *Create rectangular fine grids.* The grid generator takes all the flagged points as input and outputs a list of corners of rectangles that are the level $l+1$ grids. Near points are clustered together, and a fine grid patch spanning each cluster is formed. These clustering algorithm use the heuristic procedures described separately below.

(4) *Ensure proper nesting.* The new fine grids are checked to ensure that they are properly contained in the base level grids. If they are not, the new

grid is repeatedly subdivided until each piece does fit. Since the flagged points originally were inside the base grid, at least one cell from the boundary, the new grid containing the flagged points must be eventually lie inside as well. Since the base level grids did not move, step (2) cannot be used to ensure the proper nesting of this level. This problem only arises when the base grids are a nonconvex union of rectangles.

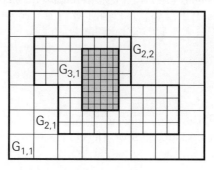

Fig. 6.4. Grid $G_{3,1}$ spans two coarser grids, but it is properly nested (from [448])

Figure 6.4 shows one grid at level 3, $G_{3,1}$, two grids at level 2, $G_{2,1}$ and $G_{2,2}$, all of them laid on the underlying uniform grid $G_{1,1}$, which covers the domain completely. Typically, the setup of a problem to be solved involves the formulation of the problem on a "physical" rectangular domain, conveniently sized and made periodic in all the directions. This domain can be discretized using the composite grids described. Level 1 is obtained through the uniform division of the domain into a regular array of computational cells, whose width and height will be assumed to be the same, h_1, for simplicity. Finer rectangular grid patches forming level $l, 2 \leq l \leq l_{\text{finest}}$, have mesh spacing:

$$h_l = \frac{h_1}{r^{l-1}}, \qquad (6.27)$$

where r is the refinement ratio used (in general, $r = 2$ or 4).

6.4.2.2 Time Integration Algorithm. To consider the application of AMR to the solution of the Maxwell equations we shall follow the algorithm developed in [45] for fluid dynamics. AMR assumes there is a basic, underlying conservative, explicit finite difference scheme of the following form:

Maxwell's equation,

$$B^{z,n+1}_{i,j,k} = B^{z,n}_{i,j,k} - \frac{c\Delta t}{\Delta x}\left(E^{y,n+\frac{1}{2}}_{i+\frac{1}{2},j,k} - E^{y,n+\frac{1}{2}}_{i-\frac{1}{2},j,k}\right) - \frac{c\Delta t}{\Delta y}\left(E^{x,n+\frac{1}{2}}_{i,j+\frac{1}{2},k} - E^{x,n+\frac{1}{2}}_{i,j-\frac{1}{2},k}\right); \qquad (6.28)$$

fluid equation,

$$U^{n+1}_{i,j,k} - U^{n}_{i,j,k} = \qquad (6.29)$$

$$-\Delta t \left(\frac{F^{x,n+\frac{1}{2}}_{i+\frac{1}{2},j,k} - F^{x,n+\frac{1}{2}}_{i-\frac{1}{2},j,k}}{\Delta x} - \frac{F^{y,n+\frac{1}{2}}_{i,j+\frac{1}{2},k} - F^{y,n+\frac{1}{2}}_{i,j-\frac{1}{2},k}}{\Delta y} - \frac{F^{z,n+\frac{1}{2}}_{i,j,k+\frac{1}{2}} - F^{z,n+\frac{1}{2}}_{i,j,k-\frac{1}{2}}}{\Delta z} \right),$$

where the equation for the other components, B^x and B^y, is derived by a cyclic replacement of z and subscripts i, j, k. The values B are cell-centered quantities. Let us consider the two-dimensional computational domain. Each cell is defined by its four corner grid points. If there are no refined regions, then (6.28–6.29), augmented by the discretized physical boundary conditions, defines the time evolution on a single grid.

With multiple grids, each grid is separately defined and has its own solution vector, so that a grid can be advanced independently of other grids, except for the determination of its boundary values. The integration steps on different grids are interleaved, so that before advancing a grid to time $t + \Delta t$ all the finer level grids have been integrated to time t. Scheme (6.28–6.29) is still initially applied on every grid at every level, but the results would be modified by a synchronization.

6.4.2.3 Boundary Conditions. Let the interior integration scheme have a stencil which is centered in space, with d points to each side. To compute the new time step, AMR provides solution values at the old time step on a border of cells of width d intersecting the physical domain. The users must supply the code to compute any additional information needed to implement the boundary conditions. For example, if boundary conditions are imposed by extrapolation, the user would provide the extrapolated values for points outside the domain.

For a grid at level l, the bordering cell values are provided using values from adjacent level l grids where they are available; otherwise, the AMR algorithm computes boundary values using bilinear interpolation from coarser level solution values. If necessary, we also interpolate linearly in time.

Our implementation partitions the required border cells at level l into rectangular boundary patches. For each rectangular piece we:

(1) find solution values from level $l-1$ grids on a slightly larger rectangular boundary piece enclosing the border cells;

(2) linearly interpolate for the border values;

(3) if there are fine grids at level l that could supply some values, overwrite the linearly interpolated values from step (2).

In step (1), most of these coarser lever values are found by intersecting the rectangular patch with level $l - 1$ grids and by filling the overlapping pieces. However, it may be necessary to go to even coarser grids to supply these level $l - 1$ values. This is done by applying (1) to (2) recursively to the smallest rectangular patch containing the unfilled cells.

6.4.2.4 Particle Refinement. In the AMPR method, we have to provide the split particle hierarchy in accordance with the grid hierarchy. Each level l is formed by a set of markers $\boldsymbol{Z}_{l,k,i} = (\boldsymbol{x}_{l,k,i}, \boldsymbol{v}_{l,k,i})$, $k = 1, 2, ..., n$, where k

denotes the index of the grid. At the coarse level $l = 1$, the unsplit particles must cover the coarse mesh. At the finer level l, $l \geq 2$, one has to define a buffer zone which surrounds the boundary grids at this level $l \geq 2$. Then one has to make a refinement of particles which occupy this mesh including the buffer cells. To create the particle at level $l \geq 2$, one has to split each particle from level $l - 1$ inside the cell with index m into r^2 (r^3) particles with the same velocity in a two-dimensional (three-dimensional) simulation. Here, r is the refinement ratio used (in general, $r = 2$ or 4). Note that we do not change the velocity under the splitting procedure in order to avoid artificial particle heating at the finer level. The finer particles are randomly distributed in space in accordance with

$$\boldsymbol{x}_{\text{new}} = \boldsymbol{x}_{\text{old}} + (\xi - 0.5)\Delta\boldsymbol{x}_m, \tag{6.30}$$

where $\Delta\boldsymbol{x}_m$ is the size of cell with index m along the coordinates x, y, z, and ξ is a set of random numbers from the uniform distribution $0 < \xi < 1$. The split particles must cover the finer mesh which belongs to the grid level l in order for the grid values of the current, charge and density at level l to be computed. At the coarse level $l = 1$, the particles update using the global time step Δt. At the finer level $l + 1$, the particles update using a time step equal to $\Delta t/r^l$ or a smaller one, and the electromagnetic field value from a finer mesh at level l. The boundary value of the electromagnetic field inside of the finer mesh may be provided from the physical boundary condition of the computational domain or by means of an interpolation from the coarse cells.

The fine time steps, Δt_{fine}, must satisfy the following CFL conditions: (a) the fluid CFL condition, namely, the particle, may intersect the faces of not more than one fine cell; (b) the whistler, helicon, Alfvén and sonic wave propagation CFL condition.

We have also to provide a procedure for the coalescence of small particles to a big one in the region with small gradients and large grid spacing (coarse grid).

At the end of global time step Δt, we have to synchronize the value of the electromagnetic field and plasma parameters in the coarse cells covered by a fine mesh and the flux (electric field, and plasma current) at the intersection between the faces of the coarse cells and the finer mesh as was done for fluid problems in [45].

Note that adaptive mesh and algorithm refinement [179] embeds a particle method within a continuum method at the finest level of an AMR hierarchy. The coupling between the particle region and the overlying continuum grid is algorithmically equivalent to that between the fine and coarse levels of AMR. Direct simulation Monte Carlo (DSMC) is used as the particle algorithm embedded within a Godunov-type compressible Navier–Stokes solver. However, the direct splitting particle procedure is not considered in this approach.

6.4.2.5 Synchronization. There are two steps in the synchronization (see, e.g., [45, 179]). In the first step, the fine grid is averaged onto the coarse grid, i.e., the conserved quantities on coarse grid cells covered by the fine grid are replaced by the average of the fine grid. The second step of the synchronization, called "refluxing", corrects for the difference in coarse and fine grid fluxes at the boundary of the fine grid. The basic approach used here is an analog of the procedure used in [8] extended to the case of nonlinear parabolic terms. During the course of the integration step, flux information is saved at the faces of the boundary of the coarse and fine grids to obtain the difference between the fluxes calculated at level l and the corresponding level $l+1$ average. The latter are the fluxes at level $l+1$ time averaged over the level l time step and spatially averaged over the area of the level l face. This time-step-weighted and area-weighted flux difference is

$$\delta F^l = \Delta t^l \left(-A^l F^{n+\frac{1}{2},l} + \frac{1}{r} \sum_{k=0}^{r-1} \sum_{\text{faces}} A^{l+1} F^{n+k+\frac{1}{2},l+1} \right), \qquad (6.31)$$

where $\boldsymbol{F}$ are the components of the convective flux (electric field) (6.28–6.29) and A is the signed area of the face of a grid cell, where the sign depends on the direction normal to the face, facing away from the fine grid [179]. The sum over faces in (6.31) is a sum over all fine grid faces that cover the coarse face.

The flux correction, δF_n^l, represents the difference between the flux used to update the coarse cells adjacent to the fine grid and the fluxes that are computed on the fine grid [179]. The solution correction may be found from

$$\delta U = \frac{\Delta t \delta F}{\Delta x \Delta y \Delta z}, \qquad (6.32)$$

and finally the coarse grid, U^{n+1}, is updated by

$$U^{n+1} = \bar{U}^{n+1} + \delta U,$$

where $\bar{U}^{n+1}$ is the coarse grid solution after averaging the fine grid solution but before computing the correction. The fine grid is updated by using a conservative scheme that interpolates the correction to the fine grid and to other finer grids contained within the fine grid. Finally, to capture the effect of the synchronization of level l and higher on level $l-1$, the flux corrections F^{l-1} are not updated until after the synchronization of level l and $l+1$ [179].

Summary

In this chapter we have tried to give a survey of the issues involved in selecting a time integration method for particle and field updating and a brief account of the main features of the most important methods. We also discussed the multiple-time and multiple-space-scale methods which may be very important for simulation of the complex multiscale systems. We hope this will be adequate to indicate the type of method likely to be the most useful in any particular case, and helpful in avoiding the choice of grossly inefficient methods.

7. Particle Loading and Injection. Boundary Conditions

7.1 Introduction

Placing particles in x, v at $t = 0$ and creating or removing particles during a run are the main subjects of this chapter. The placement involves starting with prescribed densities in space $n_0(x)$ and in velocity $f_0(v)$ and generating particle positions and velocities $(x, v)_i$. The formal process for this is called *inversion of the cumulative density* [51].

In an investigation of plasmas which have a very large spatial extent, we usually simulate only a portion of the plasma in a limited computational domain. In order to describe the interaction with the background we also use periodic, radiation or nonradiation boundary conditions for electromagnetic waves and give the possibility for re-entering and reflecting particles at the boundaries. In this chapter we shall give different methods for the creation of the initial position of particles, for the interaction of particles with the boundaries and for the generation of particles flux at the boundaries, and we shall also consider the different type of boundary conditions for electromagnetic waves.

7.2 Loading the Particles Inside the Computational Domain

7.2.1 Loading Nonuniform Distributions $f_0(v), n_0(x)$

In local simulation of plasmas (plasma beams, shocks, current sheets, etc.), we usually use the simple boundary and initial condition for particles and electromagnetic field in order to study the details of complicated nonlinear physical processes. In contrast, in global plasma simulation of the system with complex geometry (the interaction of the solar wind with planets and comets, the magnetospherical plasma dynamics, and the plasma dynamics in devices), we usually use a complex boundary and initial condition for particles and fields.

Let us suppose that we need to place particles so as to form a density $d(x)$, from $x = a$ to $x = b$. The cumulative distribution function has the form

7. Particle Loading and Injection. Boundary Conditions

$$D(x) = \frac{\int_a^x d(x')\,\mathrm{d}x'}{\int_a^b d(x')\,\mathrm{d}x'}, \tag{7.1}$$

where

$$D(a) = 0 \quad \text{and} \quad D(b) = 1. \tag{7.2}$$

Equating $D(x_s)$ to a *uniform distribution* of numbers ξ_s, $0 < \xi_s < 1$, will produce x_s corresponding to the distribution $d(x)$. In the case of the simple expression $d(x)$ the value of x_s may by found analytically; in a more complicated case it is necessary to use a numerical method to solve $x_s = x_s(\xi_s)$.

7.2.2 Loading a Maxwellian Velocity Distribution

Let us use the system of coordinates which is located in the frame of the bulk velocity, $\boldsymbol{v} = \boldsymbol{U}_{\text{bulk}}$. A normalized thermal (Maxwellian) distribution is shown in Fig. 7.1. Most of the particle are in the region out to $v = 3v_\text{T}$ (99 % within $2v_\text{T}$, so very often we do not need to place particles beyond 3 or $4v_\text{T}$). Keeping the particles with sufficiently high v may results in a strong decrease in the integration time step value due to Courant–Fridrich–Levy (CFL) condition.

Let us suppose that the density is spatially uniform with the isotropic Gaussian $f_0(\boldsymbol{v})$. Then the cumulative distribution function for the speed $v = |\boldsymbol{v}|$,

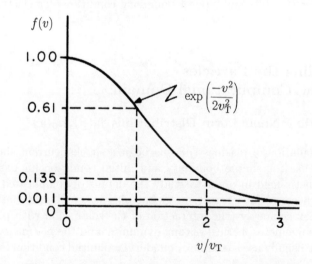

Fig. 7.1. Normalized thermal velocity distribution

7.2 Loading the Particles Inside the Computational Domain

$$\xi_s = F(\boldsymbol{v}) = \frac{\int_0^v \exp\left(-\frac{v^2}{2v_T^2}\right) d\boldsymbol{v}}{\int_0^\infty \exp\left(-\frac{v^2}{2v_T^2}\right) d\boldsymbol{v}}, \quad (7.3)$$

is set equal to a set of uniformly distributed numbers ξ_s, varying from 0 to 1, in order to obtain the values of v. In the one-dimensional case, the integration over $f(v)$ cannot be done explicitly, but it is done numerically, as in [51] (Code INIT in ES1), to produce a "quiet Maxwellian" distribution, with thermal velocity v_T. In the case of a two-dimensional isotropic thermal distribution, the integration in (7.3) can be done explicitly. The speed is $v = \sqrt{v_x^2 + v_y^2}$, angle $\theta = \arctan v_y/v_x$, $d\boldsymbol{v} = 2\pi v dv$. The inversion for speed v obtained in terms of ξ gives

$$v_s = v_T\sqrt{-2\ln \xi_s}. \quad (7.4)$$

Another set of uniform numbers ξ_m is chosen for the angles θ, over the range 0 to 2π, $\theta_m = 2\pi\xi_m$.

The text of the subroutines for the preparation of the initial array (macs3) and for the generation of particles with a Maxwellian velocity distribution (in0p3) is given in Table 7.1.

7.2.3 Loading a Ring Velocity Distribution

Let us use two systems of polar coordinates in velocity space (Fig. 7.2). The first one (dotted line) is located in the rest frame and the second one

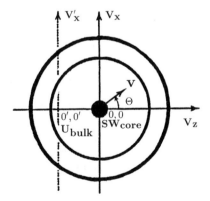

Fig. 7.2. Two systems of coordinates in velocity space. The first of them (*dotted line*) is connected with the frame at rest. The second one (*solid line*) is connected to the bulk velocity frame. The polar system and two-dimensional cut of the spherical system of coordinates are presented here. This scheme describes the ring and shell velocity distributions. The external ring corresponds to $v = V_{\max}$, while the internal ring corresponds to $v = V_{\min}$

(solid line) is located in the frame of the bulk velocity, $\boldsymbol{v} = \boldsymbol{U}_{\text{bulk}}$. Let v and θ be coordinates of this system (Fig. 7.2). We shall use the dimensionless components of velocity $v = v'/U_{\text{bulk}}$ and consider half of the ring distribution only. All particles with given v, θ are located in the intervals $V_{\text{min}} < v < V_{\text{max}}$ and $0 < \theta < \pi$.

The amount of particles within the homogeneous ring distribution, which occupy the region $[V_{\text{min}} : v, 0 : \theta]$, may be estimated by the following equation:

$$S(v, V_{\text{min}}, \theta) = \int_{V_{\text{min}}}^{v} v' dv' \int_{0}^{\theta} d\theta'$$

$$= S_v(v, V_{\text{min}}) S_\theta(\theta) = \frac{(v^2 - V_{\text{min}}^2)}{2} \theta. \quad (7.5)$$

The total amount of particles in the homogeneous ring distribution is

$$S(V_{\text{max}}, V_{\text{min}}, \pi) = S_v(V_{\text{max}}, V_{\text{min}}) S_\theta(\pi) = \frac{\pi(V_{\text{max}}^2 - V_{\text{min}}^2)}{2}. \quad (7.6)$$

The cumulative distribution function for the speed v,

$$\xi_1 = \frac{S_v(v, V_{\text{min}})}{S_v(V_{\text{max}}, V_{\text{min}})} \quad (7.7)$$

or

$$\xi_1 = \frac{v^2 - V_{\text{min}}^2}{V_{\text{max}}^2 - V_{\text{min}}^2}, \quad (7.8)$$

is set equal to a set of uniformly distributed numbers ξ_{1s}, varying from 0 to 1, in order to obtain the values of v. The value of v_s may be found by inversion of the above equation

$$v_s = \sqrt[2]{V_{\text{max}}^2 + (1 - \xi_{1s}) V_{\text{min}}^2}. \quad (7.9)$$

The cumulative distribution function for angle θ,

$$\xi_2 = \frac{S_\theta(\theta)}{S_\theta(\pi)} = \frac{\theta}{\pi}, \quad (7.10)$$

is set equal to a set of uniformly distributed numbers ξ_{2s}, varying from 0 to 1, in order to obtain the values of θ. Then θ_s may be found from

$$\theta_s = \xi_2 \pi. \quad (7.11)$$

The text of the subroutines for the preparation of the initial array (macs3) and for the generation of particles with the velocity ring distribution (in0c3) is given in Table 7.1.

7.2 Loading the Particles Inside the Computational Domain

Table 7.1. Subroutines for the generation of specific distribution. macs0: initial array; ee: initial array; anorm: initial array; macs3: initial array; rnorm: initial array; in0p3: Maxwellian velocity distribution; in0c3: velocity shell (ring) distribution; injp2: particle flux with Maxwellian velocity distribution; injc3: particle flux with velocity shell (ring) distribution; macs4: initial array; excha2: charge exchange process

```
SUBROUTINE MACS0
COMMON/MACS/U(101),IKV
IKV=101
VMAX=3
DVD=2.*VMAX/1000.
VD=-VMAX
S=0.
DO 1 J=1,1000
VD=VD+DVD
1  S=S+EXP(-VD**2.5)*DVD
DEL=S
S=0.
VD=-VMAX
DO 2 I=2,100
EM=(I-1.)/(IKV-1.)
S=S+EXP(-VD**2.5)*DVD/DEL
VD=VD+DVD
IF(EM.GT.S)GOTO 3
U(I)=VD
3  CONTINUE
U(1)=-VMAX
U(IKV)=VMAX
RETURN
END
REAL FUNCTION EE(I,J)
A=1.
EA=(A*(I-.5))/J
IF(EA.GT.1.) EA=EA-1.
EE=EA
RETURN
END
REAL FUNCTION ANORM(X)
COMMON/MACS/U(101),IKV
Y=X*(IKV-1)+1
I=Y
ANORM=U(I)*(I+1-Y)+U(I+1)*(Y-I)
RETURN
END
SUBROUTINE MACS3
INCLUDE 'commoc.f'
COMMON/MACSR/U(101),IKV
IKV=101
VMIN=0.
VMAX=VT1*3.
DVD=(VMAX-VMIN)/1000.
XX=VMIN
DO 1 I=1,1001
XX=XX+DVD*(I-0.5)+VMIN
S=S+DVD*EXP(-0.5*(XX/VT1)**2)*XX*
*(XX+0.5*(XX**2+1.))
1  CONTINUE
DEL=S
S=0.
DO 2 I=2,100
EM=(I-1.)/(IKV-1.)
S=0.
XX=VMIN
VD=VD-0.5*DVD
S=S+DVD*EXP(-0.5*(XX/VT1)**2)*XX*
*(XX+0.5*(XX**2+1.))/DEL
IF(EM.GT.S)GOTO 3
U(I)=VD
2  CONTINUE
U(1)=VMIN
U(IKV)=VMAX
PRINT 100, U
100 FORMAT(5F12.4)
RETURN
END
REAL FUNCTION RNORM(X)
REAL X
COMMON/MACSR/U(101),IKV
Y=X*(IKV-1)+1
I=Y
RNORM=U(I)*(I+1-Y)+U(I+1)*(Y-I)
RETURN
END
SUBROUTINE IN0P3
INCLUDE 'param3.f'
INCLUDE 'commoc.f'
INCLUDE 'commoc.f'
COMMON /A9/IA,IC,IB,NPC,NPCD,NWC,NPE,
*NPED,NWE,NDP,NINJ,NI23
COMMON /A11/AL2,K,ISI,ZSH,KK
COMMON /MF9/DTB,DEL4,HN2,HN1
COMMON /DIST/XCOR(NPDX),YCOR(NPDY),DDIST
DDIST=0.0
SVZ2=0.
SVX2=0.
SVY2=0.
SVXI=0.
SVZI=0.
DO 1 K=1,NPDZ
Z1=RG21-3.*DDZ2+0.5*AL2-AL2*(K-1.)/(NINJ)
K2=IA
A=F9(Z1,DTB,DEL4,HN2,HN1)
A=1.
AKSI=(2.*A-1.-DNSH)/(DNSH-1.)
AKSI=0.
DO 2 J=1,NPDY
DO 2 I=1,NPDX
N=K2+I+NPDX*(J-1)
E=EE(I,NPDX)
DELTA=RAN(IRANE)
E1=-.00001+.999*DELTA
E1=.5
E2=(E+(E1-0.5)/NPDX)
X1=DX*E2
DELTA=RAN(IRANE)
E=EE(J,NPDY)
E1=-.00001+.999*DELTA
E1=.5
E2=(E+(E1-0.5)/NPDY)
Y1=DY*E2
DELTA=RAN(IRANE)
E=.00001+.999*DELTA
E=.5
Z2=Z1+AL2*(E-.5)/(A*NINJ)
X(N)=X1
Y(N)=Y1
Z(N)=Z2
L=I/2
IF(L*2.EQ.I) GOTO 6
VT=VT1
DVX=(UX0*(1.-AKSI)+UX2*(1.+AKSI))*.5
DVY=(UY0*(1.-AKSI)+UY2*(1.+AKSI))*.5
DVZ=(UZ0*(1.-AKSI)+UZ2*(1.+AKSI))*.5
DELTA=RAN(IRANP)
E=.00001+.999*ANORM(E)+DVX
VX1=VT*ANORM(E)+DVX
DELTA=RAN(IRANP)
VY1=VT*ANORM(E)+DVY
DELTA=RAN(IRANP)
E=.00001+.999*DELTA
VZ1=VT*ANORM(E)+DVZ
U(N)=VX1
```

Table 7.1. (continued)

```
      V(N)=VY1
      W(N)=VZ1
   6  CONTINUE
      GOTO 7
      U(N)=-VX1+2.*DVX
      V(N)=-VY1+2.*DVY
      W(N)=-VZ1+2.*DVZ
   7  CONTINUE
      IA=IA+NPDX*NPDY
   2  CONTINUE
   1  CONTINUE
      RETURN
      END
      SUBROUTINE INOC3
      INCLUDE 'param3.f'
      INCLUDE 'commo6.f'
      COMMON /A9/NPS,NPSD,NWS,IA,IC,IB,NPE,NPED,NWE,NDP,NINJ
      COMMON /A11/AL2,K,IS1,ISH,KK
      COMMON /MFP/DTB,DEL4,HN2,HN1
      SINTFI=SIN(TFI)
      COSTFI=COS(TFI)
      DV3=VTT5-VTT3
      DV4=VTT6-VTT4
      Z0=RG21+3.*DDZ+0.5*AL2
      DO 1 K =1,KK
      K2=IA+IP1
      AKSI1Q0.*A.1-DNSH)/(DNSH-1.)
      AKSI=(2.*A.1-DNSH)/(DNSH-1.)
      DO 2 J=1,NIDY
      DO 2 I=1,NIDX
      N=K2+I+NIDX*(J-1)
      DELTA=RAN(IRANE)
      E=.00001+.999*DELTA
      E2=EE(I,NIDX)+(E-0.5)/NIDX
      X1=DX*E2
      DELTA=RAN(IRANE)
      E2=EE(J,NIDY)+(E-0.5)/NIDY
      Y1=DY*E2
      IF(IRING.NE.1) GOTO 6
      DELTA=RAN(IRANC)
      E=.00001+.999*DELTA
      Z1=Z0+AL2*(E-.5)/(A*NINJ)
      DVX=(UX0*(1.-AKSI)+UX2*(1.+AKSI))*.5
      DVY=(UY0*(1.-AKSI)+UY2*(1.+AKSI))*.5
      DVZ=(UZ0*(1.-AKSI)+UZ2*(1.+AKSI))*.5
      VT=(VTT3*(1.-AKSI)+VTT4*(1.+AKSI))*.5
     *+(DV3*(1.-AKSI)+DV4*(1.-AKSI))*.5*
     *(.00001+.999*DELTA)
      DV=VT*(1.-AKSI)*DVSHTT*(1.+AKSI))*.5
      DELTA=RAN(IRANC)
      E=.00001+.999*DELTA
      VX=VT*SIN(E*2.*3.14)
      WNN=DV+VT*COS(E*2.*3.14)
      VY=-WNN*SIN(TETD)
      VZ1=WNN*COS(TETD)
      VX1=VX*SINTFI+VY*COSTFI
      VY1=VY*SINTFI+VX*COSTFI
   6  IF(ISHELL.NE.1) GOTO 7
      DELTA=RAN(IRANC)
      E=.00001+.999*DELTA
      VCMIN=(VT3*(1.-AKSI)+VT4*(1.+AKSI))*.5
      VCMAX=(VT5*(1.-AKSI)+VT6*(1.+AKSI))*.5
      VR=(VCMIN**3*(1.-E)+VCMAX**3*E)**(1./3.)
      DELTA=RAN(IRANC)
      E=.00001+.999*DELTA
      COSTET=1.-2.*E
      SINTET=SQRT(1.-COSTET*COSTET)
      DELTA=RAN(IRANC)
      E=.00001+.999*DELTA
      PHI=2.*PI*E
      SINPHI=SIN(PHI)
      COSPHI=COS(PHI)
      VX1=VR*SINTET*COSPHI
      VY1=VR*SINTET*SINPHI
      VZ1=VR*COSTET
   7  CONTINUE
      X(N)=X1
      Y(N)=Y1
      Z(N)=Z1
      U(N)=VX1+DVX
      V(N)=VY1+DVY
      W(N)=VZ1+DVZ
   2  CONTINUE
      IA=IA+NIDX*NIDY
      IF(I0.LT.-RLARMP*ADJUST) GO TO 5
      Z0=Z0-AL2/(A*NINJ)
   1  CONTINUE
   5  CONTINUE
      RETURN
      END
      SUBROUTINE INJP2
      INCLUDE 'param2.f'
      INCLUDE 'commoc.f'
      INCLUDE 'commo6.f'
      COMMON/A9/IA,IC,IB,NPC,NPCD,NWC,NPE,NPED,NWE,NDP,NINJ
      COMMON/A11/AL2
      K2=IA
      DO 2 I=1,NPDX
      N=K2+I
      Z(N)=RG21+3.*AL2*AL2*(I-1.)/(NPDX*NINJ)
      X(N)=XX*E
      DELTA=RAN(IRANC)
      E=EE(I,NPDX)
      VR=RNORM(E)
      DELTA=RAN(IRANC)
      E=.00001+.999*DELTA
      A=VR/2.
      B=-1.
      IF(VR.GT.1.) C=1.5/VR-E*(1.+.5*VR+.5/VR)
      IF(VR.LT.1.) C=1.-VR+.5-E*2.
      D=B**2-4.*A*C
      COSTET=(-B-SQRT(B*B-4.*A*C))/(2.*A)
      VZ1=-1.+VR*COSTET
      SINTET=SQRT(1.-COSTET*COSTET)
      DELTA=RAN(IRANC)
      E=.00001+.999*DELTA
      PHI=E*2.*PI
      SINPHI=SIN(PHI)
      COSPHI=COS(PHI)
      VX1=VR*SINTET*COSPHI
      VY1=VR*SINTET*SINPHI
      U(N)=VX1
      V(N)=VY1
      W(N)=VZ1
   2  CONTINUE
      IF(IA.GE.IP1) STOP 3
      RETURN
      END
      SUBROUTINE INJC3
      IA=IA-IC
      INCLUDE 'param3.f'
      INCLUDE 'commo6.f'
      INCLUDE 'commoc.f'
      COMMON /A9/NPS,NPSD,NWS,IA,IC,IB,NPE,NPED,NWE,NDP,NINJ
```

7.2 Loading the Particles Inside the Computational Domain

Table 7.1. (continued)

```
      COMMON/A1/AL2
      COMMON/A7/ANK(11,201),ANKE(21,201),ANF(250),ANO(301),DN,DR
      SINTFI=SIN(TFI)
      COSTFI=COS(TFI)
      AV3=VTT5-VTT3
      K2=IA+IP1
      DO 2 J=1,NIDY
      DO 2 I=1,NIDX
      N=K2+I+NIDX*(J-1)
      DELTA=RAN(IRANE)
      E=.00001+.999*DELTA
      Z0=RG21+3.*DDZ*AL2*(E-0.5)/NINJ
      DO 21 J=1,101
      DELTA=RAN(IRANE)
      E=.00001+.999*DELTA
      E2=EE(I,NIDX)+(E-0.5)/NIDX
      X1=DX*E2
      DELTA=RAN(IRANE)
      E=.00001+.999*DELTA
      E2=EE(J,NIDY)+(E-0.5)/NIDY
      Y1=DY*E2
      Z1=Z0
      X(N)=X1
      Y(N)=Y1
      Z(N)=Z1
      IF(IRING.NE.1) GOTO 3
      DELTA=RAN(IRANC)
      M=(.00001+.999*DELTA)*200.+1
      DELTA=RAN(IRANC)
      E=.00001+.999*DELTA
      AK=E*10.+1.
      K=AK
      AK=AK-K
      VT=VTT3+AV3*E
      FI1=ANK(K,M)
      FI2=ANK(K+1,M)
      FI=FI1*(1-AK)+FI2*AK
      WNN=WTT+VT*COS(FI)
      VX=VT*SIN(FI)
      VZ1=WNN*COS(TETD)
      VX1=VX*SINTFI+VY*COSTFI
      VY1=VX*SINTFI+VX*COSTFI
      GOTO 4
    3 IF(ISHELL.NE.1) GOTO 4
      DELTA=RAN(IRANC)
      E=.00001+.999*DELTA
      VR=RNORM(E)
      DELTA=RAN(IRANC)
      E=.00001+.999*DELTA
      A=VR/2.
      B=-1.
      C=-1./(2.*VR)-E*(1.+VR/2.+1./(2.*VR))
      D=B**2-4.*A*C
      COSTET=(-B+SQRT(B**2-4.*A*C))/(2.*A)
      VZ1=1.-VB*COSTET
      SINTET=SQRT(1.-COSTET*COSTET)
      DELTA=RAN(IRANC)
      E=.00001+.999*DELTA
      PHI=E*2.*PI
      SINPHI=SIN(PHI)
      COSPHI=COS(PHI)
      VX1=VR*SINTET*COSPHI
      VY1=VR*SINTET*SINPHI
    4 CONTINUE
      U(N)=VX1
      V(N)=VY1
      W(N)=VZ1
    2 CONTINUE
      IA=IA+NIDX*NIDY
      RETURN
      END
      SUBROUTINE MACS4
      INCLUDE 'param2.f'
      COMMON/MACSW/UFI(101,101,101),URP(101,101),IKV
      COMMON/A12/DNSH,DVSH,DTSH,VT1,VT2,DPSH,VT3,VT4,VT5,VT6
      DIMENSION STORP(401)
      IKV=101
      VMAX=3.
      DRN=20./100.
      DO 1 K=1,101
      RN=(K-1.)*DRN
      DRP=VMAX/100.
      DO 1 I=1,101
      RP=(I-1.)*DRP
      S=0
      DFI=2.*PI/400.
      DO 3 J=1,400
      FI=DFI*(J-0.5)
      A=SQRT(RN**2+RP**2-2.*RN*RP*COS(FI)+.00001)
      ANU=A
      S=S+DFI*ANU
    3 STOTAL=S
      FI=-DFI*.5
      S=0.
      DO 4 J=2,100
      EM=(J-1.)/(IKV-1.)
      FI=FI+DFI
      A=SQRT(RN**2+RP**2-2.*RN*RP*COS(FI)+.00001)
      ANU=A
      S=S+DFI*ANU/STOTAL
      IF(EM.GT.S) GOTO 5
      UFI(K,I,J)=FI
      CONTINUE
    4 CONTINUE
      UFI(K,I,1)=0.
      UFI(K,I,IKV)=2.*PI
    2 CONTINUE
    1 CONTINUE
      DO 6 K=1,101
      RN=(K-1.)*DRN
      DFI=2.*PI/400.
      DRP=VMAX/400.
      S=0
      DO 7 I=1,400
      RP=(I-0.5)*DRP
      S1=0.
      DO 8 J=1,400
      FI=DFI*(J-0.5)
      A=SQRT(RN**2+RP**2-2.*RN*RP*COS(FI)+.00001)
      ANU=A
      S1=S1+ANU
    8 S1=S1*EXP(-RP**2.5)
      STORP(I)=S1
      S=S+S1
    7 CONTINUE
      STOTAL=S
      RP=-DRP*.5
      S=0.
      DO 9 I=2,100
      ITERA=0
      EM=(I-1.)/(IKV-1.)
   10 RP=RP+DRP
      ITERA=ITERA+1
      S=S+STORP(ITERA)/STOTAL
      IF(EM.GT.S) GOTO 10
      URP(K,I)=RP
    9 CONTINUE
      URP(K,1)=0.
      URP(K,IKV)=VMAX
    6 CONTINUE
      RETURN
```

Table 7.1. (continued)

```
      END
      SUBROUTINE EXCHA2(NP1,NP2,EM)
      INCLUDE 'param2.f'
      INCLUDE 'commoc.f'
      INCLUDE 'commo1.f'
      INCLUDE 'commo6.f'
      INCLUDE 'commo4.f'
      COMMON /PIOXY/ N12,IE
      COMMON /A9/ NFS,NPSD,NWS,NPC,NPCD,NWC,NPE,NPED,NWE,NDP,NINJ,N123
      COMMON/A11/AL2
      COMMON/SP/NSPEC(31),IDNP(31),DVSPEC,NSPEP(31),IDNC(31),DVSPEP
      COMMON/TERAT/ITER,TETTA,NITER
      COMMON/MACSW/UFI(101,101,101),URP(101,101)
      DRN=20./100.
      VMAX=3.
      DRP=VMAX/100.
      RK4=RG22+.1
      RK3=RG21
      RDX=1./DDX
      DDDX=1./DX
      A44=DX
      IK=MX-1
      IK1=MX+1
      HDR5=.5*DDX
      DPI=100+1
      JK=MZ-1
      RDZ=1./DDZ
      JK1=MZ+1
      RK5=RO1+.001
      RK6=RK3-.001
      RK7=RG21+4.5*AL2
      FIHK=IHK
      DO 1 N=NP1,NP2
      IF(P(N).GT.0.) GOTO 10
 6    IF (Z(N).LT.DZ+1.AND.Z(N).GT.RO1) GOTO 7
      DENPA=ANFROT
      UPA=0.
      WPA=-1.
      VTPA=0.5
 35   CONTINUE
      IB=1
      GOTO 8
 7    IB=0
      A45=(Z(N)-1.)*RDZ+1.
      J1=A45
      I1=IFIX(X(N)*RDX+1.)
      AI=I1
      AJ=J1
      IF(I1.LT.1) I1=1
      IF(I1.GT.IK) I1=IK
      IF (J1.GT.JK) J1=JK
      IF (J1.LT.1) J1=1
      N6=MZ*(I1-1)+J1
      N7=N6+1
      N8=N6+MZ
      N9=N8+1
      A5=UP(N6)
      A9=VP(N6)
      A13=WP(N6)
      A17=VTP(N6)
      A6=UP(N7)
      A10=VP(N7)
      A14=WP(N7)
      A18=VTP(N7)
      A7=UP(N8)
      A11=VP(N8)
      A15=WP(N8)
      A19=VTP(N8)
      A8=UP(N9)
      A12=VP(N9)
      A16=WP(N9)
      A20=VTP(N9)
 9    C1=A45-AJ
      C2=1.-C1
      C3=X(N)*RDX-AI+1.
      C4=1.-C3
      C5=C2*C4
      C6=C1*C4
      C7=C2*C3
      C8=C1*C3
      UPA=(A5*C5+A6*C6+A7*C7+A8*C8)
      VPA=(A9*C5+A10*C6+A11*C7+A12*C8)
      WPA=(A13*C5+A14*C6+A15*C7+A16*C8)
      VTPA=(A17*C5+A18*C6+A19*C7+A20*C8)
 8    CONTINUE
      AMODV=SQRT(U(N)**2+(W(N)-WPA)**2)
      ARTFI=ASIN(U(N)/(AMODV+0.00001))
      K=AMODV/(VTPA*DRN)+1
      E=0.0001+.9999*RAN(IEXCH)
      II=E*100+1
      VRHO=URP(K,II)
      I=VRHO/DRP+1
      E=0.0001+.9999*RAN(IEXCH)
      J=E*100+1
      PI=UFI(K,I,J)
      FITOT=ARTFI+FI
      U(N)=UPA+VTPA*VRHO*SIN(FITOT)
      E=0.0001+.9999*RAN(IEXCH)
      V(N)=VPA+VTPA*ANORM(E)
      W(N)=WPA+VTPA*VRHO*COS(FITOT)
      P(N)=-ALOG(RAN(IEXCH0))
 10   CONTINUE
 1    CONTINUE
      GOTO 14
 13   KFIN=1
 14   CONTINUE
      RETURN
      END
```

7.2.4 Loading a Shell Velocity Distribution

Let us use two systems of spherical coordinates in velocity space (Fig. 7.2). The first one (dotted line) is located in the rest frame and the second one (solid line) is located in the frame of the bulk velocity, $\boldsymbol{v} = \boldsymbol{U}_{\text{bulk}}$. Let v, θ and ϕ to be coordinates of this system, Fig. 7.2. We shall use the dimensionless components of velocity $v = v'/U_{\text{bulk}}$. All particles with given v, θ, ϕ are located in the intervals $V_{\text{min}} < v < V_{\text{max}}$, $0 < \theta < \pi$ and $0 < \phi < 2\pi$. The amount of particles with the homogeneous shell distribution, which occupy the region $[V_{\text{min}} : v, 0 : \theta, 0 : \phi]$, may be estimated by

$$S(v, V_{\text{min}}, \theta, \phi) = \int_0^\phi d\phi' \int_{V_{\text{min}}}^v v'^2 dv' \int_0^\theta \sin\theta' d\theta'$$

$$= S_\phi(\phi) S_v(v, V_{\text{min}}) S_\theta(\theta)$$

$$= \frac{(\phi - 0)(1 - \cos\theta)(v^3 - V_{\text{min}}^3)}{3}. \tag{7.12}$$

The total amount of particles with the homogeneous shell distribution is

$$S(V_{\text{max}}, V_{\text{min}}, \pi, 2\pi) = S_\phi(2\pi) S_v(V_{\text{max}}, V_{\text{min}}) S_\theta(\pi) = \frac{4\pi(V_{\text{max}}^3 - V_{\text{min}}^3)}{3}. \tag{7.13}$$

The cumulative distribution function for the speed v,

$$\xi_1 = \frac{S_v(v, V_{\text{min}})}{S_v(V_{\text{max}}, V_{\text{min}})} \tag{7.14}$$

or

$$\xi_1 = \frac{v^3 - V_{\text{min}}^3}{V_{\text{max}}^3 - V_{\text{min}}^3}, \tag{7.15}$$

is set equal to a set of uniformly distributed numbers ξ_{1s}, varying from 0 to 1, in order to obtain the values of v. The value of v_s may be found by inversion of the above equation

$$v_s = \sqrt[3]{V_{\text{max}}^3 + (1 - \xi_{1s}) V_{\text{min}}^3}. \tag{7.16}$$

The cumulative distribution function for angle θ,

$$\xi_2 = \frac{S_\theta(\theta)}{S_\theta(\pi)} = \frac{S_\theta(\theta)}{2}, \tag{7.17}$$

is set equal to a set of uniformly distributed numbers ξ_{2s}, varying from 0 to 1, in order to obtain the values of θ. Then θ may be found from

$$\cos\theta = 1 - 2\xi_2. \tag{7.18}$$

Finally, the cumulative distribution function for angle ϕ,

$$\phi = 2\pi \xi_3, \tag{7.19}$$

is a set equal a set of random numbers ξ_{3s}, varying from $0 \leq \xi_{3s} \leq 1$, in order to obtain the values of ϕ.

The text of the subroutines for the preparation of the initial array (macs3) and for the generation of particles with the velocity shell distribution (in0c3) is given in Table 7.1.

7.3 Particle Injection at Boundaries

7.3.1 Loading a Maxwellian Velocity Distribution Flux

Let us use two systems of spherical coordinates in velocity space (Fig. 7.3). The first one (dotted line) is located in the rest frame and the second one (solid line) is located in the frame of the bulk velocity, $\boldsymbol{v} = \boldsymbol{U}_{\text{bulk}}$. Let v, θ and ϕ be coordinates of the second system (Fig. 7.3). We shall use the dimensionless components of velocity $v = v'/U_{\text{bulk}}$. The z-component of the particle velocity will be $v'_z = 1 + v\cos\theta$. All particles with given v and positive component of velocity v'_z are located inside the interval $0 < \theta < \theta^*$.

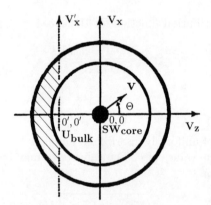

Fig. 7.3. Two systems of coordinates in velocity space. The first of them (*dotted line*) is connected with the frame at rest. The second one (*solid line*) is connected to the bulk velocity frame. The polar system and two-dimensional cut of the spherical system of coordinates are presented here. This scheme describes the ring and shell velocity distributions. The *shaded region* of the velocity distribution corresponds to the particles with negative flux. The external ring corresponds to $v = V_{\text{max}}$, while the internal ring corresponds to $v = V_{\text{min}}$

The boundary value of angle θ for particles with $v > 1$ is determined by

$$1 + v\cos\theta^* = 0 \tag{7.20}$$

or
$$\cos\theta^* = -\frac{1}{v}.$$

If the value of velocity $v \leq 1$, then $\theta^* = \pi$.

The flux of the particles with a Maxwellian distribution, which occupy the region $[0:v, 0:\theta, 0:\phi]$, may be split into the production of two parts:

$$I(v,\theta,\phi) = \int_0^\phi d\phi' \int_0^v v'^2 \exp\left(-\frac{v'^2}{2v_T^2}\right) dv' \int_0^\theta v_z(\theta')\sin\theta' d\theta'$$

$$= I_\phi(\phi)I_{v,\theta}(v,\theta). \qquad (7.21)$$

The total flux for the Maxwellian distribution is described by

$$I(V_{\max},\theta^*,2\pi) = 2\pi \int_0^{V_{\max}} v'^2 \exp\left(-\frac{v'^2}{2v_T^2}\right) F dv'$$

$$= I_\phi(2\pi)I_{v,\theta}(V_{\max},\theta^*), \qquad (7.22)$$

where

$$F(v,\theta^*) = \int_0^{\theta^*(v)} v_z(\theta')\sin\theta' d\theta'. \qquad (7.23)$$

Integration of (7.23) gives $F = 2$ for $v \leq 1$ and $F = 1 + 0.5v + 0.5/v$ for $v > 1$. The cumulative distribution function for the speed v,

$$\xi_1 = \frac{I_{v,\theta}(v,\theta^*)}{I_{v,\theta}(V_{\max},\theta^*)}, \qquad (7.24)$$

is set equal to a set of uniformly distributed numbers ξ_{1s} varying from 0 to 1, in order to obtain the values of v. To find the value of v we have to prepare the array $v = v(\xi_1)$ in advance. Then we can use linear interpolation between the nearest elements of the array. When a random value of v is found, a random value of θ may be found from the following cumulative distribution function:

$$\xi_2 = \frac{I_\theta(v,\theta)}{F}, \qquad (7.25)$$

where $I_\theta(v,\theta)$ is the flux of the particles which occupy the layer $v = \text{const}$ and satisfy the condition $\theta < \theta^*$. The function I_θ is determined by

$$I_\theta(v,\theta) = \int_0^\theta v_z(\eta)\sin\eta d\eta = -\cos\theta + 1 - \frac{v}{2}(\cos^2\theta - 1). \qquad (7.26)$$

A set of pseudo-random numbers ξ_{2s} allows us to define the values of θ from the following quadratic equation:

$$ax^2 + bx + c = 0, \quad x = \cos\theta, \qquad (7.27)$$

where

176   7. Particle Loading and Injection. Boundary Conditions

$$a = \frac{v}{2}, \quad b = 1, \quad c = -1 - \frac{v}{2} + 2\xi_2$$

for $v < 1$, and

$$a = \frac{v}{2}, \quad b = 1, \quad c = -1 - \frac{v}{2} + \xi_2\left(1 + \frac{v}{2} + \frac{1}{2v}\right)$$

for $v > 1$. The solution of (7.27) is

$$x_{1,2} = \frac{-b \pm \sqrt{b^2 - 4ac}}{2a}. \tag{7.28}$$

Finally, the cumulative distribution function for ϕ,

$$\phi = 2\pi\xi_3, \tag{7.29}$$

is set equal to a set of uniformly distributed numbers ξ_{3s} varying from 0 to 1, in order to obtain the values of ϕ.

The text of the subroutines for the preparation of the initial array (macs3) and for the generation of the particle flux with a Maxwellian velocity distribution (injp2) is given in Table 7.1.

7.3.2 Loading a Ring Velocity Distribution Flux

Let us use two systems of polar coordinates in velocity space (Fig. 7.3). The first one (dotted line) is located in the rest frame and the second one (solid line) is located in the frame of the bulk velocity, $\boldsymbol{v} = \boldsymbol{U}_{\text{bulk}}$. Let v and θ be polar coordinates (Fig. 7.3). We shall use the dimensionless velocity $v = v'/U_{\text{bulk}}$. The z-component of the particle velocity will be $v'_z = 1 + v\cos\theta$. All particles with given velocity v and positive component of velocity v'_z are located inside the interval $0 < \theta < \theta^*$. The boundary value of angle θ for particles with $v > 1$ is determined by

$$1 + v\cos\theta^* = 0 \tag{7.30}$$

or

$$\cos\theta^* = -\frac{1}{v}.$$

If the value of velocity $v \leq 1$, then $\theta^* = \pi$.

The flux of the particles with an homogeneous ring distribution, which are located at the region $[V_{\min} : v, 0 : \theta]$, may be estimated by

$$I(v, V_{\min}, \theta) = \int_{V_{\min}}^{v} v'dv' \int_0^\theta v_z(\theta')d\theta'. \tag{7.31}$$

The total flux for the homogeneous ring distribution is represented by

$$I(V_{\max}, V_{\min}, \theta^*) = \int_{V_{\min}}^{V_{\max}} v' F dv', \qquad (7.32)$$

where the function F is determined by

$$F(v, \theta^*) = \int_0^{\theta^*} v_z(\theta') d\theta', \qquad (7.33)$$

and $F = \pi$ for $v \leq 1$ and $F = \arccos(-1/v) + v \sin[\arccos(-1/v)]$ for $v > 1$. The cumulative distribution function for the speed v using random numbers $0 \leq \xi \leq 1$ for

$$\xi_1 = \frac{I(v, V_{\min}, \theta^*)}{I(V_{\max}, V_{\min}, \theta^*)} \qquad (7.34)$$

is set equal to a set of uniformly distributed numbers ξ_{1s} varying from 0 to 1 in order to obtain the values of v.

$$\xi_1 = \frac{\Phi(v) - \Phi(V_{\min})}{\Phi(V_{\max}) - \Phi(V_{\min})} \qquad (7.35)$$

for $v < 1$, where $\Phi(v) = \sqrt{v^2 - 1}/6 + v^2 \arccos(-1/v)/2 + v^3 \sin(\arccos(-1/v))/3$, and

$$\xi_1 = \frac{v^2 - V_{\min}^2}{V_{\max}^2 - V_{\min}^2} \qquad (7.36)$$

for $v > 1$. The value of v may be found from the array $v = v(\xi)$, which must be prepared in advance using linear interpolation from the nearest element in the array.

If the maximum and minimum values satisfy the condition $V_{\min} < 1 < V_{\max}$, then the value of v may be found from

$$\alpha_1 = \frac{I_{v,\theta}(v=1, V_{\min}, \theta^*)}{I(v=1, V_{\min}, \theta^*) + I(V_{\max}, v=1, \theta^*)},$$

$$\xi_1 = \alpha_1 \frac{I(v, V_{\min}, \theta^*)}{I(v=1, V_{\min}, \theta^*)}, \quad \text{for} \quad 0 \leq \xi_1 \leq \alpha_1 \quad (v < 1),$$

$$\xi_1 = \alpha_1 + \alpha_1 \frac{I(v, v=1, \theta^*)}{I(v=1, V_{\min}, v=1, \theta^*)}, \quad \text{for} \quad \alpha_1 \leq \xi_1 \leq 1 \quad (v > 1).$$

When a random value of v is found, we can write a cumulative distribution function for the value of angle θ,

$$\xi_2 = \frac{I_\theta(v_s, \theta)}{F}, \qquad (7.37)$$

where $I_\theta(v_s, \theta)$ is the flux of the particles, which occupy the layer $v = \text{const}$ and satisfy the condition $\theta < \theta^*$. I_θ is determined by

$$I_\theta(v_s, \theta) = \int_0^\theta v_z(\eta) d\eta = \theta + v \sin \theta. \qquad (7.38)$$

The values of θ may be found from a set of uniformly distributed numbers ξ_{2s}, $0 < \xi_{2s} < 1$, and
$$\theta + v\sin\theta = \pi\xi_2 \tag{7.39}$$
for $v \leq 1$ and
$$\theta + v\sin\theta = \xi_2\{\arccos(-1/v) + v\sin[\arccos(-1/v)]\} \tag{7.40}$$
for $v > 1$. To find θ those equations may be solved numerically.

The text of the subroutines for the preparation of the initial array (macs3) and for the generation of the particle flux with a velocity ring distribution (injc3) is given in Table 7.1.

7.3.3 Loading a Shell Velocity Distribution Flux

Let us use two systems of spherical coordinates in velocity space (Fig. 7.3). The first one (dotted line) is located at rest frame and the second one (solid line) is located in the frame of the bulk velocity, $\boldsymbol{v} = \boldsymbol{U}_{\text{bulk}}$. Let v, θ and ϕ be coordinates of the second spherical system of coordinates (Fig. 7.3). We shall use the dimensionless velocity $v = v'/U_{\text{bulk}}$. The z-component of the particle velocity will be $v'_z = 1 + v\cos\theta$. All particles with given velocity v and positive component of velocity v'_z are located inside the interval $0 < \theta < \theta^*$, and $0 < \phi < 2\pi$. The boundary value of angle θ for particles with $v > 1$ is determined by
$$1 + v\cos\theta^* = 0 \tag{7.41}$$
or
$$\cos\theta^* = -\frac{1}{v}.$$

If the value of velocity $v \leq 1$, then $\theta^* = \pi$.

The flux of the particles with a homogeneous shell distribution, which are located in the region $[V_{\min}:v, 0:\theta, 0:\phi]$, may be estimated by
$$I(v, V_{\min}, \theta, \phi) = \int_0^\phi d\phi' \int_{V_{\min}}^v v'^2 dv' \int_0^\theta v_z(\theta')\sin\theta' d\theta'$$
$$= I_\phi(\phi) I_{v,\theta}(v, V_{\min}, \theta). \tag{7.42}$$

The total flux for the homogeneous shell distribution is represented by
$$I(V_{\max}, V_{\min}, \theta^*, 2\pi) = I_\phi(2\pi) I_{v,\theta}(V_{\max}, V_{\min}, \theta^*) = 2\pi \int_{V_{\min}}^{V_{\max}} v'^2 F dv', \tag{7.43}$$
where the function $F = I_\theta(v, \theta^*)$ is determined by (7.23), and $F = 2$ for $v \leq 1$ and $F = 1 + 0.5v + 0.5/v$ for $v > 1$. The cumulative distribution function for the speed v using random numbers $0 \leq \xi \leq 1$ for

7.3 Particle Injection at Boundaries   179

$$\xi_1 = \frac{I_{v,\theta}(v, V_{\min}, \theta^*)}{I_{v,\theta}(V_{\max}, V_{\min}, \theta^*)} \qquad (7.44)$$

is set equal to a set of uniformly distributed numbers ξ_{1s} varying from 0 to 1 in order to obtain the values of v.

$$\xi_1 = \frac{3v^4 + 8v^3 + 6v^2 - 3V_{\min}^4 - 8V_{\min}^3 - 6V_{\min}^2}{3V_{\max}^4 + 8V_{\max}^3 + 6V_{\max}^2 - 3V_{\min}^4 - 8V_{\min}^3 - 6V_{\min}^2} \qquad (7.45)$$

for $V_{\min} > 1$, and

$$\xi_1 = \frac{v^3 - V_{\min}^3}{V_{\max}^3 - V_{\min}^3} \qquad (7.46)$$

for $V_{\max} < 1$. The value of v may be found from the array $v = v(\xi)$, which must be prepared in advance using linear interpolation from the nearest element in the array.

If the maximum and minimum values satisfy the condition $V_{\min} < 1 < V_{\max}$, then the value of v may be found from

$$\alpha_1 = \frac{I_{v,\theta}(v = 1, V_{\min}, \theta^*)}{I_{v,\theta}(v = 1, V_{\min}, \theta^*) + I_{v,\theta}(V_{\max}, v = 1, \theta^*)},$$

$$\xi_1 = \alpha_1 \frac{I_{v,\theta}(v, V_{\min}, \theta^*)}{I_{v,\theta}(v = 1, V_{\min}, \theta^*)}, \quad \text{for} \quad 0 \leq \xi_1 \leq \alpha_1 \quad (v < 1),$$

$$\xi_1 = \alpha_1 + \alpha_1 \frac{I_{v,\theta}(v, v = 1, \theta^*)}{I_{v,\theta}(v = 1, V_{\min}, \theta^*)}, \quad \text{for} \quad \alpha_1 \leq \xi_1 \leq 1 \quad (v > 1).$$

When a random value of v is found, we can write a cumulative distribution function for the value of angle θ,

$$\xi_2 = \frac{I_\theta(v_s, \theta)}{F}, \qquad (7.47)$$

where $I_\theta(v_s, \theta)$ is the flux of the particles, which occupy the layer $v = $ const and satisfy the condition $\theta < \theta^*$. I_θ is determined by

$$I_\theta(v, \theta) = \int_0^\theta v_z(\eta) \sin \eta d\eta = -\cos\theta + 1 - \frac{v}{2}(\cos^2\theta - 1). \qquad (7.48)$$

The values of θ may be found from a set of uniformly distributed numbers ξ_{2s}, $0 < \xi_{2s} < 1$, and

$$ax^2 + bx + c = 0, \quad x = \cos\theta, \qquad (7.49)$$

where

$$a = \frac{v}{2}, \quad b = 1, \quad c = -1 - \frac{v}{2} + 2\xi_2$$

for $v < 1$, and

$$a = \frac{v}{2}, \quad b = 1, \quad c = -1 - \frac{v}{2} + \xi_2\left(1 + \frac{v}{2} + \frac{1}{2v}\right)$$

for $v > 1$. The solution of (7.49) is

$$x_{1,2} = \frac{-b \pm \sqrt{b^2 - 4ac}}{2a}. \tag{7.50}$$

Finally, the cumulative distribution function for angle ϕ is

$$\phi = 2\pi\xi_3, \tag{7.51}$$

and the values of ϕ may be found from a set of uniformly distributed numbers ξ_{3s}, $0 \leq \xi_{3s} \leq 1$.

The text of the subroutines for the preparation of the initial array (macs3) and for the generation of the particle flux with a velocity shell distribution (injc3) is given in Table 7.1.

7.4 Charge Exchange Processes

The simulation of a partially ionized medium should follow the Boltzmann approach (Sect. 2.4). One of the main steps of the Boltzmann simulation is a description of the charge exchange process, namely, the frequency of generation of new atoms:

$$\nu \propto |\boldsymbol{v}_\mathrm{H} - \boldsymbol{v}_\mathrm{p}|\sigma_\mathrm{ex} \exp\left(-\frac{(\boldsymbol{v}_\mathrm{p} - \boldsymbol{U}_\mathrm{p})^2}{2v_\mathrm{T}^2}\right), \tag{7.52}$$

where $\boldsymbol{v}_\mathrm{H}$ and $\boldsymbol{v}_\mathrm{p}$ are the velocities of atoms and ions, $\boldsymbol{U}_\mathrm{p}$ is the bulk velocity of ions and v_T is the thermal velocity of ions. The cross-section σ depends on the relative velocities of the atoms and ions $|\boldsymbol{v}_\mathrm{H} - \boldsymbol{v}_\mathrm{p}|$ (see Sect. 2.4). Here we assume that ions have a Maxwellian velocity distribution function. Let us introduce the new coordinate system, which is shifted at the value of the ion bulk velocity $\boldsymbol{U}_\mathrm{p}$. The transformation to dimensionless velocity gives: $v' = v/v_\mathrm{T}$. In the two-dimensional case, let us introduce a polar system of coordinates r, ϕ in the velocity plane $v_x - v_y$, so that $\boldsymbol{r}$ is collinear with the velocity of the H atom $\boldsymbol{v}_\mathrm{H}$. Then, taking into account the weak dependence of σ on the relative velocity, (7.52) may be transformed to

$$\nu \propto \sqrt{r^2 + r_\mathrm{p}^2 - 2rr_\mathrm{p}\cos\phi}\,\exp\left(-\frac{r_\mathrm{p}^2}{2}\right). \tag{7.53}$$

The cumulative distribution function may be written as

$$D(r_\mathrm{p}, \phi) = \frac{\int_0^{r_\mathrm{p}} \int_0^{\phi} \nu(r, r'_\mathrm{p}, \phi') r'_\mathrm{p} dr'_\mathrm{p} d\phi'}{\int_0^{r_\mathrm{max}} \int_0^{2\pi} \nu(r, r'_\mathrm{p}, \phi') r'_\mathrm{p} dr'_\mathrm{p} d\phi'}. \tag{7.54}$$

7.5 Boundary Conditions for Particles and the Electromagnetic Field

However, the cumulative distribution function $D(r_p, \phi)$ may be split into two one-dimensional cumulative distribution functions. The cumulative distribution function for r_p has the form

$$\xi_1 = D(r_p) = \frac{\int_0^{r_p} \int_0^{2\pi} \nu(r, r'_p, \phi') r'_p dr'_p d\phi'}{\int_0^{r_{max}} \int_0^{2\pi} \nu(r, r'_p, \phi') r'_p dr'_p d\phi'}. \tag{7.55}$$

Equating $D(r_p)$ to a uniform distribution of numbers ξ_{1s}, $0 < \xi_{1s} < 1$, will produce the values of r_p. Equation (7.55) may be solved numerically. If the values $r_{ps} = r_p(\xi_{1s})$ are found, then one can find the value of ϕ from the next cumulative distribution function for ϕ

$$\xi_2 = D(\phi) = \frac{\int_0^{\phi} \nu(r, r'_p, \phi') d\phi'}{\int_0^{2\pi} \nu(r, r'_p, \phi') d\phi'}. \tag{7.56}$$

Equating $D(\phi)$ to a uniform distribution of numbers ξ_{2s}, $0 < \xi_{2s} < 1$, will produce the values of ϕ. The above equation $\phi = \phi(\xi_2)$ may be resolved numerically again. Returning back to the initial system of coordinates, we can write the final velocity of created atoms:

$$\theta = \arctan \frac{v_{H,y} - U_{p,y}}{v_{H,x} - U_{p,x}} + \phi, \tag{7.57}$$

$$v_x = r_p v_T \sin\theta + U_{p,x},$$
$$v_y = r_p v_T \cos\theta + U_{p,y}. \tag{7.58}$$

Tables 7.1 gives the text of the subroutines for generation the distribution (7.52) (macs4) and the charge exchange process (excha2).

7.5 Boundary Conditions for Particles and the Electromagnetic Field

7.5.1 Plasma–Vacuum Interface

Very often plasma systems include regions with vanishing plasma density. Simulation of such regions needs a very small time step in the case of the full Maxwell's equations. The reduced Maxwell's equations which are used in hybrid models need a special equation at a plasma–vacuum interface.

These equations are derived under the assumption that v_A/c is small, where v_A is the Alfvén speed. But as $n_i \to 0$, $v_A \to 0$, negating our premise. The schemes presented in Sects. 6.2.1–6.2.3 are limited in that they are sensitive to the particle density. Long before $n_i = 0$, any explicit numerical scheme will become unstable since the CFL condition is violated. In this case the CFL condition is approximately $\Delta t \leq \Delta x / v_A$, a standard result for hyperbolic systems. In certain cases, the ion response can be ignored and a parabolic system results. Then the CFL condition is $\Delta t \leq \Omega_i \Delta x^2 / 2 v_A^2$, which is typically more stringent than the hyperbolic condition [227]. To avoid this problem, $\nabla^2 \boldsymbol{E} = 0$ is solved in the low-density regions. The cutoff high- and low-density regions are always selected to satisfy the CFL condition [238, 345]. To solve the vacuum equation, a standard Gauss–Seidel method could be employed. For example, the x component of E is

$$E_{i,j,k}^x = \frac{1}{2(\Delta x^{-2} + \Delta y^{-2} + \Delta z^{-2})} \left(\frac{E_{i+1,j,k}^x + E_{i-1,j,k}^x}{\Delta x^2} \right.$$
$$\left. + \frac{E_{i,j+1,k}^x + E_{i,j-1,k}^x}{\Delta y^2} + \frac{E_{i,j,k+1}^x + E_{i,j,k-1}^x}{\Delta z^2} \right). \quad (7.59)$$

The value of the electric field is assumed to be given on the external boundary and in the plasma. For each global iteration (field calculation), the electric field is found in the low-density regions by this method.

7.5.2 Field Radiation and Absorption at the Boundaries

In space physics applications the most relevant boundary condition for the fields is that they should be able to radiate away into space and should not be reflected at the boundaries. The same problem is met when we study the electromagnetic devices such as pulsed radars and high power microwaves, transient electromagnetic waves within a rectangular waveguide, etc. To solve scattering problems embedded in an unbounded region, one needs effective radiation (Zommerfeld) boundary conditions (RBCs) or absorbing boundary conditions (ABCs) imposed at the boundary of the finite-sized computational domain; a multitude of such conditions has been derived in [51, 74, 191, 424] and they were applied to the tangential fields.

Here we discuss shortly the use of the reflectionless sponge-layer absorbing boundary condition. Suppose that the finite-difference approximation to Maxwell's equations on a Cartesian x–z grid may be written [424]

$$\mu \frac{H_{i,j+\frac{1}{2}}^{x,n+\frac{1}{2}} - H_{i,j+\frac{1}{2}}^{x,n-\frac{1}{2}}}{c \Delta t} = \delta^z(\alpha,\beta) E_{(i,j+\frac{1}{2})}^{y,n},$$

$$\mu \frac{H_{i+\frac{1}{2},j}^{z,n+\frac{1}{2}} - H_{i+\frac{1}{2},j}^{z,n-\frac{1}{2}}}{c \Delta t} = -\delta^x(\alpha,\beta) E_{(i+\frac{1}{2},j)}^{y,n},$$

$$\varepsilon \frac{E_{i,j}^{y,n+1} - E_{i,j}^{y,n}}{c \Delta t} = -\delta^x(\alpha,\beta) H_{(i,j)}^{z,n+\frac{1}{2}} + \delta^z(\alpha,\beta) H_{(i,j)}^{x,n+\frac{1}{2}}. \quad (7.60)$$

The discrete spatial derivatives are estimated as

7.5 Boundary Conditions for Particles and the Electromagnetic Field   183

$$\partial_{x(z)} \approx \delta^{x(z)}(\alpha, \beta) = \frac{1}{h}\left[\alpha\left(S_{x(z)}^{1/2} - S_{x(z)}^{-1/2}\right) + \beta\left(S_{x(z)}^{3/2} - S_{x(z)}^{-3/2}\right)\right], \quad (7.61)$$

where h is the cell size in both directions. The spatial shift operator is defined by its action, $S^l f_{(i)} = f_{i+l}$, and its subscript $x(z)$ indicates on which of the two spatial indices of the discrete field it operates. The weights α and β determine the order of accuracy of the scheme. In particular we have

$$\alpha = 1, \quad \beta = 0 \quad \rightarrow O(\Delta t^2) + O(h^2)$$

and

$$\alpha = 9/8, \quad \beta = -1/24 \quad \rightarrow O(\Delta t^2) + O(h^4).$$

The longer stencil of the spatial derivatives allows the internal scheme to be brought up to a distance of $3h/2$ away from the tangent to the boundary field variable which is to be updated with a local RBC. As a result, numerical boundary procedures are needed to update two additional tangential field nodes, one electric and one magnetic, interior to the tangential electric field boundary node. A solution to this problem can be effected using the popular local condition [232, 424]

$$B_m\left(\frac{\partial}{\partial n}, \frac{\partial}{\partial t}\right) U = \prod_{j=1}^{m}\left(\cos\theta_j^{\text{abs}} \frac{\partial}{\partial t} - c\frac{\partial}{\partial n}\right) U = 0, \quad (7.62)$$

where m is the order of "physical" accuracy, $\pm\theta_j^{\text{abs}}$ are the angles of perfect absorption, U is a tangential (electric or magnetic) field variable at the boundary and $\partial/\partial n$ is the spatial derivative in the outward direction normal to the computational domain boundary. The mesh is depicted in Fig. 7.4.

The nodes we deal with are to the right of the vertical dashed line. Assuming that the electric field everywhere is known at time level n and the magnetic field at time level $n - 1/2$, the scheme that includes the treatment of all the nodes on the vertical right-hand computational boundary and

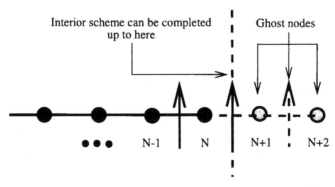

Fig. 7.4. The staggered grid along the normal to the computational domain boundary (from [424])

produces updated field variables at all locations (including the "ghost" nodes) is as follows: update H^z with (7.60) up to node $(N+\frac{1}{2},j)$ and impose $B_m H^{z,n+1/2}_{N+3/2,j} = 0$ to obtain $H^{z,n+1/2}_{N+3/2,j}$; update H^x everywhere with (7.60); update E^y with (7.60) up to node (N,j) and impose $B_m E^{y,n+1}_{N+1,j} = 0$ to obtain $E^{y,n+1}_{N+1,j}$; impose $B_m E^{y,n+1}_{N+2,j} = 0$ to obtain $E^{y,n+1}_{N+2,j}$. This scheme is possible because the operator B_m (for $m=2,3$), when applied to, for example, a tangential electric field variable, involves the node $i=\eta$ (here $\eta=N$) at time level $n+1$, the nodes $i\leq\eta$ at time levels $\leq n$, and the nodes $i\leq\eta-1$ at the $(n+1)$th time level. To compute the update of the magnetic ghost node and to implement the radiation condition at a horizontal computational boundary, the procedure described above is repeated, with the appropriate changes in the tangential fields and discrete spatial indices. The one-dimensional nature of the operators (7.62) allows for easy implementation at a corner, where the horizontal and vertical portions of the boundary meet.

In contrast to radiation, the boundary conditions for absorbing outgoing waves involves surrounding the computational domain with a wave-absorbing layer of thickness d (see [424]).

Let us consider the one-dimensional Maxwell model equations, $-\infty < x < \infty$ and $t \geq 0$:

$$\frac{\mu}{c}\frac{\partial H}{\partial t} + \sigma^* H = -\frac{\partial E}{\partial x},$$

$$\frac{\varepsilon}{c}\frac{\partial E}{\partial t} + \sigma E = -\frac{\partial H}{\partial x}, \qquad (7.63)$$

where $\sigma = \sigma^* = 0$ for $x < 0$ and $\sigma, \sigma^* > 0$ for $x \geq 0$. Equations (7.63) model an electrically and magnetically lossy "material" (occupying $x \geq 0$) with constant wavefront speed $c_\infty = c/\sqrt{(\varepsilon\mu)}$, placed adjacent to free space ($x < 0$) with wavefront speed c ($c_\infty \leq c$). Using the notation $t_c = \varepsilon/c\sigma$ and $t_c^* = \mu/c\sigma^*$, we compute the dispersion relation relevant to boundary value problems in $x > 0$. For a mode $\exp[i(\omega t - kx)]$ it is

$$-\omega^2 + i\frac{\omega}{t_c} + i\frac{\omega}{t_c^*} + \frac{1}{t_c t_c^*} + c_\infty^2 k^2 = 0, \qquad (7.64)$$

where k is the wavenumber. Upon choosing "material" properties so that $t_c = t_c^*$ (the perfectly matched layer condition), the above dispersion relation reduces to $ik = \pm(1/c_\infty)(i\omega + 1/t_c)$, i.e., left/right moving waves in this "material" travel slower than they do in $x < 0$ and decay exponentially in space at a rate that is independent of the frequency ω [424]. Further, if $\sqrt{\varepsilon/\mu} = 1$, then (7.63), written as $\boldsymbol{u}_t + A\boldsymbol{u}_x + C\boldsymbol{u} = 0$, where $\boldsymbol{u} = (B,E)^T$, exhibits the following properties: $A = R\Lambda R^{-1}$ and $C = RDR^{-1}$, where $\Lambda = \text{diag}\{-c_\infty, c_\infty\}$, $D = \text{diag}\{1/c_\infty t_c, -1/c_\infty t_c\}$, and R is the diagonalizer of the A in the region $x < 0$, where the system is $\boldsymbol{u}_t + A\boldsymbol{u}_x = 0$, i.e., $A_{x<0} = \text{diag}\{-c,c\} = R^{-1}AR$. The eigenvectors of A are preserved across $x = 0$, while its eigenvalues are reduced; the interface at $x = 0$ will be reflectionless for any propagating wave impinging upon it from the left. The waves entering this "material" will be slowed down ($c_\infty < c$) and damped independently of frequency. The region $x > 0$ can be terminated at $x = d > 0$ with any boundary condition. Appropriately setting d, t_c and c_∞ can, in principle, make the boundary condition at $x = d$ invisible to waves in the interior $x < 0$, as any outgoing wave of amplitude E_0 that has entered the layer, and has subsequently reflected from boundary condition at $x = d$, will be further attenuated while propagating back towards $x = 0$ to re-enter the computational domain with amplitude $\sim E_0 \exp(-2d/c_\infty t_c)$. The order of accuracy and stability of the interior scheme is maintained in the layer, since only time-centered, lower-order terms are used to implement it. A detailed description of multidimensional use of such sponge layers may be found in [424].

7.5.3 Boundary Conditions at the Conducting Wall

The boundary conditions have been determined by the physical constraints of a conducting wall and assumptions about the symmetry of the problem. To simplify the boundaries we confine the plasma to a domain with square, open ends (z boundaries) (see, e.g., [238, 335]). We expect that the shape of the boundary will have only a limited effect, consistent with results of [216].

Let us consider, for example, the magnetic field reconnection problem with an initial one-dimensional Harris distribution of the z-component of the magnetic field and in the absence of background plasma. On the conducting wall the normal component of the magnetic field must vanish: thus

$B_n = 0$ on the $x = $ const. boundary and on the $y = $ const. boundary.

Since at the initial moment of time only B_z is nonzero, we can assume

$B_\tau = B_z$ on the $x = $ const. boundary and on the $y = $ const. boundary.

We also assume that the walls are far enough from the plasma layer so that B_z is essentially uniform, and thus we set

$\dfrac{\partial B_z}{\partial n} = 0$ on the $x = $ const. boundary and on the $y = $ const. boundary.

Let us consider the boundary conditions for the electric field. Since there can be no tangential electric field on a perfect conductor, we have

$E_\tau = 0$ on the $x = $ const. boundary and on the $y = $ const. boundary.

The divergency equation for the electric field gives

$\dfrac{\partial E_n}{\partial n} = 0$ on the $x = $ const. boundary and on the $y = $ const. boundary.

In the considered plasma system, the particles that contact the conducting wall are lost from the system. In the z-direction we can set the axial derivative of all components of all fields equal to zero at the end of the domain, or to use the periodic boundary conditions.

In the case of the background plasma, we have to use so-called "particle-replacing" boundary conditions [353]. This condition allows the plasma to enter freely at all boundaries but does not allow the plasma to exit. The particle is replaced when it exits through one of the boundaries. The replacing particle is loaded at $x = x' + \Delta x (x = x' - \Delta x)$, $y = y'$ when the particle exits the $x =$ const. boundaries, where (x', y') is the last position outside of the simulation domain. Similarly, the replacing particle is loaded at $x = x'$, $y = y' - \Delta y$ ($y = y' + \Delta y$), when the particle exits through the $y =$ const. boundaries. When a particle moves from the last cell further into the simulation system, a new particle is placed at position $x = x' - \Delta x, y = y'$, where (x', y') is the position of the particle entering the simulation domain.

Summary

In this chapter we have tried to give a survey of the issues involved in selecting particle loading and injection algorithms. Since particle injection is used at every time step, this algorithm must be efficient and it must allow the vectorization or parallelization of the calculation, depending on the type of computer used. We also considered the difference approximation to the boundary condition for fields and particles. The boundary conditions play the same role as the difference equations inside the computational domain. Thus, one has to pay enough attention to the boundary conditions for a good formulation of the simulation model.

Part II

Applications

8. Collisionless Shock Simulation

8.1 Introduction

Research on the structure of collisionless shocks (see the corresponding literature, beginning with the first theoretical papers and reviews [166, 177, 380, 459]) is important for analyzing the interaction of the solar wind with planetary magnetospheres, for studying the acceleration of cosmic rays and for studying processes accompanying solar flares. Research on shock waves is also important for reaching an understanding of the propagation of intense disturbances in the ionosphere and magnetosphere. In this chapter we use the one-dimensional hybrid model for the simulation of processes at the front of collisionless shocks. Figure 8.1 demonstrates a scheme of typical quasiperpendicular shock.

The basic equations may be expressed in a dimensionless form by choosing a typical length L and time L/U_0, U_0 being the typical velocity. Depending on the problem, one of various velocities, for example, the bulk velocity of the incoming flow ($U_0 = U_\infty$), or the thermal velocity of ions ($U_0 = v_{T,i}$), or the Alfvén velocity ($U_0 = v_A$), may be chosen as the typical velocity.

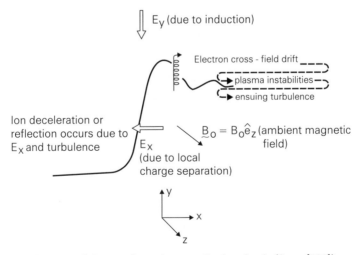

Fig. 8.1. Scheme of quasiperpendicular shock (from [580])

The basic equations are the ion equations of motion, the generalized Ohm's law, Faraday's induction equation and the adiabatic equation for the electron pressure. The ion equations (normalized) are of the form

$$\frac{d\boldsymbol{x}_{sl}}{dt} = \boldsymbol{v}_{sl}, \quad \frac{d\boldsymbol{v}_{sl}}{dt} = \frac{Z_s}{\varepsilon \tilde{M}_s}(\boldsymbol{E}^* + \boldsymbol{v}_{sl} \times \boldsymbol{B}), \quad \boldsymbol{E}^* = \boldsymbol{E} - l_d^* \boldsymbol{J}. \quad (8.1)$$

The normalized dissipation length is $l_d^* = l_d/L = 1/Re$, the magnetic Reynolds number is $Re = 4\pi U_0 L \sigma_{\text{eff}}/c^2$, and σ_{eff} is the effective conductivity. The parameter ε is the ratio of the proton Larmor radius $\varrho_{\text{cp}} = U_0/\Omega_{\text{p}} = M_A c/\omega_{\text{pi}}$ in the unperturbed field B_0 to the characteristic dimension L of the problem, $\varepsilon = \varrho_{\text{cp}}/L$, $\beta_{i,e} = p_{i,e}/(B_0^2/8\pi)$, $M_A = U_0/v_A$, $v_A = B_0/\sqrt{4\pi\varrho_0}$, and $\tilde{M}_s$ is the ratio of the ion mass to proton mass.

The electrons are considered to have a nonzero mass, so that the generalized Ohm's law is

$$\varepsilon \frac{m}{M} \frac{\partial \boldsymbol{J}_e}{\partial t} + \varepsilon \frac{m}{M}(\nabla \boldsymbol{U}_e) \boldsymbol{J}_e - \frac{\varepsilon}{M_{\text{Se}}^2 \gamma}(\nabla p_e + \partial \pi_{\alpha\beta}^e/\partial x_\beta)$$
$$- n_e \boldsymbol{E} + \boldsymbol{J}_e \times \boldsymbol{B} + l_d^* n_e \boldsymbol{J} = 0 \quad (8.2)$$

or, neglecting the pressure tensor,

$$\boldsymbol{E} = -\boldsymbol{U}_e \times \boldsymbol{B} - \frac{\varepsilon}{M_{\text{Se}}^2 \gamma n_e} \nabla p_e + l_d^* \boldsymbol{J} - \varepsilon \frac{m}{M} \frac{d\boldsymbol{U}_e}{dt}. \quad (8.3)$$

The electron $\boldsymbol{U}_e$ and ion $\boldsymbol{U}_i$ mean velocities are given by

$$\boldsymbol{U}_e = \boldsymbol{U}_i - \varepsilon \boldsymbol{J}/(M_A^2 n_e), \quad \boldsymbol{U}_i = \sum_{k=1}^{N_s} Z_k \langle N_k \boldsymbol{v}_k \rangle / \sum_{k=1}^{N_s} Z_k \langle N_k \rangle. \quad (8.4)$$

Ampere's law and the induction equation are

$$\boldsymbol{J} = \nabla \times \boldsymbol{B}, \quad \frac{\partial \boldsymbol{B}}{\partial t} + \nabla \times \boldsymbol{E} = 0, \quad (8.5)$$

$$\boldsymbol{J} = \boldsymbol{J}_e + \boldsymbol{J}_i, \quad (8.6)$$

where the dimensionless current is

$$\boldsymbol{J}_i = \sum_{k=1}^{N_s} Z_k n_k \boldsymbol{U}_k. \quad (8.7)$$

We impose the condition of quasineutrality, and we consider the electron gas to be adiabatic,

$$n_e = \sum_{k=1}^{N_s} Z_k n_k \equiv n, \quad p_e \propto n_e^\gamma. \quad (8.8)$$

8.1 Introduction

Here the dimensionless parameters may be expressed via dimensional parameters as follows:

$$\boldsymbol{U} = \boldsymbol{U}'U_0, \quad \boldsymbol{E} = \boldsymbol{E}'B_0U_0/c, \quad \boldsymbol{B} = \boldsymbol{B}'B_0, \quad p_e = p'_e p_{e0},$$

$$n = n'n_0, \quad t = t'L/U_0, \quad \boldsymbol{x} = \boldsymbol{x}'L. \tag{8.9}$$

Here $U_0 = U_\infty, B_0 = B_\infty, n_0 = n_\infty$ are the upstream values of the bulk velocity, the magnetic field and the density.

In the coordinate form the equation of motion of particles (in the 1+2/2-dimensional case) may be written as

$$\frac{d}{dt}\begin{pmatrix} v_x \\ v_y \\ v_z \end{pmatrix} = \frac{Z_s}{\varepsilon \tilde{M}_s}\begin{pmatrix} E_x^* + v_y B_z - v_z B_y \\ E_y^* + v_z B_x - v_x B_z \\ E_z^* + v_x B_y - v_y B_x \end{pmatrix}. \tag{8.10}$$

The Ohm's law equation may be written as

$$\begin{pmatrix} E_x \\ E_y \\ E_z \end{pmatrix} = -\begin{pmatrix} U_{ey}B_z - U_{ez}B_y \\ U_{ez}B_x - U_{ex}B_z \\ U_{ex}B_y - U_{ey}B_x \end{pmatrix} - \frac{\varepsilon}{M_{S_e}^2 \gamma n_e}\begin{pmatrix} 0 \\ 0 \\ \partial p_e/\partial z \end{pmatrix} + l_d^*\begin{pmatrix} J_x \\ J_y \\ J_z \end{pmatrix}$$

$$- \frac{\varepsilon m}{M}\frac{d}{dt}\begin{pmatrix} U_{ex} \\ U_{ey} \\ U_{ez} \end{pmatrix}, \tag{8.11}$$

where

$$\begin{pmatrix} U_{ex} \\ U_{ey} \\ U_{ez} \end{pmatrix} = \begin{pmatrix} U_{ix} \\ U_{iy} \\ U_{iz} \end{pmatrix} - \frac{\varepsilon}{M_A^2 n_e}\begin{pmatrix} J_x \\ J_y \\ 0 \end{pmatrix}. \tag{8.12}$$

The Maxwell equations can be written as

$$\frac{\partial}{\partial t}\begin{pmatrix} B_x \\ B_y \\ B_z \end{pmatrix} = -\begin{pmatrix} -\partial E_y/\partial z \\ \partial E_x/\partial z \\ 0 \end{pmatrix}, \quad \begin{pmatrix} J_x \\ J_y \\ J_z \end{pmatrix} = \begin{pmatrix} -\partial B_y/\partial z \\ \partial B_x/\partial z \\ 0 \end{pmatrix}. \tag{8.13}$$

The induction equation may be also written in the nondivergent form:

$$\frac{\partial}{\partial t}\begin{pmatrix} B_x \\ B_y \\ B_z \end{pmatrix} + U_{ez}\frac{\partial}{\partial z}\begin{pmatrix} B_x \\ B_y \\ B_z \end{pmatrix} - \frac{\partial}{\partial z}l_d^*(\frac{\partial}{\partial z})\begin{pmatrix} B_x \\ B_y \\ B_z \end{pmatrix} = -\begin{pmatrix} \Phi_x \\ \Phi_y \\ \Phi_z \end{pmatrix}, \tag{8.14}$$

where

$$\Phi_x = B_x\frac{\partial U_{ez}}{\partial z} - B_z\frac{\partial U_{ex}}{\partial z} + \frac{\varepsilon m}{M}\frac{\partial}{\partial z}\left(\frac{d}{dt}U_{ey}\right), \tag{8.15}$$

$$\Phi_y = B_y\frac{\partial U_{ez}}{\partial z} - B_z\frac{\partial U_{ey}}{\partial z} - \frac{\varepsilon m}{M}\frac{\partial}{\partial z}\left(\frac{d}{dt}U_{ex}\right), \tag{8.16}$$

$$\Phi_z = \left(\frac{\partial l_d^*}{\partial z}\right)\left(\frac{\partial B_z}{\partial z}\right). \tag{8.17}$$

Here, $n_e \approx n_i$ and p_e are the number density and pressure of the electrons, M_A and M_{S_e} are the Alfvén and electron sound Mach numbers:

$$M_A = M_{S_e}(\gamma\beta_e/2)^{1/2},$$

where γ is the effective specific heat ratio (in the calculations $\gamma = 2, 5/3$), and σ is the effective conductivity.

8.2 Collisionless Shocks Without Mass Loading

Analysis of experimental data obtained directly in the vicinity of the Earth's shock front, in particular, data on oblique and quasiperpendicular shocks [25, 150], has shown that there is a standing whistler at sufficiently large values of θ_{Bn}, which is the angle between the magnetic field vector and the normal to the shock front.

Hybrid simulation of the structure of oblique collisionless shocks ($m/M = 0$ [75, 303]) and kinetic simulation of quasiparallel shocks [439] have shown that a standing whistler with $\lambda = (1 - 1.5)c/\omega_{pi}$ forms at angles $\theta_{Bn} \leq 45°$. Recent fine-mesh calculations on the structure of an oblique shock [174] have established that a whistler forms at angles $\theta_{Bn} = 60°$. However, theoretical work [171, 201] indicates that whistlers can form at even much larger angles.

Mesh dissipation has been used to find the regular structure of the shock front. In resistive codes, an effective anomalous resistance is added. At large values of θ_{Bn}, fairly fine meshes must be used to deal with scales with $k_\parallel c/\omega_{pi} \gg 1$. However, mesh dissipation is not sufficient for finding a stable solution at large Mach numbers ($M_A \gg 1$), and electron inertia is used, as in some previous calculations [174, 326, 328, 330].

8.2.1 Quasiperpendicular Shocks

In this section we study the one-dimensional structure of quasiperpendicular shocks and the ion dynamics through a numerical simulation based on a hybrid code with electron inertia ($m/M \neq 0$). Fairly fine spatial meshes are used to reduce the mesh dissipation.

To study the dynamics of the front of a quasiperpendicular shock, we consider a one-dimensional flow of ions and electrons along the z-axis. We use a hydrodynamic description for the electrons, and a kinetic description for the ions (see, e.g., [244, 303, 321]). The ion distribution function $f_i(t, z, \boldsymbol{v})$ is defined in three-dimensional velocity space.

The simulation was carried out by the particle-in-cell (PIC) method with meshes of 925, 1849, and 25 000 finite-size ions. The equation of motion for the finite-size particles was integrated using the leapfrog scheme. The equation for the magnetic field was integrated by an implicit Crank–Nicolson scheme (see, e.g., [321]). The discrete ion distribution was reproduced with the help of a random-number counter. Overall energy conservation was maintained within 3% during the computation.

At the initial time, the solution describes an unperturbed supersonic flux of solar wind in the computation region ($z = [0, DZ]$). For the ions of the solar wind, the velocity distribution is specified to be Maxwellian:

$$f_i \propto \exp\left(-\frac{v_x^2 + v_y^2 + (v_z - U_z)^2}{2v_{T,i}^2}\right). \tag{8.18}$$

At the left-hand boundary, $z = 0$, we specify an unperturbed field, and we introduce an injection of ions with an unperturbed Maxwellian velocity distribution. At the right-hand boundary of the region, $z = DZ$, we impose a reflection condition for the magnetic field components,

$$\frac{\partial}{\partial z}B_{x,y} = 0, \tag{8.19}$$

and a particle reflection condition.

Calculations were carried out for plasma parameters characteristic of the solar wind at the earth's orbit: $M_A = 2, 3, 6, 10$ and 15; $\beta_e = 2$; $\beta_i = 0.5$; $\varepsilon = 3.3$; $M/m = 100, 500$ and 1840; $DZ = 10$; and $\theta_{Bn} = 72°$ [333].

At Mach numbers close to one ($M_A = 1.2$), a long-wave standing whistler with $\lambda = 0.68\varrho_{ci} = 0.8c/\omega_{pi}$ forms (Fig. 8.2). The shock front propagates very

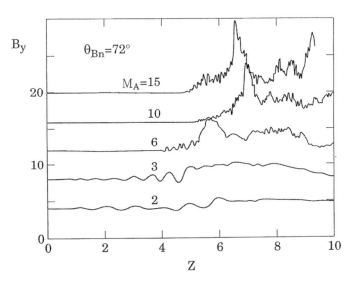

Fig. 8.2. Profile of the magnetic field component B_y for $\theta_{Bn} = 72°$ and for various Mach numbers: $M_A = 2, 3, 6, 10$ and 15 (from [333])

rapidly through the computation region. The discontinuity in the magnetic field at the front is $1.8B_0$, and the whistler amplitude is $0.3B_0$. No reflected particles are observed ahead of the ramp, although particles with $v = 2.5U_0$ moving back toward the front are observed in the transition region. There is no correlation between the profiles of the magnetic field and the density. This circumstance may be evidence of a transverse perturbation of the magnetic field in the whistler.

In the regime with $M_A = 2$, the shock front has a quasisteady front with a regular standing whistler. The wave amplitude in the whistler is $\Delta B_w = 0.8B_0$. The primary component of the magnetic field has an overshoot exceeding the value found from the Rankine–Hugoniot relation. The discontinuity in the magnetic field is $(2.4 - 2.8)B_0$. In this regime, the reflected particles move roughly half a whistler wavelength away from the ramp at a velocity $v_{i,\text{ref}} = 1.5U_0$, and the whistler formation zone has a length scale $L \approx 0.5\varrho_{ci}$.

Let us look at the case where $M_A = 6$. Figure 8.3 shows profiles of the field component B_y at times separated by an interval $0.1T_{ci}$. The overshoot in the primary component of the magnetic field in this case is $(4.4-6.4)B_0$. The whistler amplitude varies over the range $(0.7 - 1.5)B_0$. During the evolution, the structure of the front undergoes cyclic changes with a time scale of $0.3T_{ci}$. The basic steps in the cyclic changes of the magnetic field profile are as follows: (1) a steepening of the front to a length scale of $0.05\varrho_{ci}$ with a weak whistler; (2) excitation, ahead of the front, of an intense whistler, such that the front becomes very choppy with a length scale of $0.3\varrho_{ci}$; and (3) absorption (coalescence) of the intense whistler at the front, and the formation of a gently sloping front with a length scale of $0.3\varrho_{ci}$.

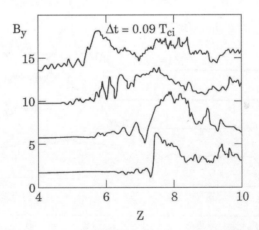

Fig. 8.3. Cyclic (in time) change in the structure of the magnetic field component B_y. The profile correspond to times separated by an interval $t = 0.09T_{ci}$ (from [333])

The plasma density is correlated with the magnetic field profile at the shock front, characterizing the compressional mode of the perturbation. The cyclic change in the profile observed here has been observed previously in a numerical simulation of quasiparallel shocks [78] and of perpendicular shocks, but without the formation of standing whistler [435].

The cyclic changes are accompanied by a cyclic change in the ion distribution function at the shock front. Figure 8.4 shows the ion velocity distributions $v_{\perp 1}$ and $v_{\perp 2}$, perpendicular to the local magnetic field, at various times and at various positions with respect to the shock front. Distribution functions (a), (b) and (c) correspond to the times $t = 0.57T_{ci}$, $0.48T_{ci}$ and $0.39T_{ci}$. The sequence of distribution functions from left to right corresponds to the pedestal, the ramp and the transition regions. The velocity range is $[-3U_0, 3U_0]$. In Fig. 8.4 we see the ions reflected from the front and those which are heated in the transition region and which form a halo around the core of the distribution function of hot ions.

In the regimes with a large Mach number, $M_A = 10 - 15$, there is an increase in the overshoot of the magnetic field: $B_{over} = (6.4 - 10)B_0$ and $(8-13)B_0$ (Fig. 8.2). The structure of the whistler is turbulent. Its amplitude is $(0.5 - 1)B_0$ and $(0.5 - 1.5)B_0$, and its wavelength is $(0.2 - 0.6)c/\omega_{pi}$ and $(0.1-0.45)c/\omega_{pi}$, respectively. The cyclic changes occur with a period of $(0.2 - 0.3)T_{ci}$ and are accompanied by a significant increase in the magnetic field in the pedestal and the overshoot. This effect is due to an increase in the flux of ions reflected from the hump of the whistler and in the overshoot. Under these conditions the whistler wavelength, $\lambda_w = (0.1 - 0.5)c/\omega_{pi}$, becomes comparable with the dissipative length, $l_d = (0.02 - 0.04)c/\omega_{pi}$, and the whistler may undergo significant damping.

Figure 8.5 shows the whistler wavelength versus the Alfvén Mach number. The wavelength is expressed in terms of the ion gyroradius (the thin arrows) and also in terms of the length scale c/ω_{pi} (the thick arrows). We recall that we have $\varrho_{ci} = M_A c/\omega_{pi}$. The thickness of the ramp of the shock front, Δ_{ramp}, is comparable to the maximum whistler wavelength, λ. We see from Fig. 8.5 that an increase in M_A leads to a decrease in the whistler wavelength and thus in the thickness of the shock front. Also shown in Fig. 8.5 is a plot of the whistler wavelength versus M_A for small values of β and M_A [525]. The dot-dashed line $\lambda = 0.5c/\omega_{pi}$ corresponds to the results of a laboratory modeling of collisionless shocks. Measurements of the thickness of the shock near the earth yield $\Delta_{ramp} = (2.5 - 3)c/\omega_{pi}$ at $\theta_{Bn} = 68 - 75°$ [146].

The nature of the standing whistler in a quasiperpendicular shock wave, as in the case of an oblique shock wave, is related to the time variation of the breaking of the shock front, according to theoretical models [170]. Since the dispersion due to the electron inertia does not stop the steepening of the shock front, this front breaks, and a whistler forms.

The whistler wavelength is comparable to the length scale c/ω_{pi}, and it falls off with distance from the front into the unperturbed flux. The coarsening

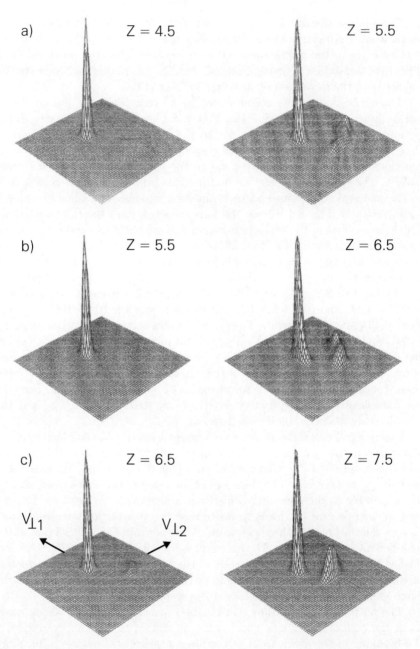

Fig. 8.4. Ion distributions with respect to the velocity perpendicular to the magnetic field in various cross-sections along the z-axis, corresponding to the three upper magnetic field profiles in Fig. 8.3: (a) $t = 0.57 T_{ci}$; (b) $0.48 T_{ci}$; (c) $0.39 T_{ci}$. (From [333])

8.2 Collisionless Shocks Without Mass Loading

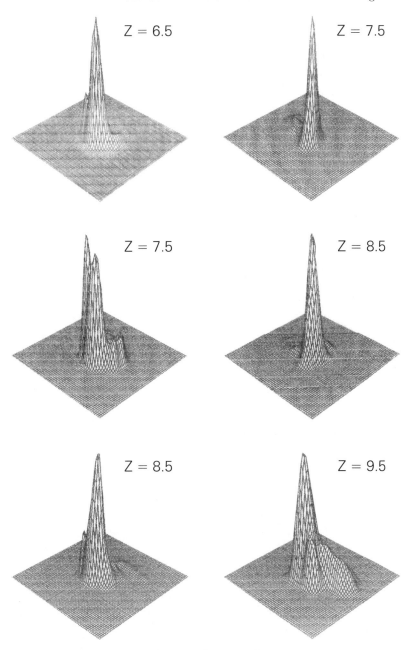

Fig. 8.4. (continued)

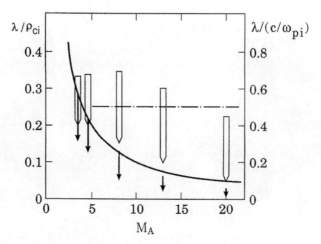

Fig. 8.5. Whistler wavelength versus the Mach number. The *thin arrows* correspond to the value of λ/ϱ_{ci}, and the *thick arrows* to the value of $\lambda(c/\omega_{ci})$. The *solid curve* is the dispersion relation for a whistler (from [525]), while the *dot-dashed line* is the observation line (from [146]). (From [333])

of the computation mesh and thus the enlargement of the dissipation length lead to the disappearance of the whistler and to a significant increase in the overshoot, i.e., to the appearance of structure in the shock front, which has been seen previously [303]. With increasing value of the Alfvén Mach number, $2 < M_A < 20$, there are decreases in both the whistler wavelength and the thickness of the shock front. At Mach numbers above a certain critical value ($M_A = 6$), we see a cyclic change in the structure of the shock front, characterized by a progressive steepening and smoothing of the magnetic field in the region of the pedestal, the ramp and the overshoot, and also by the time-varying dynamics of the whistler. The onset of the whistler gives rise to additional energy dissipation as the particles cross the shock front. Model mass ratios $M/m = 1840, 500$ and 100 lead to whistler wavelengths in the proportion $1 : 1.5 : 2.5$ with the other parameters are held constant.

8.2.2 Oblique Shocks

In the case of the simulation of oblique shocks, the shock wave profile is prescribed initially in the computational region ($z = [0, DZ]$) in accordance with the Rankine–Hugoniot relations [174]:

$$F = \frac{1}{8}\left(1 + \frac{5}{2M_A^2} + \frac{5}{\gamma M_S^2} + \sqrt{(1 + \frac{5}{2M_A^2} + \frac{5}{\gamma M_S^2})^2 + \frac{8}{M_A^2}}\right),$$

$$\frac{n_1}{n_2} = F_1, \quad B_{z2} = B_{z1}, \quad B_{\tau 1} = B_0 \sin\theta_{Bn},$$

$$B_{\tau 2} - B_{\tau 1}[1 - (B_{z1}/M_A)^2]/[F - (B_{z1}/M_A)^2],$$
$$U_{\tau 2} = B_{\tau 1}B_{z1}(1 - F)/\{[F - (B_{z1}/M_A)^2]M_A^2\},$$
$$B_{y2} = B_{\tau 2}\sin\phi, \quad B_{x2} = B_{\tau 2}\cos\phi, \quad U_{z2} = U_{z1}F,$$
$$U_{y2} = U_{\tau 2}\sin\phi, \quad U_{x2} = U_{\tau 2}\cos\phi,$$
$$v_{\text{T},\text{e}2}/v_{\text{T},\text{e}1} = v_{\text{T},\text{i}2}/v_{\text{T},\text{i}1} = (n_2/n_1)^{(\gamma-1)/2}. \tag{8.20}$$

For solar wind parameters the ion velocity distribution function before and after the shock wave is Maxwellian:

$$f_i \propto \exp\left(-\frac{v_x^2 + v_y^2 + (v_z - U_z)^2}{2v_{\text{T},i}^2}\right). \tag{8.21}$$

At the left-hand boundary, $z = 0$, the unperturbed electromagnetic field is imposed and ions are injected with the unperturbed Maxwellian velocity distribution function. The radiation boundary conditions are imposed at the right-hand boundary, $z = DZ$, for the magnetic field components along with outflow boundary conditions on the particles.

The calculations were carried out for plasma parameters typical of the solar wind at the earth's orbit: $M_A = 5, 10$ and $15; \beta_e = 2; \beta_i = 0.5; \varepsilon = 6.65; M/m = 1840; DZ = 10; \theta_{Bn} = 60°$ [174].

We begin by considering the case where $M_A = 5$. Figure 8.6 illustrates profiles of the magnetic field components $B_x(B_\perp)$ and B_y as functions of z at times $t = 0.15T_{ci}$ and $t = 0.325T_{ci}$. At the shock front the main component of the magnetic field has a value which exceeds that calculated from the usual Rankine–Hugoniot relation, i.e., there is a so-called overshoot, $B_y = 3.5B_0$. In the transition region small $(0.1-0.2)B_0$ perturbations are observed. Immediately ahead of the shock front over a region of length $0.8\varrho_{ci} = 4c/\omega_{pi}$, a right-hand polarized standing whistler is observed. The Alfvén mode corresponds to the magnetic field component $B_\perp$, and the compressible mode corresponds to B_y.

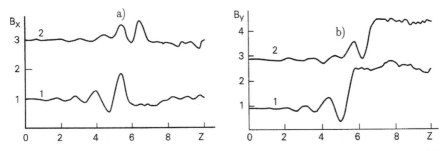

Fig. 8.6. Profiles of the components B_x (**a**) and B_y (**b**) at time $t = 0.15T_{ci}$ (1) and $t = 0.325T_{ci}$ (2) (from[174])

Four damped pulses were observed in the calculations. The amplitude of the first pulse was $\Delta B_y \approx 0.8 B_0, \Delta B_x \approx 0.8 B_0$, and the wavelength was $\lambda \approx 0.16 \varrho_{ci} \approx 0.75 c/\omega_{pi}$. The whistler wavelength and amplitude decreased as functions of distance from the front.

The density jump at the shock front was $n_2/n_1 \approx 3$. At the foot of the shock, a significant increase in the density $\Delta n \approx 0.6 n_0$ was observed, which is uncorrelated with the transverse perturbations of the magnetic field in the whistler. The small-scale perturbations of the density profile were caused by "shoot" noise. The damping of the oscillation amplitude and simultaneous decrease in the wavelength as a function of distance from the front are correlated with the degree to which reflected ions penetrate into the supersonic flow of the plasma. In this regime the whistler amplitude is comparable with the magnitude of the magnetic field in the overshoot, to which the whistler can therefore make a substantial contribution. Moving into the foot of the shock front, the ions are partly reflected from the bumps of the electromagnetic perturbation in the whistler, which weakens the ion reflection from the ramp and decreases the magnetic field amplitude of the overshoot. This is associated with a decrease in the overshoot in shock waves with small Mach numbers, in which a substantial standing whistler is present (Fig. 8.7). Note that as the quantity M/m decreases from 1840 to 500 the whistler wavelength increases by $20 - 30\%$, while for $M/m = 100$ the whistler wavelength is doubled. Therefore the results of calculations done using fully kinetic simulation [439] with $M/m = 100$ can differ from the real picture. In calculations with a fairly fine grid, which do not treat electron inertia, strong increases in the short-wavelength harmonics of the electric field and a substantial acceleration of separate groups of ions are observed for $M_A \geq 10$. Introducing electron inertia into the algorithm even with a large ratio, $M/m = 1840$, enables us to obtain a regular solution over a fairly large time interval.

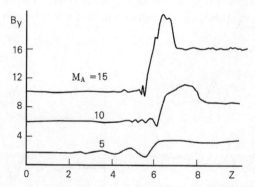

Fig. 8.7. Profiles of the component B_y for different Alfvén numbers, $M_A = 5, 10$ and 15, with $\theta_{Bn} = 60°$ (from [174])

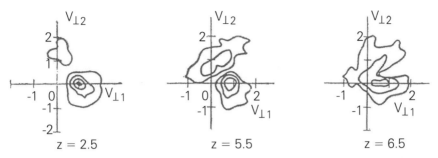

Fig. 8.8. Projection of the ion distribution function on the plane of the velocities $v_{\perp 1}$ and $v_{\perp 2}$ in the cross-sections $z = 2.5, 5.5$ and 6.5 (from [174])

Figure 8.8 shows a projection of the ion distribution function on the plane of the velocities $v_{\perp 1}$ and $v_{\perp 2}$ perpendicular to the local magnetic field for $z = 2.5, 5.5$ and 6.5 near the ramp of the shock wave (the ramp is the region with the maximum slope in the magnetic field profile). Here the sickle-shaped part of the distribution function, corresponding to the reflection of ions from the ramp, is clearly visible. The reflected particles can reach a density of $\approx 0.05 n_0$. Analysis of the ion distribution function for different values of z showed that the reflected ions have a velocity of $-(0.5-1.2)U_0$ and propagate from the front to $z = 2$ (the width of the foot is given by $\Delta_{\text{foot}} \approx 0.5 \varrho_{\text{ci}}$). In the immediate vicinity of the ramp ($z = 6$) the effective temperature jump, $T_{\text{i}2}/T_{\text{i}1} \approx 8.5$, results from the contribution of the reflected ions, while behind the shock front in the transition region the proton distribution function isotropizes ($z > 8$) and has an equivalent temperature, $T_{\text{i}3} = 3T_{\text{i}1}$.

In the case where $M_A = 10$, the overshoot in the magnetic field at the shock front is equal to $\Delta B \approx (4-5)B_0$ (Fig. 8.7). The wavelength of the standing right-hand polarized whistler varies over the range $(0.04-0.08)\varrho_{\text{ci}} = (0.4-0.8)c/\omega_{\text{pi}}$, and its amplitude has the value $(0.2-0.8)B_0$. The whistler structure is regular. The size of the shock-wave foot reaches $\Delta_{\text{foot}} \approx 0.4\varrho_{\text{ci}} = 4c/\omega_{\text{pi}}$. The increases in the ion temperature at the shock front and in the transition region are $T_{\text{i}2}/T_{\text{i}1} = 30$ and $T_{\text{i}3}/T_{\text{i}1} \approx 5$. In the case where $M_A = 15$, the magnetic field overshoot reaches a value $B_2/B_0 \approx 12-14$ (see Fig. 8.7), substantially greater than the whistler amplitude, $(0.2-1.5)B_0$. Consequently, at large M_A the presence of the standing whistler has little effect on the processes of ion reflection from the ramp of the shock wave. The wave profile is essentially time dependent and twisting changes into breaking over a time $1.2\Omega_{\text{i}}^{-1}$.

Figure 8.9 shows the whistler wavelength as a function of M_A. The wavelength is expressed in terms of both the ion gyroradius and the characteristic length c/ω_{pi}. Recall that $\varrho_{\text{ci}} = M_A c/\omega_{\text{pi}}$. The thickness of the ramp, Δ_{ramp}, is comparable with the maximum value of the whistler wavelength, λ. From Fig. 8.9 it is clear that the increase in M_A results in a decrease in the whistler wavelength and consequently in the thickness of the shock front. In order

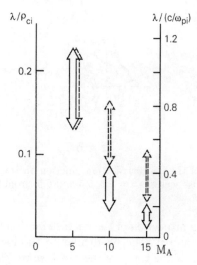

Fig. 8.9. Whistler wavelength, λ, as a function of the Alfvén number, M_A. The *solid arrows* correspond to λ/ϱ_{ci}, and the *dashed arrows* to $\lambda/(c/\omega_{pi})$. (From [174])

to explain the behavior of the standing whistler, we can attempt to account for the ion-cyclotron instability, which is associated with the anisotropy of the ion distribution function $T_{i\perp} \gg T_{i\|}$ due to ion reflection at the shock front [372, 520]. According to [520], waves satisfying the condition $k_{\|}c \sim \Omega_i$ ($\lambda \approx 2\pi c/\omega_{pi}$) have the largest growth rate. However, the calculated whistler wavelength falls into the interval $\lambda \approx (0.7 - 1.1)c/\omega_{pi}$, which is considerably smaller than the theoretical value. This effect is probably not described by the present theory.

In [170], a theory of the formation of the shock front in terms of two-fluid hydrodynamics was presented. Formation of a whistler with a wavelength on the order of the ramp thickness is related to an unsteady process, the breaking of the wave front. Because the dispersion resulting from the electron inertia does not prevent steepening of the shock front, it breaks and a whistler forms. The whistler wavelength lies in the interval $c/\omega_{pe} < \lambda < c/\omega_{pi}$ and depends on the Mach number of the incoming flow. As seen from Fig. 8.9, the qualitative conditions of this theory are in satisfactory agreement with the results of our simulations.

8.2.3 Quasiparallel Shocks

In a collisionless parallel shock the extent of the shock and its dissipation scales are not always well defined [269]. For example, fast ions reflected [202] or leaked [143] at a collisionless quasiparallel shock can generate waves and change the upstream plasma state, at variance with the concept of signal propagation in a collisional medium. In the downstream region there is no guarantee that the anomalous dissipation will result in an isotropic phase space distribution of the electrons and ions. Instead, a scattering or reflection mechanism may accomplish part of the shock heating, but a completely different mechanism may be necessary to complete it. The scales over which the transition takes place (the collisionless equivalent of the viscous layer) may be larger than the scales of the downstream region (e.g., the magnetosheath of the Earth's magnetosphere). It follows that a consistent definition and explanation of a parallel shock must include both the preshock and postshock heating regions, which can be large even using MHD scales. The simulation of quasiparallel and parallel collisionless shocks has been done in numerous papers [144, 259, 360, 433, 436, 437, 439]. In simulations the shock was created by reflecting plasma off the right-hand wall.

8.2.3.1 Low-Mach-Number Shocks. In this section we consider the results of the simulation in [437] to illustrate the transition layer of low-Mach-number parallel shock ($M_A = 2.16$), when the dissipation is provided by standing-group electromagnetic waves generated by the resonant electromagnetic ion-beam-driven instability. The transverse magnetic field at time $t\Omega_i = 100$ is shown in Fig. 8.10. In Fig. 8.10a the B_y component normalized to the upstream B_u is plotted as function of $x\omega_{pi}/c$, where c/ω_{pi} is the

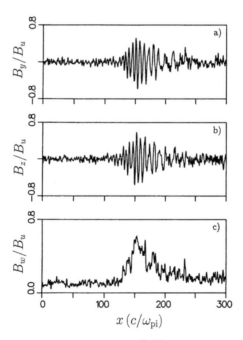

Fig. 8.10a–c. The two transverse magnetic field components and their magnitude $B_w = (B_y^2 + B_z^2)^{1/2}$ at time $t = 100\Omega_i^{-1}$. The magnetic fields are normalized to the upstream magnetic field B_u, and the position x is normalized to the upstream ion inertial length c/ω_{pi}, $M_A = 2.16$. (From [437])

upstream ion inertial length. To eliminate short-wavelength numerical noise from the plot, B_y has been smoothed by using a cubic spline fit to the data. As discussed in the preceding section, the shock is created by reflecting ions at the right-hand wall at $t = 0$ and is then allowed time to asymptotically approaches a quasisteady state. As can be seen, a coherent wave grows out of the noise at $x\omega_{\text{pi}}/c \approx 120$ and reaches its maximum value at a distance of four wavelengths ($x\omega_{\text{pi}} \approx 150$). For larger x the wave decays and increases in wavelength, and the B field magnitude asymptotically approaches a value slightly above the upstream noise level. In Fig. 8.10b the B_z component is plotted and is similar in appearance to B_y. A hodograph of $B_y - B_z$ (not shown) confirms the expectation of roughly circular polarization. If the wave is propagating in the upstream direction in the local plasma rest frame, then the wave is right-hand polarized, consistent with unstable generation by resonant, upstream flowing ($V_x < 0$) ions. In Fig. 8.10c the magnitude of the wave field, B_w, is plotted.

Consistent with the preceding figures, there is a rise in the amplitude of the wave, peaking at $B_w/B_u \approx 0.6$, followed by a slower decay. The wavenumber k of the leading edge of the wave packet in Fig. 8.10 is $\approx 1\,\omega_{\text{pi}}/c$. In Fig. 8.11 the ion phase space for the shock is plotted as a function of position. In Figure 8.11a, V_x normalized to the shock speed U_u is displayed. For simplicity, the velocity is plotted in the shock frame of reference (in which the shock front is stationary) rather than the laboratory frame (in which the plasma downstream from the shock is at rest). At $x\omega_{\text{pi}}/c \approx 150$, where the electromagnetic waves are at nearly peak amplitude (see Fig. 8.10), the ion distribution function consists of a population similar to the far upstream plasma and a diffuse population with roughly zero average velocity. From our previous discussions we expect that such a two-stream distribution will result in the unstable generation of electromagnetic waves. For $x\omega_{\text{pi}}/c > 150$, the diffuse population becomes more dense and finally merges with the upstream distribution to form a single downstream population. In Fig. 8.11b and c the V_y-x phase space and V_z-x phase space are displayed. Note that where the density increases there is no corresponding increase in the temperature.

The only observed temperature jump is in the x-direction, consistent with the theoretical prediction that the compression is one dimensional at small Mach numbers.

In Fig. 8.12 the number density, longitudinal ion pressure (P_{xx}) and ion temperature anisotropy are all plotted as a function of position. A jump in density across the region of enhanced electromagnetic wave activity is observed, followed by a slower increase over much of the remainder of the domain. A clearly defined jump in P_{xx} is seen in the region of electromagnetic waves (Fig. 8.12b), although there is some variation further downstream. The the ratio $T_{\perp i}/T_{\|i}$ (Fig. 8.12c) is about ≈ 1 upstream from the shock, but decreases strongly across the region of shock heating and wave activity to a value of ≈ 0.4.

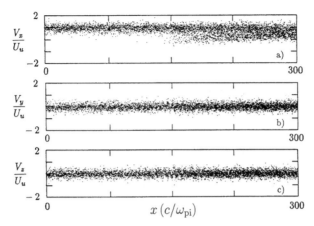

Fig. 8.11a–c. The ion phase space at time $t = 100\Omega_i^{-1}$. The velocities are normalized to the shock speed U_u, and the position x is normalized to the upstream ion inertial length, $M_A = 2.16$. (From [437])

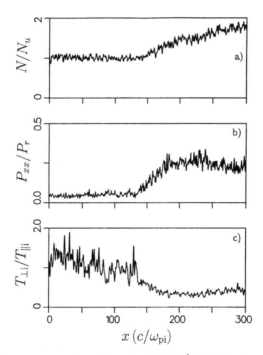

Fig. 8.12a–c. Plasma moments at time $t = 100\Omega_i^{-1}$. (a) Number density normalized to the upstream number density. (b) Field-aligned component of the ion pressure tensor (P_{xx}), normalized to the ram pressure $P_r = n_u M_i U_u^2$. (c) Perpendicular-to-parallel ion temperature ratio. The x-axis is position x normalized to the upstream ion inertial length $c/\omega_{\rm pi}$, $M_A = 2.16$. (From [437])

8.2.3.2 High-Mach-Number Shocks.

In the previous simulation studies it was shown that the structure of $\beta \approx 1$, nearly parallel shocks changes strongly as the Mach number is increased from $M_A \approx 2$ to $M_A \approx 4$ [259, 360]. At low Mach numbers the waves generated by the resonant ions can group-stand with respect to the shock. At high Mach numbers, however, the unstable spectrum shifts to longer and longer wavelengths, until it is no longer possible for the waves to group-stand [437].

A difference in structure also results because the low Mach number shocks ($M_A \approx 2$) are below or near the marginal firehose condition downstream from the shock. At higher Mach numbers, strong scattering is required to keep the downstream plasma from exceeding the firehose condition [437].

Let us consider the result of the simulation in [437]. In Fig. 8.13, the wave magnetic field is plotted as a function of position for $M_A = 5.05$ at time $t\Omega_i = 200$, again using a cubic spline to smooth the data. There are three distinct regions of wave activity. For $x\omega_{pi}/c < 300$, a long-wavelength signature can be seen embedded in the noise, increasing in magnitude as x increases. At $x\omega_{pi}/c \approx 320$, the waves are suddenly and strongly amplified and the wavenumber increases. At $x\omega_{pi}/c \approx 440$, the amplitude decreases and the wavenumber decreases.

According to the calculation of [437], the first region upstream from the shock is where the resonant ions initially generate the low-frequency whistler waves. The second region is behind the shock front and is where the waves are

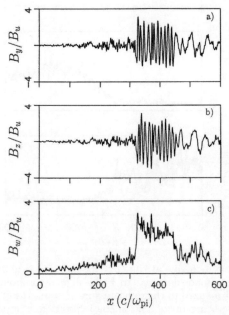

Fig. 8.13a–c. Same as Fig. 8.10, but $M_A = 5.05$ (from [437])

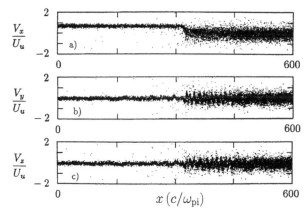

Fig. 8.14. Same as Fig. 8.11, but $M_A = 5.05$ (from [437])

compressed and amplified. The third region is where the waves finally decay, although the exact nature of the decay process is as yet not well understood. A hodograph analysis of the two components B_y and B_z showed that the waves are nearly circularly polarized.

In Fig. 8.14 the ion phase space is plotted as a function of position at time $t\Omega_i = 200$. Strong heating in V_x component occurs near $x\omega_{\rm pi}/c \approx 320$, coincident with the amplification of the upstream waves. Upstream from the shock, a few backscattered ions can be seen. The number density of these ions decreases as the distance from the shock increases, suggesting that most of the scattered ions are turned around and eventually are transported downstream from the shock. Figure 8.14b and 8.14c show strong heating of the perpendicular components of velocity, where the wave field increases. In addition, coherent oscillation can be seen downstream from the shock, consistent with the expected (bulk) motion of the ions in the wave field. The simulation shows that the energy of the ions transmitted through the shock front does not grow significantly. The irregular motion of ions is the thermalization by the large-amplitude waves. By contrast, the scattered ion is strongly accelerated, reaching energies 5–10 times the ram energy. The reason for the acceleration is the upstream waves. If conditions are right for an ion to be reflected back upstream from the shock, then because of the unstable generation of the upstream waves, the ion is likely to be turned back toward the shock because of a resonant interaction with the wave. In the wave frame the energy of the particle is conserved, but in the shock frame the particle must gain energy.

The further investigation provided in [78, 473, 476, 523, 570] has demonstrated the re-forming supercritical quasiparallel shocks. In [523], a mechanism was suggested for the generation of the large-amplitude waves seen at the shock ramp and immediately downstream, based on a resonant interaction at the interface between the incident ions and the downstream plasma.

The simulations show that during a portion of the re-formation cycle, the ramp can at times become broadened to about $10c/\omega_{\text{pi}}$, consistent with the reported measurements, but at other times it can steepen and be as narrow as c/ω_{pi} [523].

8.3 Collisionless Shocks with Mass Loading by Heavy Ions

The investigation of the "loading" of the solar wind by cometary ions in [49] showed that the loading effect is not adequate to ensure continuous transition of the supersonic plasma flow far from the comet into a subsonic flow near it, i.e., a detached shock wave must be established in the flow of the loaded solar wind before the local Mach number, calculated using the magnetosonic velocity with allowance for the pressure of the cometary ions, falls to unity from the value $M = 5-10$, typical for the solar wind. Numerical calculations of the solar wind flow past a comet [472] showed that for typical rates of evaporation of the gas from the surface of the comet's nucleus the shock wave is established at the point where the local Mach number is $M = 2$. Thus, by the time that the modified solar wind is just ahead of the shock front, its state differs considerably from the state of the unperturbed solar wind, and so we can anticipate that the jump of the plasma parameters will have a different structure at the front. It is clear that the heavy cometary ions must influence this structure.

Here it should be noted that the typical Mach numbers $M = 5-10$ for the solar wind flow at shock waves near planets significantly exceed the critical value at which hydrodynamical breaking of the shock front profile occurs [458]; this is a process which does not admit rigorous analytic description. Numerical simulation by the particle method [304] showed that in the kinetic description of the plasma the breaking corresponds to the appearance of a significant fraction of ions either reflected from the shock front or streaming out ahead of the wave front owing to the high thermal velocity of the ions behind the front [418]. At the same time, these ions acquire significant energy in the self-consistent electric field $\boldsymbol{E} = -(\boldsymbol{U} \times \boldsymbol{B})/c$ of the plasma flow ahead of the wave front, before they again fall behind the shock front and are carried away by the plasma flow. This indicates that part of the kinetic energy of the plasma flow goes over into the cyclotron gyration energy of the ions. The irreversibility of such a transition is associated with the mixing of the phases of the cyclotron gyration of the ions in the strongly inhomogeneous magnetic field at the wave front. In accordance with the classification of anomalous dissipation mechanisms at the front of collisionless shock waves, this mechanism plays the part of an anomalous viscosity. The viscosity ν is estimated as $\nu \sim l v$, where the characteristic mean free path l is found to be of the order of the Larmor radius of the reflected ions, $l \sim \varrho_{\text{ci}}$, and the

8.3 Collisionless Shocks with Mass Loading by Heavy Ions   209

"thermal" velocity is the velocity of their cyclotron gyration, i.e., it is of the order of the solar wind velocity [175, 324, 325, 332, 341, 343, 344].

The characteristic thickness of the front here is of the order of the Larmor radius of the reflected ions. The calculated Mach number at the front of the shock wave near a comet is below the critical number, and therefore the effect of the reflection of the solar wind ions from the front must be insignificant. Nevertheless, we can anticipate that the cometary ions produced in the solar wind will play the role of reflected ions in this case, and therefore, the characteristic thickness of the front will be of the order of the Larmor radius of these ions. In this case the following question remains open: Can the dissipation processes in the cometary component of the plasma by themselves ensure the required dissipation of energy at the shock front or is an even more abrupt resistive jump of the solar wind parameters necessary? We note that similar questions arise in the problem of a shock wave in space plasmas, when the significant (with respect to the energy) fraction of ultrarelativistic charged particles (cosmic rays [17]) is considered; this problem is in many respects similar to the present one. To study the shock transition layer, we shall consider here the local domain which contains upstream and downstream of the shock. The global distribution of a cometary plasma and the solar wind will be taken into account by the boundary condition (Fig. 8.15).

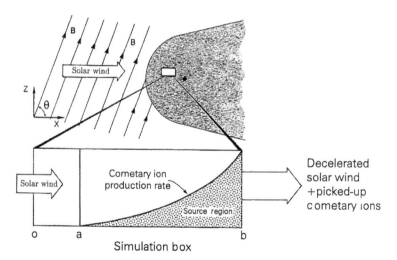

Fig. 8.15. Schematic of solar-wind–comet interaction region. Solar wind protons are continuously injected from the left-hand boundary. Cometary ion production rate increases from left to right. All particles are allowed to leave the system from the right-hand boundary. (From [406])

8.3.1 Quasiperpendicular Shocks

The detail investigation of the structure of quasiperpendicular cometary shock was done using the hybrid simulation technique of [175, 406]. For simplicity, only the case of shock propagation perpendicular to the magnetic field was considered. The basic equations used here are (8.10–8.16) (see Sect. 8.1).

At the initial time we specify a hydrodynamic shock wave with the characteristic initial front width $\delta \approx 0.5$, located in the center of the computational domain ($z = 5$). The jumps in the parameters in the shock wave are determined by the continuity relations for the mass flow, the momentum and the energy. For the values of the mean velocity, the density and the magnetic field behind the shock wave we have the following expressions:

$$\frac{n_{p1}}{n_{p2}} = \frac{n_{i1}}{n_{i2}} = \frac{U_2}{U_1} = \frac{B_1}{B_2}$$

$$= \frac{1}{3}\left(1 + \frac{2M_A^{-2} + 2M_S^{-2} + n_{i1} M_i v_{Ti1}^2}{1 + n_{i1} M_i}\right). \tag{8.22}$$

The subscripts 1 and 2 correspond to the parameters ahead of and behind the shock wave, and v_{Ti} is the thermal velocity of the ions. The velocity distribution function of the protons at the initial time is Maxwellian:

$$f_p \propto \exp\left(-\frac{v_x^2 + v_y^2 + (v_z - U_z)^2}{2v_{T,p}^2}\right). \tag{8.23}$$

The velocity distribution function of the ions is specified to be uniform on the ring

$$V_{int}^2 < (v_z - U_z)^2 + v_x^2 < V_{ext}^2 \tag{8.24}$$

in v_x–v_z space.

The temperature of the electrons, the protons and the ions behind the shock wave is calculated on the assumption that the adiabatic invariant is conserved. At the left-hand end of the region ($z = 0$), an unperturbed flow of protons and ions ($n_p U_\infty, n_i U_\infty$) is specified. At the right-hand end ($z = DZ = 10$), a subsonic flow out of the calculated region was modeled by means of a return flow of particles with velocity $v_z > 2U_2$ into the region.

The processes on the shock wave were simulated by the PIC method using a spatial grid of 500 points, with 2.5×10^5 macroprotons and 2.5×10^5 macroions. The equations of particle motion and magnetic field evolution were integrated using implicit schemes (see, e.g., [318]). The integration time step Δt satisfied the Courant-Fridrich-Levy (CFL) condition $\Delta t = \Delta z U_\infty^{-1}/10$, and the grid spacing along the z-axis satisfied the condition $\Delta z \ll \varrho_{cp2}$, where ϱ_{cp2} is the Larmor radius of the protons behind the shock wave. The discrete distribution function of the protons and ions was reproduced by means of random number generators.

8.3 Collisionless Shocks with Mass Loading by Heavy Ions

In the calculation considered here the dimensionless parameters of the plasma in front of the shock wave are typical for the solar wind at the Earth's orbit [175]:

$$M_{\rm A} = 8, \quad \beta_{\rm e} = 1, \quad \beta_{\rm p} = 0.5 \quad \text{and} \quad \varepsilon = 0.7. \tag{8.25}$$

In order to reduce the running time of the calculation, the mass of the cometary ion was assumed to be equal to five proton masses ($M_{\rm i} = 5M$). At the same time, the ratio of the mass densities of the ion and proton components was kept close to the value obtained from a numerical calculation of the dynamics of loading of the solar wind by cometary ions [472]. Figure 8.16 shows the profiles of the magnetic field, density and velocity of the plasma, the Alfvén and magnetosonic Mach numbers, and the velocity distributions of the protons and cometary ions obtained by numerical simulation. For the case shown in Fig. 8.16, the ratio of the ion and proton densities was $n_{\rm i1} M_{\rm i} = 0.30 n_{\rm p1} M$, and cyclotron gyration velocities for the cometary ions on the boundary $V_{\rm int} = 1$, $V_{\rm ext} = 1.5$ were selected. With this choice the magnetosonic Mach number was $M_{\rm MS} = 1.8$.

The profiles of the basic plasma parameters shown in Fig. 8.16 are typical for all the calculations. They clearly show the resistive jump of the electron–proton plasma component, in front of which is the foot associated with the outflow of part of the cometary ions into the region preceding the wave front. A similar picture was observed in the laboratory experiments [145]. However, unlike these experiments, the outflow of ions into the region ahead of the shock front in our case occurred not because ions are reflected under the influence of the electrostatic field at the wave front and the Lorentz force, but owing to the high cyclotron gyration velocity of the cometary ions behind the wave front. We note that the heavy cometary ions hardly feel the electrostatic electric field at the front of the electron–proton jump and therefore move immediately behind this shock with almost the same velocity as in front of it. At the same time, the larger part of the ions that cross the proton jump in the negative phase of cyclotron gyration ($v_x < 0$, see Fig. 8.16) returns to it after a certain time and re-enter the region in front of the shock. Moving in the region in front of the jump, these ions are accelerated by the electric field $\boldsymbol{E} = -\boldsymbol{U} \times \boldsymbol{B}/c$, and after dropping behind the wave front they have an energy substantially greater than their initial energy. The contours of the velocity distribution function of the accelerated ions form a crescent-shaped structure in the v_x–v_y plane, on the left above the ring of the basic distribution of ions in the foot region (Fig. 8.16). In turn, the supersonic plasma flow in front of the jump of the electron–proton plasma component is decelerated to velocities corresponding to $M_{\rm A} \approx 3-4$ under the influence of the cometary ions which have streamed forward.

It should be noted that the foot in the picture of the shock wave described above is not steady; it pulsates with a characteristic time on the order of half the cyclotron gyroperiod of the cometary ions $T_{\rm ci}$. To illustrate the pulsations

212   8. Collisionless Shock Simulation

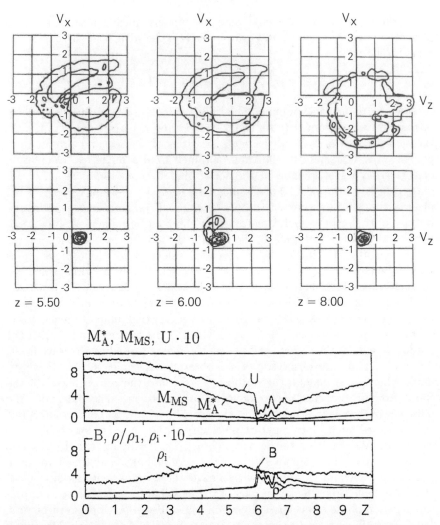

Fig. 8.16. Bottom: Profiles of the magnetic field B_y, density ϱ (normalized by the perturbed value $\varrho_1 = 1 + n_i M_i$), plasma velocity U, ion density ϱ_i, Alfvén Mach number M_A^* (calculated with allowance for the ions and normalized so that $M_A^* = M_A/\varrho$) and the magnetosonic Mach number M_{MS}. Top: The velocity distribution function for the protons and the ions. (From [175])

of the foot, Fig. 8.17 shows the profiles of the magnetic field, obtained at equal time intervals $\Delta t = 0.9 T_{ci}$. We see that, after its formation, the smooth foot again begins to accumulate in a narrow region and finally merges with the proton jump. Then the whole picture is again repeated. The pulsations are associated with the fact that the times for the cometary ions to stream into the region in front of the electron–proton jump and to return back to

8.3 Collisionless Shocks with Mass Loading by Heavy Ions   213

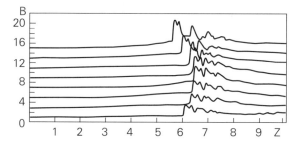

Fig. 8.17. Variation of the magnetic field profiles in time for a finite thickness ring ion distribution (from [175])

the jump add up to approximately the same total for most of these ions. In other words, they are due to the effect of the incomplete phase mixing of the outgoing ions. A further argument in favor of this conclusion is the fact that, as the spread in the cyclotron gyration velocities decreases, the local minima of the plasma velocity and maxima of the density in the region of the foot are enhanced significantly and can even give rise to the formation of a second jump there, when such a velocity spread is absent (Fig. 8.18).

Thus, the basic energy-dissipation mechanism of the directed motion of the plasma in the shock wave described above is the acceleration in the electric field of the plasma flow $\boldsymbol{E} = -\boldsymbol{U} \times \boldsymbol{B}/c$ of the cometary ions which go out ahead of the proton jump. The energy of the ion-cyclotron gyration increases and is ultimately transformed into heat due to the development of the plasma instabilities. As the measurements on the Earth's bow shock wave show [534],

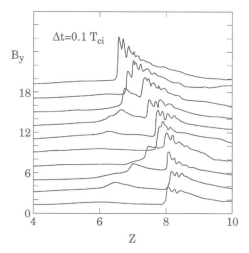

Fig. 8.18. Variation of the magnetic field profiles in time for a very thin ring ion distribution

the most effective of them is the lower-hybrid instability predicted in [459] for the case of counterstreaming ions. However, this instability develops too slowly (over times of the order of tens of cometary ion gyroperiods) to appear in our calculations. Therefore, the motion of the ions can be assumed to be partially reversible. This fact appreciably influences the structure of the shock front.

In fact, in the presence of sufficient energy dissipation at the shock front the plasma parameters change monotonically from their values in the unperturbed plasma ahead of the front to values behind the front satisfying the familiar Rankine–Hugoniot relation. In this case, the electron–proton jump develops inside the shock front, if the following condition is met: The velocity of the plasma behind the wave front must be less than the magnetosonic velocity in the electron–proton plasma component, i.e.,

$$\frac{U_2}{U_1} < \left[2\left(\frac{1}{M_A^2} + \frac{1}{M_S^2}\right)\frac{U_1}{U_2}\right]^{1/2}. \tag{8.26}$$

Here we have used the fact that the compression of the magnetic field and the proton and electron plasma components is adiabatic on the scale of the Larmor radius of the cometary ions, which characterizes the width of the shock front. Using (8.22), we rewrite this condition in the form of a bound on the ratio of the pressure of the cometary ions to the total pressure:

$$1 - \frac{M_{MS}^2(M_A^2 + M_S^2)}{M_A^2 M_S^2} = K < 1 - \frac{(M_{MS}^2 + 2)^3}{27 M_{MS}^4}, \tag{8.27}$$

where

$$\frac{1}{M_{MS}^2} = \frac{1}{M_A^2} + \frac{1}{M_S^2} + \frac{n_i M_i v_{T,i}^2}{(1 + n_i M_i)U^2} \quad \text{and} \quad \frac{1}{M_S^2} = \frac{1}{M_{Se}^2} + \frac{1}{M_{Sp}^2}.$$

This is the parameter which is usually used to evaluate the possibility of the occurrence of a narrow plasma jump inside the shock front in a plasma in which most of the pressure comes from the cosmic rays [72, 137].

For typical parameters of the supersonic plasma flow of the solar wind at the near-comet shock front, the pressure of the cometary ions dominates and condition (8.27) is not fulfilled. For example, for $M_A = 8, \beta_e = 1, \beta_p = 0.5$ and $M_{MS} = 2$, the parameter $K \approx 11/16 > K_{cr} = 0.5$. Nevertheless, the numerical calculations shows that a resistive jump develops inside the front of such a shock wave. The solution of the paradox lies in the fact that in this problem the irreversibility of the motion of the cometary ions is associated only with the mixing of their cyclotron gyration phases. Therefore, on the scale of the Larmor radius of the cometary ions, their motion can be assumed to be reversible. In such a situation the plasma parameters do not at once reach the values determined by the Rankine–Hugoniot relation, but execute several oscillations about them. At the same time the plasma pressure and the density overshoot their final values behind the front. One can imagine two simple mechanisms for such an overshoot. One of them is associated with the effects of the dispersion of waves in a weakly dissipative plasma [458]. The other treats the overshoot as a means of controlling the number of ions reflected from the shock front with the aim of achieving the necessary rate of energy dissipation at the front of a shock [304].

8.3.2 Oblique Shocks

The structure of the oblique cometary bow shock may be different from that in the perpendicular case [175, 406]. In the oblique case both macroscopic and microscopic fields are important in the coupling process [343, 406]. For the plasma parameters at the Earth's orbit this regime occurs at $20° \leq \theta_{Bn} \leq 60°$. The upper limit of this range corresponds to wave normal angles above which the electromagnetic ion-beam instabilities are stable. Here we consider the case where $\theta_{Bn} = 50°$ [406]. The simulation model used for the ion production rate as a function of distance follows from $A = (Q\nu/4\pi W_c r^2) \exp(-\nu r/W_c)$, where A is the production rate, Q is the number of molecules released from the comet per second, ν is the ionization rate, $\nu = 10^{-6}\,\mathrm{s}^{-1}$, W_c is the neutral flow velocity, $W_c = 1\,\mathrm{km/s}$, and r is the radial distance. This model makes the present simulations correspond to regions near the sun–comet line inside of the region $10^5 < r < 2 \times 10^5\,\mathrm{km}$ from the nucleus of a comet with a neutral gas production rate of $10^{30}\,\mathrm{mol/s}$ for a proton density $n_p = 10\,\mathrm{cm}^{-3}$ (Fig. 8.15).

Figure 8.19 shows the proton density and the total magnetic field at time ($t = 438\Omega_p^{-1}$). In this figure the density and the magnetic field, in contrast to those seen early ($\theta_{Bn} = 90°$), have a smoothed increase with high amplitude fluctuations; no discontinuity in density and/or magnetic field which can be associated with a shock is present. The linear characteristics of these waves have been studied in detail [183, 210, 562, 566, 579]. Since the growth rates of these waves increase with increasing density of cometary ions, the amplitude of oscillations increases with x. In Fig. 8.20 we show the time evolution of the electric (E_y) and the magnetic (B_y) fields. It is seen that the waves begin to grow at $x = 750c/\omega_{pi}$ (ω_{pi} is a proton plasma frequency), and in addition the amplitude increases monotonically with x. However, at a later time the

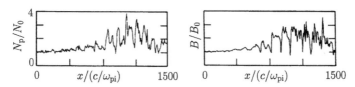

Fig. 8.19. Proton density N_p and the total magnetic field profiles at $\Omega_p t = 438$. As can be seen, no discontinuity in density and/or magnetic field which can be associated with a shock is present. (From [406])

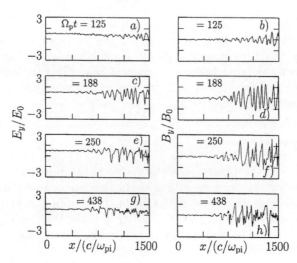

Fig. 8.20a–h. Transverse components of the electric and magnetic field at four separate times. Initially, waves are sinusoidal, their amplitude increasing with x. At later times, nonlinear effects change the waveforms. (From [406])

wave behavior becomes too complicated to be explained. It is obvious that nonlinear effects are coupled with the local plasma parameters, and hence the dispersion properties of wave result in the change of wave profiles.

It is important to note that, in spite of the fact that injection of cometary ions O^+ began at $x = 75c/\omega_{\text{pi}}$, the waves have a small amplitude up to the point $x = 750c/\omega_{\text{pi}}$. A simple explanation of this fact can be given by analysing the distribution of cometary ions, n_{O^+}, that was found, together with the density of protons n_{p} (Fig. 8.21). As can be seen, the O^+ density ($n_{O^+} \leq 0.01 n_0$) and hence the wave growth are quite small at $x \leq 375c/\omega_{\text{pi}}$, and thus one would not expect any appreciable wave amplification at these distances. Obviously, as the growth rates become larger, it would still take some time (distance) before the wave amplitudes are substantially above the background level, thus explaining the small amplitude of the waves prior to $x \approx 750c/\omega_{\text{pi}}$ [406].

Figure 8.21 clearly illustrates the fact that the coupling between the solar wind and the O^+ ions takes place through both macroscopic and microscopic electromagnetic fields. At time $\Omega_{\text{p}} t = 125$, the wave amplitudes are quite small and the solar wind deceleration is mostly due to the macroscopic coupling. At the later times, when the waves have grown (Fig. 8.21c, d), the solar wind deceleration at $x \geq 750c/\omega_{\text{pi}}$ is due to a mixture of macroscopic and microscopic coupling, while that at $x \leq 750c/\omega_{\text{pi}}$ is due to macroscopic processes alone. The result of this type of an interaction is that the proton velocity decreases in a gradual manner without any apparent discontinuity.

Figure 8.22 shows the proton phase space density plots at the moment $\Omega_{\text{p}} t = 438$. The interaction of the protons with the excited waves and the

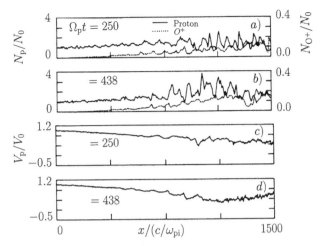

Fig. 8.21a–d. Proton density (*top*) and velocity (*bottom*) along with O^+ density at two times. Note that while the O^+ density fluctuates due to the presence of the waves, these fluctuations are not well correlated with those of the protons. Solar wind deceleration is due to macroscopic coupling, while at later times microscopic coupling is also present. (From [406])

resultant heating without a discontinuity can be seen in Fig. 8.22. The proton temperature and velocity change in a relatively smooth manner; this is one of the distinguishing characteristics between the quasiperpendicular and oblique interaction regions. The process of relaxation of the solar wind proton velocity and averaged cometary ion velocity is accompanied by the isotropization of the velocity distribution function of cometary ions. In the case of oblique

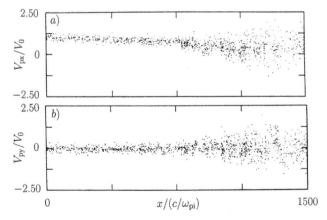

Fig. 8.22a,b. Phase space density plot of protons for $\theta_{Bn} = 50°$ and $t = 438\Omega_p^{-1}$. Solar wind heating by the excited electromagnetic waves is evident. There is no evidence of a shock. (From [406])

propagation of shock, a strong reflection of cometary ions upstream was not observed in simulations [343]. Inside of the shock transition layer, a cold beam of injected cometary ions was observed in simulations and observations of a coma [384, 385].

Since the excited waves are compressional in their linear stage, their nonlinear evolution may involve wave steepening, if the steepening rate is sufficient to overcome ion damping (see, e.g., [260]). These steepening was observed in the simulation of [406], which demonstrated the time evolution of wave steepening (Fig. 8.23). In the plasma rest frame, the excited waves have a phase velocity whose direction is toward the upstream or the left-hand side of the domain. These waves, however, are convected downstream by the solar wind. It is also evident from Fig. 8.23 that the wave-steepening process is eventually stopped due to dispersion effects, which also lead to the break-up of the shocklets.

In the models of cometary shock waves considered above, it is assumed that the velocity distribution function of pickup ions has a ring or shell distribution; however, upstream from the shock, the distribution function may have different forms: ring distributions with parallel drift relative to the solar wind, a thermalized perpendicular distribution with parallel drift, a Maxwellian distribution with no drift, and a Maxwellian distribution with the same

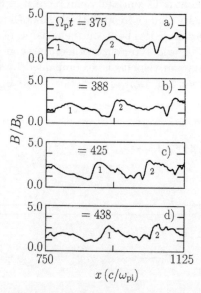

Fig. 8.23a–d. Expanded plot of the total magnetic field at four separate times showing the evolution of the two shocklets labeled 1 and 2. The phase velocity of these waves is upstream (the sun), but it is smaller than the solar wind velocity; thus, the waves are convected downstream. (From [406])

8.3.3 Quasiparallel Shocks

Determination of the structure of a quasiperpendicular cometary bow shock was carried out for $\theta_{Bn} = 90°$ in [172, 175, 319] and for $60° < \theta_{Bn} < 90°$ in [343, 405, 406]. When the interplanetary magnetic field vector and the velocity vector of the solar wind happen to align to within an angle $\theta_{BU} < 45°$, the process of loading of the solar wind with cometary ions and their being carried away is of a very complex character. As a result of cometary ions a strong MHD instability develops, linked to excitation by Alfvén [176, 460, 562, 574] as well as magnetosonic waves [176]. With the onset of the instability, particle isotropization of the velocity distribution function of the cometary ions occurs. Therefore, preceding the quasiparallel shock front, the plasma parameters have a complex structure.

To study dynamics of quasiparallel shocks, one must examine together the flow of electrons, protons, and heavy ions along the z-axis. The electrons are described in the hydrodynamic approximation, and the protons and ions by the macroscopic equation of motion. The magnetic B and electric E fields and the flow velocity U each have three components. In dimensional form the basic equations are (8.10–8.17).

At time $t = 0$ in the computational domain a hydrodynamic shock wave is set up with a characteristic initial width of the front $\delta \approx 0.5$, located in the center of the region ($z = 20$). The jumps of parameters across the shock were selected on the basis of the relation of the continuity of the mass flow, impulse and energy. For the mean velocity, number density and magnetic field, we have the following parameters across the shock [344]:

$$\frac{U_{2z}}{U_{1z}} = \frac{n_1}{n_2} = u_2, \quad \frac{B_{2y}}{B_{1y}} = \frac{\chi - b_z^2}{\chi u_2 - b_z^2}, \quad U_{2y} = \frac{b_{1y} b_z (1 - u_2)}{\chi u_2 - b_z^2}, \tag{8.28}$$

u_2 may be found from

$$a_4 u_2^4 + a_3 u_2^3 + a_2 u_2^2 + a_1 u_2 + a_0 = 0, \tag{8.29}$$

and

$$\chi = 1 + \alpha, \quad \alpha = n_i M_i / n_p M, \quad b_i = B_i / M_A,$$

$$a_4 = 4\chi^2, \quad a_3 = -\chi(C_1 + 8b_z^2),$$

$$a_2 = 2b_z^2 C_1 + C_2 + 4b_z^4 - b_{1y}^2 b_z^2,$$

$$a_1 = -\frac{C_1 b_z^4}{\chi} + C_3 - \frac{2C_2 b_z^2}{\chi} + \frac{2b_{1y}^2 b_z^2}{\chi},$$

$$a_0 = \frac{C_2 b_z^4}{\chi^2} + b_{1y} b_z^2 \left(1 - \frac{2b_z^2}{\chi}\right),$$

$$C_2 = \chi(1 + \alpha + 5v_{T1}^2 + 5\alpha v_{T,i1}^2 + 2b_{1y}^2),$$

$$C_1 = 5\left(1 + \alpha + v_{T1}^2 + \alpha v_{T,i1}^2 + \frac{b_{1y}^2}{2}\right),$$

$$C_3 = \frac{\chi}{2} b_{1y}^2 \left(1 - \frac{b_z^2}{\chi}\right)^2.$$

In this expression there are the sum of the squares of the thermal velocities of the protons and electrons v_{T1}^2 and the square of the thermal velocity of cometary ions v_{Ti1}^2, which are upstream of the shock front:

$$v_{T1}^2 = (v_{T,p1}^2 + \frac{m}{M} v_{T,e1}^2)/U_\infty^2, \quad v_{T,i1}^2 = \frac{1}{3} \int (\bm{v} - \bm{U}_i)^2 f_i d^3 \bm{v}. \quad (8.30)$$

Equation (8.29) has four roots. The first one, $u_2 = 1$, corresponds to the shockless solution; two roots are complex-conjugated; and the fourth root has the value $u_2 = 0.42$ for our parameters of incoming flow.

At the initial moment, $t = 0$, the velocity distribution function for the solar wind protons is Maxwellian:

$$f_p \propto \exp\left(-\frac{v_x^2 + v_y^2 + (v_z - U_z)^2}{2v_{T,p}^2}\right). \quad (8.31)$$

The velocity distribution function for cometary ions is a shell distribution, in agreement with calculations of the ionized cometary molecule dynamics in the solar wind [176]:

$$V_{int}^2 < v_x^2 + v_y^2 + (v_z - U_z)^2 < V_{ext}^2,$$

$$V_{int} = 0.8, \quad V_{ext} = 1.0. \quad (8.32)$$

The temperature of electrons, protons and ions beyond the shock were calculated on the assumption that adiabatic invariance is maintained. At the left-hand boundary ($z = 0$) the field is set to be unperturbed, and protons and ions are injected with an unperturbed velocity distribution function. At the right-hand boundary of the domain ($z = DZ$) the radiation condition for a component of the magnetic field is set, and the particles exit freely from the computational domain.

The simulations presented here were done using 1+2/2-dimensional hybrid code. The one-dimensional computational domain has length DZ. We use a mesh of 1000, and the number of macroprotons is $N_p = 6 \times 10^4$. The equations of motion of the particles and the evolution of the magnetic field were integrated by schemes which was used to study the interaction between the solar wind and the cometary plasma (see, e.g., [319]). The integration time step Δt satisfied the CFL condition $\Delta t = \Delta z/(4U_\infty)$; the step in the z-coordinate satisfied the condition $\Delta z \ll \varrho_{cp}$. Discrete distribution functions for protons and ions were produced through the use of a random number generator. Analysis of scheme's approximation gives the numerical viscosity with dissipation length $l_d = 1/Re = 2 \times 10^{-2}$ upstream and $l_d = 1/Re = 7 \times 10^{-3}$ downstream. This additional numerical viscosity is effective on a scale much smaller than the proton gyroradius. The average change in the total energy in the simulation is about 1%.

In variants of the calculation discussed here, the dimensionless parameters of the plasma and cometary ions were chosen typically for the solar wind at the Earth's orbit, in the coma and in the environs of the outer shock: $M_A = 8, \beta_e = 1, \beta_i = 0.5, \varepsilon = 0.35, (M_i n_i)/(M n_p) = 0.3, DZ = 40$ and $\theta_{Bn} = 30°$. In order that the calculation interval does not increase, the mass of the cometary ion is taken to be equal to five proton masses ($M_i = 5M$), thus maintaining the ratio of the mass densities of the ion and proton components near to the value that corresponds to the gas dynamical equations for loading the solar wind with cometary ions [472].

In Fig. 8.24 profiles are shown of the magnetic field, the total densities of protons and ions, and the number density of cometary ions. The complex

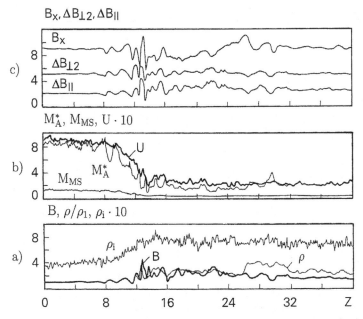

Fig. 8.24. (a) Profiles of the components of the perturbed magnetic field B, the density ϱ/ϱ_1 (normalized to the perturbed value $\varrho_1 = 1 + M_i n_i$) and the ion density ϱ_i. (b) Profiles of the plasma velocity U, Alfvén Mach number M_A (normalized such that $M_A^* = M_A/\sqrt{\varrho_1}$) and the magnetosonic Mach number M_{MS}. (c) Profiles of the transverse ($B_x, \Delta B_{\perp 2}$) and longitudinal ($\Delta B_{\parallel}$) components of the perturbed magnetic field. The position of the jump is defined to be around $z = 12.4$. (From [344])

structure of the shock and the formation of moderate proton–electron jumps can be seen. The profile of the number density of cometary ions has a diffusive structure with moderate pulsations with a wavelength $\lambda \sim \varrho_{ci}$. The jump in the mean magnetic field at the shock front is $B_2/B_1 \approx 2.5$. The density and magnetic field profiles have some differences due to the motion of particles along the field lines.

Figure 8.24b shows the profiles of models of the mean velocity and the Alfvén and magnetosonic Mach numbers. A drastic change of parameters can be seen in the region of the proton–electron jump, with a characteristic thickness $\delta \approx 4 - 5\varrho_{ci}$. The ratio of the mean velocities before and after the shock is $U_2/U_1 \approx 0.4$. The effect of velocity compression is not seen here, as opposed to in the quasiperpendicular case [175].

Particularly interesting is the presentation of the distribution along the z-axis of the perpendicular ($\Delta B_{\perp 1} = B_x, \Delta B_{\perp 2}$) and parallel ($\Delta B_\parallel$) components of the perturbed magnetic field along the shock (Fig. 8.24c). They were obtained by means of calculations from the full value of the quantities, which correspond to Hugoniot adiabat. It can be seen that in the case of a quasiparallel shock wave ($\theta_{Bn} = 30°$), intensive Alfvén oscillations are excited that have a maximum amplitude of $\Delta B_\perp \approx 3B_\infty$ on the ramp (the zone of maximal magnetic field gradient) of the shock and correspondingly $\Delta B \approx 1B_\infty$ and $\Delta B \approx 1.5B_\infty$ in the foreshock (the zone in front of the ramp) and in the subsonic region. The wavelength of these oscillations is comparable to ϱ_{ci} (in the foreshock $\lambda = 0.9\rho_{ci}$, and in the transition region $\lambda \approx 3\varrho_{ci}$); however, the frequency $\omega < \Omega_i$, and the waves have right-handed polarization in the foreshock, corresponding to the Earth's bow shock and to the zone of mass loading of the solar wind by cometary ions [176, 241, 303, 439, 566]. The distribution of $\Delta B_\parallel$ with z characterizes the excitation of magnetic oscillations. Analysis of the distribution of the pressure and temperature of the plasma showed that in the foreshock ($z < 12$) the condition $p_\parallel > p_\perp + B^2/4\pi$ is realized; therefore the perturbation of the magnetic field corresponds to the excitation of both the resonant Alfvén and the firehose instabilities. In the subsonic region the condition $p_\parallel \geq p_\perp$ holds, which corresponds to the excitation of resonant Alfvén modes [574].

The picture of the perturbation of the magnetic field obtained numerically is found to be in agreement with the data from the direct study of cometary shock waves [169, 254, 531]. We must keep in mind that, in agreement with measurements in the environs of comet Halley [169], the thickness of the diffusive shock is determined by the characteristic velocity relaxation length of the ions in the presence of diffusion by Alfvén turbulence and is approximately $5 - 10\varrho_{ci}$. Studies in the environs of comet Giacobini–Zinner [254, 531] give a frequency interval for MHD turbulence between 10^{-2} and $1\,\mathrm{Hz}$ (for the cometary ion H_2O^+), but the wavelength of magnetic field oscillations satisfies the condition $\lambda < 5\varrho_{ci}$. The jump in the magnetic field in the case of a

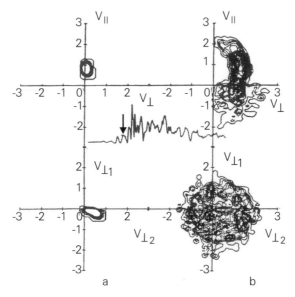

Fig. 8.25. Projection of the velocity distribution function of (a) protons and (b) ions onto the $v_{\perp 1}$–$v_{\perp 2}$ and $v_\parallel$–$v_\perp$ planes. (From [344])

quasiparallel wave amounts to 5.2 at $\theta_{Bn} = 28.7°$ [254], which approximately corresponds to our calculations.

We examine the velocity distribution function of the cometary ions. In Fig. 8.25a the projections of the velocity distribution function onto the $v_{\perp 1}$–$v_{\perp 2}$ and $v_\parallel$–$v_\perp$ velocity plane are given. In agreement with the initial conditions, the initial distribution function had the form of a shell distribution in velocity space. However, as a result of the processes which develop on the shock front, deformation of the distribution function occurs. As the wavefront is crossed, acceleration of the ions along the magnetic field occurs, on account of both the potential jump and the jump in the magnetic and inductive electric fields. The presence of Alfvén oscillations arouses additional acceleration and heating of the ions. Figure 8.25a may be looked upon as the formation of a velocity tail of ions along the magnetic field, for which $v_{\max} \approx -2U_\infty$ (for ions reflected from the front of the shock wave). Although substantial asymmetry of the distribution function of cometary ions in the $v_{\perp 1}$–$v_{\perp 2}$ velocity plane perpendicular to the magnetic field was observed by direct measurements of comet Halley [384, 385]; this was not observed in the calculations. In agreement, Fig. 8.25b has the location of partial reflection of protons ($v \approx -0.5U_\infty$) in the zone of considerable splashing of the transverse component of the magnetic field ($z = 13.33$).

Analysis of the behavior of the temperatures of ions and protons showed that on the ramp of the shock the thermal velocity ratio is as follows: $v_{T,p2}^2/v_{T,p1}^2 \approx 10 - 13$, but $v_{T,i2}^2/v_{T,i1}^2 = 1.6$. At the exit from the computational domain ($z = 40.0$), the temperatures of protons and ions tend to the value determined by the Rankine–Hugoniot relation. The large jump in the temperature of the protons on the ramp is apparently connected to the reflection of protons at the electron–proton jump and intensive heating due to interaction with the oscillations of the electromagnetic field.

8.3.4 Pickup Ion Acceleration at Shock Front. Shock Surfing

In this section, we present multiscale hybrid kinetic simulations of low-β_p ($\beta_p \leq 0.1$) supercritical perpendicular shocks (LBSPSs) which demonstrate that these shocks can significantly accelerate pickup ions (PIs) from an initial thermal shell distribution to large energies [345]. This work offers some resolution to the long-standing theoretical puzzle of how some thermal ion subpopulation acquires sufficient energy to be diffusively accelerated at a perpendicular shock wave. Accordingly, the results presented here have immediate and important implications for the acceleration of ions in astrophysical environments as diverse as supernovae shocks, cometary shocks, interplanetary shocks and stellar wind termination shocks. The simulations demonstrate self-consistently for the first time a new and fundamental dissipation mechanism for quasiperpendicular shocks, with the implication that perpendicular shocks can be very efficient in accelerating ions. A hybrid electromagnetic particle-mesh code (which includes the electron inertia term) is employed to simulate LBSPSs with a low PI density in one spatial dimension. The code allows for the analysis of the microscale proton and PI dynamics inside the foot and ramp of the shock transition layer.

The simulations were motivated by the direct observation of accelerated pickup ions near comets and planetary bow shocks and at interplanetary shocks and the expectation that the termination shock might be responsible for the acceleration of anomalous cosmic rays. Accelerated PIs have been observed directly, for example, by Ulysses at a weak corotating shock [193]. Both pickup H^+ and He^+ were observed by the solar-wind ion composition spectrometer (SWICS) to possess very hard power-law spectra which extend directly out of the expected PI distribution [539] to energies well in excess of the characteristic PI cutoff velocity $v = 2u_{SW}$ (in the spacecraft frame), where v denotes the particle velocity and u_{SW} the solar-wind flow velocity. It was observed too that $\approx 43\%$ of the pickup protons H^+ and $\approx 16\%$ of the He^+ was accelerated by the shock. The hardness of the PI spectrum was quite inconsistent with the expectations of diffusive/first-order Fermi shock acceleration. Secondly, it is generally thought that diffusive shock acceleration should favor the acceleration of heavy ions over lighter ions since the ions with large gyroradii would be "injected" more easily into the acceleration process. That neither expectation is met poses a challenge for models of particle acceleration at shock waves. Finally, particle-acceleration processes are also important in the context of laboratory shocks and shocks generated by active experiments in space.

Besides diffusive shock acceleration, two alternatives for accelerating PIs at weak quasiperpendicular shocks are shock drift acceleration [15, 76, 124, 125] and multiply-reflected-ion (MRI) acceleration or shock surfing, proposed originally in [264, 459, 461] and later in the context of PI acceleration in [297, 588]. Both approaches directly accelerate particles out of the PI "thermal" pool. However, the MRI mechanism is found to produce high energies for PIs and a very flat spectrum and to have an injection efficiency which decreases with increasing particle mass. Present models of MRI acceleration are either purely analytic [297] or quasi-analytic [588], test particle tracing [597] or test particle-mesh simulations [348], and treat the PIs as test particles and use non-self-consistent electromagnetic fields.

A set of hybrid simulations had demonstrated very effective PI acceleration by the MHD turbulence upstream of the quasiperpendicular shocks [190, 283, 306, 308]. However, this mechanism takes about several tens of gyroperiods of PIs. Shock surfing may provide preacceleration of PIs for further acceleration by the electromagnetic fluctuations. Unfortunately, self-consistent kinetic simulations of quasiperpendicular shocks (e.g., [161, 173, 301, 304, 336, 433, 435, 527]), which may include PIs [175, 190, 283, 306, 308, 321, 406], have yet to demonstrate any significant acceleration of solar wind protons and PIs by MRI-like processes in the absence of either strong wave generation or externally imposed high levels of MHD turbulence (Fig. 8.26).

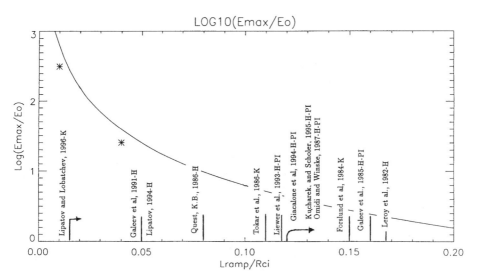

Fig. 8.26. The dependence of the maximum energy of accelerated PIs by shock surfing on the ramp thickness. ∗ denotes results from the test particle-mesh simulation [348]. The *solid line* corresponds to the estimates from [588]; H denotes the hybrid code simulations, K denotes the full kinetic code simulations. PI denotes pickup ion simulation. All of the self-consistent simulations do not resolve the fine structure of the shock ramp, therefore a shock-surfing acceleration is not effective

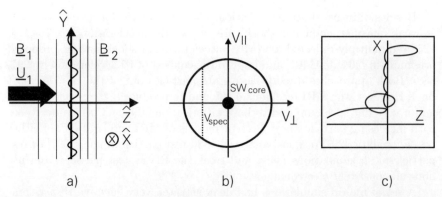

Fig. 8.27. (a) Geometry of a quasiperpendicular shock located at $z = 0$. The stream flow u is parallel to the shock normal, the magnetic field B lies along or oblique to $\hat{y}$, and the induced electric field is in the $\hat{x}$-direction. The subscripts 1 and 2 denote upstream and downstream of the shock, respectively. (b) Schematic of the idealized shell distribution assumed for pickup ions (PIs) in the fluid frame. The cold solar wind (SW) core is represented by the heavy dot at the origin, and V_{spec} corresponds to the normalized transformed specular reflection velocity, below which PIs are reflected by the electrostatic shock potential barrier. (c) Example of integrated individual PI at a perpendicular shock located at $z = 0$ with an assumed step-like electrostatic cross-shock potential. (From [348, 588])

The basic idea of particle acceleration by the multiple reflection (or surfing) of PIs at the shock ramp is easily explained. On the PI shell distribution in the shock frame, a fraction of the PI population has so little kinetic energy, that it can be reflected by the shock electrostatic potential jump inside the ramp, as illustrated in Fig. 8.27. This population of ions then drifts along the shock front surface, being multiply reflected at the shock ramp, trapped by the upstream particle Lorentz force and the electrostatic potential jump. The time spent upstream of the shock determines the maximum energy gain for a trapped PI, resulting from a balance between the particle Lorentz force and the gradient of the electrostatic potential.

For effective acceleration by shock surfing we have to satisfy the following necessary conditions:

(1) There is a balance between proton kinetic energy and a potential energy across ramp of the shock:

$$0.5 \frac{MU_\infty^2}{2} = -e\Delta\phi_0;$$

(2) The thickness of transition layer must be very small:

$$\frac{\Delta_{\text{ramp}}}{\varrho_{\text{ci}}} \leq 0.02;$$

(3) PI trapping time must be comparable with acceleration time:

8.3 Collisionless Shocks with Mass Loading by Heavy Ions

$$\tau_{\text{trap}} \approx \tau_{\text{accel}} \approx 1 T_{\text{ci}}.$$

However, there are negative factors which may reduce the efficacy of the acceleration. First of all a detached foot formation may reduce the jump in electrostatic potential. Second, a cyclic reconstruction of the shock front steepening, breaking and broading may reduce the trapping time and consequently the efficiency of acceleration.

We now present simulations of LBSPSs that exhibit significant MRI energization of PIs at the shock ramp [345]. The simulation results were obtained by using a one-dimensional hybrid kinetic code [174, 333]. Anomalous resistivity and electron inertia terms are included in this code. The simulation results are one-dimensional, as we assume spatial variation along the z-direction only, but retain all three components of the electromagnetic fields and particle velocities. The shocks studied have upstream parameters, which are expected of the solar wind in the transition layer of the termination shock: $M_A = 3 - 5$; $\beta_e = 0.1 - 2.0$; $\beta_p = 0.1 - 1.0$; $n_{\text{PI}}/n_0 = 0.001 - 0.1$; $M/m = 10^2 - 10^4$; $M_{\text{PI}}/M = 1$; $\omega_{\text{pi}}/\Omega_i = 4 \times 10^3$; $DZ = 10L$; $L = \varrho_{\text{ci}}$; $\Delta z = (0.006 - 0.0125)c/\omega_{\text{pi}}$; $\theta_{Bn} = 72°$ and $90°$; and $l_d = (0.001 - 0.25)c/\omega_{\text{pi}}$. M_A is the Alfvén Mach number, $\beta_{p(e)}$ denotes the solar wind proton(electron) plasma beta, n_{PI}/n_0 is the ratio of PI to proton density, M/m is the ratio of proton to electron mass, M_{PI}/M is the ratio of PI to proton mass, DZ is the size of computational domain, Δz is the cell size, $\varrho_{\text{ci}} = U_0/\Omega_i$ is the proton cyclotron radius ($\varrho_{\text{ci}} = \varrho_{\text{cH}^+}$), and $l_d = \nu c^2/\omega_{\text{pi}}^2 U_0$ is the resistive diffusion length [304, 435], where ν is the anomalous collision frequency.

The chosen range of the resistive diffusion length corresponds to current-driven instabilities in the foot and ramp: for the ion acoustic mode, $\nu_{\text{IA}} \approx 10^{-2}\omega_{\text{pe}}$ and $l_{d,\text{IA}} \approx 0.2c/\omega_{\text{pi}}$; for the lower-hybrid drift mode, $\nu_{\text{LH}} \approx 2.2 \times 10^{-4}\omega_{\text{pe}}$ and $l_{d,\text{LH}} \approx 3.8 \times 10^{-3}c/\omega_{\text{pi}}$; and for the modified two-stream instability, $\nu_{\text{MTS}} \approx 2.2 \times 10^{-5}\omega_{\text{pe}}$ and $l_{d,\text{MTS}} \approx 3.8 \times 10^{-4}c/\omega_{\text{pi}}$. These scales are consistent with observational and theoretical studies of instabilities at shock fronts [269, 410, 479, 560]. Initially, the proton velocity distribution function is Maxwellian, with a PI shell velocity distribution in velocity space, which we assume to have been thickened. The interior shell radius is taken to be $V_{\text{int}} = 1.1U_0$ and that of the exterior shell is $V_{\text{ext}} = 1.5U_0$, where U_0 denotes the upstream bulk velocity. Note that the effective upstream bulk velocity relative to the shock front (in the simulation) is $1.4U_0$ and the effective interior and exterior shell radii are 0.8 and 1.07, respectively. This is not a very sensitive assumption. The bulk velocity of the PIs equals the bulk velocity of the protons at the initial moment of time $\boldsymbol{U}_{\text{PI}} = \boldsymbol{U}_0$. There are $4097 - 8193$ cells in z, and initially 5×10^5 macroprotons and 5×10^5 macro-ions. Plasma is injected continuously into the left of the computational domain, and the shock is formed by reflecting the plasma off the right-hand end of the domain. The simulation time step is $5 \times 10^{-5} T_{\text{ci}}$, where $T_{\text{ci}} = 2\pi/\Omega_i$ and Ω_i is the proton gyrofrequency. Such small spatial and time scales were chosen to resolve the ramp on an electron inertial length scale and to provide an accurate calculation of PI trajectories as they are transmitted across the ramp.

Figure 8.28 illustrates a simulation of a shock with $M_A = 5$, $n_{PI}/n_0 = 0.01$, $\theta_{Bn} = 90°$ and $l_d = 0.006c/\omega_{pi}$ at time $t = 3.8T_{ci}$. Note that we show only half the computational domain. In Fig. 8.28, the velocity, magnetic field, density and electron pressure are normalized to the upstream velocity, magnetic field, proton density and electron pressure. The electric field and electrostatic potential are normalized to the upstream induced electric field $U_0 B_0/c$ and the kinetic energy of incoming protons $MU_0^2/2$. The simulation demonstrates the formation of a shock transition layer with a strong foot in the PI density profile (g) and a thin ramp, $\Delta_{\rm ramp}/\varrho_{ci} < 0.05$, in the magnetic field and electrostatic potential profiles (Fig. 8.28c and h). An additional jump forms at a distance $\delta z \approx 0.6\varrho_{ci}$ before the ramp in the electromagnetic

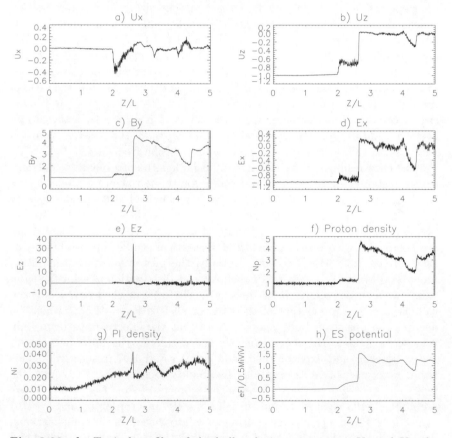

Fig. 8.28a–h. Typical profiles of the bulk velocity components U_x and U_z, the magnetic field component B_y, the electric components E_x and E_z, the proton (n_p) and PI (n_{PI}) densities, and the electrostatic potential $e\phi/MU_0^2/2$. $l_d = 0.006c/\omega_{pi}$. (From [345])

field, bulk velocity and proton density profiles. It is clearly seen that the peak in PI density corresponds to accelerated PIs and the peak is located inside the shock ramp due to temporary trapping of PIs (Fig. 8.28g).

Figure 8.29 shows the projected proton and H^+ PI distribution in the plane $v_{\perp 1}$–$v_{\perp 2}$ plane at different locations relative to the shock ramp. The panels of Fig. 8.29 are arranged in ascending order from the bottom according to position as follows: far upstream, on the shock front, just downstream of the shock and far downstream. The left (right) column illustrates the projected proton (PI) distribution in the $v_{\perp 1}$–$v_{\perp 2}$ plane at various distances from the shock ramp, where $v_{\perp 1}$ and $v_{\perp 2}$ are the velocity components perpendicular to the magnetic field. The proton distribution function has a supersonic core ahead of the ramp and a subsonic core downstream of the shock front. Before

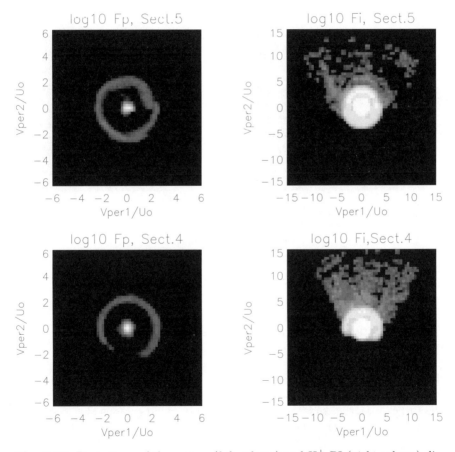

Fig. 8.29. Projections of the proton (*left column*) and H^+ PI (*right column*) distributions onto the velocity plane for spatial sections ranging from downstream to upstream. Here $v_{\perp 1} \parallel \boldsymbol{v}_z$ and $v_{\perp 2} \parallel \boldsymbol{v}_x$ are the velocity components perpendicular to the magnetic field. $l_d = 0.006c/\omega_{\mathrm{pi}}$. (From [345])

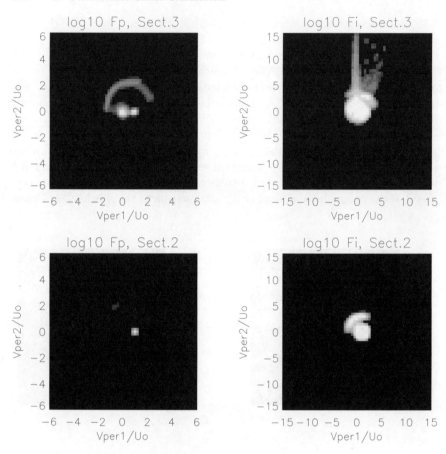

Fig. 8.29. (continued)

the ramp, we see reflected protons, while downstream the transmitted protons form a halo due to phase mixing. The bottom right panel illustrates a very typical distribution which results from ion reflection at a perpendicular shock. Such distributions are seen at virtually all quasiperpendicular shocks, both observationally (e.g., [478]) and in simulations (e.g., [304]), and contribute essentially to the formation of the ion shock foot. If the number and energy density of the reflected PIs were sufficiently high at the termination shock, the foot structure and length scales would be determined primarily by reflected PIs rather than the colder more numerous solar wind protons [306, 588]. The second panel from the bottom shows the PI distribution at the shock ramp and a strong transverse acceleration of PIs along the shock front is evident with the formation of an extended "tongue" along $v_{\perp 2}$. Finally, phase mixing occurs far downstream.

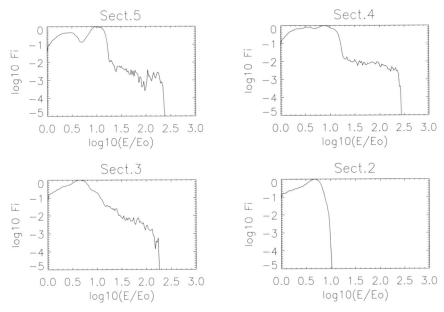

Fig. 8.30. The energy spectrum of accelerated H^+ PIs for spatial sections ranging from downstream to upstream. $l_d = 0.006c/\omega_{pi}$. (From [345])

Figure 8.30 illustrates the energy spectrum of accelerated H^+ PIs for spatial sections ranging from downstream to upstream. The PI energy spectrum has two parts, as discussed in [588]: a shell-like distribution with an energy cut-off at about $E_0 = M_{PI}U_0^2/2$ in the solar wind frame and an accelerated PI component which emerges from the shell distribution as a hard/flat power-law spectrum. The accelerated PI energy spectrum may be approximated by the power law $F_i \propto dN/N \sim (E/E_0)^{-k}$, where the energy E is calculated in the solar wind frame and N denotes the PI number density. In the present case, the index k is about 1.0–1.3 in the vicinity of the ramp. The similarity between the spectra produced by the hybrid simulation here, a test particle-mesh simulation [348], and those obtained from the quasi-analytical approach in [588] is high. The spectrum produced by the MRI mechanism is much harder than that expected from diffusive shock acceleration, which would produce an E^{-2} spectrum for the shock compression ratio used in our test particle simulations. Also, as discussed in §2 in [588], diffusive shock acceleration at a perpendicular shock imposes severe energy constraints on the particles to be accelerated, constraints which are absent for MRI acceleration. The maximum transverse energy of accelerated PIs ($\log_{10} E_{max}/E_0 = 2.3 - 2.5$) is higher than estimated in [588] for a ramp thickness is about of $L_{ramp} < 0.05\varrho_{ci}$. The simulation with a high PI density ($n_{PI}/n_0 = 0.1$) is less efficient at accelerating PIs ($\log_{10} E_{max}/E_0 = 1.7$) thanks to the formation of a strong foot and a decrease in the electrostatic potential jump. In the simulation with a large resistive diffusion length, $l_d = 0.25c/\omega_{pi}$, the maximum energy of accelerated PIs significantly decreases, since the ramp thickness is too large, and MRI acceleration becomes ineffective. For the simulation with a small resistive diffusion length, $l_d = 0.001c/\omega_{pi}$, the electrostatic jump at the ramp is comparable with the fluctuation level and no significant acceleration occurs. In the quasiperpendicular case, with the angle between the magnetic field and shock normal $\theta_{Bn} = 72°$, the formation of a whistler precursor decreases the efficacy of PI acceleration.

Further hybrid simulation (performed by the author) for $\theta_{Bn} = 90°$ with a smaller grid spacing, $\Delta z = 0.003c/\omega_{pi}$, and a smaller resistivity length, $l_d = 0.001c/\omega_{pi}$, also demonstrates significant acceleration of protons and at the same time a weaker acceleration of pickup ions (Figs. 8.31–8.34). The ac-

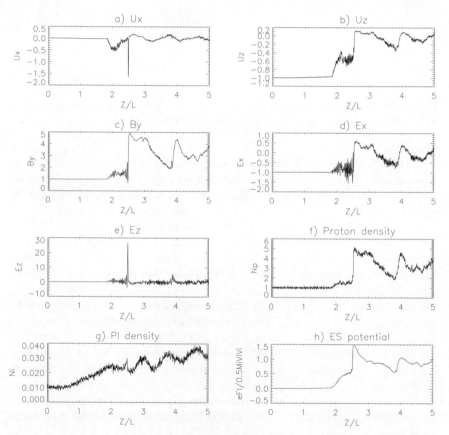

Fig. 8.31a–h. Typical profiles of the bulk velocity components U_x and U_z, the magnetic field component B_y, the electric components E_x and E_z, the proton (n_p) and PI (n_{PI}) densities, and the electrostatic potential $e\phi/MU_0^2/2$. $l_d = 0.001c/\omega_{pi}$

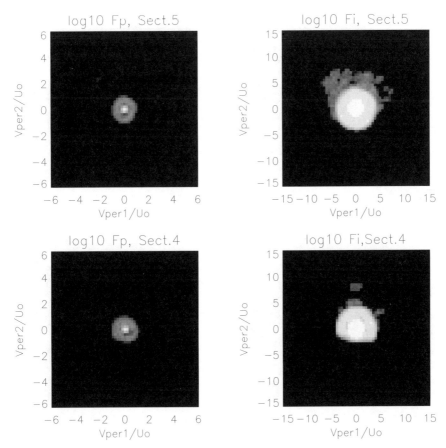

Fig. 8.32. Projections of the proton (*left column*) and H$^+$ PI (*right column*) distributions onto the velocity plane for spatial sections ranging from downstream to upstream. Here $v_{\perp 1} \parallel \boldsymbol{v}_z$ and $v_{\perp 2} \parallel \boldsymbol{v}_x$ are the velocity components perpendicular to the magnetic field. $l_d = 0.001c/\omega_{\mathrm{pi}}$

celerated protons change the profile of the electrostatic potential that results in a smaller efficacy of PI acceleration.

In conclusion, kinetic hybrid simulations of the acceleration of H$^+$ PIs at low-$\beta_\mathrm{p} \leq 0.1$ collisionless quasiperpendicular shocks (with a low PI density $[n_{\mathrm{PI}}/n_0 < 0.1]$ and an appropriate anomalous resistivity $0.006c/\omega_{\mathrm{pi}} \leq l_\mathrm{d} < 0.25c/\omega_{\mathrm{pi}}$) have demonstrated several new features, as well as providing support for the basic test-particle analysis in [297, 348, 588, 597]. Our results are as follows: (a) The energy spectrum of accelerated H$^+$ PIs at quasiperpendicular shocks may be approximated by the power law $F_i(E) \approx (E/E_0)^{-k}$, where k varies from 1.0 to 1.3. This spectrum is a little harder than that obtained by the quasi-analytical approach in [588], but both approaches give spectra which are considerably harder than those predicted by diffusive shock

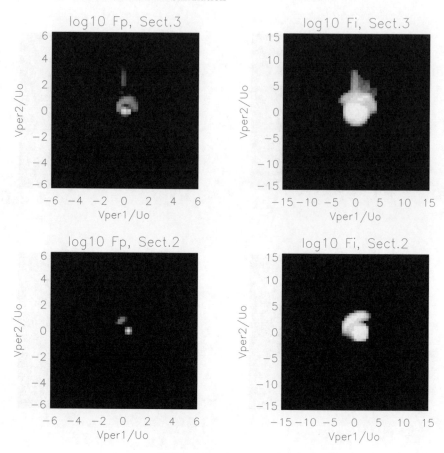

Fig. 8.32. (continued)

acceleration. (b) For MRI acceleration of H$^+$ and He$^+$ ions to be effective, a ramp thickness comparable to the electron inertial length scale is needed, whereas for heavy ions, it is sufficient to have a ramp thickness comparable to that of the ion inertial length. Thus, the key factor determining the efficacy of MRI acceleration is the existence of a strong steep ramp inside the shock transition layer. The simulation does not show MRI acceleration of H$^+$ and He$^+$ at shocks with $\beta_p > 0.1$. (c) The simulation with a very small grid spacing and a very small resistivity length demonstrates a strong acceleration of protons and a decreasing PI acceleration.

8.3 Collisionless Shocks with Mass Loading by Heavy Ions 235

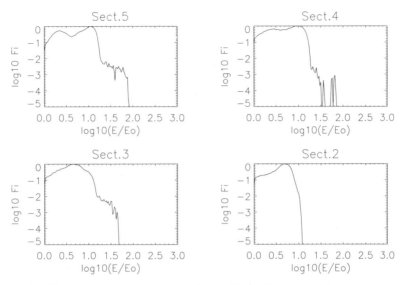

Fig. 8.33. The energy spectrum of accelerated H$^+$ PIs for spatial sections ranging from downstream to upstream. $l_\mathrm{d} = 0.001 c/\omega_\mathrm{pi}$

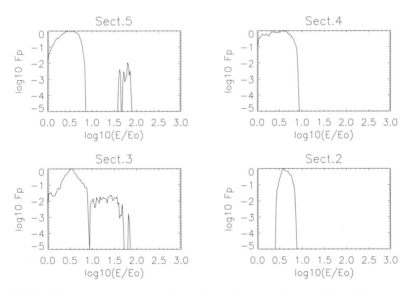

Fig. 8.34. The energy spectrum of accelerated protons for spatial sections ranging from downstream to upstream. $l_\mathrm{d} = 0.001 c/\omega_\mathrm{pi}$

Summary

In this chapter we discussed collisionless shocks by means of hybrid computer simulation. The simulations demonstrated a thin structure of the shock transition layer. Simulation also demonstrated strong wave generation inside the shock structure. Strong particle acceleration by a strong turbulence and by shock surfing under special conditions (β, anomalous resistivity) was observed. Multidimensional simulation of collisionless shock (see, e.g., [161, 321, 322, 324, 325, 326, 329, 330, 331, 348, 369, 439, 517, 520, 572, 573]) showed two- and three-dimensional wave structures which result in stronger particle heating and acceleration. However, consideration of the multidimensional simulations of the shock transition layer is not covered by this chapter.

Exercises

8.1 Derive the upstream and downstream boundary conditions for the field and protons related by the Rankine–Hugoniot relation to form a collisionless shock at rest in the simulation frame.

8.2 Derive the upstream and downstream boundary conditions for the field and pickup ions related by the Rankine–Hugoniot relation to form a collisionless shock at rest in the simulation frame.

9. Tangential Discontinuity Simulation

9.1 Introduction

The magnetopause (MP) is a critical region of geospace, since it controls the transfer of energy and momentum from the solar wind into the magnetosphere. Observations show that the MP is a finite thickness discontinuity, which separates the post-shocked solar wind (SW) from the magnetosphere. Depending on orientation (northward or southward) of the interplanetary magnetic field, the MP has been modeled as a rotational discontinuity ([200] and references therein) and as a tangential discontinuity (TD) [89, 296] respectively. While an extensive boundary layer of magnetosheath-like plasma is often observed earthward of the MP (which presumably comes from the outer magnetosphere) [139, 140, 211, 418], as shown in [414, 494], the TD model is nevertheless valuable for improving our understanding of the physical processes which occur sunward of the MP.

The structure of the MP in cases where it was modeled as a TD, has been studied using particle [40, 402] and hybrid [89, 261, 516, 519, 524, 568] computer simulations. While these simulations provide interesting physical ideas, the neglect of field-aligned dynamics casts doubt upon their applicability to MP processes. Another important shortcoming of these studies is related to their utilization of a small ion/electron mass ratio and their lack of any discussion of how the relevant physics of the problem scales with this parameter.

The transfer of energy and momentum across the dayside MP can be accomplished in two principally different ways: via reconnection processes, the Kelvin–Helmholtz instability, or the interaction of the MP with external waves (see, e.g., [252]) and via the rapid deformation of the MP, which occurs when it suffers a collision with an interplanetary discontinuity or the SW filaments. In this chapter, we consider the dynamics of the MP in the absence of external disturbances, and in Chap. 11, we shall consider the forced excitation of surface waves and other field-aligned effects occurring within the internal structure of the TD, which are a consequence of the second mechanism above.

Section 9.2 explains the model and method of calculation. In Sects. 9.3 and 9.4, we consider the structure of the MP when we model it as a TD using 2.5-dimensional hybrid code.

9.2 Formulation of the Problem and Mathematical Model

In order to study the structure of MP transition layer, we treat the MP as a TD. A TD is thin layer, characterized by the rapid change of certain plasma variables, that connects two approximately uniform plasmas and across which there is zero mass flux and zero normal component of the magnetic field. The only quantity conserved across such a layer is the total (ions, electrons and magnetic field) pressure:

$$p_{i1} + p_{e1} + B_1^2/8\pi = p_{i2} + p_{e2} + B_2^2/8\pi, \quad (9.1)$$

where subscripts 1 and 2 represent the asymptotic values on the upstream and downstream sides at the transition, respectively, and the subscript i(e) denotes ion (electron) quantities. The parameter $\beta_{i,e} = 8\pi n_{i,e} kT_{i,e}/B^2$ is the plasma beta.

We consider the case of zero initial bulk flow at the TD, which corresponds to the stagnation region of the magnetopause. The initial ion velocity distributions on both sides of the tangential discontinuity, are assumed to be Maxwellian. Across the initial tangential discontinuity the total (ion, electron and magnetic) pressure is balanced. All initial profiles of the plasma and field variables are approximated with a smoothed jump, with a finite initial thickness Δ_0. At the boundaries ($x = 0$ and $x = DX$) we choose periodic boundary conditions for the fields and particles. At the left (right) boundary $z = 0$ ($z = DZ$) we "close" the computational domain by assigning unperturbed and constant fields $B = B_1$, $E = 0$ (left) and $B = B_2, E = 0$ (right). Thus, the processes in our system do not depend on far-field variations of the SW or magnetosphere. In other problems, it may sometimes be necessary to use "open" boundary conditions for magnetic field and pressure. By "open" boundary conditions, we mean that a flux of particles across the boundaries is allowed to be nonzero. The use of "closed" boundary conditions in this paper is permissible because the initial bulk velocities of the ions and electrons are zero everywhere (thus, ruling out the existence of the Kelvin–Helmholtz instability).

For our simulation we use 2.5-dimensional hybrid codes in both one-dimensional and two-dimensional cases. The equations may be expressed in a dimensionless form by choosing a typical length L and time L/U_0, U_0 being the typical velocity. Depending on the problem, one of various velocities, for example, the bulk velocity of incoming flow, or the thermal velocity of ions, or the Alfvén velocity, may be chosen as the typical velocity. The basic equations are the ion equations of motion, the generalized Ohm's law, Faraday's induction equation and the adiabatic equation for the electron pressure. The ion equations (normalized) are of the form

$$\frac{d\boldsymbol{x}_l}{dt} = \boldsymbol{v}_l, \quad \frac{d\boldsymbol{v}_l}{dt} = (\boldsymbol{E}^* + \boldsymbol{v}_l \times \boldsymbol{B})/\varepsilon, \quad \boldsymbol{E}^* = \boldsymbol{E} - l_d^* \boldsymbol{J}. \quad (9.2)$$

The normalized dissipation length is $l_d^* = l_d/L = 1/Rc$, the magnetic Reynolds number is $Re = 4\pi v_{Ti} L \sigma_{eff}/c^2$ and σ_{eff} is the effective conductivity. The parameter ε is the ratio of the proton Larmor radius $\varrho_{ci} = v_{Ti}/\Omega_i = \sqrt{\beta_{i1}/2}c/\omega_{pi}$ in the upstream field B_0 to the characteristic dimension L of the problem, $\varepsilon = \varrho_{ci}/L$, $\beta_i = p_i/(B_0^2/8\pi)$.

The electrons are considered to have nonzero mass, so that the generalized Ohm's law is

$$\varepsilon \frac{m}{M}\frac{\partial \boldsymbol{J}_e}{\partial t} + \varepsilon \frac{m}{M}(\nabla \boldsymbol{U}_e)\boldsymbol{J}_e - \frac{\varepsilon \beta_e}{\beta_i}(\nabla p_e + \partial \pi^e_{\alpha\beta}/\partial x_\beta) - n_e \boldsymbol{E} + \boldsymbol{J}_e \times \boldsymbol{B} + l_d^* n_e \boldsymbol{J} = 0 \tag{9.3}$$

or

$$\boldsymbol{E} = -\boldsymbol{U}_e \times \boldsymbol{B} - \frac{\varepsilon \beta_e}{\beta_i n_e}\nabla p_e + l_d^* \boldsymbol{J} - \varepsilon \frac{m}{M}\frac{d\boldsymbol{U}_e}{dt}. \tag{9.4}$$

The electron $\boldsymbol{U}_e$ and ion $\boldsymbol{U}_i$ bulk velocities are given by

$$\boldsymbol{U}_e = \boldsymbol{U}_i - 2\varepsilon \boldsymbol{J}/(\beta_i n_e); \quad \boldsymbol{U}_i = \langle N_i \boldsymbol{v}_i \rangle / \langle N_i \rangle. \tag{9.5}$$

Ampere's law and the induction equation are

$$\boldsymbol{J} = \nabla \times \boldsymbol{B}; \quad \frac{\partial \boldsymbol{B}}{\partial t} + \nabla \times \boldsymbol{E} = 0, \tag{9.6}$$

$$\boldsymbol{J} = \boldsymbol{J}_e + \boldsymbol{J}_i, \tag{9.7}$$

where dimensionless current

$$\boldsymbol{J}_i = n_i \boldsymbol{U}_i. \tag{9.8}$$

We impose the condition of quasineutrality, and we consider the electron gas to be adiabatic,

$$n_e = n_i \equiv n, \quad p_e \propto n_e^\gamma. \tag{9.9}$$

Here the dimensionless parameters may be expressed via dimensional parameters on the upstream side at the transition as following

$$U = U' v_{Ti}, \quad E = E' B_0 v_{Ti}/c, \quad B = B' B_0, \quad p_e = p'_e p_{e0},$$
$$n = n' n_0, \quad t = t' L/v_{Ti}, \quad x = x' L. \tag{9.10}$$

In Sect. 9.3 we consider the one-dimensional structure of the tangential discontinuity, whereas in Sect. 9.4 we consider the two-dimensional structure of the TD.

9.3 One-Dimensional Structures

The 2.5-dimensional hybrid simulation of the TD was done in [334, 342, 568]. There are several ways to set up a time-stationary discontinuity. First, we can use a TD equilibrium obtained from simple pressure balance arguments.

In [89, 334, 342, 413], such a method was used to study TD structure and evolution in one- and two-dimensional hybrid simulations. In [524], the two-dimensional hybrid simulations were initialized with this type of equilibrium to study the Kelvin–Helmholtz instability at a TD-like discontinuity with sheared flow across the boundary. A third method is to set up the TD dynamically, through the interaction of two streaming plasmas. In [408], this technique was employed to investigate the general (Riemann) problem of how the SW plasma and magnetic field join to their counterparts in the magnetosphere. The 2.5-dimensional hybrid code can be used to simulate a one-dimensional configuration, by using only 3 grid points along the ignorable x-direction (we used 201 grid points in the z-direction).

We shall consider here the cases in which the initial ion temperatures, and thus the ion pressures (the dynamically more interesting species), are the same across the discontinuity:

$$T_{i1} = T_{i2}, \tag{9.11}$$

$$p_{e1} + B_1^2/8\pi = p_{e2} + B_2^2/8\pi. \tag{9.12}$$

In all the cases considered here, the initial location of the discontinuity is at $z = 5.0$, and the initial magnetic field is along the y-direction.

The dimensionless parameters of the plasma were assigned values typical of the SW downstream of the Earth's bow shock: $\varepsilon = 0.5$, $\Delta_0/\varrho_{ci} = 0.2$, $\Delta z/\varrho_{ci} = 0.1$, $DZ = 10$, $l_d = 0$, $\beta_{i1} = 1.0$, $\beta_{e1} = 1.0$, $n_2/n_1 = 0.2$, $B_2/B_1 = 1.65$ for the case $T_{i1} = T_{i2}$ and $\beta_{i1} = 2$, $\beta_{e1} = 2$, $\beta_{i2} = 0.89$, $n_1/n_2 = 10$, $B_2/B_1 = 1.73$ for the case with initial pressure balance $p_e + B^2/8\pi = $ const. In these particular cases, the anomalous resistivity is zero and the number of ions is 4×10^5. The y-component of the magnetic field profiles, for both cases, is shown in Fig. 9.1a for equal temperature and in Fig. 9.1b for equal pressure. The profiles of the plasma variables shown here are quasistationary and no significant change occurred prior to calculation, at $15T_{ci}$. Figure 9.1a can be broken into two principally different regions: the center of the transition layer, which is a nearly linear profile with a thickness of about ϱ_{ci}, and external to this region on both sides of the transition are regions in which, significant perturbations are generated. In general, the size of these regions varies from about one to a few ϱ_{ci}. A weak magnetic field overshoot behind and a small depression ahead of the TD are shown in Fig. 9.1a.

In the second case the parameters are similar to those of earlier TD simulations [89], except for the value of β_{e2} and B_2/B_1 (0.33 and 1.5, respectively, in the previous calculation). In our case $B_2/B_1 = 1.73$ and $\beta_{e2} = 0.014$, since we used the adiabatic electron pressure equation and a value of β_{e1}. The main difference between Fig. 9.1a and b is the displacement of the front of the TD and the formation of a significant overshoot in the magnetic field in the downstream region. The peak value of the magnetic field in the overshoot region is about $1.95B_1$, much greater than the initial value of the magnetic field downstream, e.g., $B_2 = 1.73B_1$. The thickness of this peak is about

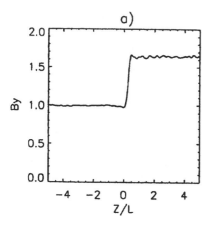 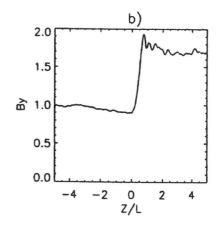

Fig. 9.1. Typical profiles of magnetic field B_y in a one-dimensional simulation of a TD with equal initial ion (**a**) temperature and (**b**) pressure across it (from [334, 342])

$0.5\varrho_{ci}$, and it is followed by multiple oscillations with wavelength $\lambda \approx 0.5\varrho_{ci}$. The size of the perturbed field region is about $4\varrho_{ci}$. In front of the TD a wide region of depression in the magnetic field is formed. The value of the magnetic field in this region is about $0.9B_1$, and the characteristic size of this region is about $6\varrho_{ci}$. This magnetic field depression is correlated with a region of reduced ion density in front of the TD with scale lengths of about $6\varrho_{ci}$. The value of ion density in this region decreases to $0.9n_1$. Significant oscillations are also present immediately downstream of the TD.

9.4 Two-Dimensional Structures

In the two-dimensional model we have more channels of energy transportation, and the structure of the TD becomes more complicated. In this section we shall consider the two-dimensional structures of the tangential discontinuities in absence of the Kelvin–Helmholtz instability.

9.4.1 Magnetic Field Oriented Perpendicular to the Simulation Plane

In this case perturbations can develop perpendicular to the magnetic field [334, 342]. In the transition region only magnetosonic-type modes are expected. The two-dimensional simulations were performed using 201×101 and 201×51 grid points and 2×10^6 ions, for equal temperatures ($T_{i1} = T_{i2}$) on both sides of the transition. The dimensionless parameters of the plasma were assigned values typical of the SW downstream of the Earth's bow shock: $\beta_{i1} = 1.0$, $\beta_{e1} = 1.0$, $\varepsilon = 0.5$, $\Delta_0/\varrho_{ci} = 0.2$, $DZ = 10$, $DX = 10$,

$n_2/n_1 = 0.2$, $B_2/B_1 = 1.65$ and $l_\mathrm{d}/\varrho_\mathrm{ci} = 0.01 - 0.02$. Typically, we used 100 particles per cell upstream of the TD and about 20 particles per cell downstream. In the one-dimensional simulations discussed in the last section, the code provided stable solutions without any additional anomalous resistivity. In the two-dimensional simulations small values of resistivity, in the range $l_\mathrm{d}/\varrho_\mathrm{ci} = 0.01 - 0.02$ are needed. In the regime with $l_\mathrm{d}/\varrho_\mathrm{ci} = 0.02$, the cuts of the two-dimensional profiles of parameters B_y and n_i in the z-direction are approximately the same as those in the corresponding one-dimensional simulations. However, the two-dimensional profiles of the magnetic field component and of the ion density have two-dimensional perturbations upstream and downstream of the TD with $k\varrho_\mathrm{ci} \approx 2$. The perturbation of the current vector field in the x-direction has $k_x \varrho_\mathrm{ci} \approx 0.6$ inside of the transition region.

In the next regime we use low anomalous resistivity, $l_\mathrm{d}/\varrho_\mathrm{ci} = 0.01$, $DZ = 10$ and $DX = 2.5$, and obtain significant perturbations in the two-dimensional profiles of the magnetic field component B_y and the density of ions. The current perturbations which propagate in the upstream region of the TD, Fig. 9.2a, may be represented as plane waves with $k\varrho_\mathrm{ci} \approx 6$. The oscillation of the magnetic field component ΔB_y has the values $\approx 0.2 B_1$, and the perturbation of density is about $\Delta n_\mathrm{i} \approx 0.1 n_\mathrm{i1}$. The effective broadening of the transition layer is caused by oblique wave emission from the front of the TD. We also conducted simulations for a thick transition layer ($\Delta_0/\varrho_\mathrm{ci} \gg 1$) of

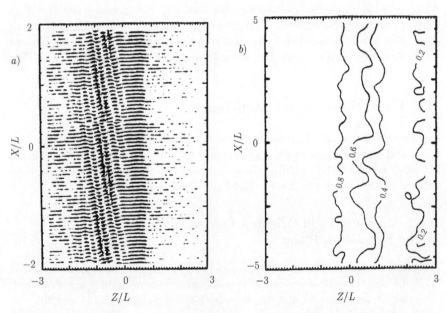

Fig. 9.2. (a) Vector field plot for the current in the x–z plane for low anomalous resistivity. (b) Contour plots of the ion density n_i for a thick TD layer. (From [334, 342])

TDs with a high anomalous resistivity: $\beta_{i1} = 1$, $\beta_{e1} = 1$, $\varepsilon = 0.1$, $n_2/n_1 = 0.2$, $\Delta_0/\varrho_{ci} = 5.0$, $DZ = 10$, $DX = 10$ and $l_d/\varrho_{ci} = 0.5$. In this case the front of the TD has strong disturbances in the contour plots of the magnetic field B_y and the ion density n_i (Fig. 9.2b). This strong disturbance of the front increases the effective thickness of the transition layer to $\Delta = 50\varrho_{ci}$. The wavelength of these surface waves is approximately $k\varrho_{ci} = 0.3$.

9.4.2 Magnetic Field in the Simulation Plane

We repeated here the same simulations as in Sect. 9.4.1 (the same plasma parameters of the plasma and computational domains), except for a different orientation of the magnetic field.

For high and low anomalous resistivities, we repeat the one-dimensional structure of transition layer. For a thick transition layer and a high anomalous resistivity, $l_d/\varrho_{ci} = 0.5, \varepsilon = 0.1, \Delta_0/\varrho_{ci} = 5.0, DZ = 10$ and $DX = 10$, there are no significant fluctuations in the contour plots of the magnetic field or ion density, in contrast to Sect. 9.4.1. Surface waves which are excited in the plane of the TD have a maximum growth rate for wave vectors which are oriented perpendicular to the computational plane; consequently, we did not see significant fluctuations in this case.

9.4.3 Analysis of the Waves at the TD and the Wave–Particle Cross-Field Transport

9.4.3.1 Waves at the TD. The dispersion equation for high-frequency electromagnetic waves with electromagnetic corrections due to the coupling of the drift wave with lower hybrid waves is easily derived from the Ampere–Poisson equations along with cold fluid equations for electrons (e.g., [561]), which yields the following:

$$1 + \frac{\omega_{pe}^2}{\Omega_e^2}\left(1 + \frac{\omega_{pe}^2}{c^2 k^2}\right) - \frac{\omega_{pe}^2}{k\Omega_e}\left(\frac{n'}{n} - \frac{B'}{B}\right)\frac{1}{\omega - kU_E} - \frac{\omega_{pi}^2}{k^2 v_i^2}Z'\left(\frac{\omega}{kv_i}\right) = 0,$$
(9.13)

where $n' = dn/dx$, and $B' = dB/dx$. The first term $(= 1)$ comes from $\nabla^2\delta\phi$ in Poisson's equation (i.e., nonneutrality); the second term is the electron polarization term with electromagnetic correction; the third term is due to the gradients; and the last term is due to the ion contribution, where Z' is the derivative of the plasma dispersion function.

In the simulation by [568], the electrons are considered to be a massless fluid, $m = 0$, and the instability is restricted to long wavelengths by the spatial resolution of the simulation ($k_{max}\Delta x \approx 1$) to $k\varrho_{ci} \leq 1$, which evidently eliminates the unstable regime. Nevertheless, the instability continues to the long-wavelength limit [221]. In this case, in the quasineutral ($\delta n_e = \delta n_i$), $m_e = 0$ limit, the dispersion equation (9.13) reduces to

9. Tangential Discontinuity Simulation

$$\beta_i + \left(\varepsilon_n \varrho_{ci} + \frac{U_e \beta_i}{v_i}\right) \frac{k\varrho_{ci}}{\omega/\Omega_i - k\varrho_{ci} U_e/v_i} - Z'\left(\frac{\omega}{kv_i}\right) = 0, \quad (9.14)$$

where $\varepsilon_n = -n_i^{-1} dn_i/dx$. The ion contribution here still assumes that ions are "unmagnetized", which is valid in the high-frequency ($\omega > \Omega_i$) limit, but less so in the long-wavelength regime. In this case one can instead assume magnetized ions, so that the last term is replaced by

$$\rightarrow 2\left(1 - I_0(\lambda)e^{-\lambda} - 2\sum_{n=1}^{\infty} I_n(\lambda)e^{-\lambda} \frac{\omega^2}{\omega^2 - n^2\Omega_i^2}\right), \quad (9.15)$$

where I_n is the modified Bessel function of order n and $\lambda = k^2 \varrho_{ci}^2/2$. Figure 9.3 compares the properties of the instability at long wavelengths with the different ion responses. The solid (dashed) curve is the real frequency ω_r for unmagnetized (magnetized) ions, while the dotted (dash-dotted) curve is the corresponding growth rate γ as a function of $k\varrho_{ci}$. For $k\varrho_{ci} \leq 3$, the instability with the magnetized ion response is stable, consistent with early analyses (e.g., [192]), while the unmagnetized ion case remains unstable.

In the simulation, from the Fourier analysis of the waves in space (e.g., Fig. 5 of [568]) and time, one obtains the following dispersion relation (expressed in terms of the local plasma parameters):

$$\frac{\omega}{\Omega_i} \approx 1.73 k\varrho_{ci}. \quad (9.16)$$

However, in the region of the TD, due to the solution of the overall Riemann problem, the ions are drifting in the simulation frame in the y-direction with

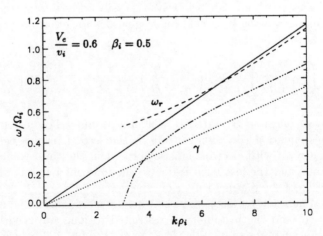

Fig. 9.3. Solution of the simplified dispersion equation (9.14–9.15) at long wavelengths, comparing ω_r for unmagnetized (*solid curve*) and magnetized (*dashed curve*) ions along with corresponding growth rates (*dotted curve*: unmagnetized ions; *dash-dotted* curve: magnetized ions). (From [568])

a velocity $U_{iy} \approx 1.6v_i$, so that (9.16) must be corrected for the Doppler shift to obtain the frequency in the ion rest frame, yielding

$$\frac{\omega}{\Omega_i} \approx 0.07k\varrho_{ci}. \tag{9.17}$$

Given the roughly 10% error in determining (9.16) and U_{iy} in the simulations and subtracting the difference between these large numbers, the agreement between (9.17) and Fig. 9.3 (where $\omega_r \approx 0.11kv_i$) is perhaps somewhat fortuitous. However, it does provide a plausibility argument to explain the existence of the waves driven by the cross-field motion of the plasma due to the gradients at the TD.

9.4.3.2 Cross-Field Diffusion Diffusion. To measure the plasma diffusion in [568], test particles were used which moved in the electric and magnetic fields obtained from the self-consistent calculations. The test ions were put in the gradient region of the TD along a narrow strip in x ($x = 105 \pm 0.5c/\omega_{pi}$) at $\Omega_i t = 0$ and randomly in y and then allowed to evolve in time. From these ions, one computes an average mean square displacement, Δx^2, which is plotted in Fig. 9.4 as a function of time. From the slope of the curve, one obtains a diffusion coefficient, $D_n = 0.2\varrho_{ci}^2\Omega_i$, which is comparable to the Bohm value, $D_B = 0.25\varrho_{ci}^2\Omega_i$. In computing the diffusion rate, the "numerical diffusion" was subtracted, which is about 40% of the above value, by redoing the calculation with B in the x–y plane that showed no wave at the TD (see also Sect. 9.4.2). The smaller diffusion in that case is due mostly to numerical fluctuations, which, given the same numerical parameters in the simulation, should be comparable in the two runs. The numerical diffusion rate, $D_{num} \approx 0.1\varrho_{ci}^2\Omega_i$, is also consistent with the magnetic field diffusion rate, $D_\eta = 0.1\varrho_{ci}^2\Omega_i$, due to the imposed numerical resistivity that damps out short-wavelength fluctuations (which can be thought of as modeling the diffusion due to high-frequency instability [e.g., the lower hybrid instability]) that are not explicitly included in the simulations [568]. The computation of the

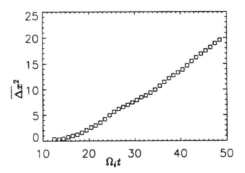

Fig. 9.4. Mean square positions of the test ions of the previous plot as a function of time, from which a net diffusion rate of $D_n = 0.2\varrho_{ci}^2\Omega_i$ is inferred (from [568])

diffusion rate from differences in the slopes of the density and/or magnetic field at various times gives the same result.

To estimate the diffusion rate theoretically, one uses the estimation of the cross-field diffusion rate

$$D \approx \frac{\varrho_{ci}^2 \Omega_i}{2} \frac{\nu_i/\Omega_i}{1 + \frac{\nu_i^2}{\Omega_i^2}}, \qquad (9.18)$$

where ν_i is the ion collision frequency for an unspecified scattering process of the ions with the waves. For $\nu_i = \Omega_i$, one obtains the largest diffusion rate, namely Bohm diffusion, with $D_B = 0.25 \varrho_{ci}^2 \Omega_i$. In [568], they used the measured diffusion rate, $D_n \approx 0.2 \varrho_{ci}^2 \Omega_i$, and concluded that in the presence of the fluctuating fields the effective ion collision frequency is essentially the ion gyrofrequency.

One can compare the measured diffusion rate with that due to high-frequency waves. For example, for waves excited by the lower hybrid drift instability [60], results from a number of simulations were used to determine an anomalous resistivity,

$$\eta_{an} = C \frac{\Omega_e}{\omega_{pe}^2} \left(\frac{m}{M}\right)^{1/2} \frac{1}{\beta_i} (\varepsilon_n \varrho_{ci})^2, \qquad (9.19)$$

where $\varepsilon_n = -n_i^{-1} dn_i/dx$ and $C \approx 3.0$ at early times (just after saturation) and $C \approx 1.0$ at late times. From the definition $\eta_{an} = 4\pi \nu_{an}/\omega_{pe}^2$, one can infer an anomalous collision frequency:

$$\frac{nu_{an}}{\Omega_i} \approx 3.4 C \frac{(\varepsilon_n \varrho_{ci})^2}{\beta_i}. \qquad (9.20)$$

In our case, $\varepsilon_n \varrho_{ci} \approx 0.8, \beta_i \approx 0.5$, so that $\nu_{an}/\Omega_i \approx 4$, while at early times, when the gradients are steeper, $\varepsilon_n \varrho_{ci} \approx 2$ and $C = 3$, giving $\nu_{an} \approx 80 \Omega_i \approx 2 \omega_{LH}$. Thus, as expected, the anomalous collision frequency due to the interaction of lower hybrid waves with ions is much larger than from low-frequency waves, although the diffusion due to high-frequency waves is generally much more localized in space [564, 568].

9.4.4 Dependence of the Final Thickness of TDs on Initial Conditions

To study the dependence of the final thickness of TDs on the initial thickness, we performed simulations with different initial thickness value, marked in Fig. 9.5 as follows: $\Delta_0/\varrho_{ci} = 0.2\,(*), 2.0\,(\triangle), 4.0\,(\diamond)$. The initial values the downstream plasma variables in the first run are the same as in (9.11) and for the next three runs as in (9.12). The anomalous resistivity is absent ($l_d = 0$) in these cases. Figure 9.5 shows the time evolution of the thicknesses of the TD for the different cases. In the first case of small initial value, the

9.4 Two-Dimensional Structures   247

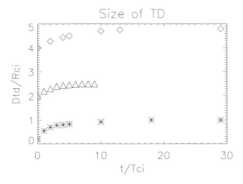

Fig. 9.5. Time evolution of the thickness of TD for different initial values of thickness: $\Delta_0/\varrho_{ci} = 0.2\,(*),\,2.0\,(\triangle),\,4.0\,(\diamond)$ (from [342])

thickness reaches the asymptotic value after $(2-3)\,T_{ci}$, while in the other 3 runs it requires about $(5-10)\,T_{ci}$. Under stationary boundary conditions $(B_{1,2} = \text{const.};\,E_{1,2} = 0)$ the final quasistationary thickness of the transition layer of TD is determined only by the initial value and may exceed it by about $1\varrho_{ci}$. With another boundary condition, for example, including the drift of the particles in the direction of the front of the TD ($E > 0$ or $E < 0$, $B = $ nonconst., $p_{i,e} = $ nonconst.), we can obtain another final thickness of TD.

The large scale widths of the TDs may correspond the SW TDs that are about up to 10 ion gyroradii. The large density and temperature gradients that are so important in the formation of TDs are not often found in the SW. The data from Pioneer 6 show that most TDs in the SW have a jump in ion density of less than 20% [491], and hence they evolve little from the initial conditions.

9.4.5 Dependence of the TD Width on Anomalous Resistivity and Numerical Viscosity

Usually the finite difference schemes provide solutions to the model equations that differ somewhat from the analytical solutions. In our case the approximation error and the effective numerical viscosity may have a significant effect on the property of the solution. In our hybrid code we have anomalous resistivity through the diffusion term with dissipation length $l_d^* = 1/Re$ and numerical viscosity length with effective dissipation length $l_{d,\text{eff}}^* \approx V_e \Delta z/2$. To study the influence of dissipation processes on the final thickness of a TD, we performed a set of simulations with different values for the anomalous resistivity and numerical viscosity. In Fig. 9.6 we gather the results of these runs in different groups. The first group (+) consists of the one- and two-dimensional regimes with $l_d/\varrho_{ci} = 0.25, \varepsilon = 0.2, \Delta z/\varrho_{ci} = 0.4$ and $l_{d,\text{eff}}/\varrho_{ci} = 0.2V_e$, where the bulk velocity of electron $V_e \ll 1$. The second group (*) consists of the two-dimensional regimes with $l_d/\varrho_{ci} = 0.05, \varepsilon =$

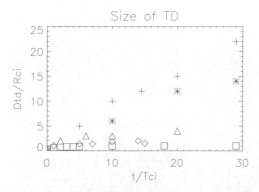

Fig. 9.6. Time evolution of the thickness of TD for different anomalous resistivity and numerical viscosity: $(+) l_d/\varrho_{ci} = 0.25, \varepsilon = 0.2, \Delta z/\varrho_{ci} = 0.4, l_{d,\text{eff}}/\varrho_{ci} = 0.2 V_e$; $(*) l_d/\varrho_{ci} = 0.05, \varepsilon = 0.2, \Delta z/\varrho_{ci} = 0.4, l_{d,\text{eff}}/\varrho_{ci} = 0.2 V_e$; $(\Delta) l_d/\varrho_{ci} = 0.02, \varepsilon = 0.5, \Delta z/\varrho_{ci} = 0.16 \div 0.2, l_{d,\text{eff}}/\varrho_{ci} = 0.1 V_e$; $(\diamond) l_d/\varrho_{ci} = 0 \div 0.04, \varepsilon = 0.5 \div 1, \Delta z/\varrho_{ci} = 0.05 \div 0.1, l_{d,\text{eff}}/\varrho_{ci} = (0.025 - 0.05) V_e$; $(\square) l_d/\varrho_{ci} = 0, \varepsilon = 0.5, \Delta z/\varrho_{ci} = 0.1, l_{d,\text{eff}}/\varrho_{ci} = 0.05 V_e$. (From [342])

$0.2, \Delta z/\varrho_{ci} = 0.4$ and $l_{d,\text{eff}}/\varrho_{ci} = 0.2 V_e$. The third group ($\Delta$) is the one- and two-dimensional regimes with $l_d/\varrho_{ci} = 0.02, \varepsilon = 0.5, \Delta z/\varrho_{ci} = 0.16 \div 0.2$ and $l_{d,\text{eff}}/\varrho_{ci} = 0.1 V_e$. The fourth group ($\diamond$) gathers one- and two-dimensional regimes with $l_d/\varrho_{ci} = 0 - 0.04, \varepsilon = 0.5 - 1, \Delta z/\varrho_{ci} = 0.05 - 0.1$ and $l_{d,\text{eff}}/\varrho_{ci} = (0.025 - 0.05) V_e$. The last group ($\square$) is the one-dimensional regime with $l_d/\varrho_{ci} = 0, \varepsilon = 0.5, \Delta z/\varrho_{ci} = 0.1$ and $l_{d,\text{eff}}/\varrho_{ci} = 0.05 V_e$. In all regimes the profiles of the plasma variables consist, as mentioned before, of two parts: linear region with the thickness of about $(1 - 2)\varrho_{ci}$ and nonlinear part with a size (in simulation) of about $(1 - 20)\rho_{ci}$. In Fig. 9.6 we present the maximum scale of the transition layer of TD.

Analysis of these results gives the following scenario for the dynamics of the transition layer of the TD. At the beginning of the simulation we have a self-consistent (in the MHD approximation) configuration of the TD. But then this system begins to evolve with characteristic space $\varrho_{ci} = \sqrt{\beta_{i1}/2} c/\omega_{pi}$ and time scales T_{ci}. During the first gyroperiod of the ions, the processes are controlled by the dynamics of the ions and the electron fluid. The numerical viscosity determined by the electron bulk velocity V_e may be significant. The time evolution of the thickness of the TD may be described as diffusion $\Delta \approx \sqrt{(l_d + l_{d,\text{eff}})t}$. But after some ion gyroperiods, the growth of the size of TD is determined by the anomalous resistivity, and it may be described by diffusion, so that $\Delta \approx \sqrt{l_d t}$.

9.4.6 The Kelvin–Helmholtz Instability at the TD

Simulations of the Kelvin–Helmholtz instability have centered mainly around MHD calculations. Initial simulations were for homogeneous configurations with velocities both parallel and perpendicular to the magnetic field [373, 578]. Later simulations included density gradients [374], arbitrary flow velocities [375] and a large computational domain [36]. Kinetic simulations have analyzed the electrostatic nature of the instability at short wavelengths [239, 394, 430, 510]. Hybrid simulations are more relevant to the MP [516, 524] and have demonstrated that the kinetic behavior of the instability is consistent with that observed in MHD simulations, i.e., initially, short wavelength perturbations grow, then evolve into vortex structures and then coalesce with longer wavelengths over time, resulting in a broad and complex mixing layer. The hybrid simulations of [524] have also shown that detached plasma structures, which form in the inhomogeneous case, indicate a strong correlation between the plasma components and the magnetic field, which are suggestive of flux-transfer events at the MP [456].

In the simulations considered above, the processes of reconnection and Kelvin–Helmholtz instability at the MP are studied separately. However, the flow accelerated by reconnection may provide the velocity shear which causes the Kelvin–Helmholtz instability on the dayside current layer [261, 469]. On the other hand, the Kelvin–Helmholtz instability could modulate the current sheet, thus creating separated and pinched regions and thereby provide the conditions for patchy reconnection at the dayside MP [286].

Now let us consider the two-dimensional hybrid simulation of the Kelvin–Helmholtz instability at the MP performed in [524]. To model the subsolar MP boundary layer, a configuration was required that contained sheared velocity flows nearly perpendicular to the magnetic field and density gradients. The initial configuration was a Vlasov equilibrium, so instability effects are not confused with any transients in the system. The simplest assumption was that the interplanetary magnetic field (IMF) is directed northward and the boundary layer is a TD. In [296], a general procedure is given for the construction of Vlasov tangential equilibria including velocity shear, which is used here (see also [85]). The essence of the method is to prescribe distribution functions for the ions (and electrons) on each side of the discontinuity (referred to as the magnetosheath and the magnetosphere) in terms of appropriate constants of motion (energy and canonical momenta). In particular, the distributions are assumed to be products of exponentials and error functions, with constants chosen such that the density of the magnetosheath ions falls off rapidly on the magnetospheric side of the discontinuity and vice versa. These distributions are integrated appropriately over velocity to obtain the charge and current densities. Poisson's and Ampere's laws are then solved to obtain the electrostatic and vector potentials and hence the electric and magnetic fields. The constants of the distributions can then be adjusted and the procedure repeated until the desired profiles of the magnetic field, flow velocity and densities of the plasma species are obtained. To initialize the simulation with the equilibrium then requires the ion distributions at each grid point to be inverted using the local values of the vector potential and the electric potential.

For their simulation, the system size is $DZ \times DX = 40c/\omega_{\rm pi} \times 40c/\omega_{\rm pi}$. Here the z-direction is normal to the boundary layer and the x-direction is along the boundary layer. They use a mesh of 200×100. The average number of macroparticles per cell is 40. The magnetosheath and magnetospheric ions are represented as separate species, using similar numbers of macroparticles weighted accordingly. In order to demonstrate the development of the instability more clearly, the calculation is done in the zero-velocity frame, with the plasma at $z > DZ/2$ moving in the positive x-direction and the plasma at $z < DZ/2$ moving in the negative x-direction. Periodic boundary conditions for the particles and fields are employed in the x-direction. In the z-direction reflecting boundary conditions are used for the ions [388], while the fields are fixed at their equilibrium values at the walls. The electrons are treated isothermally; a small resistivity, yielding a resistive diffusion length of $l_{\rm d} = \eta c^2/(4\pi U_x) = 0.025\Delta z$ is included to help smooth the fields [524]. Three simulations were presented: the first with equal densities on the two sides of the boundary, for comparison with early theory and simulations; the second with unequal densities, which is more relevant to the magnetopause; and the third with nonperiodic boundary conditions. Here we shall consider only the first case.

The temperatures and ion masses are taken as equal on the two sides of the boundary, with the electron temperature being one half of the ion temperature. These temperature ratios are arbitrary and have been chosen here more for convenience than to model any particular MP crossing. The flow speeds are equal and opposite on the two sides of the boundary with $U_x/v_{\rm A} \approx 0.165$. Figure 9.7 demonstrates color intensity plots of the density of the ion species on the right side of the interface at various times, $\Omega_i t = 55, 130, 205$ and 280 (Ω_i is a ion gyrofrequency measured in terms of the equilibrium magnetic field). At an early time ($\Omega_i t = 55$), the interaction is dominated by short-wavelength, low-amplitude perturbations that grow out of the thermal noise. These perturbations are resolved by the grid. As is characteristic of the Kelvin–Helmholtz instability, the surface waves form vortices and the initially narrow boundary is broadened. During the early stage, one can estimate the wavelength of the dominant mode to be $\lambda = 10c/\omega_{\rm pi}$ and the corresponding linear growth rate to be $\gamma = 0.1\Omega_i$. As the shear layer half thickness is $a \approx 0.75c/\omega_{\rm pi}$, one has $k_x a \approx 0.4$. The wavenumber and growth rate are thus consistent with those expected from MHD theory ($\gamma \approx 0.1\Omega_i$), even though $k_x \varrho_{\rm ci}$ is on the order of, rather than much less than, unity [430].

At a later time ($\Omega_i t = 130$), the dominant wavelength has increased, and the vortices are larger and more complex. Because of the magnetic field and density fluctuations, in part arising from the relatively small number of particles per cell as well as possibly due to ion gyroradius effects, such as occurs in the propagation of ion beams across an ambient magnetic field (e.g., [412]),

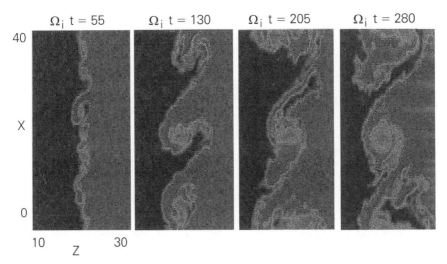

Fig. 9.7. Two-dimensional intensity plots of the density of the ion species residing initially to the right of the interface at various times (from left to right): $\Omega_i t = 55, \Omega_i t = 130, \Omega_i t = 205, \Omega_i t = 280$ for the equal density simulation. The left end of the color bar corresponds to a density of 0.0 and the right end to 12.0, where 10.0 is the initial density on the right side. (From [524])

the coherent roll up of the vortices, seen in fluid simulations, does not occur. In the regions between the vortices, the boundary remains very narrow and well defined. At times $\Omega_i t = 205$ and $\Omega_i t = 280$, the dominant wavelength continues to increase and the vortices grow in size. The mixing region also continues to expand as the boundary becomes more diffuse. Figure 9.7 shows only one ion species. In the regions where this species is absent, the other species is present, leading to approximately equal density everywhere. The large empty areas present in the vortices dominated by the second species do not persist and are quickly eliminated by the vortex motion.

One of the main consequences of the Kelvin–Helmholtz instability is the plasma transport across the TD. The transport is determined by measuring the amount of material that crosses the midplane of the shear layer. The number of ions, N_t, crossing the midplane from one side to the other is expressed as a layer thickness, δz [524]:

$$N_t = n_0 DX \delta z,$$

where DX is the system length in x and n_0 is the ion density on each side of the boundary. This quantity normalized by the ion gyroradius (ϱ_{ci} = relative ion drift speed across the layer divided by the magnetic field on either side of the boundary) is plotted in Fig. 9.8 as a function of time using solid curves. (The dashed curves correspond to the nonequal density case.) Both solid curves (representing transport in the two directions) are essentially the same, consistent with the fact that the total density remains homogeneous. The

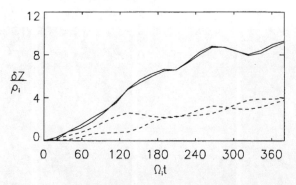

Fig. 9.8. Mass transfer versus time. The *solid lines* correspond to the transfer from one side to the other for the equal density case. ϱ_i is calculated using the equilibrium magnetic field and the relative flow speed. As discussed in the text, the number of ions transferred is expressed as a length, δZ. For the unequal density case (*dashed lines*), the magnetic field used corresponds to that on the low density side. (From [524])

small oscillations correspond to the coalescence of the instability to longer wavelength modes. The value $\delta z/\varrho_{ci}$ is approximately linear to about $\Omega_i t \approx 300$. The maximum transfer rate $(d\delta z/dt)$ corresponds to about 3% of the relative flow velocity, U_D. The transfer rate can be expressed in terms of a diffusion coefficient:

$$\frac{\partial \delta z}{\partial t} \approx \frac{1}{\varrho}\frac{\partial \varrho}{\partial t} a \approx 0.03 U_D,$$

where ϱ is the density and $a \approx \varrho_{ci}$ is initial shear layer width. Thus, the diffusion coefficient, D, is

$$D \approx \frac{1}{\varrho}\frac{\partial \varrho}{\partial t} a^2 \approx 0.03 U_D a,$$

which is about half of the Bohm rate

$$D_B \approx \frac{T_i c}{16 eB} = \frac{1}{16} v_i \varrho_{ci} \approx \frac{1}{32}\left(\frac{v_i}{U_D}\right) U_D a$$

with $T_i = 1/2 m_i v_i^2$ and $U_D/v_i \approx 2$. The rate of transport was calculated across an assumed fixed boundary. However, the interface becomes rippled in time, so that much of the transport results from a displacement of the surface. Moreover, this model does not include high-frequency waves, which can allow some diffusion of the magnetic field at small scales, and hence lead to actual diffusion of the plasma, as opposed to just dynamic mixing.

In the other hybrid simulation [516], it was conclude that the mixing occurring during the growth and nonlinear evolution of the Kelvin–Helmholtz instability is anomalously fast. In this simulation the mixing occurs in two stages, a t^2 dependence initially followed by a diffusive ($\sim t^{1/2}$) phase. In

9.4 Two-Dimensional Structures

contrast to [516], the mixing in the simulation of [524] is essentially linear in time. There are several possible explanations for this discrepancy. First, the equilibrium in simulation of [524] has a shear layer several times thinner than that in [516]. Second, the drift velocity in [516] is roughly three times larger than that in [524]. Finally, in [516] there is a single dominant wavelength in the periodic direction, whereas in [524] the dominant wavelength changes in length by about an order of magnitude over the course of simulation. Since in [524] the longest wavelength possible in the system does not dominate until the end of the simulation, it is possible that for a longer run a more diffusive $t^{1/2}$ dependence for the mixing at a later time may eventually be seen. We do not discuss here the case of unequal densities, which is more relevant to the MP, and the case with nonperiodic boundary conditions. The discussion of simulations for this case may be found in [524].

The three-dimensional simulations [519] have clarified certain aspects of the Kelvin–Helmholtz instability and have suggested other areas of interest. For uniform velocity and magnetic field in the transverse geometry, the mixing observed in three-dimensional simulations was about the same as that for two-dimensional simulations. The system is surprisingly robust to three-dimensional initial perturbations in yielding an approximately two-dimensional saturated state.

For the case where the velocity is not perpendicular to the magnetic field, approximately two-dimensional results are obtained in three dimensions as long as the condition $\boldsymbol{k} \cdot \boldsymbol{B} = 0$ is satisfied. The existence of a finite velocity component parallel to the magnetic field does not matter much in three or two-dimensional as long as the proper simulation plane is chosen for the two-dimensional simulations [519, 524]. The readers is referred to [519, 524] for details.

10. Magnetic Field Reconnection Simulation

10.1 Introduction

Magnetic reconnection plays an essential role in determining the magnetic field configuration in astrophysical and space plasmas and is responsible for the fast conversion of magnetic field energy into thermal and kinetic energy of the plasma (see references in [18]). The processes of reconnection concerning both the magnetopause and the geomagnetic tail have been studied extensively in the past by large-scale magnetohydrodynamic (MHD) simulations (e.g., [54, 295, 475]). Such simulations have confirmed some elements of Petschek's model of reconnection [423]: an inflow region, a wedge-shaped outflow jet, and a small central diffusion region (e.g., [466, 474]). In particular, slow-mode shocks bounding the outflow region have been found in the MHD simulations of reconnection. As suggested in [423], most of the energy conversion and field reversal is achieved by these slow-mode shocks.

In order to determine the actual structure and configuration of steady-state reconnection in a collisionless plasma, it is necessary to take into account the kinetic effects of the ions. Kinetic studies in the past were mainly concerned with the question of which dissipation mechanism can support magnetic reconnection by providing electric fields in the diffusion region in the case of a collisionless plasma. Some of these studies include both ion and electron dynamics (e.g., [240, 431, 506]), kinetic ions and fluid electrons with a model for the full electron pressure tensor [225], or scalar electron pressure [280], or one plasma species with the electrons represented by a charge-neutralizing fluid (e.g., [86, 131, 236, 320, 349, 350, 351, 432, 514, 592, 593, 594]).

Large-scale hybrid simulations relevant to magnetotail reconnection were performed in [280, 353]. The authors initiated reconnection in a system with antiparallel magnetic fields by a localized resistivity in the middle of the current sheet and by imposing an inflow from the top and bottom, perpendicular to the magnetic field. They found that after a transitional period, during which small-scale plasmoids emerged and were convected out of the system, a steady-state reconnection structure did build up with two pairs of thin transition layers attached to an X-point which divert and accelerate the flow. Upstream of this boundary layer, fast ion beams were found, while at the center of the current sheet, the plasma was heated and thermalized. These results are consistent with the features of fast Petschek-type reconnection. In the hybrid simulations of magnetotail reconnection [311], the reconnection was not driven but spontaneous by imposing a finite resistivity at the center of the computational domain. These simulations were started with a Harris-type neutral sheet. In [311], two slow-mode shocks bounding the outflow layer were obtained: the authors found that the ratio of normal magnetic field to normal flow velocities taken outside the reconnection layer and at the center of the reconnection layer is consistent with the Rankine–Hugoniot relations for a steady slow-mode shock, although the increase in the tangential velocity across the discontinuity was considerably less than that predicted by the jump conditions. Recently, in [386], the ion dynamics in hybrid simulations of reconnection in a double periodic system (two current sheets) were studied. The authors also started with a Harris-sheet equilibrium and initiated reconnection by imposing a localized resistivity at the center of the computational domain. In contrast to the results of [280] and [311], they found that the ions at the center of the current sheet are not fully thermalized, but rather perform modified Speiser orbits in the region close to the reconnection line, and partially have shell distributions in the region further away from the reconnection line. No indications for a slow-mode shock have been found in these simulations. In the hybrid simulation of [353], it was shown that the ion dynamics can sustain a reconnection configuration different from the one envisaged in [423]. The simulations closely resembled the collisionless merging configuration without slow-mode shocks proposed in [233].

In the hybrid simulations of [285], the reconnection in absence of anomalous resistivity was studied. The authors took into account the nongyrotropic electron pressure and the convective part of the electron inertia term. They found that the kinetic quasiviscous electron inertia associated with nongyrotropic pressure effects dominates over the electron bulk flow inertia in controlling the structure of the dissipation region around the neutral X line. The reconnection electric field based on the nongyrotropic pressure tends to reduce the current density and to relax gradients in the vicinity of the X line (similar to a localized viscosity). On the other hand, the reconnection electric field based on electron bulk flow inertia tends to require an increased current density, with gradient scales comparable to the electron skin depth.

10.2 Ion Tearing Instability

The general pattern of magnetic field reconnection in collisionless plasma has been studied in spontaneous [71, 131, 349] and induced [11] regimes. Processes in laboratory, thermonuclear [555] and space [53] plasma have been simulated mainly in the MHD approximation. Many numerical studies of the evolution of current sheets and charged particle acceleration were made in [507]. Here we consider a more narrow class of problems, that is, a kinetic description of the neutral sheet dynamics in collisionless and weakly collisional plasma. The complexity of the task is related to the inapplicability of the MHD approach, so we are forced to use the full Vlasov system of equations. These problems are essential for an appropriate understanding of the magnetospheric tail behavior and have been intensely studied in a number of theoretical [55, 116, 133, 167, 178, 427] and experimental [10] papers.

In this section a great deal of attention is also given to the theoretical analysis of the nonlinear development of tearing perturbation, which is also of general interest with regard to nonlinear wave dynamics in plasma. The simple initial state model chosen in this study allows us to compare effectively the theoretical estimates and the results of the numerical simulation, which gives us a better understanding of what the discrete approach can do for the given class of problems.

10.2.1 Formulation of the Problem and Mathematical Model

A two-dimensional neutral sheet is considers. Let the magnetic field $\boldsymbol{B}$ lie in the x–z plane and the current J be along the y-axis:

$$\boldsymbol{B} = (B_x, 0, B_z), \quad \boldsymbol{J} = (0, J_y, 0),$$
$$\boldsymbol{E} = (0, E_y, 0), \quad \boldsymbol{A} = (0, A_y, 0). \tag{10.1}$$

The stability of such two-dimensional configurations has been widely discussed during recent years with special attention to the question of the relative role of electrons and ions in the instability dynamics. In all cases of interest considered so far, ion dynamics dominates the instability development [168, 178, 471], while the electron contribution appears to be stabilizing and in some cases prohibits the instability growth [117, 178, 421]. Ion tearing mode is fully stabilized by electrons for the case of a neutral sheet with a constant normal component of magnetic field B_n, having a value sufficient to make the electron movement adiabatic [117, 300]. All these analytical results were obtained for a special magnetic field topology with $B_n = \mathrm{const.}$, and in a Wenzel–Kramers–Brillouin (WKB) approximation, and so are abandoned at comparatively small wavelengths.

If the configuration initially has an X-point, electrons could not stop the ion-tearing mode growth [167]. Moreover, as was recently shown in [198], where the exact energy principle for the two-dimensional Vlasov theory was analyzed, adiabatic electron response stabilization of the ion-tearing mode may be absent even for the two-dimensional configurations without neutral points.

In real-Earth magnetotail conditions the limitation of electron stabilization may be relaxed after taking into account even the weak pitch-angle scattering of electrons by whistler turbulence [117]. As this argument shows, the exact electron dynamics for many cases are not essential, and electrons may be treated in this limit simply as neutralizing background. As the numerical method used in this chapter deals only with one-species dynamics, it is natural to choose ions as such active types of particles. Although the final conclusion about the role of electrons and the corresponding electrostatic potential influence remains sufficiently dubious, and can be clarified only by exact two-species simulation, the arguments listed above give reason to expect the electrons to impose some threshold conditions on instability development, but since the instability had started and the magnetic islands were already formed, their future dynamics are determined by the active type of particles, i.e., ions.

Let us suppose that the ions play an active role, whereas the electrons provide the neutralization of ion charge [351].

$$|\nabla \phi| \ll \frac{1}{c} \left| \frac{\partial \mathbf{A}}{\partial t} \right|. \tag{10.2}$$

The above condition may be satisfied when the electrons and ions have the same mass and the same absolute value of charge:

$$m = M_i, \quad |q_i| = |q_e| = e/2, \tag{10.3}$$

$$\mathbf{B} = (B_x, 0, B_z), \quad \mathbf{J} = (0, J_y, 0),$$
$$\mathbf{E} = (0, E_y, 0), \quad \mathbf{A} = (0, A_y, 0). \tag{10.4}$$

After these remarks the equations of ion motion and electromagnetic field evolution can be written in a dimensionless form (see Sect. 2.2.2):

for ion motion,

$$P_y = v_y + \frac{A_y}{\varepsilon} = \text{const.}, \quad \frac{d}{dt} v_x = \frac{v_y B_z}{\varepsilon}, \quad \frac{d}{dt} v_z = -\frac{v_y B_x}{\varepsilon},$$

$$\frac{d}{dt} x = v_x, \quad \frac{d}{dt} z = v_z; \tag{10.5}$$

for vector potential,

$$\Delta_{x,z} A_y = -\frac{\langle nv_y \rangle}{\varepsilon}. \tag{10.6}$$

Taking into account (10.5), we can rewrite (10.6) as

$$\left(-\Delta_{x,z} + \frac{\langle n \rangle}{\varepsilon^2} \right) A_y = \frac{\langle n P_y \rangle}{\varepsilon}, \tag{10.7}$$

where A_y is the vector potential component; n is the plasma density; P_y and v are the canonical momentum and velocity of ions; $\varepsilon = \varrho_{ci}/L$ is the ratio of the Larmor radius of ions on the outer boundary $\varrho_{ci} = \varrho_{ci}(B_0)$ to the characteristic size of a field inhomogeneity L; and $\langle nv_y \rangle$ is the flux of particles. In (10.7) we assumed that $\langle nA_y \rangle \approx \langle n \rangle \langle A_y \rangle$. Transformation from dimensional to dimensionless variables is given by the following:

$$t = t'/\varepsilon \Omega_i, \quad r = r'L, \quad v = v'v_T, \quad n = n'n_0, \quad B = B'B_0, \tag{10.8}$$

where $\Omega_i = cB_0/Mc$ is the ion gyrofrequency, v_T is the transverse thermal velocity of ions, and n_0 is the plasma density at $t = 0, x = 0$.

Since calculations are made for the rectangular region $0 \leq x \leq DX, 0 \leq z \leq DZ$, it turned out to be worthwhile to choose an initial state similar to Harris's distribution, with the boundary condition $A_y|_{x=DZ} = 0$ taken into account:

$$B_z = \tanh x, \quad A_y = -\ln(\cosh x / \cosh DX),$$

$$f = \frac{1}{\pi^{3/2} \cosh^2 x} \exp\left(\frac{-v_x^2 - (v_y - \varepsilon)^2 - v_z^2}{2}\right). \tag{10.9}$$

On the boundaries the transverse field component is $B_x = \partial A_y/\partial z = 0$. Particles are assumed to be reflected on the boundaries of the computational domain. The symmetry condition $\partial A_y/\partial x = B_z = 0$ is assumed along the z-axis, and the dynamics of the field and the distribution function perturbations of the initial plane neutral sheet are analyzed. Our numerical scheme for solution of (10.5) and (10.7) was based on the Hamiltonian variational principle (see [305, 351] and exercises).

Shot Noise Effects The initial plasma distribution (10.9) was reproduced with a random-number counter. Particles with $v > 3v_T$ were eliminated from consideration to ensure stability of calculations in linear regimes. The initial distribution of the transverse magnetic component is $B_{ij} \propto i/\sqrt{N_p}\varepsilon(i^2 + j^2)$ (i, j are Fourier-series indices). Since a limited number of particles is used in the computations (in our case $N_p = 25\,000$ and the Euler grid is 16×32), appreciable fluctuations of macroparameters occur, resulting in effective collisions. The effective collision frequency of macroparticles (current cylinders in that case) can be estimated as in the case of Coulomb collisions. Interaction between two macroparticles is proportional to the product of their velocities v_{y1} and v_{y2} and to about $(v_T^2 + \varepsilon^2 v_T^2)$ for two statistically average particles. Further on, the expressions for Coulomb collisions can be used if basic quantities are transformed as follows:

$$e^* = \frac{ev_T}{c}\sqrt{(1+\varepsilon^2)}, \quad \lambda_D^* = \frac{\lambda_D \cdot c}{v_T\sqrt{(1+\varepsilon^2)}} = \frac{\varepsilon L}{\sqrt{(1+\varepsilon^2)}},$$

$$\omega_{pi}^* = \frac{v_T}{c}\sqrt{(1+\varepsilon^2)}\omega_{pi} = \Omega_i\sqrt{(1+\varepsilon^2)}, \quad \omega_{pi} = \sqrt{\frac{4\pi n e^2}{M}}, \tag{10.10}$$

where e^*, λ_D^* and ω_{pi}^* are the effective charge, Debye length and ion plasma frequency, respectively. According to [52] the Coulomb collision frequency $(\nu/\omega_{pi}) = (\delta/16n\lambda_D^2)$, where n is the density of charged clouds, δ is the ratio of collisional cross-sections for clouds and point particles. δ, a rapidly decreasing function of the parameter R/λ_D (R is the cloud radius), is given numerically (see Fig. 7 of [52]). As the density of macroparticles is determined by the ratio of the full number macroparticles N_p to the volume occupied by them $DZ \cdot L \cdot L$ [$n = N_p/(DZ \cdot L^2)$], in our case we have

$$\frac{\nu}{\Omega_{\rm i}} = \frac{\delta(R/\lambda_{\rm D}^*) \cdot (1+\varepsilon^2)^{3/2}}{16ne^2L^2} = \frac{10^{-5} \cdot \delta \cdot (1+\varepsilon^2)^{3/2} DZ}{\pi\varepsilon^2}. \qquad (10.11)$$

The typical value of the effective collision frequency is of the order of $10^{-5}\Omega_{\rm i}$ for $\varepsilon \approx 1$, $DZ \approx \pi$ and $N_{\rm p} = 25\,000$. Although the fluctuations of macroparticle density in a given cell are not small, the large-scale magnetic field fluctuations are determined by the full number of macroparticles, and so the global simulation can in principle be done with a relatively small number of particles.

10.2.2 Multimode Regime

Evolution of the initial distribution of the magnetic field and plasma has been studied over a wide range of parameters:

$$DX = 3, \quad \pi \le DZ \le 4\pi, \quad 0.2 \le \varepsilon \le 1.$$

As a result of tearing-instability development, large amplitudes of the transverse magnetic component are observed for harmonics with various m. The number of excited harmonics, according to the linear theory, is described by the condition

$$m = kL < 1.$$

If k is the wavenumber and DZ is the length of the computational domain, then the number of excited Fourier harmonics can be easily estimated: $N < DZ/2\pi L$. The above condition can be satisfied in this as well as in the previous numerical experiment [349].

Figure 10.1 gives the $B_{xm}(t)$ dependence for the 4-mode regime ($DZ = 4\pi$). The growth of the tearing instability is in a good agreement with the results of the linear theory [178, 351] which gives an expression for the linear instability increment as follows:

$$\frac{\gamma}{kv_{\rm T}} = \frac{1}{\sqrt{\pi}} \frac{(1-m^2)\varepsilon^{3/2}}{m}, \quad m > \sqrt{\varepsilon}$$

and

$$\frac{\gamma}{kv_{\rm T}} = \frac{1}{\sqrt{\pi}}(1-m^2)\epsilon, \quad m < \sqrt{\varepsilon},$$

$$m = kL. \qquad (10.12)$$

Figure 10.1 shows that after the linear stage of instability evolution is over, a slow algebraic growth is observed as well as distinct regular oscillations of the magnetic field amplitude. Figure 10.2 gives the density levels (dashed lines) and the magnetic field configuration (solid lines) for the stage of linear growth saturation in the above regime ($t \sim 10T_{\rm ci}$). Maximum and minimum values of the perturbed concentration along the z-axis are 1.3 and 1.4, respectively.

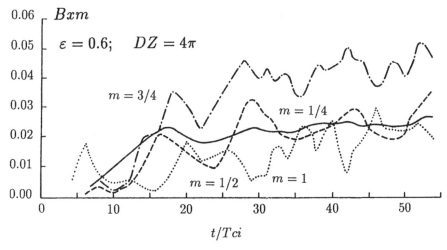

Fig. 10.1. Dynamics of the Fourier components of the magnetic field in the multimode regime for $\varepsilon = 0.6$. Time is given in number of gyroperiods (from [351])

The concentration in this regime drops drastically because it is necessary to balance the magnetic field stresses $\boldsymbol{J} \times \boldsymbol{B}$ acting over large distances DZ typical for multimode regimes. Since many modes are excited in the transverse magnetic component, strongly jagged lines of the density levels are observed. It is not easy, however, to observe how that affects the magnetic field configuration because of the low resolution of the printer used. Asymmetry in Fig. 10.2 is caused by the excitation of a few tearing modes with comparable amplitudes. It is worth remembering that Figs. 10.2–10.4 represent only one half of the period along the z-axis (full period is $2\, DZ$).

Considerable deformation of the particle distribution function (appearance of an effective anisotropy) is an essential feature of calculational results in the multimode regime. Though the distribution function observed in the

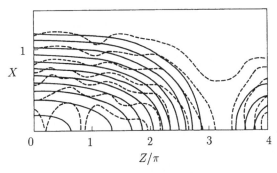

Fig. 10.2. Lines of density levels (*dashed lines*) and magnetic field lines (*solid lines*) in the multimode regime at the moment of linear growth saturation (from [351])

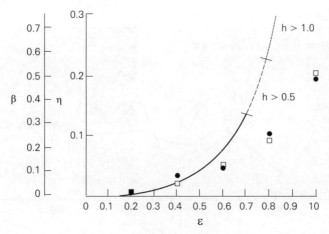

Fig. 10.3. The level of saturation of tearing linear growth versus ε in the multi-mode regime. *Squares*: numerical simulation, *solid line*: theoretical estimates, *dots*: anisotropy (from [351])

numerical experiment appreciably differs from the bimaxwellian, its deformation will nevertheless be described in terms of the parameter $\eta = (T_\| - T_\perp)/T_\|$ (where $T_\| \sim T_z$ and $T_\perp \sim T_x$). Figure 10.3 shows the theoretical ε dependence of the quasilinear saturation level $\beta^* = \sqrt{[\sum_k (b_k^2/k^2 L^2)]}$ obtained in this manner and numerically found points (squares in Fig. 10.3) obtained in the simulation experiment for different values of ε.

It is worthwhile mentioning that if $\varepsilon > 0.5$ the theoretical value β^* fairly highly exceeds the numerical value. This is due to the fact that the small-parameter ε expansion we have used becomes inapplicable for $\varepsilon \approx 1$. As has been mentioned above, an appreciable increase in the temperature anisotropy was observed in multimode regimes. Figure 10.3 also gives the ε dependence of temperature anisotropy η (dots). The $\eta = \eta(\varepsilon)$ dependence is seen to be close to the $\beta^*(\varepsilon)$ curve, which allows the conclusion $\eta \sim \beta^* \sim \varepsilon^3$, which somewhat differs from the results of Biskamp–Sagdeev–Schindler, who found $\eta \sim \beta^2/\varepsilon^2 \sim \varepsilon^4$ [55].

10.2.3 Single-Mode Regime

Since the mode with $m = 1$ has an increment $\gamma = 0$ according to the linear theory, the mode with $m = 2/3$ was used for the single-mode regime (the length of the layer is $DZ = 1.5\pi$). The linear stage of the evolution was analogous to the multimode regime (see Fig. 10.4). Stabilization was followed by a transition to a considerable algebraic growth with regular oscillations of the transverse component of the magnetic field (Fig. 10.4). The character of the algebraic growth is determined by effective collisions maintained, in

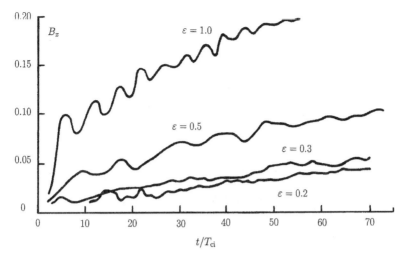

Fig. 10.4. Dynamics of the Fourier components of the magnetic field in the single-mode regime for various ε. Time is given in number of gyroperiods (from [351])

our case, by "shot noise". In that regime the distribution function was not considerably anisotropic.

Figure 10.5 shows the lines of density levels (dashed lines) and the magnetic field configuration (solid lines) in the regime of linear saturation for the single-mode case ($DZ = 1.5\pi$) at $t \approx 20 T_{\text{ci}}$. The maximum and minimum values of the perturbed concentrations along the z-axis are 1.16 and 0.65, respectively. Figure 10.5 shows that here, opposite to the multimode regime, the topology of the density level lines and the magnetic field configuration are sufficiently smooth, which corresponds to the absence of higher harmonics of

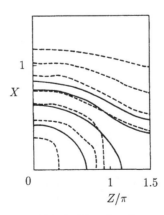

Fig. 10.5. Lines of density levels (*dashed lines*) and magnetic field lines (*solid lines*) in the single-mode regime (from [351])

the transverse magnetic field. Computations have shown that in the multimode and single-mode regimes the linear stage of tearing instability and its stabilization are observed as well as almost equal oscillations of magnetic field amplitudes, algebraic growth (stronger in the single-mode regime) and deformation of the particle distribution function (with appreciable anisotropy in the multimode regime). A more refined analysis of the numerical results will be made simultaneously with theoretical estimates in the following sections of this chapter.

10.2.4 Explosive Regime. Ion Acceleration

In [351] relatively short sheets were investigated, which explains why only the linear stage of instability was observed. To study the process of the transition between the linear and explosive stages of the instability, two sets of numerical experiments were carried out [592, 593, 594].

The first set of experiments was carried out with a fixed value of the initial magnetic field perturbation $b_0 = 0.05$ and $\varepsilon = 0.62$ for different lengths of the computational domain, DX ($DX = DX'/L, 2\pi < DX < 20\pi$). The dependence of the value of the normal magnetic field b at the end of the reconnection process on the length of the sheet DX is shown in Fig. 10.6. The critical length of the sheet DX^*, which determines the transition from the linear stage to the nonlinear stage (the region of large b) is clearly seen.

In the second set of numerical experiments with fixed values of $\varepsilon = 0.62$ and $DX = 8\pi$, the initial magnetic field perturbation b_0 was varied. Figure 10.7 shows the dependence of the normal magnetic field component at the end of the explosive stage on b_0 (circles). The quasilinear saturation amplitude is shown by triangles. The time (in gyroperiods) from the very beginning of the instability until the beginning of the explosive growth is given by squares with the scale at the right. One can see that there are three

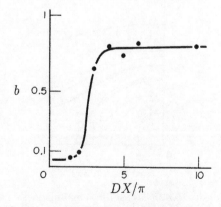

Fig. 10.6. Final nonlinear amplitude of the perturbed magnetic field b versus the length DX of the numerical simulation slab (from [594])

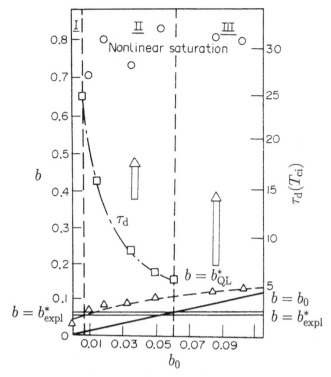

Fig. 10.7. The dependence of the final amplitude of the perturbed magnetic field b on the value of the initial perturbation b_0, obtained in numerical simulation. *Triangles*: quasilinear saturation amplitude. *Circles*: amplitude of nonlinear saturation at the final state. *Squares*: the delay time for the explosive stage onset (scale at the right). (From [594])

regimes of the instability development. The simple criterion of the transition to the explosive growth discussed in [16] and used in [514] consists of the requirement for the absence of overlapping between the regions of particle trapping $\sqrt{\varepsilon/kLb}$ in the neighboring half wavelengths of nonlinear wave perturbation, $B_\perp \sin kx$.

This gives the condition

$$b^*_{\text{expl}} = 8\varepsilon/\pi DX. \qquad (10.13)$$

On the other hand, the quasilinear saturation amplitude $b_{\text{QL}} \sim kL\varepsilon^4$ (which is in good agreement with the theoretical calculations presented in [351]) can give the intermediate state, where the metastable delay of the explosive instability growth can occur.

In the first case of $b_0 < b_{\text{QL}} < b^*_{\text{expl}}$, the quasilinear saturation state is the final state of the system evolution. In the second case of $b_0 < b_{\text{expl}} < b_{\text{QL}}$, the instability development is delayed for a while in the quasilinear saturation state, but finally the explosive process is nevertheless realized and the system passes to the state with the nonlinear amplitude $b \approx 1$. The time of delay (as is shown in Fig. 10.7), drops with the growth of b_{QL} above the threshold b^*_{expl}.

The physical cause of such a metastable character of the transition to the explosive growth is not yet theoretically understood. Finally, in the third case, when the criterion (10.13) is fulfilled directly for the initial perturbations $b^*_{expl} < b_0$, the transition to explosive development occurs immediately without any delay in the intermediate stage.

Although qualitatively the dynamics of the transition to the nonlinear stage is in agreement with (10.13), there are some quantitative discrepancies between this simple expression and the simulation results. Thus, for example, the critical length $DX^* \approx 6\pi$ (Fig. 10.6), and that is somewhat less than the estimate that comes from expression (10.13):

$$DX^* \geq \frac{8\varepsilon}{\pi b} \approx 10\pi. \tag{10.14}$$

On the other hand, sometimes the delay of the transition may still occur even in the case when the condition $b_0 > b^*_{expl}$ works. This is shown in Fig. 10.8, where the time evolution of the magnetic field amplitude for the $\varepsilon = 0.25$ and $b_0 = 0.1$ is represented. Although $b^*_{expl} \approx 0.03 < b_0$ the explosive development is delayed for about twenty gyroperiods. It seems that such retardation of the transition may be related to the complicated character of the nonlinear instability stage due to the influence of the "shot noise" effects discussed in [351].

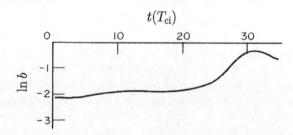

Fig. 10.8. An example of the time behavior of the perturbed magnetic field, for $\varepsilon_i = 0.25$ and $b_0 = 0.1$ (from [594])

To investigate the process of particle acceleration during the explosive stage development, numerical experiments in a range of parameters appropriate for space conditions ($0.25 \leq \varepsilon \leq 1, b_0 = 0.1, DX = 8\pi$) have been carried out. The predomination of the $kL = m = 0.25$ mode is observed (the initial amplitude was taken to be equal to $b_0 = 0.1$). During the explosive instability development, the global reconstruction of the magnetic field plasma configuration takes place, which leads to particle acceleration and the formation of rarefaction regions. In the process of numerical simulation, due to the strong variations of parameters in every calculation cell, it is expedient to study the distribution function (the number of macroparticles in energy interval ΔE) averaged over the computational domain

$$\langle f \rangle = \int_V f \mathrm{d}z \mathrm{d}x. \tag{10.15}$$

As the distribution function runs to zero as $\sqrt{E}$ with $E \to 0$, in our case it is more convenient to use the normalized distribution function $\tilde{f} = f/\sqrt{E}$ with the $\sqrt{E}$ multiplier omitted.

In the region of small E the distribution function is close to the Maxwellian with the temperature T along the x-axis somewhat higher than T_i. In the region of high energies $E > (10 - 20)T_\mathrm{i}$, a well-developed tail of accelerated particles is observed. Comparison of simulation results with the analytical expressions [594] shows good agreement for the region

$$E \geq \frac{1}{3} E_\mathrm{max}, \tag{10.16}$$

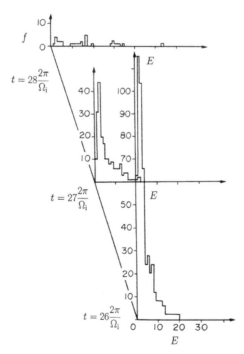

Fig. 10.9. Evolution of the energy distribution of the macroparticles in the acceleration region during the active period of explosive instability growth (from [594])

where $E_{\max}$ denotes the maximum energy of accelerated ions. Figure 10.9 shows the dynamics of the distribution of macroparticles over energies in the acceleration region for the value $\varepsilon = 0.25$ at the time moments $t = 26, 27, 28 \cdot 2\pi/\Omega_i$ corresponding to the "active" stage of the explosive instability development, whose entire duration is about $(2-3) \cdot 2\pi/\Omega_i$. In Fig. 9 in [594] the energetic particle spectra obtained in [465] are compared with the full distribution function observed in numerical simulation experiments for $\varepsilon = 0.25, 0.62$ and 0.8. As can be seen from the figure, the proposed mechanism agrees in principle with the experimental spectra, and the smaller the values of ε the better the correspondence obtained. The difference in the tail lengths are apparently related with the fact that the real ε values, which determine $E_{\max}$ in magnetotail conditions, must be somewhat smaller than in our numerical simulations. However, the shape of the spectrum is only weakly dependent on ε, so, taking into account that the exact value of $E_{\max}$ is hardly determinable from the experimental data, it is reasonable to compare the experimental results with runs for $\varepsilon = 0.25$. The linear part in the high-energy range approximately corresponds to the power-law spectrum E^{-k}, with $k = 5-6$ (see for comparison [355]).

10.3 Electron Effects on Reconnection

In Sect. 9.2 we considered Ampere's models, which include the ions as active particles and the electrons as a passive particles which provide the quasineutrality only. Now we shall use the hybrid models for magnetic field reconnection problems. The fluid approximation for electrons may be sufficient to estimate some effects of electrons on reconnection processes.

The equations may be expressed in a dimensionless form by choosing a typical length L and time L/U_0, U_0 being the typical velocity. Depending on the problem, one of various velocities, for example, the bulk velocity of the incoming flow, or the thermal velocity of ions, or the Alfvèn velocity, may be chosen as the typical velocity. The basic equations are the ion equations of motion, the generalized Ohm's law, Faraday's induction equation and the adiabatic equation for the electron pressure. We consider one sort of ions only. The ion charge equals the electron charge ($Z_i = 1$) and the ion mass equals the proton mass ($M_i = M$). The ion equations (normalized) are of the form

$$\frac{d\boldsymbol{x}_l}{dt} = \boldsymbol{v}_l, \quad \frac{d\boldsymbol{v}_l}{dt} = \frac{1}{\varepsilon}\left(\boldsymbol{E}^* + \boldsymbol{v}_l \times \boldsymbol{B}\right), \quad \boldsymbol{E}^* = \boldsymbol{E} - l_d^*\boldsymbol{J}. \qquad (10.17)$$

The normalized dissipation length is $l_d^* = l_d/L = 1/Re$; the magnetic Reynolds number is $Re = 4\pi v_{\text{Ti}} L \sigma_{\text{eff}}/c^2$; and σ_{eff} is the effective conductivity. The parameter ε is the ratio of the proton Larmor radius $\varrho_{\text{ci}} = v_{\text{Ti}}/\Omega_i = \sqrt{\beta_i/2}c/\omega_{\text{pi}}$ in the unperturbed field B_0 to the characteristic dimension L of the problem, $\varepsilon = \varrho_{\text{ci}}/L$, $\beta_i = p_i/(B_0^2/8\pi)$.

10.3 Electron Effects on Reconnection

The electrons are considered as having a nonzero mass, so that the electron and ion momentum transport equations are

$$\varepsilon \frac{m}{M}\frac{\partial \boldsymbol{J}_e}{\partial t} + \varepsilon \frac{m}{M}(\nabla \boldsymbol{U}_e)\boldsymbol{J}_e - \frac{\varepsilon \beta_e}{\beta_i}(\nabla p_e + \partial \pi^e_{\alpha\beta}/\partial x_\beta) - n_e \boldsymbol{E} + \boldsymbol{J}_e \times \boldsymbol{B} + l^*_d n_e \boldsymbol{J} = 0, \quad (10.18)$$

$$\varepsilon \frac{\partial \boldsymbol{J}_i}{\partial t} + Z_i \varepsilon \nabla \cdot \mathsf{K}_i - n_i \boldsymbol{E} - (\boldsymbol{J}_i \times \boldsymbol{B}) - l^*_d n_i \boldsymbol{J} = 0. \quad (10.19)$$

The electron $\boldsymbol{U}_e$ and ion $\boldsymbol{U}_i$ bulk velocities are given by

$$\boldsymbol{U}_e = \boldsymbol{U}_i - 2\varepsilon \boldsymbol{J}/(\beta_i n_e), \quad \boldsymbol{U}_i = \langle N_i \boldsymbol{v}_i \rangle / \langle N_i \rangle. \quad (10.20)$$

Ampere's law and the induction equation are

$$\boldsymbol{J} = \nabla \times \boldsymbol{B}, \quad \frac{\partial \boldsymbol{B}}{\partial t} + \nabla \times \boldsymbol{E} = 0, \quad (10.21)$$

$$\boldsymbol{J} = \boldsymbol{J}_e + \boldsymbol{J}_i, \quad (10.22)$$

where dimensionless ion current

$$\boldsymbol{J}_i = n_i \boldsymbol{U}_i. \quad (10.23)$$

We impose the condition of quasineutrality, and we consider the electron gas to be adiabatic:

$$n_e = n_i \equiv n, \quad p_e \propto n_e^\gamma. \quad (10.24)$$

The final equation for the electric field is of the following form:

$$\frac{2m\varepsilon^2}{M\beta_i}\nabla \times (\nabla \times \boldsymbol{E}) + n_e\left(1 + \frac{n_i m}{n_e M}\right)\boldsymbol{E} + \left[\left(1 + \frac{m}{M}\right)\boldsymbol{J}_i - \frac{2\varepsilon}{\beta_i}\nabla \times \boldsymbol{B}\right] \times \boldsymbol{B}$$

$$= -\varepsilon \frac{\beta_e}{\beta_i}\left(\nabla p_e + \partial/x_\beta \pi^e_{\alpha\beta}\right)$$

$$+ \frac{\varepsilon m}{M}[\nabla \cdot \mathsf{K}_i + (\nabla \boldsymbol{U}_e)\boldsymbol{J}_e] + (n_e l^*_d + \frac{m}{M}n_i l^*_d)\boldsymbol{J}, \quad (10.25)$$

or in the equivalent form

$$\boldsymbol{E} = -\boldsymbol{U}_e \times \boldsymbol{B} - \frac{\varepsilon \beta_e}{\beta_i n_e}\nabla \mathsf{P}_e + l^*_d \boldsymbol{J} - \varepsilon \frac{m}{M}\frac{d\boldsymbol{U}_e}{dt}. \quad (10.26)$$

Here the dimensionless parameters have been expressed via dimensional parameters as follows:

$$U = U' v_{\mathrm{Ti}}, \quad E = E' B_0 v_{\mathrm{Ti}}/c, \quad B = B' B_0, \quad p_e = p'_e p_{e0},$$

$$n = n' n_0, \quad t = t' L/v_{\mathrm{Ti}}, \quad x = x' L. \quad (10.27)$$

10.3.1 Effects of Electron Inertia and Electron Pressure Anisotropy

In order to study nongyrotropic effects, we have to use the full electron pressure tensor in the general Ohm's law [285]. Without any further approximation this equation (2.22) takes the following form

$$\boldsymbol{E} = -\boldsymbol{U}_e \times \boldsymbol{B} - \frac{\varepsilon \beta_e}{\beta_i n_e} \nabla \cdot \mathsf{P}_e - \frac{\varepsilon m}{M} \frac{\mathrm{d}\boldsymbol{U}_e}{\mathrm{d}t}. \tag{10.28}$$

Here P_e denotes the full electron pressure tensor. Traditional hybrid models usually omit the last term related to the electron bulk flow inertia and substitute the full electron pressure tensor P_e with the scalar pressure p_e, which is commonly calculated using the simple adiabatic or isothermal equation of state.

The equation for time evolution of the full electron pressure tensor may be written in the following form (see, e.g., [180, 224, 226, 285]):

$$\frac{\partial \mathsf{P}_e}{\partial t} = -\mathsf{D} - \mathsf{C}, \tag{10.29}$$

where

$$\mathsf{D} = \hat{\mathsf{D}} \mathsf{P}_e = [\boldsymbol{U}_e \cdot \nabla \mathsf{P}_e + \mathsf{P}_e \nabla \cdot \boldsymbol{U}_e + \mathsf{P}_e \cdot \nabla \boldsymbol{U}_e + (\mathsf{P}_e \cdot \nabla \boldsymbol{U}_e)^T]$$

describes convection, compression and electron fluid velocity $\boldsymbol{U}_e$ gradient effects, and

$$\mathsf{C} = \hat{\mathsf{C}} \mathsf{P}_e = \Omega_e [\mathsf{P}_e \times \hat{\boldsymbol{b}} + (\mathsf{P}_e \times \hat{\boldsymbol{b}})^T]$$

describes the cyclotron dynamics. C denotes the cyclotron tensor, and D denotes the driver tensor. Here $\Omega_e = eB/mc$, $\hat{\boldsymbol{b}} = \boldsymbol{B}/B$ is the unity vector in the direction of the magnetic field, and the superscript T denotes the transpose matrix. The only term omitted in (10.29) is the divergence of the generalized heat flux $\nabla \cdot \boldsymbol{Q}$. The simulations reported in [87] illustrated that the divergence of the generalized heat flux, which is essential around the neutral O-point, tends to zero in the vicinity of the neutral X-point.

In the simulations Harris's distribution for current sheet is given by

$$B_x(x,z) = B_0 \tanh\left(\frac{z}{L_0}\right) + \frac{\psi_0 B_0 DX}{4\pi L^2} \cos\left(2\pi \frac{(x - DX/2)}{DX}\right) z \exp\left(-\frac{z^2}{4L^2}\right),$$

$$B_z(x,z) = \psi_0 B_0 \sin\left(2\pi \frac{(x - DX/2)}{DX}\right) \exp\left(-\frac{z^2}{4L^2}\right), \tag{10.30}$$

where B_0 is the lobe magnetic field and L_0 is the halfwidth of the initial current sheet. The initial configuration of our current sheet corresponded to Harris's distribution with a small perturbation of the magnetic field localized inside the core of current sheet. The initial distribution function for ions is the Maxwellian distribution

10.3 Electron Effects on Reconnection

$$f \approx n_0 \exp\left(-\frac{(\boldsymbol{v}-\boldsymbol{U}_i)^2}{2v_{Ti}^2}\right). \tag{10.31}$$

The temperature of ions in the current layer and in the background may be different.

The number density is assumed to be given by

$$n(z) = n_p + n_b = n_0 \cosh^{-1}(z/L_0) + n_b, \tag{10.32}$$

where n_b is the lobe density (background) and n_0 is the sheet density. The initial profiles of ion bulk velocity were chosen from condition of absence of electric field so, that

$$U_{i,x} = U_{i,z} = 0,$$

$$U_{i,y} = c\frac{n_p}{n}\left(\frac{1}{M_A^2} - \frac{2\beta_e\gamma(n/n_0)^{\gamma-1}}{\beta_p}\right). \tag{10.33}$$

Now we shall discuss the results of simulations which were produced for the following parameters: (1) $m/M = 1/25$, $DX = 10c/\omega_{pi}$, $DZ = 5c/\omega_{pi}$, $L_0 = 0.5L$, $L = c/\omega_{pi}$ and $n_b/n_0 = 0.2$ [285]; (2) runs (a–f): $\beta_i = 5/6, \beta_e = 1/6$, $m/M = 1/25$, $DX = 20c/\omega_{pi}$, $DZ = 12.8c/\omega_{pi}$, $L_0 = L$, $L = c/\omega_{pi}$, $n_b/n_0 = 0.1$, $Re = 10^5$; runs (a–c): $B_y(t) = 0$; and runs (d–f): $B_y(t) \neq 0$ [335].

First we consider the simulations with nongyrotropic electrons. Figures 10.10 and 10.11 show run 1 the with nongyrotropic electrons and an electron-dominated current sheet. It can be seen from the contour plots in Fig. 10.10 that after 12 ion cyclotron times a significant amount of the magnetic flux is reconnected, i.e., magnetic reconnection has progressed to a

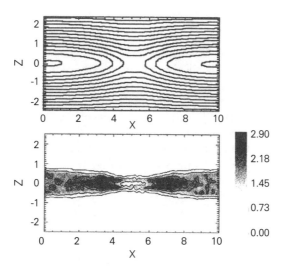

Fig. 10.10. *Top*: Magnetic field lines projection on the x–z plane; *bottom*: greyscale contour plot of the current density at the end of run 1 at $t = 12$ (from [285])

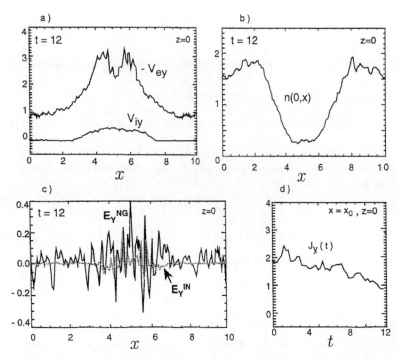

Fig. 10.11. Profiles of (a) the electron ($-U_{ey}$) and ion (V_{iy}) bulk flow velocities, (b) the number density n, and (c) the nongyrotropic E_y^{NG} (*black curve*) and inertia E_y^{IN} (*grey curve*) contributions to the reconnection electric field as a function of x along the equatorial plane $z = 0$ at the end of run 1 at $t = 12$. (d) Temporal evolution of the current density $J_y(t)$ at the neutral X-line $(0, x_0)$ for run 1. (From [285])

strongly nonlinear stage. The greyscale-coded current density distribution shown in Fig. 10.10 reveals a local current density reduction around the neutral X-line at $z = 0$, $x = x_0 = 5$. The current density gradient scale lengths around the neutral X-line are noticeably larger than the electron skin depth due to the nongyrotropic electron quasiviscous effect. The areas of increased current density are ejected away from the neutral X-line, which is consistent with the process of current filamentation. The time history of the current density J_y at the neutral X-line shown in Fig. 10.11c illustrates its gradual decrease in the course of the simulations.

Local minima in both electron flow velocity and number density around $x = 5$ can be observed in the final profiles versus x along $z = 0$ illustrated in Fig. 10.11a and b. The increase in the ion bulk flow velocity around the reconnection site can be also seen. Figure 10.11c illustrates the profiles along $z = 0$ versus x of the nongyrotropic E_y^{NG} and inertia E_y^{IN} contributions to the reconnection electric field. It can be seen that at the neutral X-line the nongyrotropic contribution to the electric field significantly exceeds the bulk

10.3 Electron Effects on Reconnection

flow inertia contribution. This is consistent with the fact that the system is evolving on the ion time scale rather than on the electron time scale, and therefore the inertia term proportional to the reconnection rate should be small. The dominant contribution to the bulk flow inertia electric field component comes from nonlinear terms (convective derivations), which are equal to zero at the reconnection site.

In the following runs (a–f) we study the effects of electron inertia on the magnetic field reconnection. In these runs the electrons are considered to be gyrotropic, i.e. the nongyrotropic off-diagonal components of the electron pressure tensor are made equal to zero.

In the runs (a–c) the magnetic field component B_y is kept equal to zero: (a) regime with nonstationary electron acceleration only; (b) regime with nonstationary and bulk electron acceleration; (c) regime with nonstationary and bulk electron acceleration, and ion bulk acceleration.

Neglecting the nongyrotropic quasiviscous effects in runs (a–c) results in drastically different dynamics in comparison with run 1. In particular, the current density J_y at the neutral X-line is increasing during the simulation, in contrast to the current density decrease for the run in Fig. 10.11d. It is seen in Fig. 10.12 (left) that after 4–5 ion gyroperiods not a lot of the magnetic flux is reconnected. A thin extended current sheet with increased current density has formed instead around the neutral X-line. The maximum value of the current density is significantly larger than for run 1, and the gradient scales are as small as the electron skin depth $c/\omega_{\rm pe}$, that is, the current density gradients are steeper than for run 1, where the electron nongyrotropic quasiviscous effects are property accounted for. The inclusion of the electron and then the ion bulk accelerations in Ohm's law results in an increase in the reconnection flux.

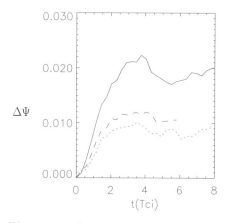
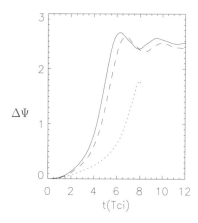

Fig. 10.12. Time evolution of reconnected flux of the magnetic field $\Delta\Psi$ ($\Delta\Psi = \Psi - \Psi_0$) for $L_0 = c/\omega_{\rm pi}$. Left: runs (a) (*dotted line*), (b) (*dashed line*) and (c) (*solid line*). Right: runs (d) (*dotted line*), (e) (*dashed line*) and (f) (*solid line*)

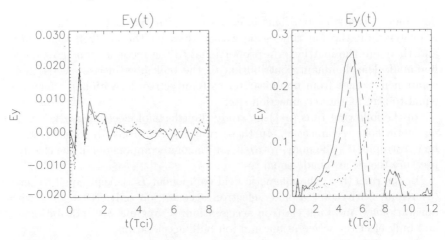

Fig. 10.13. Time evolution of the electric field E_y in the middle of the computational domain for $L_0 = c/\omega_{\text{pi}}$. Left: runs (a) (*dotted line*), (b) (*dashed line*) and (c) (*solid line*). Right: runs (d) (*dotted line*), (e) (*dashed line*) and (f) (*solid line*)

At the beginning of the reconnection ($t < 0.5 T_{\text{ci}}$) the reconnection flux has the same dynamics for runs (a–c). However, at a later time ($t > T_{\text{ci}}$) the bulk acceleration of electrons and ions begins to play a significant role. The inclusion of the electron bulk acceleration gives an extra reconnection flux of about 12% in comparison with run (a). The electron and ion bulk acceleration gives an extra reconnection flux of about 130% in comparison with run (a). The size of the magnetic islands in runs (a–c) may reach a value that is much smaller than the initial thickness of the current sheet.

Figure 10.13 demonstrates the time evolution of the electric field E_y in the middle of the computational domain. The reconnection electric field based only on the bulk electron inertia E_y^{IN} is small and strongly oscillating. The electron bulk flow velocity and current density are also oscillating, with gradient scales comparable to the electron skin depth. Run (b) corresponds to the hybrid simulation (run 2) of [285]. These results are in agreement with the full particle simulations of the driven collisionless magnetic reconnection in [237], where it was demonstrated that in the absence of the initial longitudinal magnetic field (B_y) the magnetic reconnection is supported by the electron meandering motion in a region where weak magnetic field and bulk electron inertia effects are not significant. In the hybrid simulation discussed here, kinetic effects associated with the electron meandering motion are described in terms of the evolution of the electron pressure tensor component. The spatial scale of the orbit amplitude of meandering electrons is approximately equal to half the thickness of the region, where the nongyrotropic contribution to the y-component of the electric field dominates all the other components. The overall time dependence of the electromagnetic field is approximately the same for all runs (a–c).

In the next three runs (d–f) the magnetic field is allowed to be nonzero: (d) regime with nonstationary electron acceleration only; (e) regime with nonstationary and bulk electron acceleration; (f) regime with nonstationary and bulk electron acceleration, and ion bulk acceleration. The magnetic field reconnection is drastically changed in the simulations with $B_y \neq 0$ (Fig. 10.12).

At the beginning of the reconnection ($t < 1.5T_{ci}$) the reconnection flux has the same dynamics for runs (d–f). However, at a later time ($t > 2T_{ci}$) the bulk acceleration of electrons and ions begins to play a significant role. The inclusion of the electron bulk acceleration increases significantly the speed of the reconnection by a factor of 1.5 in comparison with run (d). The electron and ion bulk accelerations together result in an increase in the reconnection speed by a factor of 1.1 in comparison with run (e). Figure 10.13 demonstrates the monotonic behavior of the electric field component E_y over time with small oscillations.

Figure 10.14 shows the typical magnetic field B (contour lines of the vector potential A_y) at time $t = 4.5T_{ci}$ for runs (d–f). One can see that the inclusion of the electron and ion bulk acceleration results in a faster magnetic field reconnection. Figure 10.15 demonstrates a strong whistler formation with a quadrupole structure.

10.3.2 Effects of Anomalous Resistivity on Reconnection

In this section we shall consider the effect of anomalous resistivity on the magnetic field reconnection processes. The different types of current instabilities which are excited in the current layer may result in effective (anomalous) resistivity. In [280, 335, 353] the maximum of resistivity diffusion length is located at the middle of the computational domain with a spatial profile given by

$$l_d = l_{d,0} \exp[-(x^2 + z^2)/\alpha]. \tag{10.34}$$

The simulations show that the presence of anomalous resistivity strongly affects the process of magnetic field reconnection. The general time evolution of reconnection includes a phase of slow reconnection, a phase of fast reconnection and a phase of saturation. The anomalous resistivity mostly affects the first phase. The second phase and saturation stage do not strongly depend on the value of the anomalous resistivity.

Figure 10.16 [three-dimensional hybrid simulation (performed by the author) with a grid of $601 \times 3 \times 151$; size of the computational domain: $DX = 240L$, $DY = 10L$, $DZ = 25.6L$; plasma parameters: $l_z = 2L$, $\delta B_z = 0.02$, $l_d = 0.2$, $M_A = 0.62$, $L = c/\omega_{pi}$, $\beta_i = 0.8$, $\beta_e = 0.2$, $\varepsilon = M_A$] shows the typical magnetic field B (contour lines of the vector potential A_y), the ion density, the y component of ion current density j_{py}, and the y component of the magnetic field B_y at $t = 30T_{ci}$. The most important difference between hybrid and MHD simulations is the occurrence of a magnetic field

276   10. Magnetic Field Reconnection Simulation

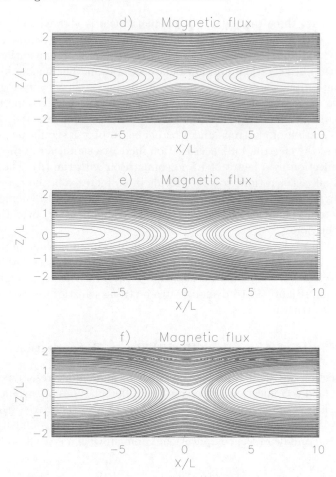

Fig. 10.14. From top to bottom, typical magnetic field B (contour lines of the vector potential A_y) at $t = 4.05T_{ci}$: (**d**) regime with nonstationary electron acceleration only; (**e**) regime with nonstationary and bulk electron acceleration; and (**f**) regime with nonstationary and bulk electron acceleration, and ion bulk acceleration

component in the invariant direction. This component is due to the Hall effect, which results into a poloidal current density j_p, i.e., a current in the x–z plane.

The Hall magnetic field component has been seen before in particle/hybrid simulations of reconnection (e.g., [225, 229, 353, 359, 386]), but the important point here is that this component is seen throughout the reconnection layer and has a magnitude of ≈ 50% of the lobe field. The smaller amplitude waves in B_y occur upstream of the separatricies. The density profile (not shown) does not show any indication of slow-mode shocks. In contrast, the high-speed flow in the reconnection wedge has a step-like profile. The magnetic

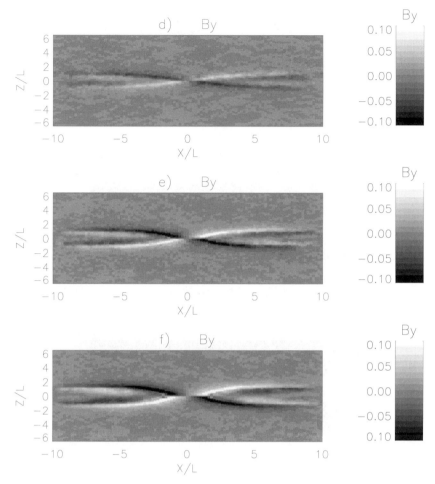

Fig. 10.15. From top to bottom, typical y-component of the magnetic field B_y at $t = 4.05T_{ci}$: (**d**) regime with nonstationary electron acceleration only; (**e**) regime with nonstationary and bulk electron acceleration; and (**f**) regime with nonstationary and bulk electron acceleration, and ion bulk acceleration

field in the invariant direction exhibits the typical quadrupolar structure. As pointed out in [225], the contours of B_y are instantaneous flow lines of the poloidal current. The Hall current loop is similar to the one shown in [386]. The current flows along the current sheet toward the X-point and returns in a thin layer at the edge of the outflow jet. The layer of outward flowing current coincides with the region of maximum vorticity. The magnetic field perturbation produced by the poloidal current system propagates essentially with an intermediate speed along the current sheet and results at later times in a steady-state Hall magnetic field along the whole reconnection layer.

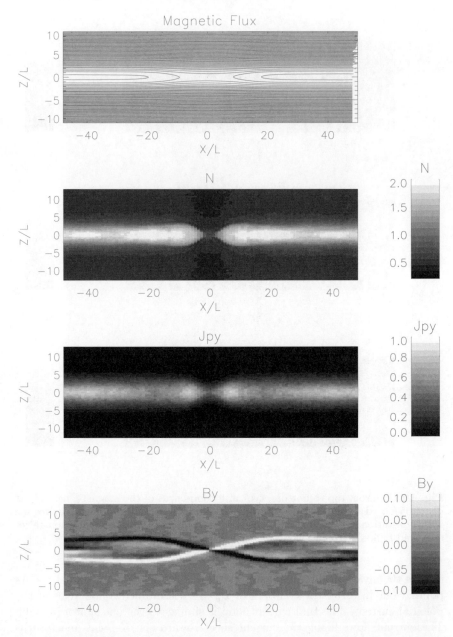

Fig. 10.16. From top to bottom at $t = 30T_{\rm ci}$: typical magnetic field B (contour lines of the vector potential A_y), the ion density, the y-component of ion current $J_{\rm py}$ and the y-component of the magnetic field B_y

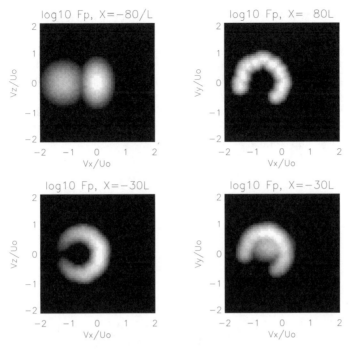

Fig. 10.17. Typical projection of the ion distribution onto the velocity planes (v_z–v_x and v_y–v_x) for spatial sections of the computational domain ($Z = 0$)

The numerous hybrid simulations (see, e.g., [164, 280, 311, 353, 359, 386]) have demonstrated the common features of the ion distribution function in current sheets. Figure 10.17 shows reduced typical distribution functions at three different positions in the center of the reconnection layer at time $t = 26T_{\text{ci}}$. It can be seen that the partial ring-type distribution is characteristic in the central current sheet of a large distance away from the X-point and dominates the distribution in the reconnection layer. The distribution function is different in the vicinity of the neutral line and in the magnetic-field pile-up region at the rear end of the plasmoid near the left end of the simulation box. The distribution in the latter region (not shown) is similar to the counterstreaming ion distribution reported in [164, 280, 311, 353]. As can be seen from Fig. 10.17, a partial shell distribution is observed near the X-point. In [353], this type of distribution was interpreted as being due to the following effects: First, the radius of curvature of the magnetic field lines near the X-point is large and is equal or exceeds the gyroradius of the lobe ions. Second, near the X-line the two-dimensional effect of the reconnection layer becomes important in that there are no ions which have been residing in the jet long enough to have, after cross-current sheet drift, v_z close to 0 and being ejected. In the v_z–v_x plane near $v_x = 0$, the two counterstreaming lobe distributions can be seen which just enter the volume determined by the small box. Particles which had reached the current sheet early and were closer to the X-point began to perform Speiser-type orbits without being ejected. The bounce motion in the z-direction exceeds the extent of the sampling box, so that the region near the current sheet is void of particles with large negative v_x and v_z close to zero ($v_z = 0$ at the turning point of the bounces).

The distribution demonstrates a strong heating of the ions inside the magnetic islands both perpendicular and parallel to magnetic field directions. The distribution function indicates that heating is stronger along the x-axis than along y- and z-axes.

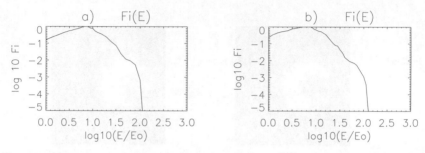

Fig. 10.18. Typical energy spectrum of accelerated ions for the following spatial sections of the computational domain: (a) $X/L = -72$; and (b) $X/L = -96$

Figure 10.18 illustrates the typical energy spectrum of accelerated ions for spatial sections ranging from $x = -DX/2$ to $x = +DX/2$. The ion energy spectrum has two parts: a bimaxwellian distribution and an accelerated component which emerges from the bimaxwellian distribution as a hard/flat power-law spectrum. The accelerated current layer ion energy spectrum may approximated by the power law $F_{c,i} \sim dN/N \approx (E/E_0)^{-k}$, where the energy E is calculated in the plasma bulk velocity frame and N denotes the ion density. In the present case, the index k is about $2-2.2$ in the vicinity of the X-point. The maximum energy of accelerated ions is $\log_{10} E_{\max}/E_0 \approx 2.2$.

The reader may refer to [225, 229, 280, 353, 359, 386] for details of the hybrid simulation of the magnetic field reconnection.

Summary

In this chapter we have discussed magnetic field reconnection and particle acceleration inside the current sheets by means of hybrid simulations. The simulations show that the Ampere models describe the linear, nonlinear and explosive processes on ion scales well, provided that the electrons do not strongly effect the ion-tearing instability. We have demonstrated that the conventional hybrid codes allow us to take into account the electron effects the electron inertia and electron pressure anisotropy. Simulation also demonstrates the strong ion acceleration with the power-law spectrum E^{-k} at the explosive stage of the reconnection process. Here we have considered the reconnection processes which start from a small initial perturbation. A more realistic situation, for example, the interaction of shock waves with the current sheets, was discussed in [77, 349]. However, to study the kinetic electron effects and charge separation effects one has to use the non-neutral full particle models.

Exercises

10.1 Derive the Lagrangian and the Hamiltonian for the magnetoinductive model (Sect. 10.2.1).

10.2 Derive the finite-difference scheme for the magnetoinductive model (Sect. 10.2.1) using the Euler–Lagrange equations.

11. Beam Dynamics Simulation

11.1 Introduction

Over the last decade, the importance of electromagnetic instabilities generated by ions backstreaming from shocks and the cometary, interplanetary or interstellar pickup ions or dust inside the solar wind or in the local interstellar medium (LISM) has been clearly established. The study of beam dynamics is also important for understanding the physical processes in the interaction of the solar wind with small space objects, in the interaction of fragments of the solar wind with the planetary magnetopause, and in the plasma environment during active experiments.

In this chapter we shall use the following model for the plasma and fields: The basic equations may be expressed in a dimensionless form by choosing a typical length L and time L/U_0, U_0 being the typical velocity. Depending on the problem, one of various velocities, for example, the bulk velocity of incoming flow ($U_0 = U_\infty$), or the thermal velocity of ions ($U_0 = v_{\mathrm{Ti}}$), or the Alfvén velocity, ($U_0 = v_{\mathrm{A}}$) may be chosen as the typical velocity. The basic equations are the ion equations of motion, the generalized Ohm's law, Faraday's induction equation and the adiabatic equation for the electron pressure. The ion equations of motion (normalized) are of the form

$$\frac{\mathrm{d}\boldsymbol{x}_{sl}}{\mathrm{d}t} = \boldsymbol{v}_{sl}, \quad \frac{\mathrm{d}\boldsymbol{v}_{sl}}{\mathrm{d}t} = \frac{Z_s}{\varepsilon \tilde{M}_s}\left(\boldsymbol{E}^* + \boldsymbol{v}_{sl} \times \boldsymbol{B}\right), \quad \boldsymbol{E}^* = \boldsymbol{E} - l_{\mathrm{d}}^* \boldsymbol{J}. \quad (11.1)$$

The normalized dissipation length is $l_{\mathrm{d}}^* = l_{\mathrm{d}}/L = 1/Re$, the magnetic Reynolds number is $Re = 4\pi U_0 L \sigma_{\mathrm{eff}}/c^2$ and σ_{eff} is the effective conductivity. The parameter ε is the ratio of the proton Larmor radius $\varrho_{\mathrm{cp}} = U_0/\Omega_{\mathrm{p}} = M_{\mathrm{A}} c/\omega_{\mathrm{pi}}$ in the unperturbed field B_0 to the characteristic dimension L of the problem, $\varepsilon = \varrho_{\mathrm{cp}}/L$, $\beta_{\mathrm{i}} = p_{\mathrm{i}}/(B_0^2/8\pi)$ and $\tilde{M}_s$ is the ratio of the ion mass to proton mass.

The electrons are considered to have a nonzero mass, so that the generalized Ohm's law is

$$\varepsilon\frac{m}{M}\frac{\partial \boldsymbol{J}_{\mathrm{e}}}{\partial t} + \varepsilon\frac{m}{M}(\nabla \boldsymbol{U}_{\mathrm{e}})\boldsymbol{J}_{\mathrm{e}} - \frac{\varepsilon}{M_{S_{\mathrm{e}}}^2 \gamma}(\nabla p_{\mathrm{e}} + \partial \pi_{\alpha\beta}^{\mathrm{e}}/\partial x_\beta) - n_{\mathrm{e}}\boldsymbol{E} + \boldsymbol{J}_{\mathrm{e}} \times \boldsymbol{B} + l_{\mathrm{d}}^* n_{\mathrm{e}} \boldsymbol{J} = 0$$
(11.2)

or
$$\boldsymbol{E} = -\boldsymbol{U}_\mathrm{e} \times \boldsymbol{B} - \frac{\varepsilon}{M_{S_\mathrm{e}}^2 \gamma n_\mathrm{e}} \nabla p_\mathrm{e} + l_\mathrm{d}^* \boldsymbol{J} - \varepsilon \frac{m}{M} \frac{\mathrm{d}\boldsymbol{U}_\mathrm{e}}{\mathrm{d}t}. \tag{11.3}$$

The electron ($\boldsymbol{U}_\mathrm{e}$) and ion ($\boldsymbol{U}_\mathrm{i}$) mean velocities are given by

$$\boldsymbol{U}_\mathrm{e} = \boldsymbol{U}_\mathrm{i} - \varepsilon \boldsymbol{J}/(M_\mathrm{A}^2 n_\mathrm{e}) \quad \text{and} \quad \boldsymbol{U}_\mathrm{i} = \sum_{k=1}^{N_\mathrm{s}} Z_k \langle N_k \boldsymbol{v}_k \rangle / \sum_{k=1}^{N_\mathrm{s}} Z_k \langle N_k \rangle. \tag{11.4}$$

Ampere's law and the induction equation are

$$\boldsymbol{J} = \nabla \times \boldsymbol{B}, \quad \frac{\partial \boldsymbol{B}}{\partial t} + \nabla \times \boldsymbol{E} = 0, \tag{11.5}$$

$$\boldsymbol{J} = \boldsymbol{J}_\mathrm{e} + \boldsymbol{J}_\mathrm{i}, \tag{11.6}$$

where the dimensionless current

$$\boldsymbol{J}_\mathrm{i} = \sum_{k=1}^{N_\mathrm{s}} Z_k n_k \boldsymbol{U}_k. \tag{11.7}$$

We impose the condition of quasineutrality and we consider the electron gas to be adiabatic

$$n_\mathrm{e} = \sum_{k=1}^{N_\mathrm{s}} Z_k n_k \equiv n, \quad p_\mathrm{e} \propto n_\mathrm{e}^\gamma. \tag{11.8}$$

Here the dimensionless parameters may be expressed using dimensional parameters as follows:

$$\boldsymbol{U} = \boldsymbol{U}' U_0, \quad \boldsymbol{E} = \boldsymbol{E}' B_0 U_0/c, \quad \boldsymbol{B} = \boldsymbol{B}' B_0, \quad p_\mathrm{e} = p'_\mathrm{e} p_{\mathrm{e}0},$$

$$n = n' n_0, \quad t = t' L/U_0, \quad \boldsymbol{x} = \boldsymbol{x}' L. \tag{11.9}$$

Here $U_0 = U_\infty, B_0 = B_\infty$ and $n_0 = n_\infty$ are the upstream value of bulk velocity, the magnetic field and the density; M_A and M_{S_e} are the Alfvén and electron sound Mach numbers:

$$M_\mathrm{A} = M_{S_\mathrm{e}} (\gamma \beta_\mathrm{e}/2)^{1/2}, \quad \beta_\mathrm{e} = p_{\mathrm{e}0}/(B_0^2/8\pi),$$

where γ is the effective specific heat ratio (in the calculations $\gamma = 2$ and $5/3$) and σ is the effective conductivity.

11.2 Cold Beam Dynamics

In this section we consider the generation of the electromagnetic waves by a cold beam in a homogeneous plasma. In many situations the solar wind interacts with newly created ions. Examples include heavy ions produced from the nuclei of comets, helium in the local interstellar medium, heavy ions emitted from the ionosphere of nonmagnetic planets, and artificially released atoms. In these cases the newly created ions are relatively cold, and collective interactions between the solar wind plasma and these cold ions result in significant coupling of two plasmas. The other examples of cold beams are the backstreaming ions in the ion foreshock and quasiparallel shocks.

11.2.1 One-Dimensional Models

11.2.1.1 Resonant Case. The excitation of electromagnetic waves by a cold beam was considered in [562, 571]. The one-dimensional simulations were carried out with a 128×4 grid, with correspondingly fewer (32 768) particles. The parameters for this case are beam density $n_\mathrm{b}/n_0 = 0.015$ and beam speed relative to the electron background $V_\mathrm{b} = 10 v_\mathrm{A}$ [571]. To achieve zero total current in the x-direction, the core ion speed is thus $V_\mathrm{c} = -(n_\mathrm{b}/n_0) V_\mathrm{b}/(1-n_\mathrm{b}/n_0) = -0.2 v_\mathrm{A}$, so that the relative ion drift speed is $V_0 = 10.2 v_\mathrm{A}$. As shown in Fig. 3 of [566], for these parameters the right-hand resonant instability dominates. Figure 11.1 (solid curves) shows time histories of various quantities: (a) the fluctuating magnetic field energy density $W_b = (1/L^2) \int \mathrm{d}x \int \mathrm{d}y [(B_x - B_0)^2 + B_y^2 + B_z^2]/B_0^2 = (\delta B/B_0)^2$; (b) V_0/v_A; (c) $T_{\mathrm{b}\perp}$; (d) $T_{\mathrm{c}\perp}$; (e) $T_{\mathrm{b}\|}$ and f) $T_{\mathrm{c}\|}$. The beam (core) parallel (perpendicular) temperatures are normalized in terms of their initial values.

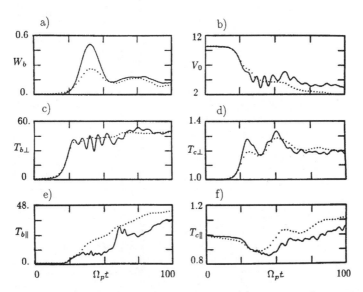

Fig. 11.1. Time histories for the resonant case: (**a**) magnetic fluctuations W_b, (**b**) drift speed V_0, and temperatures (**c**) $T_{\mathrm{b}\perp}$, (**d**) $T_{\mathrm{c}\perp}$, (**e**) $T_{\mathrm{b}\|}$, and (**f**) $T_{\mathrm{c}\|}$. *Solid* (*dotted*) *curves* are results from the one-(two-)dimensional simulation. (From [571])

The magnetic field fluctuations grow exponentially early on, then saturate at $(\delta B/B_0)_{\max} \approx 0.5$, then decay to a lower, nonlinear level at later times. The relative ion drift speed V_0 falls rapidly early on, then oscillates during the period when the field fluctuations are largest, then continues to decrease slightly. Similar oscillations are then seen in $T_{b\perp}$, and to a lesser extent $T_{b\|}$. These oscillations are thought to be due to magnetic trapping of the beam ions in the large-amplitude waves, although there is some disagreement between the actual oscillation frequency and the theoretically predicted values [181]. The parallel beam temperature does not grow so rapidly initially, although it continues to increase throughout the run. The corresponding changes in the core ion temperatures are much less: $T_{c\perp}$ increases slightly, while $T_{c\|}$ decreases during the growth and saturation phase but increases at later times [571].

Characteristics of the spectrum of waves generated are shown in Fig. 11.2. It displays profiles of one component of the transverse magnetic field B_y versus x and magnitudes of the squares of the Fourier modes $F_x = |B_y(k_x)|^2$ versus mode number $k_x = K_x/2\pi$ at various times throughout the run: (a) $\Omega_p t = 30$, corresponding to the growth stage of the instability; (b) $\Omega_p t = 45$,

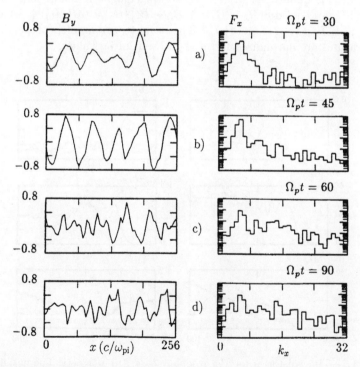

Fig. 11.2a–d. Profiles of B_y versus x and the Fourier modes of $B_y(F_x)$ versus mode number k_x at various times throughout the one-dimensional run for the resonant case (from [571])

corresponding to the decay stage; (c) $\Omega_p t = 60$ and (d) $\Omega_p t = 90$, corresponding to the nonlinear phase. The Fourier mode spectra are plotted with arbitrary but fixed scales at these four times. Throughout the run the $k_x = 5$ mode, the most unstable according to linear theory, dominates. During the growth stage, the spectrum narrows as the fastest-growing modes dominate. At early times the waveforms have a large amplitude and are fairly regular in shape. However, during the decay phase, shorter-wavelength modes are enhanced, as shown in [566]. While the $k_x = 5$ mode continues to be the largest at the end of the run, the spectrum is quite flat and the waveforms have become irregular.

11.2.1.2 Nonresonant Case. In the second simulation [571] the beam density is increased to $n_b/n_0 = 0.1$; $V_b = 10v_A$; $V_c = -1.1v_A$, so that the right-hand nonresonant instability dominates during the linear growth stage. The cell size was reduced: $\Delta x = \Delta y = 1c/\omega_{pi}$. The magnetic field fluctuations grow exponentially at early times, to much higher levels, $(\delta B/B_0)_{\max} = 2.2$, then decay to lower levels at later times (Fig. 11.3). The corresponding changes in the beam ion temperatures are about the same as in the resonant case, while those of the core ions are significantly larger (consistent with the characteristics of the nonresonant interaction). Magnetic trapping oscillations are not seen, however, in the one-dimensional run; again, the results are consistent with the nonresonant character of the waves [181, 566]. The hybrid

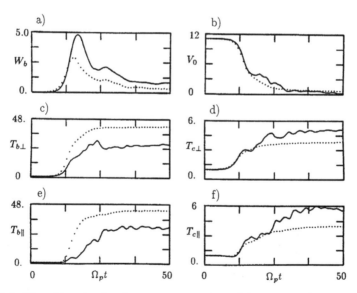

Fig. 11.3. Time histories for the nonresonant case: (**a**) magnetic fluctuations W_b, (**b**) drift speed V_0, and temperatures (**c**) $T_{b\perp}$, (**d**) $T_{c\perp}$, (**e**) $T_{b\|}$, and (**f**) $T_{c\|}$. *Solid* (*dotted*) *curves* are results from the one-(two-)dimensional simulation (from [571])

simulation of the two-pickup-ion instability in a cometary environment was done in [182].

11.2.2 Two-Dimensional Models

The two-dimensional simulation of resonant mode was done on a 128×128 grid of length $256 \times 256 c/\omega_{\mathrm{pi}}$ with 262 144 ions in [571]. Figure 11.1 (dotted curves) shows time histories of various quantities: (a) the fluctuating magnetic field energy density $W_{\mathrm{b}} = (1/L^2) \int \mathrm{d}x \int \mathrm{d}y [(B_x - B_0)^2 + B_y^2 + B_z^2]/B_0^2 = (\delta B/B_0)^2$; (b) V_0/v_{A}; (c) $T_{\mathrm{b}\perp}$; (d) $T_{\mathrm{c}\perp}$; (e) $T_{\mathrm{b}\|}$ and (f) $T_{\mathrm{c}\|}$. The beam (core) parallel (perpendicular) temperatures are normalized in terms of their initial values. The two-dimensional simulations exhibit a behavior similar to the one-dimensional simulation, but with several significant differences. First, the magnetic field fluctuations grow at a slower rate, saturate at a lower level, but then decay to about the same nonlinear level. In this case there are many more unstable modes which propagate obliquely to $\boldsymbol{B}_0$ and grow somewhat more slowly, so that it is not surprising that the composite growth rate is smaller. Second, the trapping oscillations are not visible in the two-dimensional run. Again, this is expected because all phase-dependent effects would naturally be washed away when averaging over many waves. Third, the nonlinear values of the various quantities are about the same as in the one-dimensional run, except that V_0 is slightly smaller, while $T_{\mathrm{b}\|}$ and $T_{\mathrm{c}\|}$ are slightly larger. This is primarily an artifact of breaking up the parallel energy into a mean velocity and temperature. This can be seen in Fig. 11.4, in which distribution functions for the parallel (v_x) and one transverse (v_y)

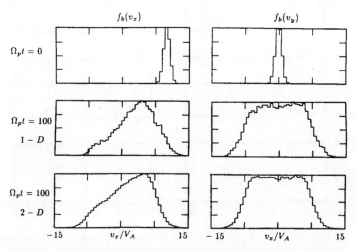

Fig. 11.4. Beam velocity distributions for the parallel (v_x) and perpendicular (v_y) components at the beginning and end of runs in one and two dimensions for the resonant case (from [571])

velocity component for the beam ions are shown at the beginning and the end of the one- and two-dimensional runs. The transverse distributions are flat-topped out to $v_y \approx V_0/2$; there is little difference between the one- and two-dimensional cases. The parallel velocity distributions are also similar. The distribution in the two-dimensional run is slightly more filled in around $v_x \approx 0$, implying a slightly smaller V_0 and larger $T_{b\|}$, as shown in Fig. 11.1. Note that no evidence of enhanced acceleration caused by the presence of a two-dimensional wave spectrum is seen.

The corresponding two-dimensional characteristics of the spectrum of waves generated are shown in Fig. 11.5 in a format similar to that in Fig. 11.2, except that now the Fourier spectra $F_x = |B_y(k_x)|^2$ averaged over k_y and $F_y = |B_y(k_y)|^2$ averaged over k_x are displayed. The profiles of B_y (averaged over y) are somewhat smaller in amplitude and not as irregular at later times. The corresponding Fourier spectra F_x shows the dominance of the $k_x = 5$ mode throughout the run, without much enhancement of shorter-wavelength modes at later times. The spectra in y indicate the continued dominance of $k_y = 0$ throughout the run, with some flattening of the spectrum at later times. Figure 11.6 shows the two-dimensional surface of B_y at various times for the resonant case. Early on a rather regular wave structure is visible, while at later times the surface is more turbulent.

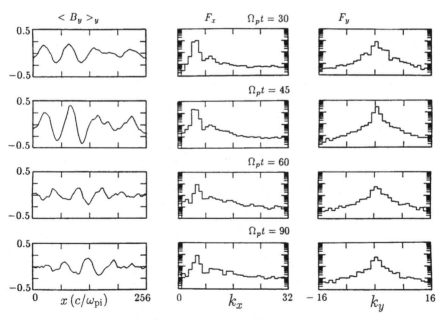

Fig. 11.5. Profile of B_y (averaged over y) versus x and the Fourier modes of B_y (F_x and F_y) versus k_x and k_y at various times throughout the two-dimensional run for the resonant case (from [571])

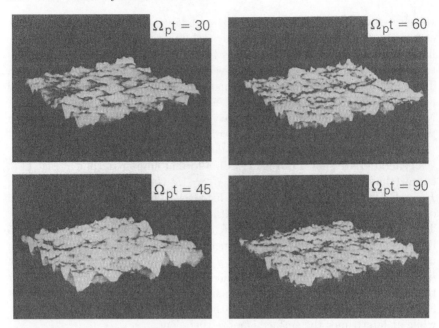

Fig. 11.6. Two-dimensional surface of B_y at various times for the resonant case. The positive x-direction points toward the right; positive y is inward (from [571])

The two-dimensional simulation of the nonresonant mode was done with a reduced cell size [571]: $\Delta x = \Delta y = 1c/\omega_{\mathrm{pi}}$. As in the resonant case, the magnetic field fluctuations in the two-dimensional simulation peak at about half the level of the corresponding one-dimensional run but now remain at a reduced level even at later times. In this case, however, large differences between both parallel and perpendicular temperatures persist at later times; the beam (core) temperatures in the two-dimensional run are about 50% larger (smaller) than the corresponding temperatures in the one-dimensional run. In two dimensions the y-averaged profiles of B_y are significantly smaller in amplitude, and the evolution toward longer waves in x is still apparent. In the y-direction the $k_y = 0$ mode persists throughout, although at later times the spectrum is enhanced and very flat beyond $k_y = \pm 2$. Finally, the waves themselves are shown in Fig. 11.7. While some regularity is apparent early on, the wave field at later times is highly turbulent.

The further two- and three-dimensional simulations [223, 518, 522] demonstrated that for relatively dense proton beams the oblique modes may play an important role in the nonlinear stage of the electromagnetic instability, even when the linearly dominant modes are the parallel ones. A very important aspect is the fact that the beam particles are resonant both with oblique and parallel modes. However, trapping in the parallel mode does not produce density fluctuations at the lowest order (at higher orders, it does [6, 7]),

11.3 Mass Loading of the Supersonic Flow by Heavy Ions

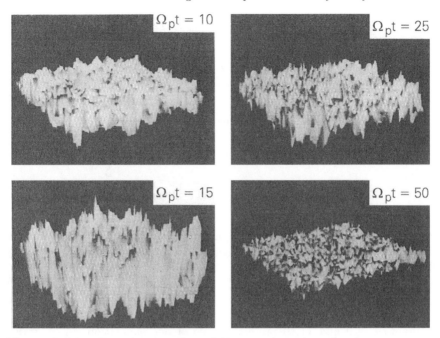

Fig. 11.7. Two-dimensional surface of B_y at various times for the nonresonant case. The positive x-direction points toward the right; positive y is inward (from [571])

but trapping in the compressible modes produces first-order density fluctuations. Thus, the trapped protons become localized in space, and the beam takes a filamentary structure; the beam filaments interact with the nearly parallel waves generated by the ambient beam protons, an interaction which curves the filaments. Such structures have not been observed in previous two-dimensional simulations of the electromagnetic ion-beam instabilities [562]. In [562] a spatial resolution of about $2c/\omega_{\mathrm{pi}}$ was used so that the oblique modes with wavelengths ≈ 10 may be not well resolved.

11.3 Mass Loading of the Supersonic Flow by Heavy Ions

11.3.1 One-Dimensional Models

In a real situation, when heavy ions injected into the solar wind have a high velocity perpendicular to the magnetic field, it is important to take into account the influence of this velocity on the excitation of instability (Figs. 8.15 and 11.8). A numerical investigation of these processes for a weak density beam was made in [574]. The model includes periodical boundary

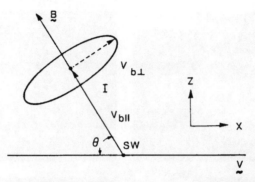

Fig. 11.8. The injection of heavy ions into the solar wind (from [574])

conditions; in addition, the initial magnetic field was oriented in the direction of initial beam velocity, the x-axis. The numerical simulation was done for the following parameters: $M_i = M, V_{b\perp} = V_{b\|} = 5.4 v_A$, $\beta_b = \beta_e = 0.01, \beta_p = 1.0$ and $\omega_{pi}/\Omega_p = 10^4$. The length scales are measured in units of c/ω_{pi} and times in Ω_p^{-1}. In order to study the various modes, the beam density and the system length L are varied.

Figures 11.9 and 11.10 present details of one simulation run in which $n_b/n_0 = 0.03$ and the length of the system is $L = 768 c/\omega_{pi}$, long enough for a number of unstable R modes to develop. In this case the density is too small to excite the N mode, and the short-wavelength H mode is not resolved (the cell size $\Delta x \sim 6c/\omega_{pi} > \lambda_H$). Figure 11.9 shows the time history of the parallel ($W_{b\|}$) and perpendicular ($W_{b\perp}$) energy density of the beam, normalized in terms of $n_0 M_b v_A^2/2$, and the magnetic field fluctuation energy

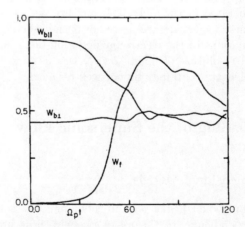

Fig. 11.9. Time histories of ring beam parallel and perpendicular energy densities, $W_{b\|}$ and $W_{b\perp}$, normalized by $n_0 M_b v_A^2/2$, and the magnetic field fluctuation energy density W_f, normalized by $B_0^2/8\pi$ (from [574])

11.3 Mass Loading of the Supersonic Flow by Heavy Ions 293

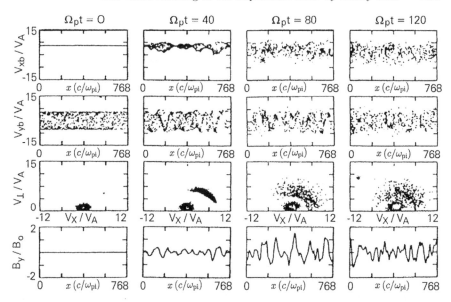

Fig. 11.10. Ring-beam phase space (v_{xb}-x, v_{yb}-x), $V_\perp$-$v_\parallel$ phase space for both ion species, and B_y profiles at various times showing the time development of the R mode ($n_b/n_0 = 0.003$ and $L = 768 c/\omega_{pi}$) (from [574])

density (normalized by $B_0^2/8\pi$) as a function of time. At early times ($0 < t < 50 \Omega_p^{-1}$) the waves grow exponentially, and the parallel beam energy decreases, while the perpendicular beam energy remains roughly constant, with about half of the beam energy ending up in the solar wind ions, principally in the perpendicular direction (not shown). After saturation ($\Omega_p t \approx 70$), the level of wave activity decreases, while the exchange of parallel and perpendicular beam energy (with the wave and solar wind ions) is much reduced [574].

The dynamical behavior of the ion ring beam is shown in Fig. 11.10. Snapshots of the beam v_{xb}-x and v_{yb}-x phase space, $v_\perp$-$v_\parallel$ phase space for both species, and profiles of B_y versus x are displayed at several times in the run. Initially, the beam is a cold, uniform ring, and the transverse components of B are zero. During the initial, linear phase ($\Omega_p t = 40$), waves in B_y (and B_z) grow and perturb the beam, as seen in the v_{xb}-x and v_{yb}-x phase space.

At this stage, pitch-angle scattering occurs, as seen by the narrow velocity shell in $v_\perp$-x phase space for the beam ions. The behavior later on ($\Omega_p t \approx 80$) is more complicated. Because of wave–wave interaction (parametric decay), the regular motion of the beam in the waves is disrupted, leading to a thermalization of the beam, as shown by the spreading of the velocity shell of the beam and the irregular waveforms in B_y. At still later times ($\Omega_p t \approx 120$) the disruption of the beam is complete, with many beam particles strongly scattered from their initial velocities, a more smeared out velocity distribution, and the persistence of some level of wave activity. The process by which the disruption occurs is similar to the case where the beam consists of cold protons, as discussed more fully in [566].

To investigate the behavior of the short-wavelength H mode, in [574] a much shorter system ($L = 10c/\omega_{\mathrm{pi}}$) was used, which allowed many wavelengths of the H mode to grow, but excluded the R mode. Figure 11.11 shows the results of such a calculation for $n_{\mathrm{b}}/n_0 = 0.1$, the v_y–v_x phase space for both ion components and B_y profiles at various times. Initially, the beam ions form a narrow ring, which spreads out along x in time (shown at $\Omega_{\mathrm{p}}t \approx 6$, corresponding to saturation of the wave amplitudes) along with the growth of very regular waveforms. At later times ($\Omega_{\mathrm{p}}t = 9$) the spreading of the beam is slightly greater, which corresponds to a second and larger maximum in the wave energy density, but unlike the first example, no disruption of the beam occurs. The corresponding picture of the waves at this time shows the persistence of the sinusoidal wave pattern, but with slightly longer wavelengths dominating. The overall behavior is reminiscent of the Weibel instability [302, 382], in that the instability evolves in time such as to relax the initial anisotropy, while the dominant wavelength increases. In the case of a low initial density $n_{\mathrm{b}}/n_0 = 0.01$, the initial thermal spread of the beam Δv_x is comparable to γ/k, so the instability grows very little.

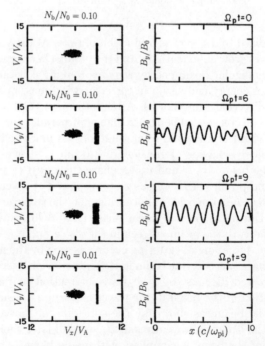

Fig. 11.11. v_y–v_x phase space and B_y profiles at various times showing time evolution of the H mode ($L = 10c/\omega_{\mathrm{pi}}$ and $n_{\mathrm{b}}/n_0 = 0.1$ and 0.01) (from [574])

11.3 Mass Loading of the Supersonic Flow by Heavy Ions

The simulation of the long system ($L = 60c/\omega_{\text{pi}}, n_{\text{b}}/n_0 = 0.1$) with a different size of grid spacing shows that the H mode does not strongly affect the spreading of the beam or the coupling of the ring-beam ions to the solar wind. Taking into account the temperature of the ion beam does not affect the excitation of the nonresonance (fire hose-like) mode.

11.3.2 Two-Dimensional Models

The periodical models considered above contain a limited class of solutions. A two-dimensional simulation of the process of mass loading of the solar wind with heavy cometary ions with nonperiodical boundary conditions was done in [176, 327]. These investigations are a natural generalization of the results of [460], in which the analysis of relaxation of the cometary ions was done on the basis of the quasilinear theory, which is valid for a small density of cometary ions only. According to gas dynamical calculations [49, 472], the maximum density of cometary ions ahead of the shock wave is $M_i n_i / M n_p = 0.185$, so for simplicity we consider a bounded domain into which a small number of cometary ions with a given initial ring-shaped distribution function is injected. We assume $M_i n_i / M n_p = 0.1$.

11.3.2.1 Mathematical Model and Formulation of Problem.
To study the relaxation of a beam of cometary ions to the average parameter values of the solar wind, we consider the combined flow of electrons, protons, and heavy cometary ions in the (z, x) two-dimensional formulation with $\partial/\partial y = 0$ (Fig. 11.12). In dimensionless form, the basic equations for our hybrid model are (11.1–11.8).

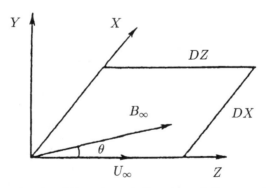

Fig. 11.12. Computational domain and directions of the unperturbed magnetic field and of the solar-wind velocity

The following spatially uniform velocity distribution of the particles is specified to prevail at the initial time in the computational domain, which is a rectangle with sides DZ and DX. The distribution for the protons of the solar wind is Maxwellian,

$$f_{\rm p} \sim \exp\left(-\frac{(v_z - U_z)^2 + v_x^2 + v_y^2}{2v_{\rm Tp}^2}\right), \tag{11.10}$$

and the distribution function for the cometary ions depends on the angle θ_{BU} between the vectors $\boldsymbol{B}_\infty$ and $\boldsymbol{U}_\infty$:

$$f_{\rm i} \sim \delta((\boldsymbol{v}_\perp - \boldsymbol{V}_{\rm drift})^2 - \boldsymbol{V}_\perp^{*2}) \cdot \delta(v_\parallel),$$

$$\boldsymbol{V}_{\rm dr} = -\left[(\boldsymbol{U}_\infty \times \hat{\boldsymbol{b}}) \times \hat{\boldsymbol{b}}\right], \quad |\boldsymbol{V}_\perp^*| = |\boldsymbol{V}_{\rm drift}| = U_\infty \sin\theta_{BU};$$

$$\hat{\boldsymbol{b}} = \boldsymbol{B}_\infty / B_\infty. \tag{11.11}$$

This distribution function is specified to be uniform on a narrow ring in a velocity plane which is perpendicular to the field $\boldsymbol{B}$:

$$v_{\rm Ti}^2 = (\boldsymbol{v}_{\perp 1} - \boldsymbol{V}_{\rm drift})^2 + \boldsymbol{v}_{\perp 2}^2 = V_\perp^{*2}. \tag{11.12}$$

The velocity $\boldsymbol{v}_\perp$, perpendicular to the magnetic field, has the components $\boldsymbol{v}_{\perp 1}$ and $\boldsymbol{v}_{\perp 2}$. Along the magnetic field $\boldsymbol{B}$, the average ion velocity $V_\parallel$ is zero, and the cometary ions acquire a transverse velocity as a consequence of the entrainment of ionized molecules by the flow of the solar wind [172, 319].

At the boundary $x = 0, DX$, a periodic condition is imposed on the fields, and when particles cross the boundary moving outside the domain, they are re-introduced into the region at its opposite boundary. At the left boundary, $z = 0$, the unperturbed field is specified, and protons and ions are injected with an unperturbed velocity distribution. At the right boundary of the region, $z = DZ$, a radiation condition is imposed on the component of the magnetic field, and the particles escape freely from the computational domain.

The simulation is carried out by the particle-in-cell method on a 255×17 or 501×4 spatial mesh with 3×10^5 macroprotons and 3×10^5 macro-ions. The equation of motion of the particles and the evolution equation of the magnetic field are integrated by the implicit schemes, which have been used in the calculation of the interaction of the solar wind with a cometary plasma [319]. The time step of the integration, Δt, satisfies the CFL condition $\Delta t = \Delta z/(4U_\infty)$; the step along the coordinate z satisfies the condition $\Delta z \ll \varrho_{\rm cp}$. The discrete distribution functions of the protons and ions are generated by a random number generator. The analysis of a numerical viscosity for $u_z = 1, u_x = 0.1$ gives the following value of the dissipative length $l_{\rm d}^*$ and effective Reynolds number Re:

$$l_{{\rm d},z}^* = \frac{1}{Re_z} = \frac{u_z \Delta z + u_z^2 \Delta t}{2} = 0.02, \quad l_{{\rm d},x}^* = \frac{1}{Re_x} = \frac{u_x \Delta x + u_x^2 \Delta t}{2} = 0.005. \tag{11.13}$$

11.3.2.2 Results and Conclusions. In the versions of the calculations which we will be discussing here, the dimensionless parameters of the plasma and the cometary ions were assigned values typical of the solar wind at the Earth's orbit and in the outer part of the coma, quite far from the outer shock waves: $M_A = 8$, $\beta_e = 1$, $\beta_p = 0.5$, $\varepsilon = 0.06$, $M_i n_i / M n_p = 0.1$, $DZ = 10L$, $DX = 1.5L$, $L = 1.33 \times 10^2 c/\omega_{pi}$ and $\theta_{BU} = 20°$. To reduce the calculation time, we took the mass of the cometary ion to be five times the mass of the proton ($M_i = 5M$). With this choice, the ratio of the mass densities of the ions and protons remained close to the value corresponding to gas dynamical calculations on the mass loading of the solar wind by cometary ions [472]. We first look at the results of the two-dimensional simulation.

Figure 11.13a shows some typical profiles of the components of the perturbed magnetic field, $B_\parallel, B_{\perp 1} = B_y$ (the value of $B_\parallel$ is close to ΔB_z), corresponding to the time $t = 5.2 T_{ci}$. We can see that a wave with the components $B_{\perp 2}, B_{\perp 1}$ ($B_{\perp 1}$ is not shown) is excited. These field components are

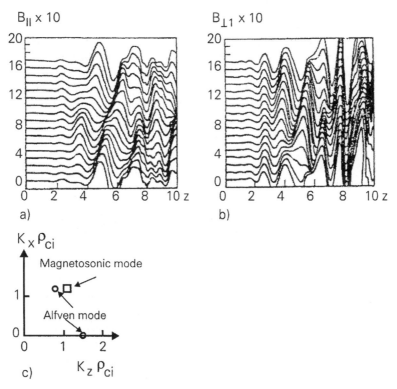

Fig. 11.13. Profiles of the components of the perturbed magnetic field, (**a**) $B_\parallel = \Delta B_z$ and (**b**) $B_{\perp 1} = B_y$, in the (z, x) two-dimensional region at the time $t = 5.2 T_{ci}$. Each profile corresponds to the field distribution along an $x = $ const. row of the computational mesh and is shifted 1 unit along the ordinate axis with respect to the adjacent profile. (**c**) Spectrum of generated waves. (From [176])

perpendicular to the main field $\boldsymbol{B}$. Because of the particular excitation conditions (the perturbations grow beginning at the amplitude of the shot noise), however, the wave pulse is nonuniform; it is modulated in the transverse (x) direction. This perturbation structure is associated with the excitation of a wave with a longitudinal magnetic field component $B_\| \approx \Delta B_z$.

A discrete Fourier analysis of the two-dimensional picture of the perturbation of the magnetic field (Fig. 11.13b) shows that a wide spectrum of waves with various harmonics $K_{n,m}$ is excited. Most of the energy of the transverse waves is in a mode which is independent of z, $K_{8,0}$ ($k_z \varrho_{ci} = 1.5$). The two-dimensional transverse oscillations have their maximum amplitude at $K_{4,1}$ ($k_z \varrho_{ci} = 0.75, k_x \varrho_{ci} = 1.2$). The magnetosonic oscillations are concentrated primarily in the $K_{4,1}$ and $K_{5,1}$ modes ($k_z \varrho_{ci} = 0.94, k_x \varrho_{ci} = 1.2$), although some of their energy is in a mode which is independent of z: $K_{8,0}$.

Analysis of the time evolution of the z-profile of the perturbed magnetic field B_y with $x = 0.5DX$ yielded the following results: Over a time $t = 1T_{ci}$, the amplitude of the transverse component of the magnetic field of the pulse reaches saturation at a value of about 0.6. The wavelength of the pulse satisfies the relation $k_z \varrho_{ci} = 1.3$ and corresponds to the resonant mode in [574] with a growth rate $\gamma \approx 0.3\Omega_i$. It propagates along the magnetic field with a phase velocity $v_{ph} \approx 0.1-0.2 U_\infty$ in a coordinate system moving with the solar wind. According to an analysis of the dynamics of transverse components of an isolated pulse, the perturbation has a right-hand polarization in the coordinate system moving with the solar wind. Over the instability rise time, the pulse drifts a distance $\approx 2\pi \varrho_{ci}$ along the z-axis. Figure 11.14a shows $(v_\|, v_{\perp 1})$ and $(v_\|, v_{\perp 2})$ sections through the ion velocity distribution for the value $z = 8.33$.

As the instability of the resonant mode develops, the distribution function becomes nearly isotropic; at $z = 0$ it is in the form of a ring oriented perpendicular to the magnetic field, while at $z = 8.33$ it has roughly the shape of a part of a spherical shell of finite thickness. A parameter which is an important characteristic of the extent to which the cometary ions are entrained is the difference between the macroscopic velocities of the solar wind and the flux of cometary ions along the z-axis; this parameter is shown in Fig. 11.14b. For the given calculation parameter values, we observe an increase in the velocity of the ion flux: U_{iz} increases to $0.6 U_\infty$, where U_∞ is the macroscopic velocity of the protons of the unperturbed solar wind.

According to the results of the calculations, variations in the modulus of the magnetic field joined with perturbations of large amplitude in both the transverse components, $B_{\perp 1} \approx B_y$ and $B_{\perp 2} \approx \Delta B_x$, and the longitudinal component, $B_\| \approx \Delta B_z$. Since the proton are magnetized, the condition $n_p \sim B$ holds quite well. The density of cometary ions with a Larmor radius ϱ_{ci} substantially smaller than the perturbation wavelength does not follow the magnetic field; because of nonlinear effects, it exhibits significant amplitude surges, which are also modulated in the transverse (x) direction.

11.3 Mass Loading of the Supersonic Flow by Heavy Ions

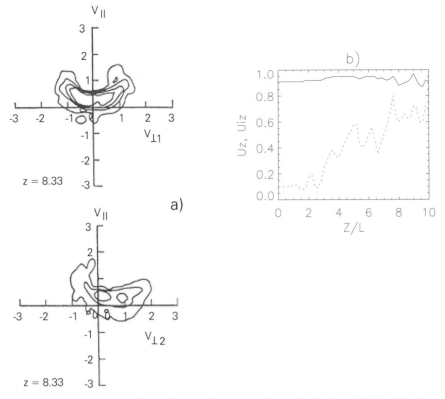

Fig. 11.14. (a) $(v_\parallel, v_{\perp 1})$ and $(v_\parallel, v_{\perp 2})$ sections through the ion velocity distribution, averaged over x, for $z = 8.33$. (b) Relaxation of the velocities U_z (*solid line*) and U_{iz} (*dotted line*) along the z-axis in the two-dimensional case. (From [176])

Similar calculations with $\partial/\partial x = 0$ lead to one-dimensional structures with large saturation amplitudes and a nearly isotropic distribution function. The apparent reason for these results is a decrease in the phase space of the wave vectors of the waves which are excited in the one-dimensional case and, correspondingly, an increase in the energy density of the waves.

On the other hand, to eliminate effects of the right-hand boundary on the saturation amplitude, we carried out a quasi-one-dimensional calculation on a 501×4 mesh with a physical length DZ containing $72\varrho_{ci}$ ($\varepsilon = 0.03$).

Let us examine the results of this quasi-one-dimensional calculation. As the z-dimension of the calculation region is increased, there are increases in both the saturation amplitude of the instability of the transverse component of the magnetic field, $B^*_\perp = 0.8 - 0.9$ [$B^*_\perp = 2$ for the level of mass loading $(M_i n_i)/(M n_p) = 0.3$, in agreement with the one-dimensional calculations of [562], for which the value of $B^*_\perp = 4 - 5$ for $(M_i n_i)/(M n_p) = 1.6$, but the elevated density of nonentrained ions does not correspond to the actual

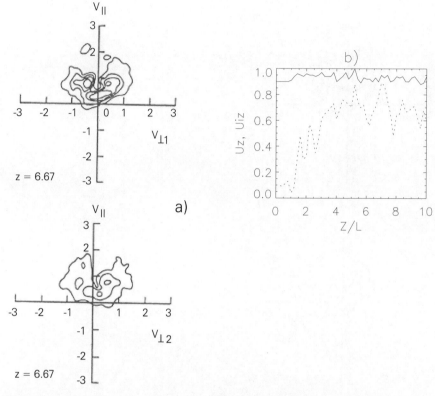

Fig. 11.15. (a) $(v_\parallel, v_{\perp 1})$ and $(v_\parallel, v_{\perp 2})$ sections through the ion velocity distribution, averaged over x, for $z = 6.67$. (b) Relaxation of the velocities U_z (*solid line*) and U_{iz} (*dotted line*) along the z-axis in the quasi-one-dimensional case. (From [176])

values for the supersonic part of the solar wind] and the extent to which the velocity distribution with respect to the velocity of the solar wind becomes isotropic (Fig. 11.15a). It can be seen that some of the ions reach a velocity $v_z \approx 2U_\infty$ at distance $z = 6.67$.

Figure 11.15b shows the bulk velocities of the solar wind, U_z, and pickup ions, U_{iz}, as functions of z at a time $t = 7.09 T_{ci}$. It can be seen from this that in the interval $0 < z < DZ/2$ there is a rapid increase in U_{iz}, up to a value of $(0.9 - 1.0)U_\infty$, which corresponds to a pronounced anisotropy of the temperature distribution $(T_\parallel/T_\perp \gg 1)$ and the growth of a resonant mode. Later on, at $z > DZ/2$, the anisotropy fades and waves with smaller values of k and correspondingly lower growth rates are apparently excited.

It thus follows from the results of these calculations that over a scale $\Delta z \sim 2\pi \varrho_{ci}$ newly formed cometary ions may be entrained essentially completely by the solar wind, and the distribution function of the beam of cometary ions simultaneously becomes nearly isotropic. However, we can estimate the fraction of cometary ions which are injected with the initial distribution function into the solar wind over a distance Δz:

$$\frac{\Delta n_{\rm i}}{n_{\rm p}} = \frac{G\nu}{4\pi n_{\rm p} W_{\rm c} U_\infty} \int_z^{z+\Delta z} \frac{{\rm e}^{-\nu r/W_{\rm c}}}{r^2} {\rm d}r \approx \frac{G\nu}{4\pi n_{\rm p} W_{\rm c} U_\infty} \cdot \frac{{\rm e}^{-\nu z/W_{\rm c}}}{z^2} \Delta z. \quad (11.14)$$

If the background plasma is assumed to consist of solar-wind protons and cometary ions which have undergone relaxation, then our beam would be the flux of newly ionized $n_{\rm i}$.

Let us take some parameter values corresponding to the Halley comet. The velocity and density of the solar wind are $U = 550\,{\rm km/s}$ and $n_{\rm p} = 10\,{\rm cm}^{-3}$; the evaporation rate of the nucleus is $G = 10^{30}\,{\rm mol/s}$; the ionization rate of the cometary is $\nu = 10^{-6}\,{\rm s}^{-1}$; the initial velocity of the cometary ions is $W_{\rm c} = 1\,{\rm km\,s}^{-1}$; and the distance to the outer shock wave is $R_{\rm sh} = 0.5 \cdot 10^6\,{\rm km}$. With $z = 10^6\,{\rm km}$ and $\Delta z = 10^5\,{\rm km}$, we then find $n_{\rm i}/n_{\rm p} = 3 \times 10^{-3}$. At this beam density, according to [574], for example, only the resonant mode would be excited ($\omega << \Omega_{\rm p}$). In contrast, the nonresonant mode ($\omega << \Omega_{\rm p}$) associated with the hydrodynamical fire-hose instability (the case when $p_\parallel/p_\perp >> 1$) would not be excited at such low beam densities because of the high threshold for the instability [574] in the case with $\theta_{BU} \approx 20° - 30°$ and because of the low growth rate $\gamma/\Omega_{\rm p} \sim (n_{\rm i} U_\parallel)/(n_{\rm p} v_{\rm A})$.

11.4 Finite Size Beam (Plasma Cloud) Dynamics

Finite size beam and plasma cloud dynamics addresses many plasma phenomena: the interaction of the solar wind with high speed filaments; weak comets; asteroids and other space objects and nonmagnetic planets; and plasma clouds released during the active experiments in space and laboratory plasmas. Plasma cloud dynamics is also related to the penetration of the solar-wind fragments across the magnetopause, the generation of waves by a tether in the ionosphere [94], and the motion of ionized clouds in the ionosphere [43] and in the laboratory plasma [542].

11.4.1 Generation of Low-Frequency Waves by Three-Dimensional and 2.5-Dimensional Beams in a Homogeneous Background

The interaction of ion beams oblique to an ambient magnetic field into a background plasma is a problem with a long history of investigation, for example, in [158, 183, 197, 379, 422, 521, 566]. It has been speculated that large-scale ($\lambda \sim c/\omega_{\rm pi}$) electromagnetic instabilities driven by ions streaming along the unperturbed magnetic field are related to the formation of parallel and quasiparallel collisionless shocks in plasma. In [521] it was found that a finite-size plasma cloud traveling parallel to an ambient magnetic field in the presence of a uniform background plasma causes a right-handed electromagnetic instability. The instability develops when the relative velocity of the plasma cloud to the background plasma is somewhat larger than the local Alfvén velocity. The instability serves to couple the cloud ions with the background ions.

In this section we present the results of a study of whistler generation by a finite-size beam propagating in an homogeneous background plasma [334]. The plasma and field variables are taken as uniform in the x-, y- and z-directions. In all the cases considered here the initial magnetic field is along the x-direction and the direction of the beam is the z-direction. The dimensionless parameters of the plasma were assigned values typical of the solar wind downstream of the Earth's bow shock: $\beta_{i1} = 1.0, \beta_{e1} = 0 - 1.0, \varepsilon = \varrho_{ci}/L = 1, h_z/\varrho_{ci} = h_x/\varrho_{ci} = 0.1, DX = 20 - 80L, DY = 0.3 - 8.0L$ and $DZ = 5 - 10L$. The size of the beam is $1\varrho_{ci} \times 1\varrho_{ci} \times 1\rho_{ci}$, and it is centered at the point $(x = 0, y = 0, z = -2)$. In the simulations considered here, the time scale is much smaller than T_{ci}, so that in the first approximation we may fix the position of the background ions and update the position of the beam ions only. The number of beam ions is 10^5. One-dimensional profiles (cuts through the simulation plane) of the components of the magnetic field (B_z and B_y) and the components of the electric field (E_y and E_z) along the x-direction at time $t = 0.03T_{ci}$ are shown in Fig. 11.16.

Figures 11.16a and b are for the case $n_b/n_p = 1$, $V_b = 2$, $m/M = 0$ and $l_d = 0$. The figure shows the formation and propagation of a right-hand-polarized whistler wave along the magnetic field and a left-hand-polarized whistler in the opposite direction. Two-dimensional sections of the same profiles, through the x–y and x–z planes, are shown in Fig. 11.17.

These simulations show that the envelope of the generated whistler depends on the value of the density, the beam velocity, the anomalous resistivity, β_e and the mass ratio m/M. The quasi-two-dimensional regime with $DX = 80L$, $DY << \varrho_{ci}$ and $DZ = 10L$ was also simulated. This simulation with a 2.5-dimensional hybrid code yielded the same results as three-dimensional hybrid codes with both implicit and explicit calculation of the electric field.

11.4.2 Interaction of the 2.5-Dimensional Beam with Tangential Discontinuities

In this section we consider the transfer of energy and momentum across the dayside magnetopause (MP) via the rapid deformation of the MP, which occurs when the latter suffers a collision with the solar-wind filaments. We consider the forced excitation of surface waves and other field-aligned effects occurring within the internal structure of the tangential discontinuity (TD). Three-dimensional MHD simulations of the interaction of the interplanetary discontinuity with the MP (Sudden Commencement) [312, 536, 537] have shown that the pressure jump, associated with the incident disturbance, can

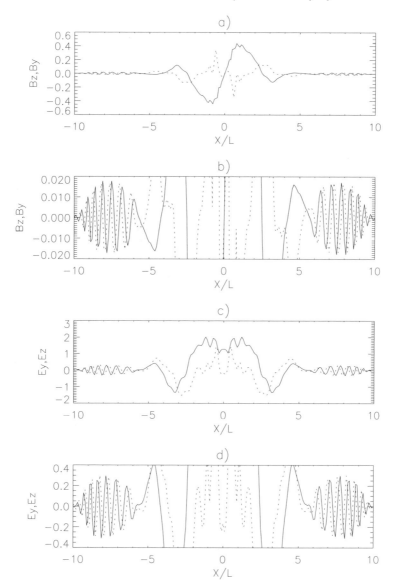

Fig. 11.16. Beam propagation in an homogeneous plasma. One-dimensional sections of (**a,b**) the magnetic field components B_z (*solid line*), B_y (*dotted line*) and (**c,d**) electric field components E_y (*solid line*), E_z (*dotted line*) at $t = 0.03 T_{ci}$ (from [334])

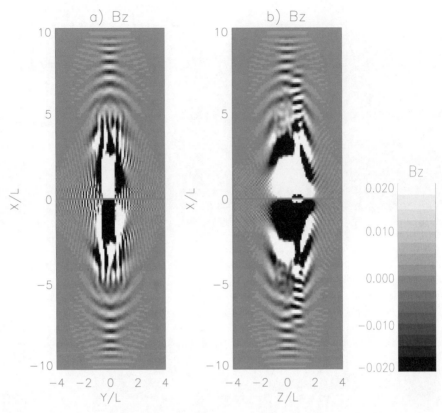

Fig. 11.17. The two-dimensional sections of (**a,b**) the magnetic field component B_z and (**c,d**) the electric field component E_z at $t = 0.03T_{ci}$ (from [334])

excite magnetosonic impulses at the boundary of the magnetosphere. A portion of the energy associated with this impulse is transformed into Alfvén waves, which are generated when the resultant magnetosonic impulse passes through the noon–midnight region of magnetosphere. The propagation of these Alfvén waves is associated with a current along the geomagnetic field.

The objective of this section is to understand the impulsive local excitation of waves, whose sources are either in the vicinity of or internal to the structure of the MP. We consider the generation of whistler waves which are generated when a supersonic beam collides with a TD (i.e., the MP). Thus, we hope to model some processes associated with dust in the solar wind (e.g., from small comets and asteroids [271]) and also processes associated with the collision of these "clouds" with TDs (e.g., collisions of the MP with solar-wind fragments, ionized atmosphere and dust from space objects and comets) [341].

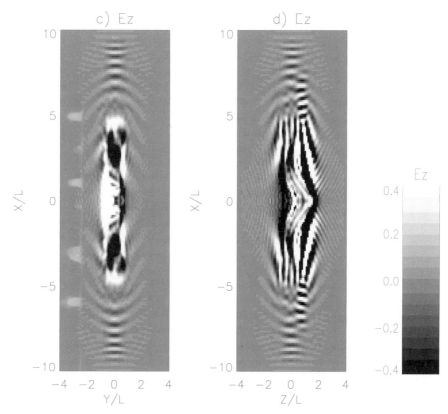

Fig. 11.17. (continued)

Previous two-dimensional MHD [121] and hybrid [470] simulations of this process did not include the field-aligned effects, because of the particular choice of field orientation by the authors. In our simulations the initial magnetic field lies in the simulation plane, and its direction is parallel to the front of TD, along x-axis. In our 2.5-dimensional simulations, we used 201×51 grid points and 2×10^6 ions for the background plasma and 10^4 ions for the beam [334, 342]. The dimensionless parameters of the plasma were assigned values typical of the solar wind downstream of the Earth's bow shock: $\beta_{i1} = 1.0, \beta_{e1} = 1.0, \varepsilon = \varrho_{ci}/L = 0.5, DX = DZ = 10L, \Delta x/\varrho_{ci} = 0.4, \Delta z/\varrho_{ci} = 0.1$. The beam parameters are $V_b = 5, n_b/n_1 = 0.5$ and the size of the beam is $1\varrho_{ci} \times 1\varrho_{ci}$. At the start the beam is at $x = DX/2, z = DZ/2 - 1\varrho_{ci}$, thereafter it propagates in the direction of the TD. The simulations show that the penetration of the beam through the TD is accompanied by its expansion in the direction of the magnetic field. Upstream of the TD, the beam generates right-hand-polarized whistler waves, which propagate along the magnetic field. At the TD, a fast-mode magnetosonic wave detaches from the beam and propagates ahead of the beam, since the phase velocity of magnetosonic waves is much larger on the far ("earthward") side of the TD. Figure 11.18 shows (a) the density levels and (b) the current distribution in the computational plane at $t = 0.15 T_{\rm ci}$, when the beam is centered on the TD. This current is associated with whistler propagation along the magnetic field.

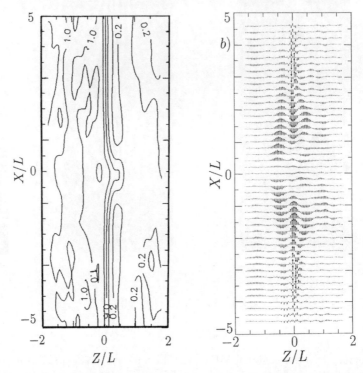

Fig. 11.18. Wave generation by a finite size beam inside of the TD. Contour plots of (**a**) the ion density and (**b**) the two-dimensional distribution of total current in the x–z plane at $t = 0.10 T_{\rm ci}$ (from [334, 342])

Summary

In this chapter we have discussed the dynamics of the beams and plasma clouds in background plasma. The simulations show that the hybrid codes allow us to study the low-frequency waves (Alfvén waves, helicons and whistlers) which are generated by finite size beams. The hybrid codes allow us also to study the generation of low-frequency waves in the region of mass loading of the plasma flow (the solar wind, the local interstellar medium, etc.) by using pickup ions. The study of the interaction of the plasma clouds with a magnetopause allows us to estimate the mass, momentum and energy flux transfer across the magnetopause. The finite electron mass effects may result in modulation of the wave envelope of the generated whistler. However, a realistic fully three-dimensional self-consistent simulation of beam dynamics has yet to be done.

12. Interaction of the Solar Wind with Astrophysical Objects

12.1 Introduction

This chapter is devoted to the global interaction of the solar wind with non-magnetic planets, comets and the local interstellar medium (LISM). The hybrid simulation allows us to predict the different plasma structures in the planetary and cometary environments: bow shock, magnetopause, magnetotail etc. These models must be compared with observational data to improve our knowledge about the plasma environment of these objects. The global hybrid MHD–Boltzmann simulation of the solar wind–local interstellar medium will be considered also. Although the main structure of the outer heliosphere is still the subject of theoretical research, these models allow us to predict the global structure of the heliosphere and the penetration of the neutral component from the LISM through the heliosphere.

12.2 Interaction of the Solar Wind with Strong Comets

In the case of a planet, the solar wind encounters a localized obstruction (the magnetosphere or the ionosphere of the planet), but in the case of a comet the influence of the obstruction is felt at much greater distances. This is because the gravitational field of the comet's nucleus is negligible, so that the neutral gas which evaporates from the surface of the nucleus under the influence of solar radiation expands freely into interplanetary space. At a distance of one astronomical unit from the Sun, i.e., at the Earth's orbit, the rate of expansion is of the order of 1 km/s, and therefore the neutral gas manages to escape to millions of kilometers from the comet's nucleus before it is ionized under the influence of the solar radiation and solar-wind electrons or owing to charge exchange (the characteristic times of all these processes are of the order of 10^6 s).

The problem of the interaction of the solar wind with a comet involves a large set of physical processes. The mathematical complexity of the study of these phenomena is due to the different spatial and temporal scales of the phenomena occurring in the coma and tail of the comet.

During the last two decades many works have appeared devoted to hydrodynamic evaluations and numerical calculations of the plasma flow in the vicinity of a comet under different assumptions with respect to the model of the material evaporating from the surface of the comet under the action of solar radiation [27, 28, 49, 68, 245, 365, 472, 546]. As is well known (see, e.g., [246]), depending on the rate of evaporation of cometary material, the physics of the interaction of the solar wind with a comet is analogous to the interaction of the solar wind with the Moon [314, 496, 577] (very low evaporation rate) and Venus [38, 318, 497] (high evaporation rate) or to the interaction of the interstellar gas with the solar wind [26, 547].

12.2.1 Formulation of the Problem and Mathematical Model

The pattern of the flow around a comet, the location of the external shock wave and the configuration of the magnetic field at the head of the comet for a specified shape of the surface of contact discontinuity were obtained in axially symmetric, three-dimensional, magnetohydrodynamic calculations in [472]. However, since the gyroradius of the heavy cometary ions, calculated from the velocity and magnetic field of the solar wind, can be 10^4 km, the hydrodynamic equations lose their validity in the vicinity of the shock wave and the contact surface. Moreover, the cometary ions acquire in the acceleration process energy significantly greater than the energy of the solar-wind protons, which leads to a two-phase flow of the solar wind and cometary plasma with different thermodynamic parameters. An attempt is made in this work to study the flow structure in the coma of a comet with consideration of the effects of the finite Larmor radius of the solar-wind protons and the cometary ions. The combined flow of the solar wind and the ionized components of the cometary gas is examined within coaxial spheres of radius R_∞ and R_0 (Fig. 12.1). The inner sphere R_0 is surrounded by a thin "film" of radius R_1, simulating the ionopause.

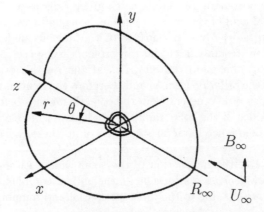

Fig. 12.1. Geometry of the calculation region

12.2 Interaction of the Solar Wind with Strong Comets

We use the equations of motion of the particles to describe the motion of the macroprotons of the solar wind and the cometary macro-ions, while the electrons are described in the hydrodynamics approximation. This approach (separation of the ions for the kinematic description) was developed for the study of thermonuclear (see, e.g., [598]) and cosmic plasma (see, e.g., [304, 318]). The principal equations are written in the following dimensionless form:

The ion equations (normalized) are of the form

$$\frac{d\boldsymbol{x}_{sl}}{dt} = \boldsymbol{v}_{sl}, \quad \frac{d\boldsymbol{v}_{sl}}{dt} = \frac{Z_s}{\varepsilon \tilde{M}_s}(\boldsymbol{E}^* + \boldsymbol{v}_{sl} \times \boldsymbol{B}), \quad \boldsymbol{E}^* = \boldsymbol{E} - l_d^*\boldsymbol{J}. \quad (12.1)$$

The normalized dissipation length is $l_d^* = l_d/L = 1/Re$, the magnetic Reynolds number is $Re = 4\pi v_{Ti} L \sigma_{\text{eff}}$ and σ_{eff} is the effective conductivity. The parameter ε is the ratio of the proton Larmor radius $\varrho_{cp} = U_0/\Omega_p$ in the unperturbed field B_∞ to the characteristic dimension L of the problem, $\varepsilon = \varrho_{cp}/L$; $\tilde{M}_s$ is the ratio of the ion mass to proton mass. The electrons are considered to have zero mass, so that the generalized Ohm's law is

$$\boldsymbol{E} = -\boldsymbol{U}_e \times \boldsymbol{B} - \frac{\varepsilon}{M_{Se}^2 \gamma n_e}\nabla p_e + l_d^*\boldsymbol{J}. \quad (12.2)$$

The electron $\boldsymbol{U}_e$ and ion $\boldsymbol{U}_i$ mean velocities are given by

$$\boldsymbol{U}_e = \boldsymbol{U}_i - \varepsilon \boldsymbol{J}/(M_A^2 n_e), \quad \boldsymbol{U}_i = \sum_{k=1}^{N_s} Z_k \langle N_k \boldsymbol{v}_k \rangle / \sum_{k=1}^{N_s} Z_k \langle N_k \rangle. \quad (12.3)$$

Ampere's law and the induction equation are

$$\boldsymbol{J} = \nabla \times \boldsymbol{B}, \quad \frac{\partial \boldsymbol{B}}{\partial t} + \nabla \times \boldsymbol{E} = 0, \quad (12.4)$$

$$\boldsymbol{J} = \boldsymbol{J}_e + \boldsymbol{J}_i, \quad (12.5)$$

where dimensionless ion current is

$$\boldsymbol{J}_i = \sum_{k=1}^{N_s} Z_k n_k \boldsymbol{U}_k. \quad (12.6)$$

We impose the condition of quasineutrality and we consider the electron gas to be adiabatic

$$n_e = \sum_{k=1}^{N_s} Z_k n_k \equiv n, \quad p_e = n_e^\gamma, \quad (12.7)$$

where the effective adiabatic index $\gamma = 5/3$.

The production rate of ions (cometary material) is

$$G = \frac{\nu L}{W_c} \frac{Q}{n_\infty L^2 U_\infty} \frac{1}{4\pi r^2} \exp\left(-\frac{\nu L r}{W_c}\right). \quad (12.8)$$

Here the dimensionless parameters may be expressed via dimensional parameters as follows:

$$U = U' U_\infty, \quad E = E' B_\infty U_\infty / c, \quad B = B' B_\infty, \quad p_e = p'_e p_{e\infty},$$

$$n = n' n_\infty, \quad t = t' L / U_\infty, \quad x = x' L. \tag{12.9}$$

The simplified form of Ohm's law is used to model the "ionosphere" in the region between the spheres R_1 and R_0 (Fig. 12.1),

$$E = \frac{\nabla \times B}{Re}. \tag{12.10}$$

It is assumed that the "ionosphere" is initially embedded in the homogeneous interplanetary magnetic field. At the outer boundary of the computational domain $r = R_\infty$ towards the Sun ($\theta > \pi/2$, in the spherical coordinate system with $z \parallel U_\infty$), the parameters of the oncoming plasma flux and field correspond to the interplanetary field, such that $U_\infty = \hat{e}_z U_z$, $B_\infty = \hat{e}_y B_y$, where $\hat{e}$ is the unit vector, while the condition of free escape of particles and field perturbations is used at the tail part of the outer boundary ($\theta < \pi/2$), which permits a sufficiently accurate examination of the formation of the plasma wake. The magnetic field is assumed to be unperturbed at the inner boundary of the computational domain ($r = R_0$). The next condition of radiation of the magnetic field perturbation is used in the tail part of the computational domain

$$\frac{\partial}{\partial t} b + U_n \frac{\partial}{\partial n} b = 0. \tag{12.11}$$

The collisionless particle-mesh method is used in this work to calculate the dynamics of the protons of the solar wind and the ion component of the cometary plasma. The oncoming flux is approximated by a cold beam of protons with a directed velocity U_∞, while the cometary ions are approximated by a beam with a radial velocity W_c. A grid, logarithmic in radius $r = \exp[j/j_{\max} \ln(R_\infty)]$ and uniform over angles θ, in the planes $\phi = 0, \phi = \pi/2$ ($z, x; z, y$) of the spherical coordinate system was used to calculate the hydrodynamic parameters and fields; $j_{\max}$ is the number of nodes of the grid at the outer boundary $r = R_\infty$.

Since in flow problem the zone of strong deformation of the flux is localized to the vicinity of the origin ($r \sim R_0$), it was necessary in the calculations to maintain good statistics in the cells of small size [$\Delta r \sim r \ln(R_\infty)/j_{\max}$]; macroprotons with a weight function depending on the distance from the z-axis were used for this, so that the concentration of macroparticles has a large value in the vicinity of the z-axis and a small value far away from it. To generate the macro-ions, pseudorandom number generators were used, which permit the production of ionized cometary ions to be approximated according to (12.8). The macro-ions were generated with the same weight function, which provided a high concentration in the vicinity of $r \sim R_0$ and, consequently, sufficiently good statistics in the cells of small volume. The number of simulated particles was 10^5 for protons and 10^5 for ions. Each macroparticle had four coordinates: θ, r, V_θ and V_r. Storage of the data was on disk, with unpacking and packing for the arrangement of two coordinates of the particle into one standard byte of disk memory being carried out with the exchange between the external and internal memory. When a particle leaves the calculation region, its place in the memory is occupied by a new particle.

As experience in numerical modeling of three-dimensional plasma flow in the vicinity of the model ionosphere of Venus [318] showed, the averaged hydrodynamic parameters have a mirror symmetry with respect to the z–x and z–y planes. Moreover, although the actual motion of the individual particles is three-dimensional, the axially symmetric approximation of the averaged hydrodynamic flow,

$$\boldsymbol{U} = \boldsymbol{U}(r,\theta), \quad n = n(r,\theta),$$

can be used to solve the induction equation (12.4).

The dependence on ϕ is

$$B_r = b_r \sin\phi, \quad B_\theta = b_\theta \sin\phi, \quad B_\phi = b_\phi \cos\phi, \qquad (12.12)$$

where B_r, B_θ, B_ϕ are the components of the magnetic field in the spherical coordinate system and b_r, b_θ, b_ϕ are the components of the first term of the expansion, which was used to solve the field equations in the first approximation. The induction equation can then be represented in the following form:

$$\frac{\partial b_i}{\partial t} + L_1 b_i - L_2 b_i = f_i, \qquad (12.13)$$

where

$$L_1 b_i = U_k \frac{\partial b_i}{\partial x_k} + b_i \left(\frac{\partial}{\partial x_k} \alpha_k U_k + \beta_i \right)$$

and

$$L_2 b_i = \frac{\partial}{\partial x_k} \left(\gamma_{kj} \frac{\partial}{\partial x_j} b_i \right), \quad b_i = \{b_r, b_\theta, b_\phi\}$$

$\alpha_k, \beta_i, \gamma_{kj}$ are variable coefficients, summation is carried out over the indices, and f_i is the term corresponding to the pressure gradient in (12.2).

The concrete form of the induction equation that we have to solve is the following:

$$\frac{\partial}{\partial t} b_r = -U_r \frac{\partial}{\partial r} b_r - \frac{U_\theta}{r} \frac{\partial}{\partial \theta} b_r + \left(\frac{b_\theta}{r} \frac{\partial}{\partial \theta} U_r - \frac{U_\theta \cos\theta}{r \sin\theta} b_r - \frac{b_r}{r} \frac{\partial}{\partial \theta} U_\theta - \frac{2U_r}{r} b_r \right)$$

$$+ \frac{1}{Re} \left[\Delta_m b_r - \frac{2}{r^2} \left(b_r + \frac{1}{\sin\theta} \frac{\partial}{\partial \theta} (\sin\theta \cdot b_\theta) - \frac{m}{\sin\theta} b_\phi \right) \right], \qquad (12.14)$$

$$\frac{\partial}{\partial t}b_\theta = -U_r \frac{\partial}{\partial r}b_\theta - \frac{U_\theta}{r}\frac{\partial}{\partial \theta}b_\theta$$

$$-\left[\frac{U_\theta}{r}b_r + \frac{U_\theta \cos\theta}{r\sin\theta}b_\theta - b_r\frac{\partial}{\partial r}U_\theta + b_\theta\left(\frac{\partial U_r}{\partial r} + \frac{U_r}{r}\right)\right]$$

$$+\frac{1}{Re}\left[\Delta_m b_\theta + \frac{2}{r^2}\left(\frac{\partial b_r}{\partial \theta} - \frac{b_\theta}{2\sin^2\theta} + \frac{\cos\theta}{\sin^2\theta}b_\phi\right)\right], \quad (12.15)$$

$$\frac{\partial}{\partial t}b_\phi = -U_r\frac{\partial}{\partial r}b_\phi - \frac{U_\theta}{r}\frac{\partial}{\partial \theta}b_\phi - b_\phi\left(\frac{U_r}{r} + \frac{\partial}{\partial r}U_r + \frac{1}{r}\frac{\partial}{\partial \theta}U_\theta\right)$$

$$+\frac{\varepsilon}{M_{S_e}^2 \gamma r}\left[\frac{\partial}{\partial r}\left(\frac{1}{n}\frac{\partial}{\partial \theta}p_e\right) - \frac{\partial}{\partial \theta}\left(\frac{1}{n}\frac{\partial}{\partial r}p_e\right)\right]$$

$$+\frac{\varepsilon}{M_A^2 r}\left(\frac{\partial}{\partial r}r(\boldsymbol{J}\times\boldsymbol{B})_\theta - \frac{\partial}{\partial \theta}(\boldsymbol{J}\times\boldsymbol{B})_r\right)$$

$$+\frac{1}{Re}\left[\Delta_m b_\phi + \frac{2}{r^2\sin\theta}\left(b_r + \frac{b_\theta}{\tan\theta} - \frac{b_\phi}{2\sin\theta}\right)\right], \quad (12.16)$$

where

$$\Delta_m = \frac{1}{r}\frac{\partial}{\partial r}\left(r^2\frac{\partial}{\partial r}\right) + \frac{1}{r^2\sin\theta}\frac{\partial}{\partial \theta}\left(\sin\theta\frac{\partial}{\partial \theta}\right) - \frac{m^2}{r^2\sin^2\theta}.$$

The symmetry condition

$$b_r = \frac{\partial}{\partial \theta}b_\theta = \frac{\partial}{\partial \theta}b_\phi = 0 \quad (12.17)$$

is used on the z-axis.

Two grids (59 × 32), uniform over θ, were used to solve (12.8) together with (12.1–12.7), with the grid for the calculation of n and W shifted by $\Delta t/2, \Delta r/2$ and $\Delta\theta/2$ with respect to the grid for the calculation of B. The Crank–Nicolson method was used to solve (12.8); the difference operators L_1 and L_2 have the approximation $O(\Delta r^2, \Delta \theta^2)$, respectively. The equations of motion of the protons and ions (12.1) were approximated by the implicit scheme with an accuracy $O(\Delta t^2)$:

$$\frac{v_r^{n+1} - v_r^n}{\Delta t} = \frac{v_\theta^{n+1} + v_\theta^n}{2}\frac{v_\theta^{(s)}}{r} + \frac{v_\theta^{n+1} + v_\theta^n}{2}\frac{B_\phi}{\varepsilon \tilde{M}_i} + \frac{E_r}{\varepsilon \tilde{M}_i}, \quad (12.18)$$

$$\frac{v_\theta^{n+1} - v_\theta^n}{\Delta t} = -\frac{v_r^{n+1} + v_r^n}{2}\frac{v_\theta^{(s)}}{r} - \frac{v_r^{n+1} + v_r^n}{2}\frac{B_\phi}{\varepsilon \tilde{M}_i} + \frac{E_\theta}{\varepsilon \tilde{M}_i}, \quad (12.19)$$

$$\frac{r^{n+1} - r^n}{\Delta t} = \frac{v_r^{n+1} + v_r^n}{2}, \quad \frac{\theta^{n+1} - \theta^n}{\Delta t} = \frac{v_\theta^{n+1}/r^{n+1} + v_\theta^n/r^n}{2}, \quad (12.20)$$

where superscripts $n+1$ and n denote the value of v at the new and old time steps, s is the number of the iteration and $\tilde{M}_i = M_i/M$ is the dimensionless mass of the particle.

The algorithm described above was realized by the modification of the program for modeling the flow around Venus [318]. Numerous control calculations, in which both the parameters of the discrete model (number of macroparticles and nodes of the Euler grid) and the shape of the computational domain and boundary conditions at the inner boundary were varied, were made to verify the stability of the results obtained.

12.2.2 Structure of the Region of Mass Loading by Cometary Ions

The calculations were carried out with the input data specified in the following [172, 319]. Parameters of the solar wind: $M_A = M_S = 10, M_{MS} = (M_A^{-2} + M_S^{-2})^{-1/2} = 7, \beta_p = 0, \beta_e = 1, \gamma = 5/3, n_\infty = 5\,\text{cm}^{-3}$, $U_{SW} = 350\,\text{km/s}$, $R_0 = 1L, R_\infty = 10L$ and $L = 5 \times 10^3\,\text{km}$ (the size of computational domain in the z-direction equals $20L$), $Re = 10^2$. Parameters of the cometary plasma: $Q = (3-5) \times 10^{27}\,\text{mol/s}$, $W_c = 1\,\text{km/s}$, $M_i = 23.3M, \nu = 10^{-6}\,\text{s}^{-1}$ and $\varepsilon = 0.7$. The hydrodynamic parameters above correspond to the hydrodynamic calculations in [472]. However, to simplify the calculations, the effective mass of the cometary ion M_i was taken as $5M$, with a simultaneous increase in the evaporation rate of the comet Q by $23.3/5$ times.

General Flow Pattern in the Vicinity of the Comet Let us examine the general pattern obtained with numerical modeling for an evaporation rate of the cometary material of $Q = 5 \times 10^{27}\,\text{mol/s}$. The density distribution of the solar-wind protons in the interaction zone is given in Fig. 12.2a.

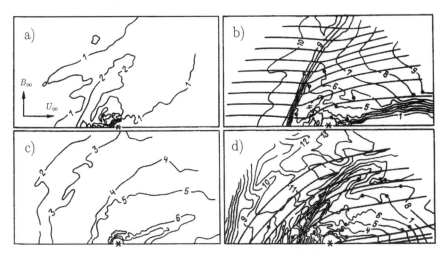

Fig. 12.2. Contour plots (*thin lines*) of the (**a**) solar-wind-proton and (**c**) cometary-ion densities; streamlines (*thick lines*) and contour plots of (**b**) proton and (**d**) cometary-ion velocities. ∗ – position of cometary kernel (from [319])

Here, as in Figs. 12.3a,c,d and Fig. 12.4a,b the numbers on the contour lines indicate the values of the distribution under consideration. A sharp increase in the density at the shock wave front and in the region of the magnetic barrier is seen. The distribution of the proton velocity and streamlines is given in Fig. 12.2b. Here, as in Fig. 12.2d and Fig. 12.3b, the numbers on the contour lines indicate the values of the quantity $10 \cdot |U|$. The crowding of the velocity contour lines corresponds to the position of the shock wave. Behind the shock wave front the streamlines diverge under the action of the pressure of the heavy cometary ions. The density distribution of the cometary ions is given in Fig. 12.2c. For convenience, the numbering of the contour lines is as follows: the ion density, $n_i = 0.001 \times 10^{(j-1)/2}$ for odd j and $n_i = 0.005 \times 10^{j/2-1}$ for even j. It can be seen that the cometary ions are effectively dragged into the region of the tail, forming an extended structure whose characteristic width is determined by the Larmor radius of the cometary ions ϱ_{ci}. The distribution of the velocity contours of the cometary ions and streamlines is given in Fig. 12.2d. Since the cometary ions have a small initial velocity W_c, then, upon entering the moving electromagnetic field coupled with the solar-wind plasma, it acquires a velocity comparable to the undisturbed U_∞. The ion energy $M_i U_\infty^2 / 2$ is significantly greater than the energy of the protons in the oncoming flux, and so the so-called two-phase flow is formed.

The presence of ions with a transverse temperature significantly greater than the temperature along the magnetic field can lead to the development of the hose instability examined in [378], but the examination of such effects with consideration of ion flow along the field lines can be carried out only in the completely three-dimensional flow model. Thanks to the large Larmor radius of the cometary ions $\varrho_{ci}/L = 3.5$ determining the characteristic velocity relaxation length, a difference in the streamlines of the cometary ions and the solar-wind protons is observed at the head of the coma of the comet. The ion streamlines directly on the flow axis are practically orthogonal to the proton streamlines.

The distribution of the total plasma density and the hydrodynamic distribution of the contours and streamlines are given in Fig. 12.3a,b. We recall that the hydrodynamic parameter (density and mass-averaged [bulk] velocity) are defined in our problem in the following manner:

$$\varrho = M n_p + M_i n_i \quad \text{and} \quad \boldsymbol{U} = (\boldsymbol{U}_p M n_p + \boldsymbol{U}_i M_i n_i)/\varrho. \tag{12.21}$$

12.2.3 Induced Magnetosphere, Bow Wave and Magnetic Barrier

The distribution of the magnitude and configuration of the magnetic field lines (in the vertical plane) are given in Fig. 12.3c,d.

The field discontinuity at the shock wave front and the field enhancement in the region of formation of the magnetic barrier (vicinity of a spherical

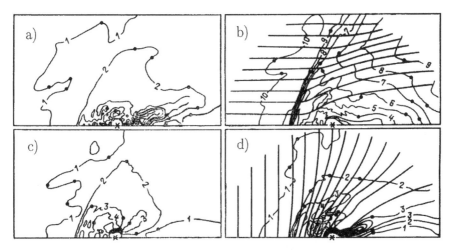

Fig. 12.3. Contour plots (*thin lines*) of (**a**) hydrodynamic density, (**b**) magnitude of hydrodynamic velocity, (**c**) magnetic field in x–z plane and (**d**) magnetic field in y–z plane. *Thick lines* in (**b**) and (**d**): streamlines and magnetic field lines. ∗ – position of cometary kernel (from [319])

obstacle of finite conductivity) are clearly seen. The deformation of the magnetic field lines at the shock wave front, the "linkage" of the field lines in the magnetic barrier region and the elongation of the field lines along the tail of the induced magnetosphere can be seen in Fig. 12.3d.

The magnitude of the magnetic field in the tail region is $2B_\infty$ in the horizontal and $(2-3)B_\infty$ in the vertical. The configuration of the field lines is on the whole in qualitative agreement with the results of the laboratory simulation in [426].

The distribution of the Alfvén and magnetosonic Mach numbers is given in Fig. 12.4a,b. We recall that the magnetosonic Mach number is calculated from the Alfvén and sound Mach numbers by $M_{\mathrm{MS}} = (M_{\mathrm{A}}^{-2} + M_{\mathrm{S}}^{-2})^{-1/2}$. In our system M_{S} is determined by the total pressure of the solar-wind electrons p_{e}, protons $p_{\mathrm{p}} = n_{\mathrm{p}} M \langle (\boldsymbol{v}_{\mathrm{p}} - \boldsymbol{U})^2 \rangle / 2$ and ions $p_{\mathrm{i}} = n_{\mathrm{i}} M_{\mathrm{i}} \langle (\boldsymbol{v}_{\mathrm{i}} - \boldsymbol{U})^2 \rangle / 2$. The

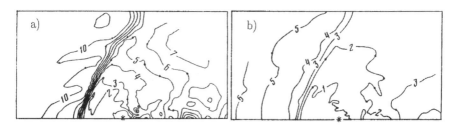

Fig. 12.4. Contour plots of (**a**) Alfvén and (**b**) magnetosonic Mach numbers (from [319])

sharp drop in the value of the Mach number near the shock wave front is seen in the figure. A transition region with $M_{MS}, M_A < 1$ is located behind the shock wave front, and the sonic line ($M_{MS} = 1$) is seen at its flanks. In the system under consideration, the distance from the bow shock to the comet is 1.6 times greater than the corresponding value in MHD models [472]. This elevated value is related to three factors: (a) the replacement of the actual three-dimensional flow by a two-dimensional flow; (b) the insufficiently small size of the inner boundary of the calculation region; and (c) the effect of the large Larmor radius of the cometary ion.

Structure of the External Shock Wave and Transition Region The distributions of the primary parameters along the z-axis (spherical coordinate system) parallel to the undisturbed velocity vector U_∞ were constructed to analyze the physical processes accompanying the flow around the comet. The results of the calculations of the system examined above with the molecular evaporation rate $Q = 5 \times 10^{27}$ mol/s are shown in Fig. 12.5.

The structure of the shock front and the transition region changes significantly when the solar wind is loaded by the cometary ions. The distributions of the parameters of the solar wind and cometary ions are given in Fig. 12.5a,b. These distributions (for example, n_p, B, M_{MS} and M_A) show that the thickness of the shock front is already determined by the Larmor

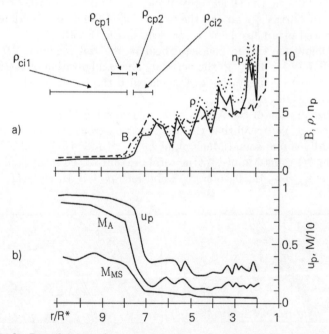

Fig. 12.5a,b. Distribution of parameters along the z-axis with cometary ion loading; $\varrho_{cp1}, \varrho_{ci1}$: Larmor radius of protons and cometary ions in magnetic field B_∞, calculated from velocity U_∞ (from [319])

radius of the hot cometary ions ϱ_{ci2}, calculated from the directed velocity and average magnetic field directly behind the shock wave. The structure of the transition region is also characterized by oscillations of the parameters (see Fig. 12.5a,b) with a characteristic spatial scale of ϱ_{ci2}. With a decrease in the evaporation rate of cometary material to $Q = 3 \times 10^{27}$ mol/s, the size of the transition region and, consequently, the number of oscillations of the parameters decrease, and their distortion under the action of the magnetic barrier increases.

The specifics of the shock wave and transition region observed in the calculations are typical of comet-like objects. For a relatively small evaporation rate from the surface of the comet, $Q < 10^{33}$ mol/s, the position of the shock wave front is determined approximately by the radius of critical loading, $R_{Sh} > R_{Sr}$, where R_{Sr} is the radius of critical loading for which the plasma density reaches the critical value $(\varrho_0 + \Delta\varrho)/\varrho_0 \approx 1 + (\Delta\varrho/\varrho_0)_{cr} = \gamma^2/(\gamma^2 - 1)$; the magnetosonic Mach number $M_{MS} = 1$ according to [49], while the Mach number at the shock wave $M_{MS} \approx 2$. The relative density of the heavy ions at the shock front $\Delta\varrho/\varrho < (\Delta\varrho/\varrho_0)_{cr} \approx 1/(\gamma^2 - 1) \approx 0.3$, and the transverse velocity $U_\perp \approx U_\infty$. For large values of the evaporation rate, $Q > 10^{33}$ mol/s, the position of the shock wave front is controlled by the obstacle (in this case, the discontinuity surface at which there is a balance between the pressures of the solar wind $\varrho_\infty U_\infty^2$ and the cometary ions $\varrho_i U_i^2$) and the Mach number at the shock wave is large ($M_{MS} \gg 1$), while the relative content of the energetic cometary ions is small [$\Delta\varrho/\varrho \ll (\Delta\varrho/\varrho)_{cr}$]. In this limiting case, the thickness of the shock wave and the structure of the transition region will evidently be determined only by the gyroradius of the solar-wind protons.

Of course, in a real situation, for example, in the case of comet Halley, the composition of the cometary material has a wide spectrum of components, and, consequently, one can expect the appearance of more complex structures of shock fronts, but the defining characteristics (width of the shock front and scale of the oscillations) will evidently be determined by the average mass of the cometary ions:

$$M_i = \sum_{j=1}^{N_s} n_j M_j / \sum_{j=1}^{N_s} n_j, \tag{12.22}$$

where N_s is the number of ion types in the simulation. One should also note the presence of cometary ions and protons with velocities of $(2-4)U_\infty$, accelerated at the shock wave front and carried by the plasma flow into the tail region.

12.3 Interaction of the Solar Wind with Weak Comets and Related Objects

12.3.1 Formulation of the Problem and Mathematical Model

Bi-ion fluid simulations [56, 467] have demonstrated several distinct features which occur in mass-loading zones associated with the interaction of the solar wind with comets whose rate of gas production is small, specifically, bi-ion magneto-acoustic structuring and pulsations. Because of the limited applicability of the bi-ion fluid model, hybrid codes are used in order to check the results. The simulations were performed for gas production rates $5 \times 10^{24}\,\text{mol/s} \leq Q \leq 10^{28}\,\text{mol/s}$ and for interplanetary magnetic field oriented perpendicular to the simulation plane. In the case of $Q \leq 8.4 \times 10^{26}\,\text{mol/s}$ the cometary atmosphere forms a strong cycloid-type tail. When $Q \geq 10^{27}\,\text{mol/s}$, the cometary atmosphere forms a cone-type tail and structuring of the coma occurs. The results of these simulations apply to other weak solar-wind mass-loading sources, e.g., the active magnetospheric particle tracer explorer (AMPTE) releases [411], possibly Phobos and Deimos and possibly even Pluto. The results presented here may be important for studies of the dynamics of the ionized environment near the "Solar Probe" spacecraft in the future.

The observations of comets Halley, Giacobini–Zinner and Grigg–Skjellerup have demonstrated that the transition region between a comet and the solar wind is distinct from that in the planetary case. This region has been shown to be dominated by turbulence, mainly at MHD frequencies corresponding to the gyrofrequencies of the water-group ions. Consequently, the magnetic moments of the cometary ions are no longer adiabatic invariants, and their interaction with the solar-wind plasma is essentially kinetic [175, 176, 325, 338, 339, 407, 468, 530]. On the other hand, planetary bow shocks are collisionless in nature, and the kinetic behavior of the ions is again an essential feature. In both the cometary and planetary cases, it is thus essential to incorporate the plasma kinetic behavior into any study, because of its leading role in dissipation and acceleration processes. The need for a kinetic description is further accentuated in the cometary case by the large gyroradius of the cometary ions. Hybrid simulation of the interaction of the comet Shoemaker–Levy 9 with Jupiter [340, 341] has also demonstrated the importance of the kinetic description of solar-wind protons, cometary ions and dust. Thus, kinetic simulation remains very important for the study of the interaction of the solar wind with weak comets (Rosetta mission) and related objects.

Hybrid simulations of the interaction of the solar wind with gas-producing bodies have been performed using a 2.5-dimensional hybrid code, which was adapted to a Cartesian system of coordinates [176]. In the case of weak comets, the morphology of the interaction (shock, tail, etc.) is in principle three-dimensional. However, we can approximate the three-dimensional structure of the interaction zone by 2.5-dimensional simulations using different orientations of the interplanetary magnetic field. Initially, the computational domain only contains the supersonic solar wind. The ionized component of heavy cometary particles is described by the following distribution function: $G = Q\nu e^{-\nu R/W_c}/4\pi(r^2 + r_0^2)W_c$, where Q is the rate of evaporation of gas from the surface of the comet, ν is the rate of ionization of cometary moleculars, r_0 is the radius of cometary nucleus and W_c is the initial velocity of cometary molecules. In all present simulations the interplanetary electromagnetic field is directed perpendicular to the bulk velocity of the solar wind [338, 339]. Since the proton and heavy ion gyroradii must be resolved, we require a grid-point spacing less than 1 gyroradii in order to avoid numerical dispersion and dissipation. On the other hand, we also require good statistics and, therefore, a sufficiently large number of particles per cell (i.e., low "shot" noise). For this reason a homogeneous mesh is superior to an adaptive mesh, because it is easily to satisfy both conditions mentioned above. Our computational domain had dimensions of 201×201 (in 2.5-dimensional simulations) and $101 \times 51 \times 101$ (in three-dimensional simulations) grid points and we used $4 \times 10^5 - 10 \times 10^6$ macroparticles for both the solar-wind protons and the heavy ions.

12.3.2 Interaction of the Solar Wind with Very Weak Comets

12.3.2.1 Test Particle Regime of the Heavy Ion Dynamics.
Let us consider the results of hybrid simulation of the interaction of the solar wind with very weak comets [338]. The simulations were performed for the following solar-wind parameters in a near earth orbit: $M_A = 5, \beta_p = 0.5, \beta_e = 2, B = 5\,\text{nT}, n_{SW} = 5\,\text{cm}^{-3}, U_{SW} = 350\,\text{km/s}, W_c = 1\,\text{km/s}, Q = 8.4 \times 10^{26}\,\text{mol/s}$ and $r_0 = 10\,\text{km}$. The heavy-ion to solar-wind-proton mass ratio was $M_i/M = 10$. The characteristic scale $L = c/\omega_{pi} = 140\,\text{km}$, the proton (heavy ion) gyroradius $\varrho_{cp} = U_{SW}/\Omega_p = M_A L = 700\,\text{km}$ ($\rho_{ci} = 7000\,\text{km}$), the effective radius of the cometary atmosphere $R_{eff} = 5L$, and the Reynolds number Re = 20. Figures 12.6, 12.7 and 12.8 show the two-dimensional sections of the cometary density, solar-wind density and magnetic field profiles produced in three-dimensional simulations. In this case, in the region outside the vicinity of the nucleus, the solar wind is not strongly perturbed and the dynamics of the heavy ions is the same as if they were test particles, i.e., they move approximately along a cycloidal path. The global picture of the interaction is fully asymmetrical. In the region below the solar-wind–comet line, there is a Mach cone which is not influenced by the heavy ions. In contrast, in the region above the solar-wind line, we see significant mass loading of the solar wind by the heavy ions in the part of the cometary tail where the total mass injected along a streamline has a maximum. The jumps of proton density and magnetic field at the flanks are $n_{p2}/n_{p1} = 1.4 - 1.5$ and $B_2/B_1 = 1.5$ in the upper portion of the figure

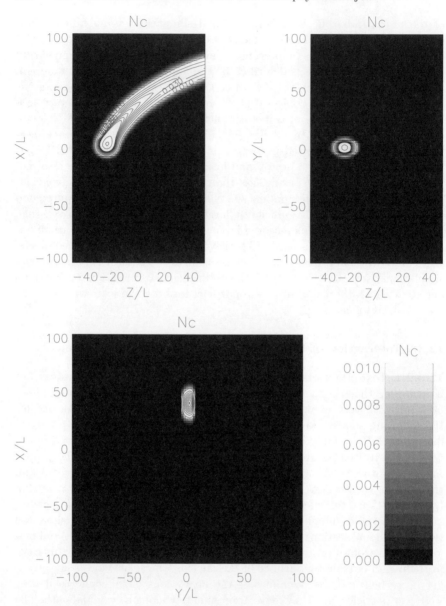

Fig. 12.6. Two-dimensional section of the cometary ion density profile. $Q = 8.4 \times 10^{26}$ mol/s, $M_i/M = 10$, $R_{\text{eff}} = 5L$ (from [338])

and $n_{p2}/n_{p1} = 2.5 - 1.9$ and $B_2/B_1 = 1.9$ in the lower portion. The structure of the Mach cone transition region in the lower part of the figure is approximately planar, with a maximum jump in the vicinity of the cometary nucleus.

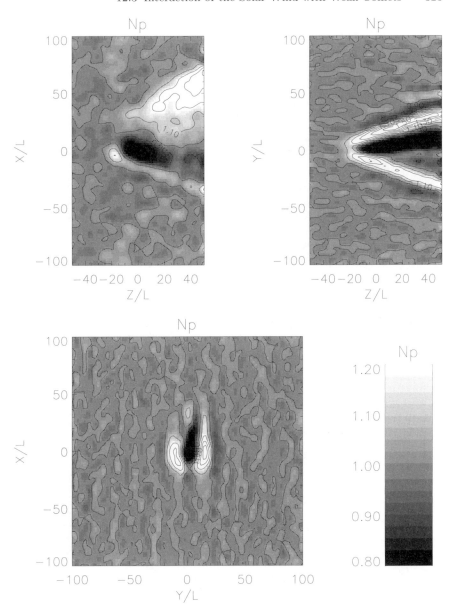

Fig. 12.7. Two-dimensional section of the solar wind density profile. $Q = 8.4 \times 10^{26}$ mol/s, $M_i/M = 10$, $R_{\text{eff}} = 5L$ (from [338])

12.3.2.2 Splitting of the Cometary Tail Structure. The structure and orientation of the cometary tail depends upon the heavy-ion production rate, the gyroradii of the protons ϱ_{cp} and of the heavy ions ϱ_{ci}, and the mass ratio M_i/M. In the following simulations we increased the gas production

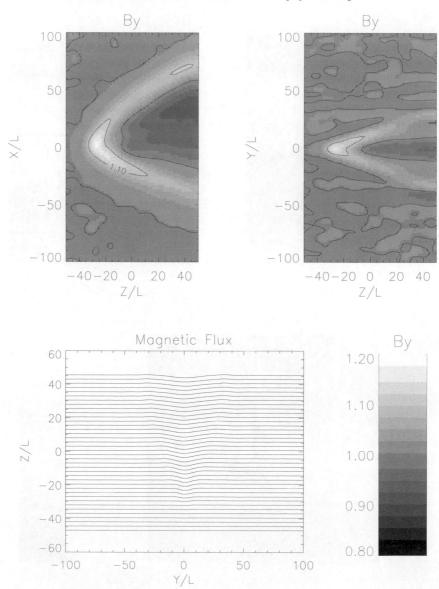

Fig. 12.8. Two-dimensional section of the magnetic field component B_y profile and field lines in y–z plane. $Q = 8.4 \times 10^{26}$ mol/s, $M_i/M = 10$, $R_{\text{eff}} = 5L$ (from [338])

rate to $Q = 2.5 \times 10^{27}$ mol/s, but otherwise they are the same as those discussed in the preceding discussion. In this case, a more symmetrical Mach cone is formed, but the primary difference is the appearance of a split in the cometary tail structure. Figures 12.9, 12.10 and 12.11 show the two-

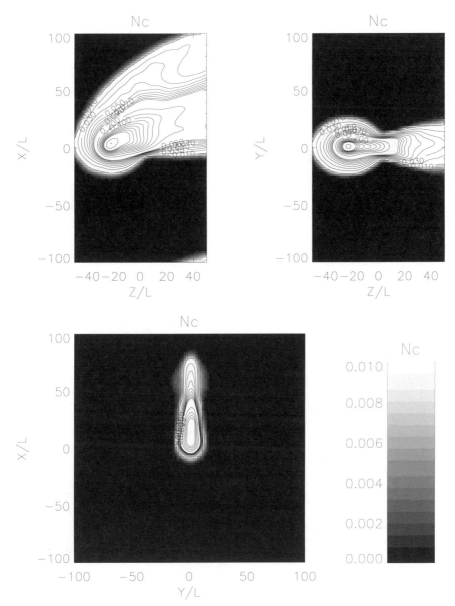

Fig. 12.9. Two-dimensional section of the cometary ion density profile. $Q = 2.5 \times 10^{27}$ mol/s, $M_i/M = 10$, $R_{\text{eff}} = 25L$ (from [338])

dimensional sections of the cometary density, solar-wind density and magnetic field profiles produced in three-dimensional simulations for the parameters of the cometary atmosphere.

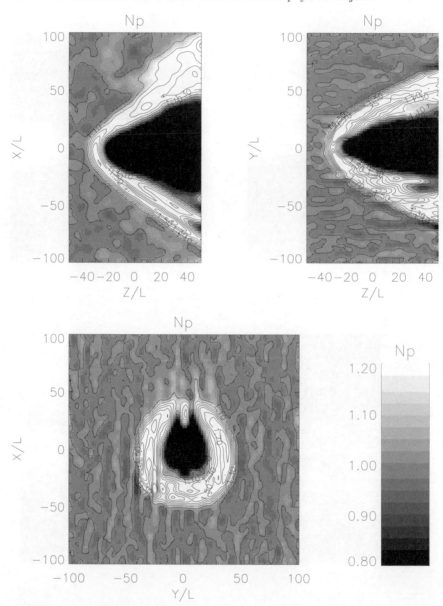

Fig. 12.10. Two-dimensional section of the solar wind density profile. $Q = 2.5 \times 10^{27}$ mol/s, $M_i/M = 10$, $R_{\text{eff}} = 25L$ (from [338])

We also present also the result of a 2.5-dimensional simulation with a very high grid resolution to demonstrate the fine structures in the cometary tail (Fig. 12.12). Figure 12.12 shows three heavy-ion density structures in the tail. In this regime we have a strong deceleration of the bulk velocity and

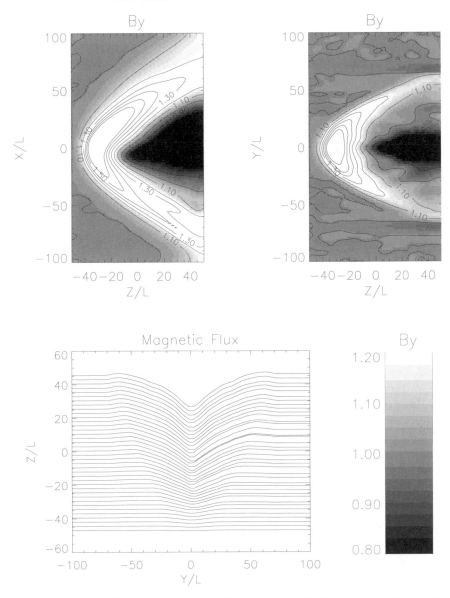

Fig. 12.11. Two-dimensional section of the magnetic field component B_y profile and field lines in y–z plane. $Q = 2.5 \times 10^{27}$ mol/s, $M_i/M = 10$, $R_{\text{eff}} = 25L$ (from [338])

a decrease in the inductive electric field, because of significant mass loading outside the nucleus. The electrostatic field has a strong radial component. A portion of the heavy ions, from a lobe (subsolar) region outside the nucleus, move as test particles, and they form a weak wake oblique to the solar-wind

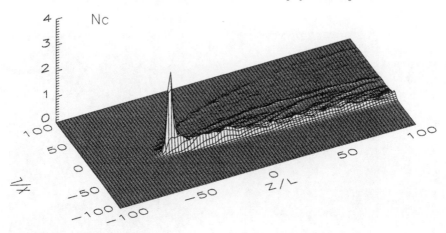

Fig. 12.12. Distribution of cometary ion density in the 2.5-dimensional simulation, $Q = 1.6 \times 10^{26}$ mol/s, $M_i/M = 10$, $R_{\text{eff}} = 30L$

flow (Fig. 12.12). Another amount of the heavy ions, from both flank regions outside the nucleus, moves initially along the radial direction, and then they drift in the direction slightly oblique to the solar-wind flow, accompanying gyrorotation with a small gyroradius because of the weak inductive electric field. The jump in the magnetic field at the upper flank of the Mach cone is larger than in the previous case.

The heavy cometary ion tail has an asymmetric structure, and its direction is slightly oblique to the solar-wind–cometary-nucleus line. The strong fluctuations in the density contour plot, with a wavelength of the order of ϱ_{ci}, may result from both the incomplete phase mixing of the heavy ions and a Kelvin–Helmholtz instability at the boundary between the solar-wind protons and the heavy ions.

Let us consider the dependence of the tail-splitting structure on the gas production rate using one-dimensional section ($z = DZ/4$). For $\bar{Q} \ll 1.0$ ($\bar{Q} = Q \times 10^{-26}$ s/mol), the cometary heavy ions form a thin tail which follows the test-particle trajectory. For a high gas production tail, $\bar{Q} \approx 1.0$, the cometary ions form a cone-like tail with a significant halo that corresponds to the test-particle dynamics (Fig. 12.13a). With a gas production rate of $1.3 < \bar{Q} < 2.0$, the cometary tail includes a strong double-tail structure with two maxima in the ion density profile at points $x = 22.2$ and $x = 6.94$ and a weak oblique wake with a maximum in ion density profile at the point $x = 69.4$ (Fig. 12.13b–c). It is very important that all three peaks in the ion density profile are stable in location for the gas production range $1.3 < \bar{Q} < 2.0$. When the gas production rate increases from 1.3 to 2.0, the upper maximum of the double-tail structure decreases, whereas the bottom maximum increases (Fig. 12.13b–c). In the region inside the double-tail structure, there are strong pulsations of the density due to a phase mixing

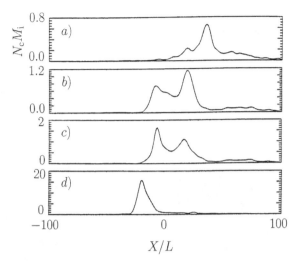

Fig. 12.13. One-dimensional sections of the cometary ion density in the $x(z = D_z/4)$ direction

of the ion rotational motion. Figure 12.4 demonstrates these pulsations in the cometary density profile. The thin structure of this double-tail system includes additional beams of heavy ions, the sources of which are distributed inside the bottom tail.

12.3.3 Interaction of the Solar Wind with Weak Comets

In the case of relatively high gas production ($Q \geq 5 \times 10^{27}$ mol/s), the zone of strong mass loading of the solar wind by heavy cometary ions is much larger than in the previous cases (note that all other parameters are the same as in the previous cases) (Figs. 12.14, 12.15 and 12.16). The bulk velocity in the vicinity of the nucleus decreases significantly, thus decreasing the gyroradius of the cometary ions. The heavy ions form a wide region of enhanced density (Fig. 12.14). The high-density heavy-ion tail has an asymmetric structure, and its direction is slightly oblique to the solar-wind–cometary-nucleus line. The strong fluctuations in the density contour plot, with a wavelength of the order of ϱ_{ci}, may result from both the incomplete phase mixing of the heavy ions and a Kelvin–Helmholtz instability at the boundary between the solar-wind protons and the heavy ions. The cometary bow shock (Figs. 12.15 and 12.16) is located at a distance of $1\varrho_{ci}$ and has a quasisymmetric structure.

12.3.4 Interaction of the Solar Wind with Pluto

The characteristic size of the interaction region between the solar wind and Pluto is also much smaller than the pick-up radius of the heavy planetary ions.

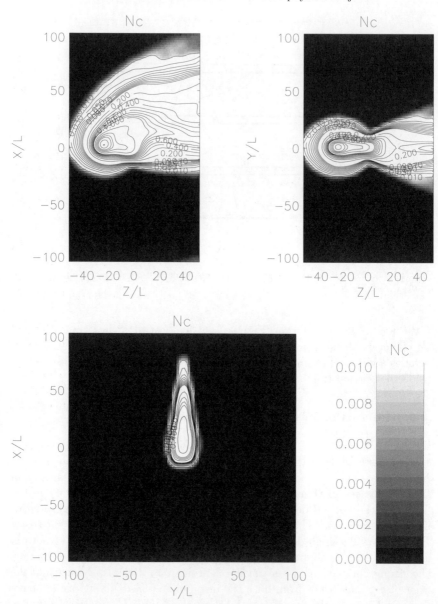

Fig. 12.14. Two-dimensional section of the cometary ion density profile. $Q = 5 \times 10^{27}$ mol/s, $M_i/M = 10$, $R_{\text{eff}} = 25L$ (from [338])

This is due to the weak interplanetary magnetic field ($B \approx 0.1\,\text{nT}$) and the strongly reduced ionization rate of planetary ions ($\nu \approx 10^{-9}\,\text{s}^{-1}$) at such large distances from the Sun ($\approx 30\,\text{AU}$) [22]. In the simulations the parameters of the solar wind and planetary ions are assumed to be [338, 339]: $M_A = 18, \beta_p =$

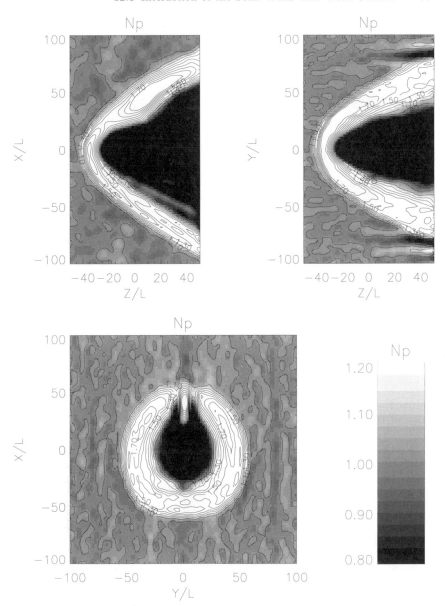

Fig. 12.15. Two-dimensional section of the solar wind density profile. $Q = 5 \times 10^{27}$ mol/s, $M_i/M = 10$, $R_{\text{eff}} = 25L$ (from [338])

$0.5, \beta_e = 1, B = 0.1\,\text{nT}, n_{\text{SW}} = 0.01\,\text{cm}^{-3}, W_{\text{SW}} = 400\,\text{km/s}, L = 2300\,\text{km}, M_i/M = 10, Q = 10^{27} - 5 \times 10^{28}\,\text{mol/s}, r_0 = 1000\,\text{km}, W_c = 1\,\text{km/s}$ and $\varrho_{\text{cp}} = M_A L$. In the case of low gas production ($10^{27}\,\text{mol/s} \leq Q \leq 10^{28}\,\text{mol/s}$), the outermost planetary ions are forced to move approximately in cycloidal

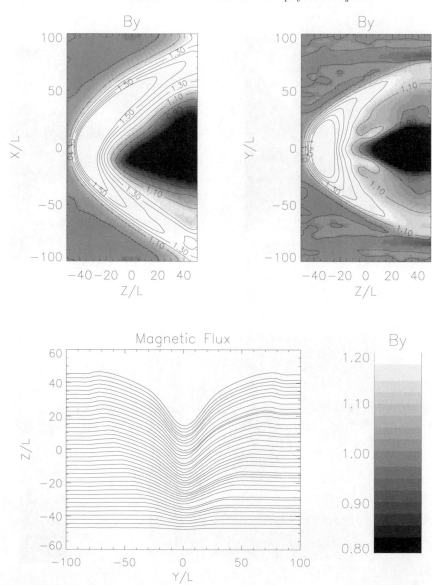

Fig. 12.16. Two-dimensional section of the magnetic field component B_y profile and field lines in y–z plane. $Q = 5 \times 10^{27}$ mol/s, $M_i/M = 10$, $R_{\text{eff}} = 25L$ (from [338])

orbits. The typical distribution of the plasma parameters are closed to the test-particle regime (Figs. 12.9, 12.10 and 12.11). The heavy-ion distribution is nongyrotropic in accordance with test-particle calculations [267]. The Mach cone has an asymmetric structure, and it possesses a flank an the solar-wind–Pluto lines (Figs. 12.10 and 12.11). The jumps in proton density and magnetic field are $n_{p2}/n_{p1} = 3 - 3.7$ and $B_2/B_1 = 2.5 - 2.8$. In the region above the solar-wind–Pluto line, the proton density profile is not significantly perturbed (Fig. 12.10). The magnetic field profile increases by about 10% at the forward part of the front above the solar-wind–Pluto line, with a 30% decrease in the vicinity of the solar-wind–Pluto line (Fig. 12.11). A detached bow shock is not formed in this case. In simulations with high gas production rate ($Q = 5 \times 10^{28}$ mol/s), a quasisymmetric strong detached bow shock is formed. Analysis of the electric field perturbation shows that upstream of the cometary bow shock the Alfvén and magnetosonic waves and whistler are generated in accordance with the study of the plasma cloud dynamics (see Figs. 11.16–11.17).

12.4 Interaction of the Solar Wind with Venus

12.4.1 Formulation of the Problem and Mathematical Model

On the basis of magnetic field and plasma measurements in the vicinity of Mars [134, 205, 533] and Venus [135, 204, 533], the presence of collisionless bow shock waves near the planets, a weak magnetic field near Mars of $\sim 50\,\gamma$ and no appreciable field near Venus were established.

Gas dynamical models of streamline flow in the atmospheres of Mars and Venus were developed in [38, 498]. It was shown that the outer boundary of the atmosphere consists of a tangential discontinuity and that the drag pressure in the transition zone is in equilibrium with the atmospheric pressure. Also well known are quasihydrodynamic intensity estimates for the induced magnetic field associated with diamagnetic photo-ions and conduction in the ionosphere, neglecting the back influence of the field on flow in the transition zone [100].

Direct numerical simulation of the interaction of the solar-wind plasma with the ionosphere [315, 316] has shown that for subsonic streamline flow an induced magnetosphere may exist. However, subsequent calculations for the supersonic case showed that the plasma and solar-wind flows were locked at the ionosphere [318]. This fact showed that it is necessary to allow for interaction processes at the ionopause in the general picture of streaming flow.

Due to the conditions $\varrho_{ce} \ll \delta_i \sim \varrho_{ci}$ and $\Omega_e \gg (d/dt) \ln B$, in the solar wind, electrons will be described by the equations of MHD and ions by a Vlasov equation (see Sect. 12.2.1). Here ϱ_{ce} and ϱ_{ci} are the electron and ion Larmor radii, Ω_e and Ω_i are the gyrofrequencies, and δ_i is the depth of the ionosphere.

Since in the present section we are mainly interested in processes in the shock transition layer and in the upper ionosphere, we consider a simplified model of the ionosphere:

- the charged particle distribution in the ionosphere is stationary;
- there is no ionospheric convection in the first approximation;
- "breakaway" of the magnetic field from the ionosphere is determined by the effective conductivity, such that $Re_i = 10 - 10^5 \ll Re_i^*$ (Re_i^* is the Reynolds number with Coulomb collisions);
- the ionosphere has the form of a spherical shell of thickness δ_i.

In this approximation, the generalized Ohm's law takes the form

$$\boldsymbol{E} = -\frac{\varepsilon}{n_{\text{isp}}} \boldsymbol{J} \times \boldsymbol{B} - \frac{\epsilon}{M_{\text{Se}}^2 \gamma n_e} \nabla p_e + l_{\text{d,i}}^* \boldsymbol{J}, \tag{12.23}$$

where n_{isp} and p_e are the number density and electron pressure in the ionosphere.

It is assumed that at $t = 0$ the ionosphere is immersed in a homogeneous interplanetary magnetic field and that the plasma parameters in the vicinity of the ionosphere are also unperturbed. At the outer boundary of the calculation zone, $r = R_k$, the parameters of the incoming plasma flux and field have their corresponding interplanetary values, with $U_\infty \perp B_\infty$. Homogeneous second-order boundary conditions are obtained for the magnetic field at the boundary of the computational domain near the terminator ($\theta = \pi/2$). The magnetic field is considered to be unperturbed at the inner boundary of the ionosphere $r = 1 - \delta_i$ (Fig. 12.17, dashed line).

In this section we employed the CIC ("cloud in cell") or PM ("particle-mesh") method to calculate the solar-wind ion dynamics. The incoming flux is approximated by a cold ion beam. Since the field equations (Sect. 12.2.1) are being solved in the planes $\phi = 0, \pi(z, x; z, y)$, it turns out to be advisable to inject the plasma in the vicinity of these planes. Such an initial beam distribution provides a full approximation to the beam parameters in the planes $\phi = 0, \pi/2$ with a minimal number of macroparticles $N = 100 \times 15 \times 7$ [317]. The solution of the field equations to the first approximation uses the ϕ dependence as (12.12). The solution of (12.1), (12.2) and (12.4) together with (12.8) and (12.10) uses two uniform grids in r and θ (31×31), the grid used in calculating n and $\boldsymbol{U}$ being shifted by $\Delta t/2$, $\Delta r/2$ and $\Delta\theta/2$ with respect to that for $\boldsymbol{B}$. Use is made of the symmetry condition $b_r = \partial b_\theta/\partial\theta = 0, b_\phi = b_\theta$ along the z-axis. Equations (12.13) are solved by Crank–Nicolson method, where the approximation errors to operators L_1 and L_2 are $O(\Delta r, \Delta\theta)$ and $O(\Delta r^2, \Delta\theta^2)$, respectively. The equation for ion motion is approximated by an implicit method with accuracy $O(\Delta t^2)$.

12.4.2 Results and Conclusions

Calculations were performed for the following parameters of the incoming flux and ionosphere [318]: $M_A = M_S = 10, U_\infty/\Omega_i L = 0.086, \gamma = 5/3, R_k = 1.35, n_i = 10^2 - 10^4, \delta_i = 0.1$ and $Re_i = 10 - 10^5$. As a result of the encounter between the plasma flux and the ionosphere, electrons and ions penetrate to

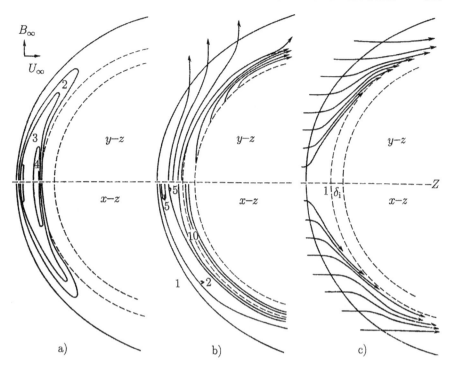

Fig. 12.17. (a) Density distribution in the y–z and x–z planes; (b) magnetic lines of force in the y–z plane and level lines in the x–z plane and (c) streamlines in the y–z and x–z planes. The numbers give the integer part of the magnitude of the density and magnetic field. (From [318])

a depth of the order of the gyroradius ϱ_{ce}, ϱ_{ci}, so that the electric field in the ionosphere is small $\boldsymbol{E}_i \ll \boldsymbol{B}_i$.

Since in the initial magnetic field the dimensionless ion Larmor radius $\varrho_{ci}/L = U_\infty/\Omega_i L = 0.086$ is comparable to the depth of the ionosphere δ_i, the ions which constitute the flux layer fill the ionosphere. The electrons, with gyroradius ϱ_{ce} three orders of magnitude smaller than ϱ_{ci}, practically fail to penetrate the ionosphere. The ions penetrating the ionosphere transfer momentum to those within the ionosphere in a layer of depth $l \sim \varrho_{ci}$. The finite ionic Larmor radius gives rise to a viscous interaction between the transition zone plasma and the ionosphere. This effect essentially distinguishes the interaction model being considered from gas dynamical models [38, 498].

Let us consider in detail the streamline plot for the conditions $Re_i = 10^3$ and $n_i = 10^2$ at time $t = 0.85$. Figure 11.18a shows the solar-wind plasma density distribution in the transition region and ionosphere for the equatorial and meridional planes. The sharp rise in density immediately behind the shock front is apparent. The density perturbation decreases as $\theta \to \pi/2$ far from z-axis.

Figure 12.17b gives the distribution of the magnetic field magnitude in the equatorial plane and the configuration of the magnetic lines of force in the meridional plane. Immediately behind the shock front, the magnetic field increases somewhat, just as the density does. Near the ionopause, crowding of the lines of force is also observed, the plasma is no longer frozen in and the magnetic field does not correspond to the density. Finite conductivity causes diffusion of the magnetic field into the ionosphere. The characteristic stagnation zone width with a strong magnetic field equals the ion gyroradius in the transition zone. Growth of the transverse component of the magnetic field behind the shock front and the ionopause is also observed in laboratory [12] and in space [135] experiments.

Plasma streamlines in the equatorial and meridional planes are shown in Fig. 12.17c. It is apparent that the streamlines pass around the strong magnetic field zone. In the regions of comparatively weak magnetic field, the plasma flows into the ionopause, partially dragging the interplanetary field behind it. Calculated results show an asymmetry in the streamline plots for the equatorial and meridional planes. Moreover, due to the influence of the finite ion Larmor radius, a shift of the zero flow line in the equatorial plane is observed.

Maximum magnetic field strengths at the ionopause of $B_i = 2, 4, 10$ and 12 correspond to regions in the ionosphere with magnetic Reynolds number $Re_i = 10, 10^2, 2 \times 10^2$ and 10^3. It is apparent that as Re_i decreases magnetic field diffusion through the ionosphere increases, while pumping of the field at the ionopause is also approximated as $V_i \approx Re_i/\delta$ (where δ is a characteristic diffusion scale).

Figure 12.18 shows the variation with time of the width of the transition zone at the subsolar point L_{trans}, as a function of Re_i. The calculations showed the delay in the formation of particle reflection and of a shock wave resulting from strong field diffusion in the ionosphere, and, consequently, of a larger ion Larmor radius. The width of the transition zone in our model is only very slightly larger than the values obtained from gas dynamic models [38, 498]. This result is basically connected with not assuming that the ionopause is a tangential discontinuity, and consequently, with the enhancement of viscosity effects.

In this section we used the ionosphere with effective conductivity to simulate the mass loading of the solar wind by planetary pickup ions. This model provides the study of some main effects: formation of the bow shock and of the induced magnetosphere and magnetic barrier. To study the wave generation and pickup dynamics one has to use bi-ion hybrid models. Application of such models to the solar-wind–Venus and solar-wind–Mars interactions was done in [64, 65]. However, we do not consider such simulations here.

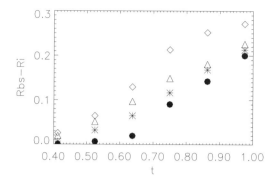

Fig. 12.18. Variation in the width $L_{bs} - R_i$ of the transition zone as a function of time for different ionospheric Reynolds numbers: (•) $Re_i = 10$; (∗) $Re_i = 10^2$; (△) $Re_i = 2 \times 10^2$; (◇) $Re_i = 10^3$ (from [318])

12.5 Interaction of the Solar Wind with the Moon

Measurements of the structure of the interplanetary magnetic field and plasma in the vicinity of the Moon, taken by Explorer 35 [390, 489], have established that in the interaction of the solar wind with the Moon no bow shock wave is formed and there is a cavity free of plasma on the night side. The magnetic field in the cavity is greater than the unperturbed interplanetary field (see Fig. 1.3). Further investigation of the solar-wind flow around the Moon was continued partially by the spacecrafts WIND (solar-wind plasma measurements) and Lunar Prospector. This observation data allow us to further study the plasma and field in the day-side environment and the dynamics processes in the plasma wake.

Theoretical studies of a stationary, two-dimensional interaction between the solar wind and a nonconducting lunar model are well known. In [496, 577], the equations of gas dynamics were used to describe the solar wind, whereas in [551, 552] a guiding-center model was used to describe the solar wind protons. It was assumed that the guiding centers move along rectilinear trajectories.

The recent observations and theoretical models indicate that electrostatic instabilities may play an important role in plasma dynamics in the wake (see, e.g., [153]). The kinetic instabilities which are associated with a nonuniform distribution of ions may result in plasma density and electromagnetic perturbations in the wake. However, in this section we shall consider the quasineutral and quasistationary model only [313, 314].

According to [492], the Moon has a conducting core ($\sigma = 10^{-3}\,\Omega^{-1}\,\mathrm{m}^{-1}$), surrounded by a poorly conducting layer ($\sigma = 10^{-7}\,\Omega^{-1}\,\mathrm{m}^{-1}$). Hence, it is useful to study the influence of the conducting core of the Moon on the structure of the wake.

12.5.1 Formulation of the Problem and Mathematical Model

On the day-time part of the lunar surface, the plasma particles are absorbed, and the perturbation region forms a thin boundary layer of thickness $\delta \sim c/\omega_{\rm pe}$ [391]. On the night-time side the perturbation region is bounded by a surface of weak perturbations forming a cone with half-apex angle $\sin^2\theta = 1/M_{\rm A}^2 + 1/M_{\rm S}^2$ ($M_{\rm A}$ and $M_{\rm S}$ are the Alfvén and sound Mach numbers) [552].

The solar-wind parameters are such that $\varrho_{\rm i,e} \ll R_{\rm M}$, $\Omega_{\rm i} \gg (\partial/\partial t)\ln B$, $v_{\rm Ti} \ll U \ll v_{\rm Te}$ ($\varrho_{\rm i,e}$ is the ion or electron gyroradius, $R_{\rm M}$ is the radius of the Moon, B is the absolute value of the magnetic field, and v and U are the thermal and bulk velocities of particles); therefore, to describe the plasma it is convenient to use the drift approximation for the ions (see Sect. 2.2.7) and to assume a Maxwell–Boltzmann distribution for the electrons [9, 92, 490]:

$$f_{\rm e} \propto \exp\left[-\frac{m(v_\parallel - U_\parallel)^2}{2kT_{\rm e}} - \frac{\mu_{\rm e} B}{kT_{\rm e}} - \frac{e}{kT_{\rm e}}\int_\infty^r d\boldsymbol{x}\left(\boldsymbol{E} + \frac{\boldsymbol{v}_{\rm e}\times\boldsymbol{B}}{c}\right)\right]. \quad (12.24)$$

The dimensionless forms of the plasma and field equations are as follows: The equation for the distribution function $f(t, \boldsymbol{x}, V_\parallel)$ is

$$\frac{\partial}{\partial t}f + (\boldsymbol{U}_\perp + \hat{\boldsymbol{e}}_1 v_\parallel)\nabla f + \dot{v}_\parallel \frac{\partial}{\partial v_\parallel}f = 0, \quad (12.25)$$

where $v_\parallel$ and $\boldsymbol{U}_\perp$ are the particle velocities along and transverse to the magnetic field; the equation for the transverse and longitudinal motion of the ions is

$$\boldsymbol{U}_\perp = \left(\boldsymbol{E} - \frac{\varepsilon\beta_{\rm i}\nabla B}{2M_{\rm A}^2} - \varepsilon v_\parallel^2\frac{\partial \hat{\boldsymbol{e}}_1}{\partial x_\parallel}\right)\times\frac{\hat{\boldsymbol{e}}_1}{B}, \quad (12.26)$$

$$\frac{\rm d}{{\rm d}t}v_\parallel = \frac{E_\parallel}{\varepsilon} - \frac{\beta_{\rm i}}{2M_{\rm A}^2}\frac{\partial B}{\partial x_\parallel} + \boldsymbol{U}_E\times\frac{\rm d}{{\rm d}t}\hat{\boldsymbol{e}}_1, \quad (12.27)$$

$$\frac{{\rm d}\boldsymbol{x}}{{\rm d}t} = \boldsymbol{U}_\perp + v_\parallel\hat{\boldsymbol{e}}_1, \quad \boldsymbol{U}_E = \frac{\boldsymbol{E}\times\hat{\boldsymbol{e}}_1}{B}, \quad (12.28)$$

where $E_\parallel$ is a component of the electric field along the magnetic field and $\hat{\boldsymbol{e}}_1 = \boldsymbol{B}/|\boldsymbol{B}|$.

The conservation of the magnetic moment is

$$\mu_{\rm i,e} = \frac{p_{\rm i,e\perp}}{nB} = \frac{\beta_{\rm i,e}B_\infty}{8\pi n_\infty} = {\rm const.}, \quad \varepsilon = \frac{\varrho_{\rm ci}}{R_{\rm M}}, \quad \varrho_{\rm ci} = \frac{v_{\perp\infty}}{\Omega_{\rm i}}. \quad (12.29)$$

The Maxwell equations in the plasma and in the Moon are

$$\frac{\partial \boldsymbol{B}}{\partial t} = -\nabla\times\boldsymbol{E}, \quad (12.30)$$

$$\nabla\cdot\boldsymbol{E} = \frac{\omega_{\rm pe}^2}{\omega_{\rm pi}}\bar{\varrho}, \quad \nabla\cdot\boldsymbol{B} = 0. \quad (12.31)$$

Ampere's law, taking the drift current into account, has the following form:

12.5 Interaction of the Solar Wind with the Moon

In the solar wind,

$$\nabla \times \boldsymbol{B} = \frac{4\pi}{B^2}\left[-p_\perp(\nabla B \times \hat{\boldsymbol{e}}_1) + p_\parallel B k^2 \boldsymbol{R}_{\text{cur}} \times \hat{\boldsymbol{e}}_1 - \varrho B \dot{\boldsymbol{v}} \times \hat{\boldsymbol{e}}_1\right]$$

$$-4\pi \nabla \times \left(\frac{p_\perp \hat{\boldsymbol{e}}_1}{B}\right), \quad (12.32)$$

$$\nabla \times \boldsymbol{B} = \left(\frac{\beta}{2}(-\nabla n + \chi n \cdot \boldsymbol{k}) - \frac{M_A^2 n \dot{U}}{B}\right) \times \frac{\hat{\boldsymbol{e}}_1}{1 + \frac{\beta n}{2B}} \quad (12.33)$$

Inside of the Moon,

$$\nabla \times \boldsymbol{B} = Re\sigma \boldsymbol{E}, \quad (12.34)$$

where

$$p_\perp = B\int d^3 v \mu F_i, \quad p_\parallel = M_i \int d^3 v (v_\parallel - U_\parallel)^2 F_i, \quad (12.35)$$

$$\boldsymbol{k} = \hat{\boldsymbol{e}}_1 \times (\nabla \times \hat{\boldsymbol{e}}_1), \quad p_\parallel \propto \varrho^3/B^2. \quad (12.36)$$

$\boldsymbol{k}$ is the curvature vector of the magnetic field, $\boldsymbol{R}_{\text{cur}}$ is the radius of curvature of the magnetic field and $|\boldsymbol{k}| = R_{\text{cur}}^{-1}$; $n, p_\parallel$ and $p_\perp$ are the density and longitudinal and transverse pressures of the plasma. The plasma parameters are relative to the unperturbed solar-wind parameters:

$$\bar{t} = tU_\infty/R_{\text{M}}, \quad \bar{\boldsymbol{x}} = \boldsymbol{x}/R_{\text{M}}, \quad \bar{U} = U/U_\infty, \quad \bar{\nabla} = \nabla/R_{\text{M}}.$$

The basic dimensionless parameters of the problem are as follows: $M_A = U_\infty\sqrt{4\pi\varrho}/B_\infty$ is Alfvén Mach number, $Re = 4\pi U_\infty R_{\text{M}}\sigma/c^2$ is the magnetic Reynolds number, $S = U_\infty/v_\parallel$ is the longitudinal Mach number, $\chi = T_\parallel/T_\perp$ is the temperature anisotropy, $\Omega_i R_{\text{M}}/U_\infty = 1/\varepsilon$ is the magnetization of the plasma and θ_{BU} is the angle between the velocity and the magnetic field.

The following boundary conditions are imposed: At the outer boundary of the perturbation region, $[B_n] = 0, [E_\tau] = 0$ and $[f] = 0$. When calculating the steady-state flow, we assume the Moon is nonconducting. In calculating the interaction between the solar-wind discontinuity and the Moon, it is assumed in the first approximation that the Moon has a perfectly conducting core of radius $a = 0.85 R_{\text{M}}$ surrounded by a nonconducting shell. At the core boundary we impose the condition $B_n = 0$.

The computational domain of the wake is bounded by a cross-section, on which is imposed Sommerfeld's radiation condition [526]. In the present problem this condition is implied by (12.30) and has the form

$$(\partial/\partial t)\boldsymbol{B} = -\nabla \times (\boldsymbol{U} \times \boldsymbol{B}).$$

Equations (12.30–12.31) with the quasineutrality of the plasma taken into account lead to the simplification $\boldsymbol{E} \approx \boldsymbol{E}_{\text{boun}}(t, \boldsymbol{x})$. Henceforth we shall only consider the case $\boldsymbol{U}_\infty \perp \boldsymbol{B}_\infty$ [313, 314].

12.5.2 Method of Solution

To solve the system (12.26–12.29), we use the discrete guiding-center-in-cell model [313]. Calculation of plasma parameters is made for four sections: $\phi = 0, \pi/6, \pi/3, \pi/2$. We used a 10×30 mesh (Figs. 12.19 and 12.20).

When solving the quasistationary problem, we neglect the conductivity of the nucleus. Let us introduce the vector potential $\boldsymbol{B} = \nabla \times \boldsymbol{A}$, which satisfies the calibration condition $\nabla \boldsymbol{A} = 0$; thus, a solution of the equation

$$\Delta A_{x,y,z} = -J_{x,y,z} \tag{12.37}$$

may be found using a Fourier transform,

$$A_x = \sum_{m=1}^{m_k} A_{xm} \cos(2m\phi) + A_{x0}, \quad A_y = \sum_{m=1}^{m_k} A_{ym} \sin(2m\phi),$$

$$A_z = \sum_{m=1}^{m_k} A_{zm} \cos[(2m-1)\phi],$$

and (12.37) reduces to

$$\Delta_m A_{x,y,zm} = -J_{x,y,zm}, \quad m = 1, 2, ..., \tag{12.38}$$

where

$$\Delta_m = \frac{1}{r^2} \frac{\partial}{\partial r} \left(r^2 \frac{\partial}{\partial r} \right) + \frac{1}{r^2 \sin \theta} \frac{\partial}{\partial \theta} \left(\sin \theta \frac{\partial}{\partial \theta} \right) - \frac{m^2}{r^2 \sin^2 \theta}$$

and where the following boundary conditions apply: $\partial A_m/\partial\theta|_{\theta=0} = 0$ and $\boldsymbol{A}|_{\text{bound}} = \boldsymbol{A}_\infty(z - U_{SW}t)$. The right-hand parts of the system (12.38) are found from the nodes of grids using the least-squares fitting for fixed values of r and θ. The solution of (12.38) was found by using an implicit iterative method. The splitting of the operator was done on the basis of additive schemes which provide the total approximation to (12.38) [463].

Fig. 12.19. The topology of computational meshes in (a) stationary and (b) non-stationary regimes (from [313, 322])

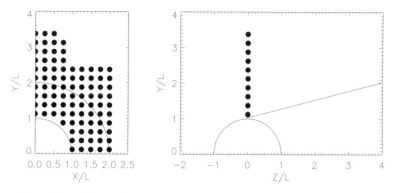

Fig. 12.20. The scheme for guiding-centers injection into the computational domain. • denotes the locations of the injected guiding centers (from [313, 322])

In the case of a propagation of a discontinuity of the solar wind along the lunar wake, the perturbed component $\boldsymbol{b} = \boldsymbol{B} - \boldsymbol{B}_0$ ($\boldsymbol{B}_0$ is a field in the stationary wake) was found from

$$\nabla \times (\nabla \times \boldsymbol{b}) = \nabla \times \left(\frac{4\pi}{c}\boldsymbol{J}\right), \quad \nabla \cdot \boldsymbol{b} = 0, \qquad (12.39)$$

where $\boldsymbol{J} = \boldsymbol{J}_{\text{drift}}$ in the plasma and $\boldsymbol{J} = 0$ in the nonconducting layer of the Moon. The following boundary conditions were applied: at the surface of the Moon $r = R_M$, the reflection condition for transversal components of the magnetic field on the day-side portion of a surface is

$$b_r = b_{r\infty}|_{\theta<\pi/2}, \quad \frac{\partial}{\partial r}(rb_{\theta,\phi})|_{\theta<\pi/2} = 0;$$

at the surface of the nucleus $r = a$, the reflection condition for transversal components of the magnetic field is

$$b_r = 0, \quad \frac{\partial}{\partial r}(rb_{\theta,\phi}) = 0;$$

and at the external boundary and $\theta > \pi/2$,

$$\boldsymbol{b} = \boldsymbol{b}_\infty(z - U_{\text{SW}}t).$$

Using the symmetry condition, we seek a solution to (12.39) in the form of a series:

$$b_r = \sum_{m=1}^{m_k} a_m \sin(2m-1)\phi, \quad b_\theta = \sum_{m=1}^{m_k} b_m \sin(2m-1)\phi,$$

$$b_\phi = \sum_{m=1}^{m_k} c_m \cos(2m-1)\phi, \quad m = 1, 3, \ldots \qquad (12.40)$$

Putting (12.40) into (12.39), we obtain a system of equations in partial derivatives of a_m, b_m, and c_m:

$$(\Delta b)_{r,\theta,\phi,m} = -(\nabla \times J)_{r,\theta,\phi,m}, \quad m = 1, 2, \dots \qquad (12.41)$$

To solve (12.40) and (12.41), we constructed a second-order difference scheme based on the alternative direction method [463]. The topology of the computational domains and meshes used in stationary and nonstationary regimes are demonstrated in Fig. 12.19. Figure 12.20 shows the scheme for guiding-centers injection into the computational domain.

12.5.3 Results and Conclusions

The flow around the Moon was calculated for the following solar-wind parameters: $M_A = 10, S_i = 10, S_e = 0.23, \xi_{i,e} = 2, U_\infty/\Omega_i R_M = 0.3$. The unperturbed ion and electron distribution function was assumed to be Maxwellian.

Figure 12.21 gives the steady-state picture of the magnetic field B_y/B_∞ and plasma density n/n_∞ in the lunar wake in the planes of the equator $(x-z)$ and principal meridian $(y-z)$. The drift perturbations of the magnetic field are essentially three-dimensional; the magnetic-field profile is monotonic in

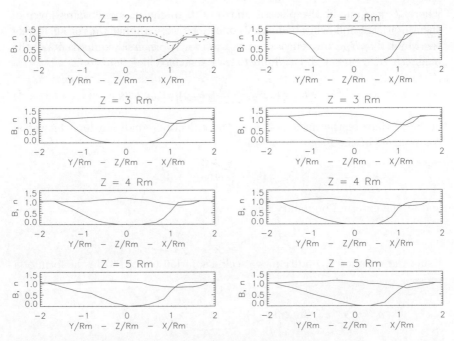

Fig. 12.21. Distribution of the relative plasma density n/n_∞ and magnetic field B_y/B_∞: left: in a stationary lunar wake; right: in the wake on passing through the discontinuity (from [314])

the y–z plane and periodic in the x–z plane. The amplitude of the magnetic-field perturbation is mainly determined by the parameter β and is $\approx 14\%$ for $\beta \approx 1$, which corresponds roughly to the two-dimensional calculations from [390].

The experimental profile of the plasma density and field obtained on Explorer 35 [390, 489] is shown in Fig. 12.21 (left) (dashed lines). We see that the observed profile is similar to the calculated one. However, at the outer edge of the wake the calculations do not give any additional perturbations of the plasma and field, which is apparently connected with using plasma macroparticles that are too large.

In calculating the perturbed plasma-wake pattern it is assumed that a weak plane discontinuity propagates in the solar wind with the velocity of the fast magnetoacoustic wave $u = U_\infty(1 + \sqrt{M_A^{-2} + M_S^{-2}})$. The parameters of the discontinuity are chosen as follows: The thickness of the discontinuity is $2\lambda \approx 2R_M$ and the relative enhancement of the magnetic field is $\Delta B/B_\infty = 0.25$, which varies according to a linear law across the discontinuity. It is assumed that the lunar core is perfectly conducting and its radius is $a = 1500$ km.

Figure 12.21 (right) gives the pattern of the perturbations in the magnetic field and plasma density in the wave at the time $\bar{t} = tU_\infty/R_M = 5$ in the planes of the equator $(x$–$z)$ and the principal meridian $(y$–$z)$. The calculation shows that the perturbation of the plasma density in the wake is mostly determined by the parameters of the discontinuity. The perturbation of the magnetic field in the wake is determined both by the parameters of the discontinuity and by the drift currents and currents in the lunar core.

Figure 12.22 gives the change in the magnetic field on the z-axis at the times $\bar{t} = 0$–7 (at $\bar{t} = 0$ the forward front of the discontinuity is located near the subsolar point of the lunar surface). The maximum enhancement of the magnetic-field discontinuity $\Delta B/B_\infty$ is reached in practice for $\bar{t} > 10$ and

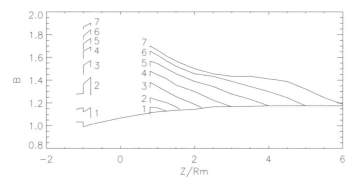

Fig. 12.22. Change in the magnetic field on the z-axis at the times $\bar{t} = 0-7$ (from [314])

equals 3.3 at the subsolar point and 2.1 at the antisolar point. This result agrees with estimates of the electromagnetic induction in a sphere placed in an asymmetric cavity with conducting walls [477, 535]. The magnetic field induced by the currents in the core has a maximum value at the equator and quickly falls off along the wake. At a distance of $2.5R_M$ the conducting core has practically no effect on the wake structure.

12.6 Interaction of Neutral Interstellar Atoms with the Heliosphere

12.6.1 Formulation of the Problem and Mathematical Model

The global interaction of the local interstellar medium (LISM) with the solar wind is a fundamental problem of heliospheric physics. It requires the solution of a highly nonlinear coupled set of integro–MHD–Boltzmann equations, which describe the dynamics of the ionized solar wind and LISM together with interstellar and heliospheric neutral atoms. To the first order, the plasma and neutral atoms are coupled by resonant charge exchange, although other coupling processes are present. The characteristic scale of the ionized components is determined usually by the typical ion gyroradius, which is much less than characteristic global heliospheric scales of interest. By contrast, the mean free path of neutral particles is comparable to characteristic heliospheric scales such as the distance separating the bow shock and the heliopause or the heliopause and the termination shock, or even the radial extent of the supersonic solar wind [347]. Consequently, the Knudsen number $Kn = \lambda/L$ (λ is the mean free path of neutral particles and L is a characteristic heliospheric scale), which is a measure of the distribution relaxation distance, satisfies $Kn \approx 1$. Thus, it is difficult to assume that the neutral H distribution can relax to a Maxwellian distribution, and one needs ideally to solve a Boltzmann equation for the neutral component in which charge exchange and photo-ionization processes are included [347].

Several approaches for including the interstellar neutral component self-consistently in models describing the solar-wind–LISM interaction have been formulated. In [29, 30], a Monte-Carlo technique was used to solve the stationary neutral Boltzmann equation; in [147, 148, 409, 443] the method of characteristics was used to solve the Boltzmann equation directly; in [590], an integral method developed in [213] was used; while in [26, 262, 307, 310, 420, 501, 553, 557, 591], hydrodynamic models were developed to describe the neutral H component.

Here we use a Boltzmann particle-mesh approach to study the dynamics of the neutral component, whereas the ionized solar wind and LISM are described in a gas dynamical approximation (Sect. 2.4.1). Neutral atoms experience charge exchange with the ionized component of the LISM and solar wind and

12.6 The Interaction of Neutral Interstellar Atoms

may also be affected by photo-ionization. The charge exchange cross-section σ_{ex} adopted here is the empirical fit of [156]:

$$\sigma_{\text{ex}} = \left(2.1 \times 10^{-7} - 9.2 \times 10^{-9} \ln v\right)^2, \qquad (12.42)$$

where σ_{ex} is measured in cm^2 and the particle velocity v is measured in centimeters per second.

Here the average velocity of protons relative to an H atom with velocity $\boldsymbol{v}_i$ is [443]

$$V_{\text{rel,p}} = v_{\text{T,p}} \left[\frac{\exp(-\omega_{\text{p}}^2)}{\sqrt{\pi}} + \left(\omega_{\text{p}} + \frac{1}{2\omega_{\text{p}}}\right) \text{erf}(\omega_{\text{p}}) \right],$$

$$\omega_{\text{p}} = \frac{1}{v_{\text{T,p}}} |\boldsymbol{v}_i - \boldsymbol{U}_{\text{p}}| \equiv \Delta u / v_{\text{T,p}}. \qquad (12.43)$$

erf(x) denotes the usual error function, $v_{\text{T,p}} = \sqrt{(2kT_{\text{p}}/M)}$ is the thermal velocity of the proton, k is Boltzmann's constant and $\boldsymbol{U}_{\text{p}}$ is the bulk flow velocity of the plasma.

The photo-ionization rate of neutral atoms may be expressed as

$$\beta_{\text{ph}} = \beta_{\text{ph},E} (r_E/r)^2, \qquad (12.44)$$

where $\beta_{\text{ph},E}$ is the photoionization rate at 1 AU ($r = r_E$).

If the particle satisfies (2.164), then we have to exchange the velocity of this neutral macroparticle with the velocity of a proton from the ionized component of the LISM or the solar wind. This is accomplished using a random number generator for the probability (or the frequency) of charge exchange of the atom with velocity $\boldsymbol{v}$ and the proton with velocity $\boldsymbol{v}_{\text{p}}$ [358],

$$\nu(\boldsymbol{v}_{\text{p}}) \sim |\boldsymbol{v} - \boldsymbol{v}_{\text{p}}| \sigma_{\text{ex}} \exp\left(-\frac{(\boldsymbol{v}_{\text{p}} - \boldsymbol{U}_{\text{p}})^2}{2v_{\text{T,p}}^2}\right). \qquad (12.45)$$

In the present simulations, we do not take the cross-section σ into account in (12.45) because of the weak dependence on $|v - v_{\text{p}}|$ (as was done in the work of [358]).

In our simulation we use the plasma distribution in the computational domain from the solution of the hydrodynamical equations (2.166–2.168). We have assumed that $M_{\text{n}} = M$ (M_{n} the neutral mass).

The two-dimensional computational domain has dimensions $DX = 1600$ AU and $DY = 2000$ AU. We use a 401×401 mesh, and the number of macroparticles (macroatoms) is $N_A = 8 \times 10^6$. The time step Δt satisfies the condition $v_{\text{max}} \Delta t \leq \min(\Delta x, \Delta y)$. The atoms which are created in the solar wind impose stringent limits on the value of Δt. At the left boundary ($x = -600$ AU), the neutral LISM H flux is assumed to be the Maxwellian distribution:

$$f = n_\infty (\pi v_T^2)^{-3/2} \exp\left(-\frac{(\boldsymbol{v} - \boldsymbol{U})^2}{2v_T^2}\right), \tag{12.46}$$

and v_T and $\boldsymbol{U}$ are the neutral thermal and bulk velocities. For the right boundary ($x = 1000\,\mathrm{AU}$), we assume that neutral particles escape freely. On the side boundaries ($y = \pm 1000\,\mathrm{AU}$), a cyclical boundary condition is imposed.

The Boltzmann code described here is based on existing 2.5-dimensional and three-dimensional particle codes developed for numerous hybrid plasma simulations, a summary of which can be found in [334].

12.6.2 Results and Conclusions

In simulating the evolution of the interstellar neutral H distribution as it traverses the heliosphere, only charge exchange is considered for the present. Photo-ionization of atoms, solar radiation pressure and gravity are excluded. We should emphasize again that we do not present a self-consistent Boltzmann simulation in which we update the fluid plasma in response to the evolving neutral distribution. Instead, we use a gas dynamics simulation for the plasma (in which charge-exchange processes were calculated self-consistently within the multifluid description of [591]), and we then use the plasma distribution obtained as the prescribed background for our Boltzmann simulation. Two sets of LISM parameters were adopted. For the neutral H component: $n_{H\infty} = 0.14\,\mathrm{cm^{-3}}$, $U_{H\infty} = 26\,\mathrm{km/s}$ and either $T_{H\infty} = 8000\,\mathrm{K}$ or $10\,900\,\mathrm{K}$. For the ionized interstellar component: $n_{H^+\infty} = 0.07\,\mathrm{cm^{-3}}$, $U_{H^+\infty} = 26\,\mathrm{km/s}$ and either $T_{H^+\infty} = 8000\,\mathrm{K}$ or $10\,900\,\mathrm{K}$. The solar wind parameters at 1 AU were $n_{p,E} = 5.0\,\mathrm{cm^{-3}}$, $U_{p,E} = 400\,\mathrm{km/s}$, $T_E = 10^5\,\mathrm{K}$ and $M_{S,E} = 7.6$. Both sets of parameters yield a characteristic two-shock model for the solar-wind–LISM interaction [591].

12.6.2.1 Global Distribution of the Macroscopic Parameters of the H Component.
Before discussing the detailed neutral velocity distributions, we describe the global density, temperature and velocity characteristics of neutral hydrogen throughout the solar-wind–LISM interaction region. Upstream of the detached bow shock, charge exchange does not greatly change the velocity distribution function of the neutral component. Nonetheless, a small increase in the neutral density ahead of the bow shock occurs thanks to atoms created in the solar wind which stream radially outwards. Between the bow shock and the heliopause, the neutral component is strongly affected by charge exchange with the (weak) shock-heated, LISM-ionized component. Figure 12.23 shows the two-dimensional distribution of the H density in the two-dimensional plane. Figure 12.24 illustrates one-dimensional line-of-sight profiles along and orthogonal to the stagnation axis for the H density n_H, the H bulk velocity U_{bulk} and the effective temperature $T_H \propto \langle(\boldsymbol{v} - \boldsymbol{U}_{\mathrm{bulk}})^2\rangle/3$. Ahead of the bow shock, the neutral density and temperature increase significantly up to values of $n_{H,\max} \approx 1.9 n_{H\infty}$ (Figs. 12.23

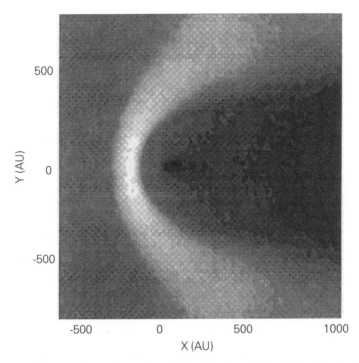

Fig. 12.23. A typical two-dimensional distribution of the H density in the two-dimensional Boltzmann simulation ($T_{H\infty} = 10\,900\,\text{K}$) (from [347])

and 12.24a) and $T_H \approx (2-3)T_{H\infty}$ (Fig. 12.24c), while the bulk velocity decreases ($U_{\text{bulk}} \approx 0.35 - 0.6 U_\infty$ along the stagnation axis; Fig. 12.24b).

At the heliopause ($x = 130\,\text{AU}$) the density of H decreases to $1.1 n_{H\infty}$ (Fig. 12.24a), the temperature increases, $T_H \approx 9 T_\infty$ (Fig. 12.24c), and the bulk velocity decreases to $0.55 U_{H\infty}$ (Fig. 12.24b).

Between the termination shock and the heliopause, the density of H atoms decreases to $\approx 0.8 - 0.9 n_{H\infty}$ while the bulk velocity is reduced to $U_{\text{bulk}} = (0.42 - 0.7) U_{H\infty}$ (Fig. 12.24b). The temperature of H atoms varies significantly from a heliopause value of $T_H \approx 9 T_{H,\infty}$ to $T_H \approx (6-7) T_{H,\infty}$ at the termination shock. Inside the supersonic solar wind, an additional peak in the atom temperature profile appears; this is due to atoms created in the solar wind. Figure 12.25 illustrates the streamlines of neutral hydrogen in the two-dimensional plane. The bow shock front, the heliopause, and the lobe of the termination shock can be discerned. In the tail region, the termination shock does not divert the plasma flow very strongly, and so it is difficult to distinguish the downstream termination shock front in Fig. 12.25. However, in this region, immediately downstream of the Sun, neutral H is accelerated monotonically to a value of $U_{\text{bulk}} \approx 1.2 U_\infty$ at the right-hand boundary of the computational domain ($x = 1000\,\text{AU}$). Associated with the velocity increase is an increase in the H density to $n_H \approx 0.5 n_{H\infty}$ up to about $x \approx 200$ AU, after which the density begins to decrease slightly, reaching $n_H \approx 0.4 n_{H\infty}$ by the right-hand boundary ($x = 1000$ AU). Similarly, the temperature increases to $T \approx (7-8) T_{H\infty}$ before cooling again.

Recent observational data from the Ulysses Neutral Gas Experiment GAS [575] indicate that the temperature of LISM H atoms may be $T_\infty = 8000$ K rather than $T_\infty = 10\,900$ K. Accordingly, simulations using $T_{H\infty} = 8000$ K were performed. The global plasma configuration using the 8000 K temperature is a little different from that obtained from the 10 900 K simulations in that one has a slightly stronger bow shock and a smaller separation distance between the bow shock and heliopause. As a result, the increase in the H density downstream of the bow shock is about 16% higher than with $T_\infty = 10\,900$ K, as illustrated explicitly in Fig. 5 of [347]. The thickness of the H density profile at the heliopause is also much smaller than that of the high $T_{H,\infty}$ case. An analysis of the density profile reveals that diffusion of H atoms through the bow shock and heliopause is smaller for the $T_\infty = 8000$ K case than for the higher LISM H temperature model. In the supersonic solar wind, the H density profile does not depend strongly on the assumed LISM H temperature.

The results of our Boltzmann simulation can be compared with corresponding results obtained in a four-fluid simulation using the same LISM ($T_\infty = 8000$ K) and solar-wind parameters [591]. Figure 12.26 shows one-dimensional sections of the H atom density profile along the upstream stagnation axis $x(y=0)$ for the Boltzmann (solid line) and four-fluid (dotted line) simulations. The four-fluid model gives a slightly different distribution of

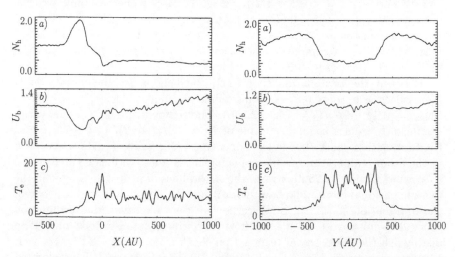

Fig. 12.24. One-dimensional central sections of (**a**) H density, (**b**) the x component of the bulk velocity U_{bulk}, and (**c**) the temperature T_H in the $x(y=0)$ direction (*left*) and in the $y(x=0)$ direction (*right*) ($T_{H\infty} = 10\,900$ K) (from [347])

12.6 The Interaction of Neutral Interstellar Atoms   349

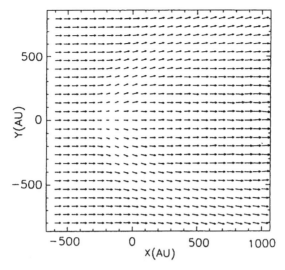

Fig. 12.25. Vector field plot for the bulk velocity of the neutral H component ($T_{H\infty} = 10\,900\,\mathrm{K}$) (from [347])

H density. Far upstream of the bow shock, the Boltzmann code and the four-fluid model give approximately the same distribution except for "shot noise" fluctuations in the Boltzmann simulation. Just ahead of the bow shock, the Boltzmann solution demonstrates strong diffusion of the neutral H compared to the four-fluid solution. The amplitude of the hydrogen wall is comparable in the two models and the filtration is similar, with a slightly smaller density crossing the termination shock in the four-fluid simulation. At distances $r \leq 30\,\mathrm{AU}$ from the Sun, the neutral H number density in the Boltzmann simulation is about 30% higher than in the four-fluid model. However, the resolution in our Boltzmann simulation in the vicinity of Sun becomes ina-

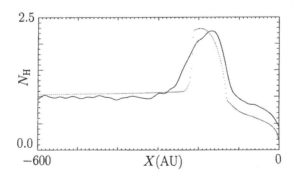

Fig. 12.26. One-dimensional central sections of the H density profile along the x-axis for a Boltzmann simulation (*solid line*) and a four-fluid model (*dotted line*) ($T_\infty = 8000\,\mathrm{K}$) (from [347])

dequate since $\Delta z = 4\,\text{AU}$ and $\Delta x = 5\,\text{AU}$. This may result in a numerical reduction of the charge-exchange process, especially in regions where the plasma density increases markedly. This may also result in the incorrect estimation of the value of the atom temperature at distances $r \leq 100\,\text{AU}$. Much better statistics are needed for those regions of our simulation where the neutral number density is small. In a future simulation, we plan to use either a variable particle size or a particle splitting procedure.

Although the parameters of our simulation differ from those of the simulation of [29, 30], we can make a qualitative comparison. In [29], the following parameters for the incident LISM flow are used: $n_{H\infty} = 0.14\,\text{cm}^{-3}$, $n_{H^+\infty} = 0.07\,\text{cm}^{-3}$ and $U_\infty = 25\,\text{km/s}$; however, the exact value of T_∞ is not specified.

The maximum density of H atoms ($n_{H\text{max}} = 1.78 n_{H\infty}$) is located approximately at $x = -200\,\text{AU}$, whereas the density at $x = -100\,\text{AU}$ has a value $n_H = 0.7 n_{H\infty}$. In [31], by using $n_{p\infty} = 0.1\,\text{cm}^{-3}$, $n_{H\infty} = 0.2\,\text{cm}^{-3}$ and $T_\infty = 7000\,\text{K}$ for the incident LISM flow, and including electron impact ionization, it was found that the maximum H atoms density ($n_{H\text{max}} = 1.75 n_{H\infty}$) is located approximately at $x = -200\,\text{AU}$, whereas the density at $x = -100\,\text{AU}$ has a value $n_H = 0.5 n_{H\infty}$. Our simulation gives $n_{H\text{max}} \approx 2.2 n_{H\infty}$ at $x = -170\,\text{AU}$ and $n_H \approx 0.93 n_{H\infty}$ at $x = -100\,\text{AU}$ in the case where $T_{H\infty} = 8000\,\text{K}$. In our simulation with $T_{H\infty} = 10\,900\,\text{K}$, we have a maximum density $n_{H\text{max}} \approx 1.9 n_{H\infty}$ at a distance $x = -200\,\text{AU}$ and $n_H \approx 0.93 n_{H\infty}$ at $x = -100\,\text{AU}$. In the tail region, our simulation gives an asymptotic value of $n_H \approx 0.5 n_{H\infty}$ for the density, which corresponds to the model of [29]. In our simulations, the neutral atom temperature at the bow shock is the same as in the Baranov and Malama simulation. Inside the supersonic solar wind, however, and in the tail region, the atom temperature is higher in our simulation than in the Baranov and Malama simulation. This difference may result from our solar-wind gas dynamical temperature distribution differing from that of the self-consistent Baranov and Malama Boltzmann gas dynamical solution.

12.6.2.2 Velocity Distribution Function of the H Component.

Consider now the velocity distribution of the neutral H component rather than the various moments discussed in the previous subsection. The typical velocity distribution function consists of several fairly distinct components: a large core, which corresponds to atoms originating in the LISM, slightly broadened by charge exchange with bow-shock-heated LISM material; a beam of hot neutral atoms originating from the subsonic solar wind; and a cooler beam of neutral H produced in the supersonic solar wind. Figure 12.27 shows the projection of the H velocity distribution function for different sections along the $x(y = 0)$ axis, ordered from top to bottom according to the following radial distances:

- Outside the bow shock, the H velocity distribution function consists of a strong core of slightly heated atoms originating in the LISM and a weak

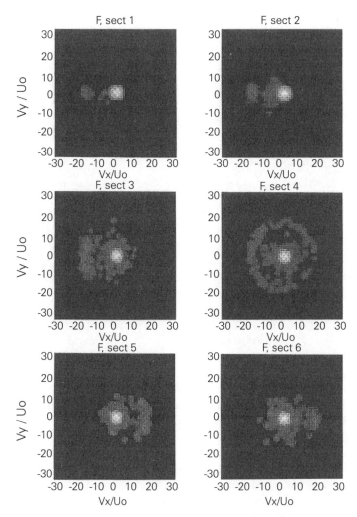

Fig. 12.27. Projection of the H velocity distribution function in the v_x–v_y plane for different sections along the $x(y=0)$ axis. From top to bottom: $x = -300, -200, -100, 0, +100, +200, +300, +400, +500$ and $+600$ AU. (From [347])

beam of neutral atoms produced by charge exchange in the heliosheath and the supersonic solar wind. Typical velocities in the beam are about $18 U_\infty = 468$ km/s (Fig. 12.27, section 1, $x = -300$ AU).

- Inside the bow shock, the H velocity distribution function consists of a core of heated atoms that are created by charge exchange between inflowing neutral atoms and the bow-shock-heated plasma together with a tongue-like distribution produced by charge exchange in the heliosheath and supersonic solar wind. The core is labeled component 1 and the secon-

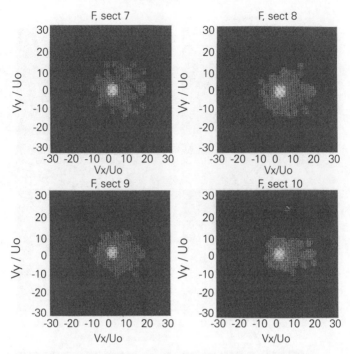

Fig. 12.27. (continued)

dary or tongue component is called component 2 in [591]. The distribution is displayed in Fig. 12.27, section 2 for $x = -200\,\mathrm{AU}$.

- Just downstream of the termination shock in the heliosheath, the H velocity distribution function consists of a component 1 core and a strong tongue-like beam (component 2) together with an halo of heated atoms produced by charge exchange between H atoms and the supersonic solar wind (called component 3 or the "splash" component). The velocity of halo atoms is comparable with the thermal velocity of supersonic solar wind protons (illustrated in Fig. 12.27, section 3 for $x = -100\,\mathrm{AU}$). The component 2 distribution is wider in this section than those in sections 1 and 2 because the distribution function is an average value over the volume element $100\,\mathrm{AU} \times 100\,\mathrm{AU}$ in x–y plane.
- In the vicinity of the Sun, the H velocity distribution function consists of the component 1 core, the heated component 2 neutrals, and component 3 atoms (illustrated in Fig. 12.27, section 4 for $x = 0\,\mathrm{AU}$). Since the velocity distribution is smoothed, section 4 illustrates a halo distribution of component 3 atoms rather than a tongue-type distribution.
- In the heliotail region just downstream of the termination shock, the H atom velocity distribution function has a core of LISM atoms, a halo of component 2 heated atoms (with the thermal velocity now slightly smaller

than found in the equivalent upstream heliosheath region) and a tongue of atoms produced in the supersonic solar wind (component 3). The distribution is shown in Fig. 12.23, section 5, where $x = 100$ AU, and Fig. 12.27, section 6, for $x = 200$ AU. In the deeper heliotail region, the effective thermal velocity of neutral atoms increases again (Fig. 12.27, section 7, $x = 300$ AU and Fig. 12.27, section 8, $x = 400$ AU), but as the heliotail cools [591], the effective thermal velocity of neutral atoms also decreases (Fig. 12.27, section 9, $x = 500$ AU, and Fig. 12.27, section 10, $x = 600$ AU). Further development of this model may be found in [346, 383, 587, 589].

Summary

In this chapter we have discussed the simulation of the interaction of the solar wind with the astrophysical objects. The global hybrid model describes well the interaction of the solar wind with nonmagnetic planets and comets. The simulations demonstrate the formation of the collisionless bow shock, induced magnetosphere, magnetic barrier and other plasma structures. The hybrid model also describes well the thin structure of the cometary tail which has not been described to date using bi-ion fluid models. The results of the simulation of the interaction of the solar wind with weak comets may also be used for the study of the plasma phenomena near small bodies like Phobos, Deimos, Pluto etc., and also in AMPTE releases. The hybrid fluid–Boltzmann model describes well enough the dynamics of the external heliosphere, whereas the Boltzmann particle-mesh model does not approximate well the dynamics of neutral components near the Sun. To reduce the approximation error one has to use the multiscale methods (adaptive grids, e.g., the AMPR method) for simulation of the neutral component. The hybrid ion-guiding-center–electron-fluid model describes well low-frequency perturbations of the electromagnetic field and plasma near the Moon and in the plasma wake. However, to study the high-frequency and small-scale phenomena one has to take into account the non-neutrality and electron plasma oscillations. For these problems one has to use the full particle models.

Exercises

12.1 Find the interstellar oxygen-to-hydrogen density ratio.

12.2 Derive the algorithm for source of cometary ions with the following distribution:
$$G = \frac{Q\nu}{4\pi W_c (r^2 + a^2)} \exp\left(-\frac{\nu r}{W_c}\right),$$
where Q, ν and W_c denote gas production rate, the photo-ionization rate, and the initial velocity of the cometary ions; a is an effective radius of the cometary nucleus to cancel a singularity in the above expression.

13. Appendix

13.1 Coordinate Form of Maxwell's Equations and the Electron Pressure Equations

In this section we provide the Cartesian, cylindrical and spherical coordinate forms for Maxwell's equations and the electron pressure equations.

13.1.1 Cartesian Coordinates

Ohm's law:

$$\begin{pmatrix} E_x \\ E_y \\ E_z \end{pmatrix} = -\frac{1}{c} \begin{pmatrix} U_{ey}B_z - U_{ez}B_y \\ U_{ez}B_x - U_{ex}B_z \\ U_{ex}B_y - U_{ey}B_x \end{pmatrix} - \frac{1}{en_e} \begin{pmatrix} \frac{\partial p_e}{\partial x} \\ \frac{\partial p_e}{\partial y} \\ \frac{\partial p_e}{\partial z} \end{pmatrix} + \frac{1}{\sigma_{\text{eff}}} \begin{pmatrix} J_x \\ J_y \\ J_z \end{pmatrix}$$

$$-\frac{m}{e}\frac{d}{dt}\begin{pmatrix} U_{ex} \\ U_{ey} \\ U_{ez} \end{pmatrix}. \qquad (13.1)$$

Faraday's law:

$$\frac{\partial}{\partial t}\begin{pmatrix} B_x \\ B_y \\ B_z \end{pmatrix} = -c \begin{pmatrix} \frac{\partial E_z}{\partial y} - \frac{\partial E_y}{\partial z} \\ \frac{\partial E_x}{\partial z} - \frac{\partial E_z}{\partial x} \\ \frac{\partial E_y}{\partial x} - \frac{\partial E_x}{\partial y} \end{pmatrix}. \qquad (13.2)$$

Ampere's law:

$$\begin{pmatrix} J_x \\ J_y \\ J_z \end{pmatrix} = \frac{c}{4\pi} \begin{pmatrix} \frac{\partial B_z}{\partial y} - \frac{\partial B_y}{\partial z} \\ \frac{\partial B_x}{\partial z} - \frac{\partial B_z}{\partial x} \\ \frac{\partial B_y}{\partial x} - \frac{\partial B_x}{\partial y} \end{pmatrix}. \qquad (13.3)$$

13. Appendix

The inductive equation may be written in the following nondivergent form:

$$\frac{\partial}{\partial t}\begin{pmatrix}B_x\\B_y\\B_z\end{pmatrix} + \left(U_{ex}\frac{\partial}{\partial x} + U_{ey}\frac{\partial}{\partial y} + U_{ez}\frac{\partial}{\partial z}\right)\begin{pmatrix}B_x\\B_y\\B_z\end{pmatrix}$$

$$-\left(\frac{\partial}{\partial x}D\frac{\partial}{\partial x} + \frac{\partial}{\partial y}D\frac{\partial}{\partial y} + \frac{\partial}{\partial z}D\frac{\partial}{\partial z}\right)\begin{pmatrix}B_x\\B_y\\B_z\end{pmatrix} = -\begin{pmatrix}\Phi_x\\\Phi_y\\\Phi_z\end{pmatrix}, \quad (13.4)$$

where

$$\Phi_x = B_x\left(\frac{\partial U_{ey}}{\partial y} + \frac{\partial U_{ez}}{\partial z}\right) - B_y\frac{\partial U_{ex}}{\partial y} - B_z\frac{\partial U_{ex}}{\partial z}$$

$$-\frac{c}{e}\left(\frac{\partial}{\partial y}\frac{1}{n_e}\frac{\partial p_e}{\partial z} - \frac{\partial}{\partial z}\frac{1}{n_e}\frac{\partial p_e}{\partial y}\right)$$

$$+\left(\frac{\partial D}{\partial y}\right)\left(\frac{\partial B_y}{\partial x}\right) + \left(\frac{\partial D}{\partial x}\right)\left(\frac{\partial B_x}{\partial x}\right) + \left(\frac{\partial D}{\partial z}\right)\left(\frac{\partial B_z}{\partial x}\right),$$

$$\Phi_y = -B_x\frac{\partial U_{ey}}{\partial x} + B_y\left(\frac{\partial U_{ex}}{\partial x} + \frac{\partial U_{ez}}{\partial z}\right) - B_z\frac{\partial U_{ey}}{\partial z}$$

$$-\frac{c}{e}\left(\frac{\partial}{\partial z}\frac{1}{n_e}\frac{\partial p_e}{\partial x} - \frac{\partial}{\partial x}\frac{1}{n_e}\frac{\partial p_e}{\partial z}\right)$$

$$+\left(\frac{\partial D}{\partial z}\right)\left(\frac{\partial B_z}{\partial y}\right) + \left(\frac{\partial D}{\partial y}\right)\left(\frac{\partial B_y}{\partial y}\right) + \left(\frac{\partial D}{\partial x}\right)\left(\frac{\partial B_x}{\partial y}\right),$$

$$\Phi_z = -B_x\frac{\partial U_{ez}}{\partial x} + B_z\left(\frac{\partial U_{ex}}{\partial x} + \frac{\partial U_{ey}}{\partial y}\right) - B_y\frac{\partial U_{ez}}{\partial y}$$

$$-\frac{c}{e}\left(\frac{\partial}{\partial x}\frac{1}{n_e}\frac{\partial p_e}{\partial y} - \frac{\partial}{\partial y}\frac{1}{n_e}\frac{\partial p_e}{\partial x}\right)$$

$$+\left(\frac{\partial D}{\partial x}\right)\left(\frac{\partial B_x}{\partial z}\right) + \left(\frac{\partial D}{\partial z}\right)\left(\frac{\partial B_z}{\partial z}\right) + \left(\frac{\partial D}{\partial y}\right)\left(\frac{\partial B_y}{\partial z}\right),$$

and the magnetic field diffusion is

$$D = \frac{c^2}{4\pi\sigma_{\text{eff}}}.$$

The time evolution of the electron pressure is described by the following equation:

$$\frac{\partial p_e}{\partial t} + \left(\frac{\partial p_e U_{ex}}{\partial x} + \frac{\partial p_e U_{ey}}{\partial y} + \frac{\partial p_e U_{ez}}{\partial z}\right) + \gamma p_e\left(\frac{\partial U_{ex}}{\partial x} + \frac{\partial U_{ey}}{\partial y} + \frac{\partial U_{ez}}{\partial z}\right) =$$

$$= \frac{(\gamma - 1)}{\sigma_{\text{eff}}}J^2. \quad (13.5)$$

In above equation, we neglected the viscosity stress tensor π^e and the heat flux q_e.

13.1.2 Cylindrical Coordinates

Ohm's law:

$$\begin{pmatrix} E_r \\ E_\phi \\ E_z \end{pmatrix} = -\frac{1}{c} \begin{pmatrix} U_{e\phi} B_z - U_{ez} B_\phi \\ U_{ez} B_r - U_{e\varrho} B_z \\ U_{e\varrho} B_\phi - U_{e\phi} B_r \end{pmatrix} - \frac{1}{en_e} \begin{pmatrix} \dfrac{\partial p_e}{\partial r} \\ \dfrac{1}{r}\dfrac{\partial p_e}{\partial \phi} \\ \dfrac{\partial p_e}{\partial z} \end{pmatrix} + \frac{1}{\sigma_{\text{eff}}} \begin{pmatrix} J_r \\ J_\phi \\ J_z \end{pmatrix}$$

$$-\frac{m}{e}\frac{\mathrm{d}}{\mathrm{d}t}\begin{pmatrix} U_{er} \\ U_{e\phi} \\ U_{ez} \end{pmatrix}. \qquad (13.6)$$

Faraday's law:

$$\frac{\partial}{\partial t}\begin{pmatrix} B_r \\ B_\phi \\ B_z \end{pmatrix} = -c \begin{pmatrix} \dfrac{1}{r}\dfrac{\partial E_z}{\partial \phi} - \dfrac{\partial E_\phi}{\partial z} \\ \dfrac{\partial E_r}{\partial z} - \dfrac{\partial E_z}{\partial r} \\ \dfrac{1}{r}\left(\dfrac{\partial (rE_\phi)}{\partial r} - \dfrac{\partial E_r}{\partial \phi}\right) \end{pmatrix}. \qquad (13.7)$$

Ampere's law:

$$\begin{pmatrix} J_r \\ J_\phi \\ J_z \end{pmatrix} = \frac{c}{4\pi}\begin{pmatrix} \dfrac{1}{r}\dfrac{\partial B_z}{\partial \phi} - \dfrac{\partial B_\phi}{\partial z} \\ \dfrac{\partial B_r}{\partial z} - \dfrac{\partial B_z}{\partial r} \\ \dfrac{1}{r}\left(\dfrac{\partial (rB_\phi)}{\partial r} - \dfrac{\partial B_r}{\partial \phi}\right) \end{pmatrix}. \qquad (13.8)$$

The inductive equation may be written in the following nondivergent form:

$$\frac{\partial}{\partial t}\begin{pmatrix} B_r \\ B_\phi \\ B_z \end{pmatrix} + \left(\frac{U_{er}}{r}\frac{\partial}{\partial r}r + \frac{U_{e\phi}}{r}\frac{\partial}{\partial \phi} + U_{ez}\frac{\partial}{\partial z}\right)\begin{pmatrix} B_r \\ B_\phi \\ B_z \end{pmatrix}$$

$$-\left(\frac{\partial}{\partial r}\frac{D}{r}\frac{\partial}{\partial r}r + \frac{1}{r^2}\frac{\partial}{\partial \phi}D\frac{\partial}{\partial \phi} + \frac{\partial}{\partial z}D\frac{\partial}{\partial z}\right)\begin{pmatrix} B_r \\ B_\phi \\ B_z \end{pmatrix} = -\begin{pmatrix} \Phi_r \\ \Phi_\phi \\ \Phi_z \end{pmatrix}, \qquad (13.9)$$

where

$$\Phi_r = B_r\left(\frac{1}{r}\frac{\partial U_{e\phi}}{\partial \phi} + \frac{\partial U_{ez}}{\partial z}\right) - \frac{B_\phi}{r}\frac{\partial U_{e\phi}}{\partial \phi} - B_z\frac{\partial U_{er}}{\partial z}$$

$$-\frac{c}{er}\left(\frac{\partial}{\partial\phi}\frac{1}{n_e}\frac{\partial p_e}{\partial z} - \frac{\partial}{\partial z}\frac{1}{n_e}\frac{\partial p_e}{\partial\phi}\right) + \frac{1}{r^2}\left(\frac{\partial D}{\partial\phi}\right)\left(\frac{\partial r B_\phi}{\partial r}\right)$$

$$-\left(\frac{\partial D}{\partial r}\right)\left(\frac{\partial B_z}{\partial z}\right) + \left(\frac{\partial D}{\partial z}\right)\left(\frac{\partial B_z}{\partial r}\right) - r\left(\frac{\partial}{\partial r}\frac{D}{r^2}\right)\left(\frac{\partial B_\phi}{\partial\phi}\right),$$

$$\Phi_\phi = -B_r\left(\frac{\partial U_{e\phi}}{\partial r} - \frac{U_{e\phi}}{r}\right) + B_\phi\left(\frac{\partial U_{er}}{\partial r} + \frac{\partial U_{ez}}{\partial z} - \frac{U_{er}}{r}\right) - B_z\frac{\partial U_{e\phi}}{\partial z}$$

$$-\frac{c}{e}\left(\frac{\partial}{\partial z}\frac{1}{n_e}\frac{\partial p_e}{\partial r} - \frac{\partial}{\partial r}\frac{1}{n_e}\frac{\partial p_e}{\partial z}\right)$$

$$-\frac{1}{r}\left(\frac{\partial D}{\partial\phi}\right)\left(\frac{\partial B_z}{\partial z}\right) + r\left(\frac{\partial}{\partial r}\frac{D}{r^2}\right)\left(\frac{\partial B_r}{\partial\phi}\right) - \frac{1}{r^2}\left(\frac{\partial D}{\partial\phi}\right)\left(\frac{\partial r B_\phi}{\partial r}\right),$$

$$\Phi_z = -\varrho B_r\frac{\partial U_{ez}}{\partial r} + B_z\left(\frac{\partial U_{er}}{\partial r} + \frac{1}{r}\frac{\partial U_{e\phi}}{\partial\phi}\right) - \frac{B_\phi}{r}\frac{\partial U_{ez}}{\partial\phi}$$

$$-\frac{c}{er}\left(\frac{\partial}{\partial r}\frac{1}{n_e}\frac{\partial p_e}{\partial\phi} - \frac{\partial}{\partial\phi}\frac{1}{n_e}\frac{\partial p_e}{\partial r}\right)$$

$$+\frac{1}{r}\left(\frac{\partial r D}{\partial r}\right)\left(\frac{\partial B_r}{\partial z}\right) - \frac{1}{r}\left(\frac{\partial D}{\partial z}\right)\left(\frac{\partial \varrho B_r}{\partial \varrho}\right) - \frac{D}{r}\frac{\partial B_r}{\partial z}$$

$$-\frac{1}{r}\left(\frac{\partial D}{\partial z}\right)\left(\frac{\partial B_\phi}{\partial\phi}\right) + \frac{1}{r}\left(\frac{\partial D}{\partial\phi}\right)\left(\frac{\partial B_\phi}{\partial z}\right) + B_\phi\frac{\partial}{\partial r}\frac{D}{r}.$$

The time evolution of the electron pressure is described by the following equation:

$$\frac{\partial p_e}{\partial t} + \left(\frac{1}{r}\frac{\partial(r p_e U_{er})}{\partial r} + \frac{1}{r}\frac{\partial(p_e U_{e\phi})}{\partial\phi} + \frac{\partial(p_e U_{ez})}{\partial z}\right)$$

$$+\gamma p_e\left(\frac{1}{r}\frac{\partial(r U_{er})}{\partial r} + \frac{1}{r}\frac{\partial(U_{e\phi})}{\partial\phi} + \frac{\partial U_{ez}}{\partial z}\right) = \frac{(\gamma-1)}{\sigma_{\text{eff}}}J^2. \quad (13.10)$$

In above equation, we neglected of the viscosity stress tensor π^e and the heat flux q_e.

13.1.3 Spherical Coordinates

Ohm's law:

$$\begin{pmatrix}E_r \\ E_\theta \\ E_\phi\end{pmatrix} = -\frac{1}{c}\begin{pmatrix}U_{e\theta}B_\phi - U_{e\phi}B_\theta \\ U_{e\phi}B_r - U_{er}B_\phi \\ U_{er}B_\theta - U_{e\theta}B_r\end{pmatrix} - \frac{1}{en_e}\begin{pmatrix}\frac{\partial p_e}{\partial r} \\ \frac{1}{r}\frac{\partial p_e}{\partial\theta} \\ \frac{1}{r\sin\theta}\frac{\partial p_e}{\partial\phi}\end{pmatrix} + \frac{1}{\sigma_{\text{eff}}}\begin{pmatrix}J_r \\ J_\theta \\ J_\phi\end{pmatrix}$$

13.1 Maxwell's Equations and Electron Pressure Equations

$$-\frac{m}{e}\frac{d}{dt}\begin{pmatrix} U_{er} \\ U_{e\theta} \\ U_{e\phi} \end{pmatrix}. \qquad (13.11)$$

Faraday's law:

$$\frac{\partial}{\partial t}\begin{pmatrix} B_r \\ B_\theta \\ B_\phi \end{pmatrix} = -c \begin{pmatrix} \frac{1}{r\sin\theta}\left(\frac{\partial(E_\phi \sin\theta)}{\partial \theta} - \frac{\partial E_\theta}{\partial \phi}\right) \\ \frac{1}{r}\left(\frac{1}{\sin\theta}\frac{\partial E_r}{\partial \phi} - \frac{\partial(rE_\phi)}{\partial r}\right) \\ \frac{1}{r}\left(\frac{\partial(rE_\theta)}{\partial r} - \frac{\partial E_r}{\partial \theta}\right) \end{pmatrix}. \qquad (13.12)$$

Ampere's law:

$$\begin{pmatrix} J_r \\ J_\theta \\ J_\phi \end{pmatrix} = \frac{c}{4\pi} \begin{pmatrix} \frac{1}{r\sin\theta}\left(\frac{\partial(B_\phi \sin\theta)}{\partial \theta} - \frac{\partial B_\theta}{\partial \phi}\right) \\ \frac{1}{r}\left(\frac{1}{\sin\theta}\frac{\partial B_r}{\partial \phi} - \frac{\partial(rB_\phi)}{\partial r}\right) \\ \frac{1}{r}\left(\frac{\partial(rB_\theta)}{\partial r} - \frac{\partial B_r}{\partial \theta}\right) \end{pmatrix}. \qquad (13.13)$$

The inductive equation may be written in the following nondivergent form:

$$\frac{\partial}{\partial t}\begin{pmatrix} B_r \\ B_\theta \\ B_\phi \end{pmatrix} + \left(\frac{U_{er}}{r}\frac{\partial}{\partial r}r + \frac{U_{e\theta}}{r}\frac{\partial}{\partial \theta} + \frac{U_{e\phi}\sin\theta}{r}\frac{\partial}{\partial \phi}\right)\begin{pmatrix} B_r \\ B_\theta \\ B_\phi \end{pmatrix}$$
$$-\left(\frac{1}{r}\frac{\partial}{\partial r}D\frac{\partial}{\partial r}r + \frac{1}{r^2}\frac{\partial}{\partial \theta}D\frac{\partial}{\partial \theta} + \frac{1}{r^2\sin^2\theta}\frac{\partial}{\partial \phi}D\frac{\partial}{\partial \phi}\right)\begin{pmatrix} B_r \\ B_\theta \\ B_\phi \end{pmatrix} = -\begin{pmatrix} \Phi_r \\ \Phi_\theta \\ \Phi_\phi \end{pmatrix},$$
$$(13.14)$$

where

$$\Phi_r = \frac{B_r}{r}\left(2U_{er} + \frac{\partial U_{e\theta}}{\partial \theta} + U_{e\theta}\cot\theta - \frac{1}{\sin\theta}\frac{\partial U_{e\phi}}{\partial \phi}\right) - \frac{B_\theta}{r}\frac{\partial U_{er}}{\partial \theta} - \frac{B_\phi}{r\sin\theta}\frac{\partial U_{er}}{\partial \phi}$$

$$-\frac{c}{er^2\sin\theta}\left(\frac{\partial}{\partial \theta}\frac{1}{n_e}\frac{\partial p_e}{\partial \phi} - \frac{\partial}{\partial \phi}\frac{1}{n_e}\frac{\partial p_e}{\partial \theta}\right)$$

$$+rB_r\frac{\partial}{\partial r}\frac{D}{r^2} + \frac{D\cot\theta}{r^2}\frac{\partial B_r}{\partial \theta} + \left(\frac{\partial}{\partial r}\frac{D}{r}\right)\left(\cot\theta B_\theta + \frac{\partial B_\theta}{\partial \theta}\right)$$

$$-\frac{D}{r^2}\frac{\partial B_\theta}{\partial \theta} - \frac{DB_\theta\cot\theta}{r^2} - \frac{1}{r^2}\left(\frac{\partial D}{\partial \theta}\right)B_\theta - \frac{1}{r}\left(\frac{\partial D}{\partial \theta}\right)\left(\frac{\partial B_\theta}{\partial r}\right)$$

$$-\frac{1}{r^2\sin\theta}\left(\frac{\partial D}{\partial\phi}\right)\left(\frac{\partial rB_\phi}{\partial r}\right) - \frac{2D}{r^2\sin\theta}\frac{\partial B_\phi}{\partial\phi} + \frac{1}{r\sin\theta}\left(\frac{\partial D}{\partial r}\right)\left(\frac{\partial B_\phi}{\partial\phi}\right),$$

$$\Phi_\theta = -B_r\left(\frac{\partial U_{e\theta}}{\partial r} - \frac{U_{e\theta}}{r}\right) + B_\theta\left(\frac{\partial U_{er}}{\partial r} + \frac{1}{r\sin\theta}\frac{\partial U_{e\phi}}{\partial\phi} + \frac{U_{e\theta}\cot\theta}{r}\right)$$

$$-\frac{B_\phi}{r\sin\theta}\frac{\partial U_{e\theta}}{\partial\phi} - \frac{c}{er\sin\theta}\left(\frac{\partial}{\partial\phi}\frac{1}{n_e}\frac{\partial p_e}{\partial r} - \frac{\partial}{\partial r}\frac{1}{n_e}\frac{\partial p_e}{\partial\phi}\right)$$

$$-\frac{1}{r}\left(\frac{\partial D}{\partial r}\right)\left(\frac{\partial B_r}{\partial\theta}\right) + \frac{1}{r^3}\left(\frac{\partial D}{\partial\theta}\right)\left(\frac{\partial r^2 B_r}{\partial r}\right) + \frac{2D}{r^2}\frac{\partial B_r}{\partial\theta} - \frac{B_\theta D}{r^2\sin^2\theta}$$

$$+\frac{B_\theta\cot\theta}{r^2}\left(\frac{\partial D}{\partial\theta}\right) + \frac{D\cot\theta}{r^2}\frac{\partial B_\theta}{\partial\theta} - \frac{1}{r^2\sin\theta}\left(\frac{\partial D}{\partial\phi}\right)\left(\frac{\partial B_\phi}{\partial\theta}\right)$$

$$+\frac{1}{r^2\sin\theta}\left(\frac{\partial D}{\partial\theta}\right)\left(\frac{\partial B_\phi}{\partial\phi}\right) - \frac{2D\cos\theta}{r^2\sin^2\theta}\frac{\partial B_\phi}{\partial\phi} - \frac{B_\phi\cos\theta}{r^2\sin^2\theta}\frac{\partial D}{\partial\phi},$$

$$\Phi_\phi = -B_r\left(\frac{\partial U_{e\phi}}{\partial r} - \frac{U_{e\phi}}{r}\right) - \frac{B_\theta}{r}\left(\frac{\partial U_{e\phi}}{\partial\theta} - U_{e\phi}\cot\theta\right) + B_\phi\left(\frac{\partial U_{er}}{\partial r} + \frac{1}{r}\frac{\partial U_{e\theta}}{\partial\theta}\right)$$

$$-\frac{c}{er}\left(\frac{\partial}{\partial r}\frac{1}{n_e}\frac{\partial p_e}{\partial\theta} - \frac{\partial}{\partial\theta}\frac{1}{n_e}\frac{\partial p_e}{\partial r}\right)$$

$$-\frac{1}{r^2\sin\theta}\left(\frac{\partial D}{\partial r}\right)\left(\frac{\partial B_r}{\partial\phi}\right) + \frac{2B_r}{r^2\sin\theta}\frac{\partial D}{\partial\phi} + \frac{1}{r\sin\theta}\left(\frac{\partial D}{\partial\phi}\right)\left(\frac{\partial B_r}{\partial r}\right)$$

$$+\frac{2D}{r^2\sin\theta}\frac{\partial B_r}{\partial\phi} + \frac{D\cot\theta}{r^2}\frac{\partial B_\theta}{\partial\theta} + \frac{B_\theta\cos\theta}{r^2\sin^2\theta}\frac{\partial D}{\partial\phi}$$

$$+\frac{1}{r^2\sin\theta}\left(\frac{\partial D}{\partial\phi}\right)\left(\frac{\partial B_\theta}{\partial\theta}\right) + \frac{D\cos\theta}{r^2\sin^2\theta}\frac{\partial B_\theta}{\partial\phi}$$

$$+\frac{D}{r^2\sin\theta}\frac{\partial^2 B_\theta}{\partial\phi\partial\theta} + \frac{B_\phi}{r^2}\frac{\partial D\cot\theta}{\partial\theta}.$$

The time evolution of the electron pressure is described by the following equation:

$$\frac{\partial p_e}{\partial t} + \left(\frac{1}{r^2}\frac{\partial(r^2 p_e U_{er})}{\partial r} + \frac{1}{r\sin\theta}\frac{\partial(p_e U_{e\theta}\sin\theta)}{\partial\theta} + \frac{1}{r\sin\theta}\frac{\partial p_e U_{e\phi}}{\partial\phi}\right)$$

$$+\gamma p_e\left(\frac{1}{r^2}\frac{\partial(r^2 U_{er})}{\partial r} + \frac{1}{r\sin\theta}\frac{\partial(U_{e\theta}\sin\theta)}{\partial\theta} + \frac{1}{r\sin\theta}\frac{\partial U_{e\phi}}{\partial\phi}\right)$$

$$= \frac{(\gamma-1)}{\sigma_{\text{eff}}}J^2. \tag{13.15}$$

In above equation, we neglected the viscosity stress tensor π^e and the heat flux q_e.

13.2 Solving One-Dimensional Difference Equations

13.2.1 Three-Point Difference Equation with Nonperiodic Boundary Conditions: Forward-Elimination–Backward-Substitution Method

Let us consider the three-point equations (see, e.g., [364, 463])

$$A_i y_{i-1} - C_i y_i + B_i y_{i+1} = -F_i, \quad i = 1, 2, ..., N-1 \qquad (13.16)$$

with boundary conditions

$$y_0 = \chi_1 y_1 + \mu_1, \quad y_N = \chi_2 y_{N-1} + \mu_2.$$

Assume that the coefficients A_i, B_i and C_i satisfy the conditions

$$A_i \neq 0, \quad B_i \neq 0, \quad C_i \geq A_i + B_i, \quad i = 1, 2, ..., N-1,$$

where the last inequality holds at least for one index i.

Assume that there is a relation

$$y_i = \alpha_{i+1} y_{i+1} + \beta_{i+1}, \quad i = 0, ..., N-1 \qquad (13.17)$$

with unknown coefficients α_i and β_i. A substitution of the expression $y_{i-1} = \alpha_i y_i + \beta_i$ into (13.16) gives:

$$(A_i \alpha_i - C_i) y_i + A_i \beta_i + B_i y_{i+1} = -F_i.$$

Using (13.17) one can write

$$[(A_i \alpha_i - C_i)\alpha_{i+1} + B_i] y_{i+1} + A_i \beta_i + (A_i \alpha_i - C_i)\beta_{i+1} = -F_i.$$

This equation must be satisfied for any y_i, if

$$(A_i \alpha_i - C_i)\alpha_{i+1} + B_i = 0, \quad A_i \beta_i + (A_i \alpha_i - C_i)\beta_{i+1} + F_i = 0.$$

The recurrence relation may be obtained for α_{i+1}:

$$\alpha_{i+1} = \frac{B_i}{C_i - \alpha_i A_i}, \quad i = 1, 2, ..., N-1. \qquad (13.18)$$

Let us suppose that the denominator of (13.18) is not equal zero, then we can obtain the recurrence relation for the calculation of β_i:

$$\beta_{i+1} = \frac{A_i \beta_i + F_i}{C_i - \alpha_i A_i}, \quad i = 1, 2, ..., N-1. \qquad (13.19)$$

The left boundary condition, $y_0 = \chi_1 y_1 + \mu_1$, and the recurrence relation at point $i = 0$, $y_0 = \alpha_1 y_1 + \beta_1$, give

$$\alpha_1 = \chi_1, \quad \beta_1 = \mu_1. \tag{13.20}$$

Thus, one has a Cauchy problem for the determination of α_i and β_i, the formula (13.19). Now one has to determine the boundary value y_N from the following relations:

$$y_N = \chi_2 y_{N-1} + \mu_2, \quad y_{N-1} = \alpha_N y_N + \beta_N;$$

alternatively, if $1 - \alpha_N \chi_2 \neq 0$, one has

$$y_N = \frac{\mu_2 + \chi_2 \beta_N}{1 - \alpha_N \chi_2}. \tag{13.21}$$

One has again the Cauchy problem for the determination of y_i. Thus, the common loop for solving (13.16) may be written as follows:

$$(\rightarrow)\alpha_{i+1} = \frac{B_i}{C_i - \alpha_i A_i}, \quad i = 1, 2, ..., N-1, \alpha_1 = \chi_1, \tag{13.22}$$

$$(\rightarrow)\beta_{i+1} = \frac{A_i \beta_i + F_i}{C_i - \alpha_i A_i}, \quad i = 1, 2, ..., N-1, \beta_1 = \mu_1, \tag{13.23}$$

$$y_N = \frac{\mu_2 + \chi_2 \beta_N}{1 - \alpha_N \chi_2}, \tag{13.24}$$

$$(\leftarrow)y_i = \alpha_{i+1} y_{i+1} + \beta_{i+1}, \quad i = N-1, N-2, ..., 1, 0. \tag{13.25}$$

The direction of the arrow indicates the direction of calculation.

13.2.2 Three-Point Difference Equation with Periodic Boundary Conditions: Forward-Elimination–Backward-Substitution Method

The cyclic sweep method is used to obtain the periodic solution of the finite-difference equation or (system of equations). Such a problem may occur in the calculation in cylindrical and spherical coordinates.

Let us consider the system of three-point equations (see, e.g., [364, 463]):

$$A_1 y_N - C_1 y_1 + B_1 y_2 = -F_1,$$
$$A_i y_{i-1} - C_i y_i + B_i y_{i+1} = -F_i, \quad i = 2, 3, ..., N-1, \tag{13.26}$$
$$A_N y_{N-1} - C_N y_N + B_N y_1 = -F_N.$$

Such a problem occurs when searching for the periodic ($y_{i+N} = y_i$) solution of the system of three-point equations,

$$A_i y_{i-1} - C_i y_i + B_i y_{i+1} = -F_i, \quad i = \pm 0, \pm 1, ...,$$

with the periodic boundary conditions

$$A_{i+N} = A_i, \quad B_{i+N} = B_i, \quad C_{i+N} = C_i, \quad F_{i+N} = F_i.$$

Assume that the coefficients A_i, B_i and C_i satisfy the conditions

$$A_i > 0, \quad B_i > 0, \quad C_i > A_i + B_i. \tag{13.27}$$

The final loop for solving (13.26) is the following:

$$\alpha_{i+1} = \frac{B_i}{C_i - \alpha_i A_i}, \quad \beta_{i+1} = \frac{A_i \beta_i + F_i}{C_i - \alpha_i A_i}, \quad \gamma_{i+1} = \frac{A_i \gamma_i}{C_i - A_i \alpha_i},$$

$$i = 2, 3, ..., N, \tag{13.28}$$

where

$$\alpha_2 = B_1/C_1, \quad \beta_2 = F_1/C_1, \quad \gamma_2 = A_1/C_1;$$

$$p_i = \alpha_{i+1} p_{i+1} + \beta_{i+1}, \quad q_i = \alpha_{i+1} q_{i+1} + \gamma_{i+1}, \quad i = N-2, ..., 1, \tag{13.29}$$

where

$$p_{N-1} = \beta_N, \quad q_{N-1} = \alpha_N + \gamma_N;$$

and finally

$$y_i = p_i + y_N q_i, \quad i = 1, 2, ..., N-1, \tag{13.30}$$

where

$$y_N = \frac{\beta_{N+1} + \alpha_{N+1} p_1}{1 - \alpha_{N+1} q_1 - \gamma_{N+1}}.$$

The stability of this method is provided under the condition $1 - \alpha_{N+1} q_1 - \gamma_{N+1} \neq 0$. Indeed, one can see from (13.28) that $\alpha_i < 1$, $\gamma_i > 0$, $\alpha_2 + \gamma_2 < 1$. Assume that $\alpha_i + \gamma_i < 1$, then

$$\alpha_{i+1} + \gamma_{i+1} = \frac{B_i + A_i \gamma_i}{C_i - A_i \alpha_i} < \frac{B_i + A_i - A_i \alpha_i}{C_i - A_i \alpha_i} < 1. \tag{13.31}$$

Thus, one can find that $q_{N-1} < 1$, $q_i < 1$ and finally $1 - \alpha_{N+1} q_1 - \gamma_{N+1} > 0$.

13.2.3 Five-Point Difference Equation: Forward-Elimination–Backward-Substitution Method

Let us consider the following system of the five-point equations:

$$C_0 y_0 - D_0 y_1 + E_0 y_2 = F_0,$$

$$-B_1 y_0 + C_1 y_1 - D_1 y_2 + E_1 y_3 = F_1,$$

$$A_i y_{i-2} - B_i y_{i-1} + C_i y_i - D_i y_{i+1} + E_i y_{i+2} = F_i, \quad 2 \leq i \leq N-2, \tag{13.32}$$

$$A_{N-1} y_{N-3} - B_{N-1} y_{N-2} + C_{N-1} y_{N-1} - D_{N-1} y_N = F_{N-1},$$

$$A_N y_{N-2} - B_N y_{N-1} + C_N y_N = F_N,$$

which can occur in the finite-difference approximation to the fourth-order difference equations.

The algorithm for five-point forward-elimination–backward-substitution method is the following:

$$\alpha_{i+1} = [C_i - (B_i - A_i\beta_{i-1})\beta_i - A_i\alpha_{i-1}]^{-1} E_i, \quad i = 2, 3, ..., N-2 \quad (13.33)$$

and

$$\alpha_1 = C_0^{-1} E_0, \quad \alpha_2 = (C_1 - B_1\beta_1)^{-1} E_1,$$

$$\beta_{i+1} = [C_i - (B_i - A_i\beta_{i-1})\beta_i - A_i\alpha_{i-1}]^{-1} [D_i - (B_i - A_i\beta_{i-1})\alpha_i],$$

$$i = 2, 3, ..., N-1 \quad (13.34)$$

and

$$\beta_1 = C_0^{-1} D_0, \quad \beta_2 = (C_1 - B_1\beta_1)^{-1} (D_1 - B_1\alpha_1),$$

$$\gamma_{i+1} = [C_i - (B_i - A_i\beta_{i-1})\beta_i - A_i\alpha_{i-1}]^{-1} [F_i - (B_i - A_i\beta_{i-1})\gamma_i - A_i\gamma_{i-1}],$$

$$i = 2, 3, ..., N, \quad (13.35)$$

where

$$\gamma_1 = C_0^{-1} F_0, \quad \gamma_2 = (C_1 - B_1\beta_1)^{-1} (F_1 + B_1\gamma_1).$$

The final solution is obtained by the formulae

$$y_i = -\alpha_{i+1} y_{i+2} + \beta_{i+1} y_{i+1} + \gamma_{i+1}, \quad i = N-2, N-3, ..., 0, \quad (13.36)$$

where

$$y_N = \gamma_{N+1}, \quad y_{N-1} = \beta_N \gamma_{N+1} + \gamma_N.$$

The stability and applicability follows from satisfaction of the following conditions:

$$|A_i|, |B_i|, |D_i|, |E_i| > 0, \quad |C_i| \geq |A_i| + |B_i| + |D_i| + |E_i|, \quad (13.37)$$

$$2 \leq i \leq N-2, \quad |C_0| \geq |E_0| + |D_0|, \quad |C_N| \geq |A_N| + |B_N|,$$

$$|C_1| \geq |B_1| + |D_1| + |E_1|, \quad |C_{N-1}| \geq |A_{N-1}| + |B_{N-1}| + |D_{N-1}|,$$

where one of these inequalities must be strict.

14. Solutions

2.1 Select the macroatom with index j from (2.189) or (2.190). Then consider the charge exchange between a macroatom with weight γ_i and a proton with weight α_j. For convenience we can split the weight α_j in two parts $\alpha_j = \alpha_j^{(1)} + \alpha_j^{(2)}$. The first term is due to charge exchange with the atom i, and second term is a rest part of ion j.

If the value of γ_i satisfies the condition $\gamma_i < \alpha_j$, then a macroatom is totally charge exchanged with a macroproton and

$$\alpha_j^{(1)} := \gamma_i, \quad \alpha_j^{(2)} := \alpha_j^{(1)} - \gamma_i.$$

If the value of γ_i satisfies the condition $\gamma_i > \alpha_j$, then only part of a macroatom is charge exchanged with a macroproton and

$$\alpha_j^{(1)} := \alpha_j, \quad \alpha_j^{(2)} = 0.$$

Then we have to select a new macro-ion with index j' from (2.189) or (2.190). The rest portion of the macroatom $\gamma_i - \alpha_j$ must be charge exchanged with this macroproton.

3.1 Let $\Delta t \Rightarrow -\Delta t$, $n+1/2 \Rightarrow n-1/2$ and $n-1/2 \Rightarrow n+1/2$ in this scheme. Then (3.28–3.29) become

$$\frac{v^{n-1/2} - v^{n+1/2}}{-\Delta t} = \frac{q}{M}\left(E(x^n) + \frac{v^{n-1/2} + v^{n+1/2}}{2c} \times B(x^n)\right), \quad (14.1)$$

$$\frac{x^{n-1} - x^n}{-\Delta t} = v^{n-1/2}. \quad (14.2)$$

We can see that this scheme is equal to the scheme (3.28–3.29). The scheme (3.28–3.29) is the second-order finite-difference approximation to the equation of particle motion since all variables are centered at time level $n + 1/2$.

3.2 Let $\Delta t \Rightarrow -\Delta t$, $n+1/2 \Rightarrow n-1/2$ and $n-1/2 \Rightarrow n+1/2$ in this scheme. Then (3.32–3.34) become

$$\frac{v^{n-1} - v^n}{-\Delta t} = \frac{q}{M}\left(E^{n-1/2} + \frac{v^{n-1} + v^n}{2c} \times B^{n-1/2}\right), \quad (14.3)$$

$$\frac{x^{n-1} - x^n}{-\Delta t} = \frac{v^{n-1} + v^n}{2}, \tag{14.4}$$

where

$$E^{n-1/2} = E\left(\frac{x^{n-1} + x^n}{2}\right), \quad B^{n-1/2} = B\left(\frac{x^{n-1} + x^n}{2}\right). \tag{14.5}$$

We can see that this scheme equals to the scheme (3.32–3.34). The scheme (3.32–3.34) is the second-order finite-difference approximation to the equation of particle motion since all variables are centered at time level $n + 1/2$.

3.3 Let $\Delta t \Rightarrow -\Delta t$, $n+1/2 \Rightarrow n-1/2$ and $n-1/2 \Rightarrow n+1/2$ in this scheme. Then (3.38–3.41) become

$$v_1 = v^n - c\boldsymbol{E} \times \boldsymbol{B}/B^2; \tag{14.6}$$

$$\frac{(v_2 - v_1)}{-\Delta t} = \frac{q}{M}\boldsymbol{E}_\parallel + (v_1 + v_2) \times \boldsymbol{\Omega}/2; \tag{14.7}$$

$$v^{n-1} = v_2 + c\boldsymbol{E} \times \boldsymbol{B}/B^2; \tag{14.8}$$

$$\frac{x^{n-1} - x^n}{-\Delta t} = \frac{v^n + v^{n+1}}{2}. \tag{14.9}$$

We can see that this scheme is equal to the scheme (3.38–3.41). The scheme (3.38–3.41) is the second-order finite-difference approximation to the equation of particle motion since all variables are centered at time level $n + 1/2$.

3.4 Let $\Delta t \Rightarrow -\Delta t$, $n+1/2 \Rightarrow n-1/2$ and $n-1/2 \Rightarrow n+1/2$ in this scheme. Then (3.42–3.45) become

$$\frac{v_1 - v^n}{-\Delta t/2} = q\boldsymbol{E}^n/M, \tag{14.10}$$

$$\frac{v_2 - v_1}{-\Delta t} = (v_1 + v_2) \times \frac{\boldsymbol{\Omega}^n}{2}, \tag{14.11}$$

$$\frac{v^{n-1} - v_2}{-\Delta t/2} = q\boldsymbol{E}^n/M, \tag{14.12}$$

$$\frac{x^{n-1} - x^n}{-\Delta t} = \frac{v^n + v^{n-1}}{2}. \tag{14.13}$$

We can see that this scheme is equal to the scheme (3.42–3.45). The scheme (3.42–3.45) is the second-order finite-difference approximation to the equation of particle motion since all variables are centered at time level $n + 1/2$.

3.5 In coordinate form the dimensionless equation of motion of particles may be written as

$$\frac{d}{dt}\begin{pmatrix} v_x \\ v_y \\ v_z \end{pmatrix} = \frac{1}{\varepsilon \tilde{M}_s}\begin{pmatrix} E_x + v_y B_z - v_z B_y \\ E_y + v_z B_x - v_x B_z \\ E_z + v_x B_y - v_y B_x \end{pmatrix}. \quad (14.14)$$

Here $\boldsymbol{E}^*$ is the effective electric field. The parameter ε is the ratio of the proton Larmor radius $\varrho_{ci} = U_0/\Omega_i$ in the unperturbed field B_1 to the characteristic dimension L of the problem, $\varepsilon = \varrho_{ci}/L$, U_0 is the characteristic velocity of the problem and $\tilde{M}_s$ is the ratio of ion to proton masses.

We can write the following finite-difference scheme:

$$v_x^{n+1} - v_x^n = \tilde{E}_x + (v_y^{n+1} + v_y^n)\tilde{B}_z - (v_z^{n+1} + v_z^n)\tilde{B}_y,$$

$$v_y^{n+1} - v_y^n = \tilde{E}_y + (v_z^{n+1} + v_z^n)\tilde{B}_x - (v_x^{n+1} + v_x^n)\tilde{B}_z, \quad (14.15)$$

$$v_z^{n+1} - v_z^n = \tilde{E}_z + (v_x^{n+1} + v_x^n)\tilde{B}_y - (v_y^{n+1} + v_y^n)\tilde{B}_x,$$

where

$$\tilde{\boldsymbol{B}} = 0.5\frac{\Delta t}{\varepsilon \tilde{M}_s}\boldsymbol{B}^{n+1/2}, \quad \tilde{\boldsymbol{E}} = \frac{\Delta t}{\varepsilon \tilde{M}_s}\boldsymbol{E}^{n+1/2}.$$

Finally, we have

$$v_x^{n+1} = \Delta_x/\Delta, \quad v_y^{n+1} = \Delta_y/\Delta, \quad v_z^{n+1} = \Delta_z/\Delta, \quad (14.16)$$

where

$$\Delta = 1 + \tilde{B}_x^2 + \tilde{B}_y^2 + \tilde{B}_z^2,$$

$$\Delta_x = F_x + \tilde{B}_x \tilde{B}_z F_z + \tilde{B}_x \tilde{B}_y F_y - \tilde{B}_y F_z + \tilde{B}_z F_y + \tilde{B}_x^2 F_x,$$

$$\Delta_y = F_y + \tilde{B}_y \tilde{B}_x F_x + \tilde{B}_y \tilde{B}_z F_z - \tilde{B}_z F_x + \tilde{B}_x F_z + \tilde{B}_y^2 F_y,$$

$$\Delta_z = F_z + \tilde{B}_z \tilde{B}_y F_y + \tilde{B}_z \tilde{B}_x F_x - \tilde{B}_x F_y + \tilde{B}_z F_x + \tilde{B}_z^2 F_z, \quad (14.17)$$

where

$$F_x = v_x^n + \tilde{E}_x + v_y^n \tilde{B}_z - v_z^n \tilde{B}_y,$$

$$F_y = v_y^n + \tilde{E}_y + v_z^n \tilde{B}_x - v_x^n \tilde{B}_z,$$

$$F_x = v_z^n + \tilde{E}_z + v_x^n \tilde{B}_y - v_y^n \tilde{B}_x.$$

5.1 We have the second Maxwell equation

$$\boldsymbol{B}^{n+\theta} = \boldsymbol{B}^n - \theta \Delta t \nabla \times \boldsymbol{E}^{n+\theta}. \tag{14.18}$$

The equation for updating of the electric field is

$$d\nabla \times (\nabla \times \boldsymbol{E}^{n+\theta}) + A\boldsymbol{E}^{n+\theta} + (\nabla \times \boldsymbol{E}^{n+\theta}) \times \boldsymbol{I}$$
$$+ g((\nabla \times (\nabla \times \boldsymbol{E}^{n+\theta})) \times \boldsymbol{B}^n) = \boldsymbol{Q}, \tag{14.19}$$

where

$$d = \frac{m}{M}\frac{\varepsilon^2}{M_A^2} + \theta n_e l_d^* \Delta t, \quad A = n_e, \quad g = \frac{\theta \Delta t \varepsilon}{M_A^2}, \tag{14.20}$$

$$\boldsymbol{I} = \frac{\theta \varepsilon \Delta t}{M_A^2}\left(\frac{M_A^2}{\varepsilon}\sum_{k=1}^{N_s}\boldsymbol{J}_k^{n+1/2} - \nabla \times \boldsymbol{B}^n\right), \tag{14.21}$$

$$\boldsymbol{Q} = -\left(\sum_{k=1}^{N_s}\boldsymbol{J}_k^{n+\frac{1}{2}} - \frac{\varepsilon}{M_A^2}\nabla \times \boldsymbol{B}^n\right) \times \boldsymbol{B}^n - \frac{\varepsilon \beta_e}{2M_A^2}\nabla p_e + n_e l_d^* \nabla \times \boldsymbol{B}^n. \tag{14.22}$$

The transformation from dimensional variables to dimensionless variables is described by

$$t = t'L/U_\infty, \quad r = r'L, \quad \boldsymbol{u} = \boldsymbol{u}'U_\infty,$$
$$\boldsymbol{B} = \boldsymbol{B}'B_\infty, \quad \boldsymbol{E} = \boldsymbol{E}'U_\infty B_\infty/c, \quad p = p'p_\infty, \tag{14.23}$$

where M_A is the Alfvén Mach number; $\varepsilon = \varrho_{ci}/L$ is the ratio of the ion gyroradius to the characteristic length L; and U_∞, B_∞ and n_∞ are the characteristic velocity, the magnetic field and the density of incoming flow. The dissipation length equals $l_d^* = 1/Re = \eta c^2/(4\pi L U_\infty)$, where η is the anomalous resistivity.

5.2 This equation is evaluated at a time level between the n and $n+1$ levels. The electric and magnetic fields at time level $n + 1/2$ are given by

$$\boldsymbol{E}^{n+1/2} = \frac{1}{2}(\boldsymbol{E}^{n+1} + \boldsymbol{E}^n), \tag{14.24}$$

$$\boldsymbol{B}^{n+1/2} = \frac{1}{2}(\boldsymbol{B}^{n+1} + \boldsymbol{B}^n). \tag{14.25}$$

In the predictor step one needs to calculate the electric $\boldsymbol{E}$ and magnetic $\boldsymbol{B}$ fields at time level $n + 1/2$; for this purpose we used the second Maxwell equation, that gives

$$\boldsymbol{B}^{n+1} = \boldsymbol{B}^n - c\Delta t \nabla \times \boldsymbol{E}^{n+\frac{1}{2}}. \tag{14.26}$$

The evaluation of (5.41) at time level $n + 1/2$ in combination with (14.26) results in the equation for $\boldsymbol{E}^{n+1/2}$, i.e. the electric field at time level $(n+1/2)$

$$d\nabla \times (\nabla \times \boldsymbol{E}^{n+\frac{1}{2}}) + A\boldsymbol{E}^{n+\frac{1}{2}} = \boldsymbol{Q}^{n+\frac{1}{2}}, \tag{14.27}$$

where

$$d = c^2, \quad A = \omega_{\text{pe}}^2 \left(1 + \sum_{k=1}^{N_s} \frac{n_k Z_k^2 m}{n_e M_k}\right), \tag{14.28}$$

and

$$\boldsymbol{Q}^{n+\frac{1}{2}} = -\frac{e}{m} \left\{ \frac{4\pi}{c} \sum_{k=1}^{N_s} \left(1 + \frac{Z_k m}{M_k}\right) \boldsymbol{J}_k^{n+\frac{1}{2}} - \nabla \times \boldsymbol{B}^{n+\frac{1}{2}} \right\} \times \boldsymbol{B}^{n+\frac{1}{2}}$$

$$-\frac{4\pi e}{m} \{\nabla p_e + \partial/\partial x_\beta \pi_{\alpha\beta}^e\} + 4\pi \left(\sum_{k=1}^{N_s} Z_k e \nabla \cdot \mathsf{K}_k + (\nabla \boldsymbol{U}_e) \boldsymbol{J}_e \right)$$

$$-\frac{4\pi e}{m} \left(\boldsymbol{R}_e - \sum_{k=1}^{N_s} \frac{Z_k m}{M_k} \boldsymbol{R}_k \right). \tag{14.29}$$

5.3 Let us write the Crank–Nicolson scheme for (5.168) as

$$\frac{\boldsymbol{B}^{n+1} - \boldsymbol{B}^n}{\Delta t} + U_x \frac{\partial \boldsymbol{B}^{n+\frac{1}{2}}}{\partial x} - \frac{\partial}{\partial x} l_{\text{d}}^* \frac{\partial}{\partial x} \boldsymbol{B}^{n+\frac{1}{2}} = -\boldsymbol{\Phi}^{n+\frac{1}{2}}, \tag{14.30}$$

here $\boldsymbol{B}^{n+1}$ and $\boldsymbol{B}^n$ are the values of the magnetic field at time levels $n+1$ and n, and $\boldsymbol{B}^{n+\frac{1}{2}} = 0.5(\boldsymbol{B}^{n+1} + \boldsymbol{B}^n)$. Substitution of expression $\boldsymbol{B}^{n+\frac{1}{2}}$ into (14.30) finally gives

$$\frac{\boldsymbol{B}_i^{n+1} - \boldsymbol{B}_i^n}{\Delta t} + \frac{|U_{x,i}| + U_{x,i}}{4} \frac{\boldsymbol{B}_i^{n+1} - \boldsymbol{B}_{i-1}^{n+1}}{h_{i-\frac{1}{2}}} + \frac{|U_{x,i}| + U_{x,i}}{4} \frac{\boldsymbol{B}_i^n - \boldsymbol{B}_{i-1}^n}{h_{i-\frac{1}{2}}}$$

$$-\frac{|U_{x,i}| - U_{x,i}}{4} \frac{\boldsymbol{B}_{i+1}^{n+1} - \boldsymbol{B}_i^{n+1}}{h_{i+\frac{1}{2}}} - \frac{|U_{x,i}| - U_{x,i}}{4} \frac{\boldsymbol{B}_{i+1}^n - \boldsymbol{B}_i^n}{h_{i+\frac{1}{2}}}$$

$$-\frac{1}{2h_i}\left(l^*_{\text{d},i+\frac{1}{2}} \frac{\boldsymbol{B}_{i+1}^{n+1} - \boldsymbol{B}_i^{n+1}}{h_{i+\frac{1}{2}}} - l^*_{\text{d},i-\frac{1}{2}} \frac{\boldsymbol{B}_i^{n+1} - \boldsymbol{B}_{i-1}^{n+1}}{h_{i-\frac{1}{2}}}\right)$$

$$-\frac{1}{2h_i}\left(l^*_{\text{d},i+\frac{1}{2}} \frac{\boldsymbol{B}_{i+1}^n - \boldsymbol{B}_i^n}{h_{i+\frac{1}{2}}} - l^*_{\text{d},i-\frac{1}{2}} \frac{\boldsymbol{B}_i^n - \boldsymbol{B}_{i-1}^n}{h_{i-\frac{1}{2}}}\right) = -\boldsymbol{\Phi}^{n+\frac{1}{2}}, \tag{14.31}$$

where $h_{i+\frac{1}{2}} = x_{i+1} - x_i$, $h_{i-\frac{1}{2}} = x_i - x_{i-1}$ and $h_i = (x_{i+1} - x_{i-1})/2$.

The coefficients of the final three-point equation

$$A_i \boldsymbol{B}_{i-1} - C_i \boldsymbol{B}_i + B_i \boldsymbol{B}_{i-1} = -F_i$$

are the following:

$$A_i = \frac{|U_{x,i}| + U_{x,i}}{4h_{i-\frac{1}{2}}} + \frac{l^*_{\text{d},i-\frac{1}{2}}}{h_{i-\frac{1}{2}} h_i}, \quad B_i = \frac{|U_{x,i}| - U_{x,i}}{4h_{i+\frac{1}{2}}} + \frac{l^*_{\text{d},i+\frac{1}{2}}}{h_{i+\frac{1}{2}} h_i},$$

$$C_i = \frac{1}{\Delta t} + A_i + B_i, \quad F_i = \left[A_i \boldsymbol{B}_{i-1}^n - \left(A_i + B_i - \frac{1}{\Delta t}\right)\boldsymbol{B}_i^n + B_i \boldsymbol{B}_{i+1}^n\right].$$
(14.32)

The stability conditions for this scheme, $A_i \neq 0$, $B_i \neq 0$ and $C_i \geq A_i + B_i$, do not have any limit for the Courant–Fridrich–Levy number, CFL $= \Delta t U/h$.

5.4 For this case we may use the alternative direction implicit scheme: first, in the x-direction we have

$$\frac{\boldsymbol{B}^{n+\frac{1}{2}} - \boldsymbol{B}^n}{\Delta t/2} + U_x \frac{\partial \boldsymbol{B}^{n+\frac{1}{2}}}{\partial x} - \frac{\partial}{\partial x} l_d^* \frac{\partial \boldsymbol{B}^{n+\frac{1}{2}}}{\partial x} + U_y \frac{\partial \boldsymbol{B}^n}{\partial y} - \frac{\partial}{\partial y} l_d^* \frac{\partial \boldsymbol{B}^n}{\partial y} = -\boldsymbol{\Phi};$$
(14.33)

second, in the y-direction we have

$$\frac{\boldsymbol{B}^{n+1} - \boldsymbol{B}^{n+\frac{1}{2}}}{\Delta t/2} + U_y \frac{\partial \boldsymbol{B}^{n+1}}{\partial y} - \frac{\partial}{\partial y} l_d^* \frac{\partial \boldsymbol{B}^{n+1}}{\partial y} + U_x \frac{\partial \boldsymbol{B}^{n+\frac{1}{2}}}{\partial x} - \frac{\partial}{\partial x} l_d^* \frac{\partial \boldsymbol{B}^{n+\frac{1}{2}}}{\partial x} = -\boldsymbol{\Phi}.$$
(14.34)

The finite-difference approximation to the operators in this equation may be performed by the same way as in (14.31):
in the x-direction,

$$\frac{\boldsymbol{B}_{i,j}^{n+\frac{1}{2}} - \boldsymbol{B}_{i,j}^n}{\Delta t/2} + \frac{|U_{x,i,j}| + U_{x,i,j}}{2} \frac{\boldsymbol{B}_{i,j}^{n+\frac{1}{2}} - \boldsymbol{B}_{i-1,j}^{n+\frac{1}{2}}}{h_{x,i-\frac{1}{2}}}$$

$$- \frac{|U_{x,i,j}| - U_{x,i,j}}{2} \frac{\boldsymbol{B}_{i+1,j}^{n+\frac{1}{2}} - \boldsymbol{B}_{i,j}^{n+\frac{1}{2}}}{h_{x,i+\frac{1}{2}}}$$

$$- \frac{1}{h_{x,i}} \left(l_{d,i+\frac{1}{2},j}^* \frac{\boldsymbol{B}_{i+1,j}^{n+\frac{1}{2}} - \boldsymbol{B}_{i,j}^{n+\frac{1}{2}}}{h_{x,i+\frac{1}{2}}} - l_{d,i-\frac{1}{2},j}^* \frac{\boldsymbol{B}_{i,j}^{n+\frac{1}{2}} - \boldsymbol{B}_{i-1,j}^{n+\frac{1}{2}}}{h_{x,i-\frac{1}{2}}} \right)$$

$$+ \frac{|U_{y,i,j}| + U_{y,i,j}}{2} \frac{\boldsymbol{B}_{i,j}^n - \boldsymbol{B}_{i,j-1}^n}{h_{y,j-\frac{1}{2}}} - \frac{|U_{y,i,j}| - U_{y,i,j}}{2} \frac{\boldsymbol{B}_{i,j+1}^n - \boldsymbol{B}_{i,j}^n}{h_{y,j+\frac{1}{2}}}$$

$$- \frac{1}{h_{y,j}} \left(l_{d,i,j+\frac{1}{2}}^* \frac{\boldsymbol{B}_{i,j+1}^n - \boldsymbol{B}_{i,j}^n}{h_{y,j+\frac{1}{2}}} - l_{d,i,j-\frac{1}{2}}^* \frac{\boldsymbol{B}_{i,j}^n - \boldsymbol{B}_{i,j-1}^n}{h_{y,j-\frac{1}{2}}} \right) = -\boldsymbol{\Phi}_{ij}; \quad (14.35)$$

in the y-direction,

$$\frac{\boldsymbol{B}_{i,j}^{n+1} - \boldsymbol{B}_{i,j}^{n+\frac{1}{2}}}{\Delta t/2} + \frac{|U_{y,i,j}| + U_{y,i,j}}{2} \frac{\boldsymbol{B}_{i,j}^{n+1} - \boldsymbol{B}_{i,j-1}^{n+1}}{h_{y,j-\frac{1}{2}}}$$

$$- \frac{|U_{y,i,j}| - U_{y,i,j}}{2} \frac{\boldsymbol{B}_{i,j+1}^{n+1} - \boldsymbol{B}_{i,j}^{n+1}}{h_{y,j+\frac{1}{2}}}$$

$$-\frac{1}{h_{y,j}}\left(l^*_{\text{d},i,j+\frac{1}{2}}\frac{B^{n+1}_{i,j+1}-B^{n+1}_{i,j}}{h_{y,j+\frac{1}{2}}} - l^*_{\text{d},i,j-\frac{1}{2}}\frac{B^{n+1}_{i,j}-B^{n+1}_{i,j-1}}{h_{y,j-\frac{1}{2}}}\right)$$

$$+\frac{|U_{x,i,j}|+U_{x,i,j}}{2}\frac{B^{n+\frac{1}{2}}_{i,j}-B^{n+\frac{1}{2}}_{i-1,j}}{h_{x,i-\frac{1}{2}}} - \frac{|U_{x,i,j}|-U_{x,i,j}}{2}\frac{B^{n+\frac{1}{2}}_{i+1,j}-B^{n+\frac{1}{2}}_{i,j}}{h_{x,i+\frac{1}{2}}}$$

$$-\frac{1}{h_{x,i}}\left(l^*_{\text{d},i+\frac{1}{2},j}\frac{B^{n+\frac{1}{2}}_{i+1,j}-B^{n+\frac{1}{2}}_{i,j}}{h_{x,i+\frac{1}{2}}} - l^*_{\text{d},i-\frac{1}{2},j}\frac{B^{n+\frac{1}{2}}_{i,j}-B^{n+\frac{1}{2}}_{i-1,j}}{h_{x,i-\frac{1}{2}}}\right) = -\Phi_{ij}. \tag{14.36}$$

5.5 In the one-dimensional case it is possible to use the implicit numerical scheme for the magnetic field induction equation [41, 328, 330, 333]:

$$\frac{\partial B_x}{\partial t} = -\frac{\varepsilon B_z}{M_A^2}\frac{\partial}{\partial z}\left(\frac{J_x}{n}\right) - \frac{\partial}{\partial z}U_z B_x + B_z\frac{\partial}{\partial z}U_x + \frac{1}{Re}\frac{\partial}{\partial z}\left(\frac{J_y}{n}\right)$$

$$+\frac{m\varepsilon^2}{MM_A^2}\frac{\partial}{\partial z}\frac{\text{d}}{\text{d}t}\left(\frac{1}{n}\frac{\partial}{\partial z}B_x\right) - \frac{\varepsilon m}{M}\frac{\partial}{\partial z}\frac{\text{d}}{\text{d}t}U_{iy} \tag{14.37}$$

and

$$\frac{\partial B_y}{\partial t} = -\frac{\varepsilon B_z}{M_A^2}\frac{\partial}{\partial z}\left(\frac{J_y}{n}\right) - \frac{\partial}{\partial z}U_z B_y + B_z\frac{\partial}{\partial z}U_y - \frac{1}{Re}\frac{\partial}{\partial z}\left(\frac{J_x}{n}\right)$$

$$+\frac{m\varepsilon^2}{MM_A^2}\frac{\partial}{\partial z}\frac{\text{d}}{\text{d}t}\left(\frac{1}{n}\frac{\partial}{\partial z}B_y\right) - \frac{\varepsilon m}{M}\frac{\partial}{\partial z}\frac{\text{d}}{\text{d}t}U_{ix}. \tag{14.38}$$

Define the new variables Q_x and Q_y as [41, 328]

$$Q_x = B_x - \frac{m\varepsilon^2}{MM_A^2}\frac{\partial}{\partial z}\left(\frac{1}{n}\frac{\partial B_x}{\partial z}\right), \tag{14.39}$$

$$Q_y = B_y - \frac{m\varepsilon^2}{MM_A^2}\frac{\partial}{\partial z}\left(\frac{1}{n}\frac{\partial B_y}{\partial z}\right). \tag{14.40}$$

Then we can transform (14.37–14.38) to

$$\frac{\partial Q_x}{\partial t} = \frac{MB_z}{m\varepsilon}(B_y - Q_y) - \frac{\partial}{\partial z}U_z Q_x + B_z\frac{\partial}{\partial z}U_x + \frac{MM_A^2}{m\varepsilon^2 Re}(B_x - Q_x)$$

$$-\frac{\varepsilon m}{M}\frac{\partial}{\partial z}\frac{\text{d}}{\text{d}t}U_{iy}, \tag{14.41}$$

$$\frac{\partial Q_y}{\partial t} = -\frac{MB_z}{m\varepsilon}(B_x - Q_x) - \frac{\partial}{\partial z}U_z Q_y + B_z\frac{\partial}{\partial z}U_y + \frac{MM_A^2}{m\varepsilon^2 Re}(B_y - Q_y)$$

$$+\frac{\varepsilon m}{M}\frac{\partial}{\partial z}\frac{\text{d}}{\text{d}t}U_{ix}. \tag{14.42}$$

First, (14.41–14.42) are solved for Q_x and Q_y using the first-order approximation of convective terms, taking into account the direction of ion bulk velocity. Second, the value of the magnetic field can be found from (14.39) and (14.40) using Q_x and Q_y as given.

5.6 The electron inertia term due to (5.164) may be represented as

$$\frac{m}{ne^2}\frac{\partial}{\partial t}\boldsymbol{J} - \frac{m}{e}(\boldsymbol{U}_e \cdot \nabla)\boldsymbol{U}_e. \tag{14.43}$$

The assumption that

$$\frac{\partial}{\partial t}\boldsymbol{J}_i \ll \frac{\partial}{\partial t}\boldsymbol{J} \approx \frac{\partial}{\partial t}\boldsymbol{J}_e \tag{14.44}$$

gives

$$\frac{m}{ne^2}\frac{\partial}{\partial t}\boldsymbol{J}_e - \frac{m}{e}(\boldsymbol{U}_e \cdot \nabla)\boldsymbol{U}_e \tag{14.45}$$

and, if the ion density satisfies the condition

$$\frac{\partial}{\partial t}\ln n_i \ll 1, \tag{14.46}$$

one can rewrite (14.43), finally, as

$$-\frac{m}{e}\left(\frac{\partial}{\partial t}\boldsymbol{U}_e + (\boldsymbol{U}_e \cdot \nabla)\boldsymbol{U}_e\right) = -\frac{m}{e}\frac{\mathrm{d}}{\mathrm{d}t}\boldsymbol{U}_e. \tag{14.47}$$

Equation (14.47) approximates the electron inertia term under assumptions (14.44) and (14.46).

5.7 The electron inertia term due to (5.167) may be represented as

$$-\frac{m}{ne^2}\left(\frac{1}{ne}[(\boldsymbol{J}\nabla)\boldsymbol{J} + \nabla J^2] - \frac{m}{ne^2}\frac{\partial}{\partial t}\boldsymbol{J}\right). \tag{14.48}$$

The assumption that

$$\frac{\partial}{\partial t}\boldsymbol{J}_i \ll \frac{\partial}{\partial t}\boldsymbol{J} \approx \frac{\partial}{\partial t}\boldsymbol{J}_e \tag{14.49}$$

and

$$\frac{\partial}{\partial t}\ln n_i \ll 1 \tag{14.50}$$

gives, finally,

$$-\frac{m}{e}\left(\frac{\partial}{\partial t}\boldsymbol{U}_e + (\boldsymbol{U}_e\nabla)\boldsymbol{U}_e - \nabla U_e^2\right). \tag{14.51}$$

The first two terms of (14.51) give an approximation to the electron inertia term

$$-\frac{m}{e}\left(\frac{\mathrm{d}}{\mathrm{d}t}\boldsymbol{U}_e\right). \tag{14.52}$$

However, one has to make an additional assumption about the third term for a complete approximation of (14.52).

5.8 We can use the shifted mesh for the electron pressure $p_{i\pm1/2}$ and add a small resistive term in order to satisfy the stability condition for the sweep method:

$$\frac{p^{n+1}-p^n}{\Delta t}+U_x\frac{\partial p^{n+\frac{1}{2}}}{\partial x}+\gamma p^{n+\frac{1}{2}}\frac{\partial U_x}{\partial x}=2(\gamma-1)J^2/\beta_e Re,$$

$$J^2=\left(\frac{\partial B_z}{\partial x}\right)^2+\left(\frac{\partial B_y}{\partial x}\right)^2+\nu\frac{\partial^2 p_e^{n+1}}{\partial x^2}. \qquad (14.53)$$

The above equation is written in dimensional form. Here Re is the Reynolds number, and ν is the effective scheme's viscosity. Substitution of the expression $p^{n+\frac{1}{2}}=0.5(p^{n+1}+p^n)$ into (14.53) gives

$$\frac{p_i^{n+1}-p_i^n}{\Delta t}+\frac{|U_{x,i}|+U_{x,i}}{4}\frac{p_i^{n+1}-p_{i-1}^{n+1}}{h_{i-\frac{1}{2}}}+\frac{|U_{x,i}|+U_{x,i}}{4}\frac{p_i^n-p_{i-1}^n}{h_{i-\frac{1}{2}}}$$

$$-\frac{|U_{x,i}|-U_{x,i}}{4}\frac{p_{i+1}^{n+1}-p_i^{n+1}}{h_{i+\frac{1}{2}}}-\frac{|U_{x,i}|-U_{x,i}}{4}\frac{p_{i+1}^n-p_i^n}{h_{i+\frac{1}{2}}}$$

$$+\frac{\gamma}{4h_i}(p^{n+1}+p^n)(U_{x,i+1}-U_{x,i-1})=2(\gamma-1)J^2/\beta_e Re, \qquad (14.54)$$

where $h_{i+\frac{1}{2}}=x_{i+1}-x_i$, $h_{i-\frac{1}{2}}=x_i-x_{i-1}$ and $h_i=(x_{i+1}-x_{i-1})/2$. For the coefficients of the three-point equation

$$A_i p_{i-1}-C_i p_i+B_i p_{i-1}=-F_i$$

we have

$$A_i=\frac{|U_{x,i}|+U_{x,i}}{4h_{i-\frac{1}{2}}}+\frac{\nu}{h_{i-\frac{1}{2}}h_i}, \quad B_i=\frac{|U_{x,i}|-U_{x,i}}{4h_{i+\frac{1}{2}}}+\frac{\nu}{h_{i+\frac{1}{2}}h_i},$$

$$C_i=\frac{1}{\Delta t}+A_i+B_i+\frac{\gamma}{4h_i}(U_{x,i+1}-U_{x,i-1}),$$

$$F_i=\left[A_i p_{i-1}^n-\left(A_i+B_i-\frac{1}{\Delta t}+\frac{\gamma}{4h_i}(U_{x,i+1}-U_{x,i-1})\right)p_i^n+B_i p_{i+1}^n\right]$$

$$-\frac{2(\gamma-1)J^2}{\beta_e Re}. \qquad (14.55)$$

The stability condition of this method, $A_i\neq 0$, $B_i\neq 0$ and $C_i\geq A_i+B_i$, gives the next limitation for the Courant–Fridrich–Levy number, CFL $=\Delta t U/h$:

$$\text{CFL}\leq\frac{4}{\gamma}\left|\frac{U_i}{U_{i+1}-U_{i-1}}\right|, \quad i=1,...,N. \qquad (14.56)$$

5.9 We can use the shifted mesh for the electron pressure $p_{i\pm\frac{1}{2},j\pm\frac{1}{2}}$. For this case we may use the alternative direction implicit scheme: first, in the x-direction we have

$$\frac{p^{n+1}-p^n}{\Delta t/2}+U_x\frac{\partial p^n}{\partial x}+U_y\frac{\partial p^n}{\partial y}=-\gamma p^{n+\frac{1}{2}}\left(\frac{\partial U_x}{\partial x}+\frac{\partial U_y}{\partial y}\right)+\frac{2(\gamma-1)}{\beta_e Re}J^2; \tag{14.57}$$

second, in the y-direction we have

$$\frac{p^{n+1}-p^{n+\frac{1}{2}}}{\Delta t/2}+U_x\frac{\partial p^{n+\frac{1}{2}}}{\partial x}+U_y\frac{\partial p^{n+1}}{\partial y}=-\gamma p^{n+\frac{1}{2}}\left(\frac{\partial U_x}{\partial x}+\frac{\partial U_y}{\partial y}\right)+\frac{2(\gamma-1)}{\beta_e Re}J^2, \tag{14.58}$$

where $p^{n+1/2}=0.5(p^{n+1}+p^n)$. The finite-difference approximation of the operators in this equation may be made in the same way as in (14.54).

5.10 The semi-discretization form of the advection equation is as follows:

$$\frac{du_n}{dt}+cA^{-1}\Delta u_n/h=0, \tag{14.59}$$

where A and Δ are operators from [528]:

$$A=A_0-0.25s\Delta_0,\quad \Delta=0.5(\Delta_0-s\Delta_2),\quad s=\operatorname{sgn}c, \tag{14.60}$$

and

$$A_0f=\frac{1}{6}f_{i-1}+\frac{2}{3}f_i+\frac{1}{6}f_{i+1},\quad \Delta_0=f_{i+1}-f_{i-1},\quad \Delta_2=f_{i+1}-2f_i+f_{i-1}. \tag{14.61}$$

Looking for the solution of (14.59) in the form $U(t)\exp ikhn$, $k=\mathrm{const}$, one obtains the ordinary differential equation

$$\frac{dU}{dt}+\left[\frac{c}{h}W_0(kh)+\mathrm{i}\frac{c}{h}W_1(kh)\right]U=0, \tag{14.62}$$

where

$$W_0(\alpha)=96s\Phi(\alpha)\sin^4\alpha/2,\quad W_1(\alpha)=24\Phi(\alpha)\sin\alpha(7-\cos\alpha),$$

$$\Phi(\alpha)=\left[(8+4\cos\alpha)^2+35\sin^2\alpha\right]^{-1}.$$

Comparing (14.60) with the exact equation $U'+ickU=0$, one may obtain for the numerical phase velocity c^*

$$c^*=cW_1(\alpha)/\alpha,\quad \alpha=kh,$$

where $c=\mathrm{const}$ is recognized as the exact phase velocity of the harmonics with the wavenumber k.

The negative term $cW_0(kh)/h$ in (14.62) describes the dissipation. For $kh\ll 1$ one can see a small phase error and dissipation d

$$c^* = c\left[1 + O(\alpha^4)\right], \quad ch^{-1}W_0(\alpha) = ch^{-1}d = O(\alpha^3), \quad W_{0\max} = W_0(\pi) = 6.$$

In the high-frequency region ($\pi/2 < kh \leq \pi$), where the phase velocity strongly deviates from the exact value, the dissipation rapidly increases with kh, thus providing the efficient damping of spurious oscillations. Therefore, the compact differencing $A^{-1}\Delta$ contains a built-in filter which does not perturb the physically relevant components of solutions and extinguishes those components with the mesh size h [528].

5.11 Let us introduce a new variable $q = \partial u/\partial x$:

$$\frac{\partial u}{\partial t} + \frac{\partial(\phi - \epsilon q)}{\partial x} = f,$$

$$\frac{\partial u}{\partial x} - q = 0. \qquad (14.63)$$

For the first equation of (14.63) one can use the same operators A and Δ as in (14.60) with $s = \operatorname{sgn}\phi'(u^n)$. In order to approximate the second equation of (14.63), one can apply CUD-3 formulas with the opposite orientation of the operators, that is by setting $s = -\operatorname{sgn}\phi'(u^n)$. One may write

$$A\langle\frac{\partial u}{\partial t}\rangle^{n+1} + \Delta(\phi - \epsilon q)^{n+1}/h = Af^{n+1}, \qquad (14.64)$$

$$\tilde{A}q^{n+1} = \tilde{\Delta}u^{n+1}/h,$$

where $\tilde{A} = A^*q = A_0 + 0.25\operatorname{sgn}\phi'(u^n)\Delta_0$, $\tilde{\Delta}u = 0.5(\Delta_0 + \operatorname{sgn}\phi'(u^n)\Delta_2)$.

Considering the case $\phi(u,x) = au$, $a = \text{const}$, $\epsilon = \text{const}$, $\langle\partial u/\partial t\rangle_i^{n+1} = (u_i^{n+1} - u_i^n)/\Delta t$ the unconditional stability of the scheme (14.64) can be easily provided by the spectral method. One can verify that the truncation error of scheme (14.64) is $O(\Delta t^k + h^3)$ on the solution of (5.173) if the derivative $\partial u/\partial t$ is approximated with an error of $O(\Delta t^k)$ [528].

8.1 Let us consider a one-dimensional computational domain. Suppose the proton velocity distribution function is a Maxwellian distribution at the boundaries. At the upstream boundary we have the unperturbed electromagnetic field, $\boldsymbol{B} = \boldsymbol{B}_1$, $\boldsymbol{E} = \boldsymbol{E}_1$. In the upstream we have a supersonic flow, so the incoming flux is easily estimated as

$$F_{\text{up,inc}} = \int_0^\infty dv_x \int_{-\infty}^\infty dv_y dv_z v_x f_{\text{p}}(\boldsymbol{v}, \boldsymbol{U}_{\text{p1}}, V_{\text{Tp1}}) = n_{\text{p1}}U_{\text{p1}}. \qquad (14.65)$$

At the downstream boundary the electromagnetic field is determined by the Rankine–Hugoniot relation, $\boldsymbol{B} = \boldsymbol{B}_2$, $\boldsymbol{E} = \boldsymbol{E}_2$ (see Sect. 8.2). The leaving $F_{\text{d,l}}$ and incoming $F_{\text{d,inc}}$ fluxes must satisfy the relation

$$F_{\text{d,l}} - F_{\text{d,inc}} = n_{\text{p2}}U_{\text{p2}x}, \qquad (14.66)$$

where

$$F_{d,l} = \int_0^\infty dv_x \int_{-\infty}^\infty dv_y dv_z v_x f_p(\boldsymbol{v}, \boldsymbol{U}_{p2}, V_{Tp2}) \tag{14.67}$$

and

$$F_{d,\text{inc}} = \int_{-\infty}^0 dv_x \int_{-\infty}^\infty dv_y dv_z v_x f_p(\boldsymbol{v}, \boldsymbol{U}_{p2}, V_{Tp2}). \tag{14.68}$$

Equation (14.66) indicates that in order to keep a quasistationary shock front at rest one has to provide the corresponding incoming flux at the downstream boundary. The numerical algorithms for the generation of the fluxes $F_{\text{up,inc}}$ and $F_{l,\text{inc}}$ are considered in Sects. 7.2 and 7.3.

We can estimate the flux $F_{d,l}$ as

$$F_{d,l} = \int_0^\infty v_x f_p(v_x) dv_x = n_2 U_2$$

$$\cdot \left(1 + \frac{1}{2} \left\{ \frac{\exp(-MU_2^2/2kT_{p2})}{\pi^{1/2} U_2 (M/2kT_{p2})^{1/2}} - \text{erfc}\left[U_2 \left(\frac{M}{2kT_{p2}}\right)^{1/2} \right] \right\} \right)$$

assuming $f_p = f_p(v_x)$ [304]. At each time step a fraction $(1 - n_2 U_2/F_{d,l})$ of the particles escaping the system are re-admitted into the downstream boundary, with negative random x velocities approximating the negative wing of f_p; the remaining fraction $n_2 U_2/F_{d,l}$ of protons are injected into the upstream boundary. The net flux leaving the system at the downstream boundary is thus exactly $n_2 U_2$ and balances the incoming upstream flux at the left end [304]. This procedure keeps the total number of particles conserved, and there are no sources of mass, momentum or energy in the system.

8.2 Let us consider a one-dimensional computational domain. Suppose the pickup ion velocity distribution function is a ring distribution at the boundaries. The boundary conditions are the same as in Exercise 8.1. At the upstream boundary the incoming and leaving fluxes satisfy the relation

$$F_{\text{up,inc}} - F_{\text{up,l}} = n_{\text{PI1}} U_{\text{PI1}}, \tag{14.69}$$

where

$$F_{\text{up,inc}} = \int_0^\infty dv_x \int_{-\infty}^\infty dv_y dv_z v_x f_i(\boldsymbol{v}, \boldsymbol{U}_1, V_{\max 1}, V_{\min 1}) \tag{14.70}$$

and

$$F_{\text{up,l}} = \int_{-\infty}^0 dv_x \int_{-\infty}^\infty dv_y dv_z v_x f_i(\boldsymbol{v}, \boldsymbol{U}_1, V_{\max 1}, V_{\min 1}). \tag{14.71}$$

At the downstream boundary the electromagnetic field is determined by the Rankine–Hugoniot relation, $\boldsymbol{B} = \boldsymbol{B}_2$, $\boldsymbol{E} = \boldsymbol{E}_2$ (see Sect. 8.2). The leaving $F_{d,l}$ and incoming $F_{d,\text{inc}}$ fluxes must satisfy the relation

$$F_{\text{d,l}} - F_{\text{d,inc}} = n_{\text{PI2}} U_{\text{PI2}x}, \tag{14.72}$$

where

$$F_{\text{d,l}} = \int_0^\infty dv_x \int_{-\infty}^\infty dv_y dv_z v_x f_{\text{i}}(\boldsymbol{v}, \boldsymbol{U}_{\text{PI2}}, V_{\max 2}, V_{\min 2}) \tag{14.73}$$

and

$$F_{\text{d,inc}} = \int_{-\infty}^0 dv_x \int_{-\infty}^\infty dv_y dv_z v_x f_{\text{i}}(\boldsymbol{v}, \boldsymbol{U}_2, V_{\max 2}, V_{\min 2}). \tag{14.74}$$

Equations (14.69) and (14.72) indicate that in order to keep a quasistationary shock front at rest one has to provide the corresponding incoming flux at the downstream boundary. The numerical algorithms for the generation of the fluxes $F_{\text{up,inc}}$ and $F_{\text{l,inc}}$ are considered in Sects. 7.3.1–7.3.3.

10.1 The numerical simulation of the system of equations is accomplished using the CIC method based on the Hamilton variation principle. In our case (with no scalar potential, $\boldsymbol{A} = \boldsymbol{A}_y$) the Lagrangian according to [305] can be written as follows:

$$L = \int d^3\boldsymbol{r}' d^3\boldsymbol{v}' \cdot f(\boldsymbol{r}', \boldsymbol{v}', 0) \left\{ \frac{M \cdot \dot{\boldsymbol{R}}^2}{2} + \frac{Q \cdot \dot{\boldsymbol{R}} \cdot A_y(R, t)}{c} \right\}$$

$$- \int d^2\boldsymbol{r} \frac{B^2}{8\pi}, \tag{14.75}$$

where $M, Q, \boldsymbol{R} = \boldsymbol{R}(\boldsymbol{r}', \boldsymbol{v}', t), \boldsymbol{v} = \boldsymbol{R}(\boldsymbol{r}', \boldsymbol{v}', t)$ are the mass, charge, coordinates and velocity of a particle whose initial position and velocity are $\boldsymbol{r}'$ and $\boldsymbol{v}'$, respectively.

To obtain specific approximation schemes an assumption is made that the initial distribution function is:

$$f(\boldsymbol{r}', \boldsymbol{v}', 0) = \sum_{l=1}^{N_p} \delta(\boldsymbol{r}' - \boldsymbol{r}_l) \delta(\boldsymbol{v}' - \boldsymbol{v}_l). \tag{14.76}$$

The simplest dependence may be chosen for functions A_y and $\boldsymbol{R}$:

$$A_y = \sum_{n,m=1}^{N_g} \beta_{m,n}(t) a_{m,n}(\boldsymbol{r}),$$

$$\boldsymbol{R} = \sum_{l=1}^{N_p} (\gamma_{lx}(t) \boldsymbol{R}_{lx} + \gamma_{ly}(t) \boldsymbol{R}_{ly} + \gamma_{lz}(t) \boldsymbol{R}_{lz}), \tag{14.77}$$

where $\beta_{m,n}, \gamma_e$ are time-dependent coefficients, and

$$\boldsymbol{R}_{lx}(\boldsymbol{r}',\boldsymbol{v}') = \begin{cases} \boldsymbol{i}, & \text{if } \boldsymbol{r}' = \boldsymbol{r}_l \text{ and } \boldsymbol{v}' = \boldsymbol{v}_l, \\ 0, & \text{otherwise}, \end{cases}$$

$$\boldsymbol{R}_{ly}(\boldsymbol{r}',\boldsymbol{v}') = \begin{cases} \boldsymbol{j}, & \text{if } \boldsymbol{r}' = \boldsymbol{r}_l \text{ and } \boldsymbol{v}' = \boldsymbol{v}_l, \\ 0, & \text{otherwise}, \end{cases}$$

$$\boldsymbol{R}_{lz}(\boldsymbol{r}',\boldsymbol{v}') = \begin{cases} \boldsymbol{k}, & \text{if } \boldsymbol{r}' = \boldsymbol{r}_l \text{ and } \boldsymbol{v}' = \boldsymbol{v}_l, \\ 0, & \text{otherwise}. \end{cases}$$

The Lagrangian and Hamiltonian are given as

$$L = \sum_{l=1}^{N_{pa}} \left\{ \frac{M}{2} [\dot{\gamma}_{lx}^2 + \dot{\gamma}_{ly}^2 + \dot{\gamma}_{lz}^2] + \frac{Q}{c} \dot{\gamma}_{ly} \sum_{m,n=1}^{N_g} \beta_{mn} a_{mn}(\gamma_l) \right\}$$

$$- \int \frac{d^2\boldsymbol{r}}{8\pi} \left\{ \left[\sum_{m,n=1}^{N_g} \beta_{mn} \frac{\partial}{\partial x} a_{mn} \right]^2 + \left[\sum_{m,n=1}^{N_g} \beta_{mn} \frac{\partial}{\partial z} a_{mn} \right]^2 \right\}, \qquad (14.78)$$

$$H = \sum_{l=1}^{N_p} \frac{M}{2} [\dot{\gamma}_{lx}^2 + \dot{\gamma}_{ly}^2 + \dot{\gamma}_{lz}^2]$$

$$+ \int \frac{d^2\boldsymbol{r}}{8\pi} \left\{ \left[\sum_{m,n=1}^{N_g} \dot{\beta}_{mn} a_{mn} \right]^2 \right.$$

$$+ \left. \left[\sum_{m,n=1}^{N_g} \beta_{mn} \frac{\partial}{\partial x} a_{mn} \right]^2 + \left[\sum_{m,n=1}^{N_g} \beta_{mn} \frac{\partial}{\partial z} a_{mn} \right]^2 \right\}. \qquad (14.79)$$

The theorem of energy conservation states, in our case, that the sum of the kinetic energy of the particles, of the transverse electric component and the magnetic energy are constant.

10.2 Substitution of (14.78–14.79) into the Euler–Lagrange equations [305]

$$\frac{d}{dt}\left(\frac{\partial L}{\partial \dot{\gamma}_{kl}}\right) - \frac{\partial L}{\partial \gamma_{kl}} = 0,$$

$$\frac{d}{dt}\left(\frac{\partial L}{\partial \dot{\beta}_m}\right) - \frac{\partial L}{\partial \beta_m} = 0,$$

$$\frac{\partial L}{\partial \alpha_n} = 0,$$

gives an approximation to the system of equations of particle motion

$$M\dot{\gamma}_{ly} + \frac{Q}{c} \sum_{m,n=1}^{N_g} \beta_{mn} a_{mn}(\gamma_l) = \text{const.} = c_l,$$

$$M\ddot{\gamma}_{ly} - \frac{Q}{c}\dot{\gamma}_{ly}\sum_{m,n=1}^{N_g}\beta_{mn}\frac{\partial}{\partial x}a_{mn} = 0, \qquad (14.80)$$

$$M\ddot{\gamma}_{lz} - \frac{Q}{c}\dot{\gamma}_{ly}\sum_{m,n=1}^{N_g}\beta_{mn}\frac{\partial}{\partial z}a_{mn} = 0,$$

and of the field dynamics

$$\frac{1}{4\pi}\sum_{m,n=1}^{N_g}\beta_{mn}\int d^2\mathbf{r}\left(\frac{\partial}{\partial x}a_{mn}\frac{\partial}{\partial x}a_{ij} + \frac{\partial}{\partial z}a_{mn}\frac{\partial}{\partial z}a_{ij}\right)$$

$$= \frac{Q}{Mc}\sum_{l=1}^{N_p}c_l a_{ij}(\gamma_l) - \frac{Q^2}{c^2 M}\sum_{l=1}^{N_p}\sum_{m,n=1}^{N_g}\beta_{mn}a_{mn}(\gamma_l)a_{ij}(\gamma_l). \qquad (14.81)$$

The scheme (14.80–14.81) was accomplished on the basis of the method of finite elements [364, 398] and the discrete Fourier transform method. To derive field equation (14.81) iterative schemes of the following type were used:

$$A\beta_{mn}^{s+1} = B\beta_{mn}^s + J_{mn}^s.$$

The operator is reversed by the factorization method (see, for example, [463]), and it is of an algebraic type if the Fourier method is used. In this case the norm of the step operator, $T = A^{-1}[B]$ met the stability requirement $||T|| \leq 1$ for any ε. Since for small ε the characteristic timescale of the magnetic variations is longer than the particle gyroperiod, it was expedient to use two time grids with different time steps to calculate the field and the trajectory of particles. This approach reduces the time required for calculating one gyroperiod by a factor of 5–10 (depending on the number of corrections) as compared to the usual leapfrog scheme traditionally used in simulation methods. Variation of the total energy of the system was used to control the accuracy of the calculations. In our computations the relative energy change was about 1.5×10^{-3}.

12.1 Since O atoms mainly exchange their charge with H$^+$ ions, the following charge-exchange equilibrium (charge-exchange production and losses of O atoms cancel!) is required at the unperturbed VLISM [149, 248, 346, 499]:

$$n_H n_O^* \langle \sigma_{ex}^{HO^+}\rangle V_{rel}(H, O^+) = n_O n_H^* \langle \sigma_{ex}^{OH^+}\rangle V_{rel}(O, H^+).$$

Since in the unperturbed very local interstellar medium (VLISM) ions and atoms are considered to have the same temperature, the mean relative velocities on the two sides of above equation cancel each other and one obtains

$$(n_O^*/n_O) = \xi_O = \xi_H \frac{\langle \sigma_{ex}^{OH^+}\rangle}{\langle \sigma_{ex}^{HO^+}\rangle},$$

which means that $\xi_O = (8/9)\xi_H$ [499]. Here, $\xi_H = n_H^*/n_H$ and $\xi_O = n_O^*/n_O$ denote the fractional ionization of hydrogen and oxygen.

References

1. R. Abgrall: J. Comput. Phys. **114**, 45 (1994)
2. M.N. Acuna, J.E.P. Connerney, N.F. Ness, R.P. Lin, D. Mitchell, C.W. Carlson, J. McFadden, K.A. Anderson, H. Réme, C. Mazelle, D. Vignes, P. Wasilewski, P. Cloutier: Science **284**, 790 (1999)
3. J.C. Adam: 'Modern development in particle simulation'. In: *Computer Simulation of Space Plasmas*, ed. by H. Matsumoto, T. Sato (Terra Scientific, Tokyo 1984) p. 117
4. Y. Adam: J. Comput. Phys. **24**, 10 (1977)
5. J.C. Adam, A. Gourdin-Serveniere, A.B. Langdon: J. Comput. Phys. **47**, 229 (1982)
6. K. Akimoto, D. Winske, T.G. Onsager, M.F. Thomsen, S.P. Gary: J. Geophys. Res. **96**, 17599 (1991)
7. K. Akimoto, D. Winske, S.P. Gary, M.F. Thomsen: J. Geophys. Res. **98**, 1419 (1993)
8. A.S. Almgren, J.B. Bell, P. Colella, L.H. Howell, M.L. Welcomm: J. Comput. Phys. **142**, 1 (1998)
9. Ya. L. Al'pert, A.V. Gurevich, L.P. Pitaevskii: *Space Physics with Artificial Satellites*, (Consultant Bureau, New York 1965: Science, Moscow 1964) (in Russian)
10. A.T. Altintsev, V.I. Krasov: Sov. J. Technich. Phys. **44**, 2629 (1974)
11. K. Amano, T.J. Tsuda: Geomagn. Geoelectr. **29**, 9 (1979)
12. Yu.V. Andrijanov, I.M. Podgorny: 'Effect of the frozen-in magnetic field on the formation of Venus plasma shell boundary: experimental configuration'. In: *Solar-wind Interaction with the Planets Mercury, Venus, and Mars. The Proceeding of a bilateral seminar of the US-USSR Joint Working group on Near-Earth Space, the Moon, and Planets, held at the Space Research Institute of Academy of Sciences of the USSR, Moscow, November 17–21, 1975*, ed. by N.F. Ness (NASA, SP-397 Washington, D.C. 1976) p. 101
13. T.P. Armstrong, R.C. Harding, G. Knorr, D. Montgomery: 'Solution of Vlasov's Equation by Transform Methods'. In: *Method in Computational Physics. Plasma Physics, Vol. 9*, ed. by B. Alder, S. Fernbach, M. Rotenberg (Academic Press, New York, London 1970) p. 30
14. T.P. Armstrong, D.C. Montgomery: J. Plasma Phys. **1**, 425 (1967)
15. T.P. Armstrong, M.E. Pesses, R.B. Decker: 'Shock drift acceleration'. In: *Collisionless Shocks in the Heliosphere: Reviews of Current Research, Geophysical Monograph Series, Vol. 35*, ed. by B.T. Tsurutani, R.G. Stone (AGU, Washington, D.C. 1985) p. 271
16. M. Ashour-Abdalla, F.M. Coronity, A.A. Galeev: Geophys. Res. Lett. **5**, 707 (1978)
17. W.I. Axford: In: *Plasma Astrophysics*, ed. by T.D. Guyenne, G. Levy (ESA SP-161, Noordwijk 1981) p. 425

18. W.I. Axford: 'Magnetic field reconnection'. In: *Reconnection in Space and Laboratory Plasmas, Geophysical Monograph Series, Vol. 30*, ed. by E.W. Hones Jr. (AGU, Washington, D.C. 1984) pp. 4-14
19. A.Y. Aydemir: Phys. Plasmas **1**, 822 (1994)
20. H. Babovsky, F. Gropengiesser, H. Neunzert, J. Struckmeier, B. Wiesen: 'Low-Discrepancy Method for the Boltzmann Equation'. In: *Rarefied Gas Dynamics: Theoretical and Computational Techniques, 118*, ed. by E.P. Muntz, D.P. Weaver, D.H. Campbell (AIAA, Washington 1989) p. 85
21. G. Backus: J. Math. Phys. **1**, 178 (1960)
22. F. Bagenal, R.L. McNutt: Geophys. Res. Lett. **16**, 1229 (1989)
23. V.B. Balakin: Sov. J. Computational and Mathematical Physics **10**(6), 1512 (1970)
24. S.D. Bale, C.J. Owen, J.L. Bougeret, K. Goetz, et al.: Geophys. Res. Lett. **24**, 1427 (1997)
25. M.A. Balikhin, T.M. Vinogradova, L.D. Volkomirskaya, et al.: 'Nonlinear wave dynamics in the front of the strong shock'. In: *Collisionless Shocks, Proceedings of International Symposium, Balatonfüred, Hungary, June 1987*, ed. by K. Szegö (CRIP, Budapest 1987) p. 172
26. V.B. Baranov, K.V. Krasnobaev, A.G. Kulikovsky: Sov. Phys. Dokl. **15**, 791 (1971)
27. V.B. Baranov, M.G. Lebedev: Sov. Astron. J. **7**(3), 378 (1981)
28. V.B. Baranov, M.G. Lebedev: Astrophys. Space Sci. **147**, 69 (1988)
29. V.B. Baranov, Y.G. Malama: J. Geophys. Res. **98**, 15157 (1993)
30. V.B. Baranov, Y.G. Malama: J. Geophys. Res. **100**, 14755 (1995)
31. V.B. Baranov, Y.G. Malama: Space Sci. Rev. **78**, 305 (1996)
32. D.C. Barnes, T. Kamimura: Res. Rep.-Nagoya Univ. Inst. Plasma Phys., IPPJ-570 (1982)
33. D.C. Barnes, T. Kamimura, J.N. Leboeuf, T. Tajima: J. Comput. Phys. **52**, 480 (1983)
34. D.C. Barnes, R.A. Nebel, C.E. Seyler: Bull. Am. Phys. Soc. **37**, 1557 (1992)
35. M.L. Begue, A. Ghizzo, P. Bertrand: J. Comput. Phys. **151**, 458 (1999)
36. G. Belmont, G. Chanteur: Phys. Scr. **40**, 124 (1989)
37. O.M. Belotserkovskii: *Numerical Simulation of the Continuous Medium Mechanics* (Nauka Press, Moscow 1994) (in Russian)
38. O.M. Belotserkovskii, T.K. Breuss, A.M. Krymskii, V.Ya. Mitnitskii: Geophys. Res. Lett. **14**(5), 503 (1987)
39. E.V. Belova, R.E. Denton, A.A. Chan: J. Comput. Phys. **136**, 324 (1997)
40. J. Berchem, H. Okuda: J. Geophys. Res. **95**, 8133 (1990)
41. Yu.A. Berezin: *Simulation of Nonlinear Wave Processes* (Nauka Press, Novosibirsk 1982) (in Russian)
42. Yu.A. Berezin, V.A. Vshivkov: *Particle Method in Collisionless Plasma Dynamics* (Nauka Press, Novosibirsk 1980) (in Russian)
43. Yu.A. Berezin, V.A. Vshivkov, G.I. Dudnikova, M.P. Fedoruk: Sov. J. Plasma Phys. **18**(12), 1567(1992)
44. I.S. Berezin, N.P. Zhidkov: *Numerical Methods, Vols. 1-2* (Fizmatgiz, Moscow 1966)
45. M.J. Berger, P. Colella: J. Comput. Phys. **82**(1), 64 (1989)
46. M.J. Berger, J. Oliger: J. Comput. Phys. **53**, 484 (1984)
47. H.L. Berk, D. Book: Phys. Fluids **12**, 649 (1969)
48. H.L. Berk, K.V. Roberts: 'The Water-Bag Model'. In: *Method in Computational Physics. Plasma Physics, Vol. 9*, ed. by B. Alder, S. Fernbach, M. Rotenberg (Academic Press, New York, London 1970) p. 88
49. L. Biermann, B. Brosowski, H.U. Schmidt: Solar Phys. **1**, 254 (1967)

50. G.A. Bird: *Molecular Gas Dynamics and the Direct Simulation of Gas Flows* (Clarendon Press, Oxford 1994)
51. C.K. Birdsall, A.B. Langdon: *Plasma Physics via Computer Simulation*, The Adam Hilger Series on Plasma Physics (Adam Hilger, Bristol, Philadelphia, New York 1991; McGraw-Hill, New York 1985)
52. C.K. Birdsall, A.B. Langdon, H. Okuda: 'Finite-Size Particle Physics Applied to Plasma Simulation'. In: *Methods in Computational Physics. Plasma Physics, Vol. 9*, ed. by B. Alder, S. Fernbach, M. Rotenberg (Academic Press, New York and London 1970) p. 241
53. J. Birn: J. Geophys. Res. **85**, 1244 (1980)
54. J. Birn, M. Hesse, K. Schindler: J. Geophys. Res. **101**, 12939 (1996)
55. D. Biskamp, R.Z. Sagdeev, K. Schindler: Cosmic Electrodyn. **1**, 297 (1970)
56. A. Bogdanov, K. Sauer, K. Baumgärtel, K. Srivastava: Planet. Space Sci. **44**, 519 (1996)
57. J.P. Boris: 'Relativistic Plasma Simulation-Optimization of a Hybrid Code'. In: *Proc. Fourth Conf. Numer. Sim. of Plasmas*, ed. by J.P. Boris, R.A. Shanny (NRL, Washington, D.C. 1970) p. 3–67
58. J.P. Boris, D.L. Book: 'Solution of Continuity Equations by the Method of Flux-Corrected Transport'. In: *Methods in Computational Physics. Controlled Fusion, Vol. 16*, ed. by B. Alder, S. Fernbach, M. Rotenberg (Academic Press, New York and London 1976) p. 85–130
59. J.U. Brackbill, D.W. Forslund: J. Comput. Phys. **46**, 271 (1982)
60. J.U. Brackbill, D.W. Forslund, K.B. Quest, D. Winske: Phys. Fluids **27**, 2682 (1984)
61. J.U. Brackbill, D.W. Forslund: 'Simulation of Low- Frequency Electromagnetic Phenomena in Plasma'. In: *Multiple Time Scales. Computational Techniques, Vol. 3*, ed. by J.U. Brackbill, B.I. Cohen (Academic Press, New York 1985) p. 272
62. J.U. Brackbill, G. Lapenta: J. Comput. Phys. **114**, 77 (1994)
63. S.L. Braginskii: 'Transport processes in a plasma'. In: *Reviews of Plasma Physics*, ed. by M.A. Leontovich (Consultants Bureau, New York 1965) p. 205
64. S.A. Brecht, J.R. Ferrante: J. Geophys. Res. **96**(A7), 11209 (1991)
65. S.A. Brecht, J.R. Ferrante, J.G. Luhmann: J. Geophys. Res. **98**(A2), 1345 (1993)
66. S.A. Brecht, V.A. Thomas: Comput. Phys. Comm. **48**, 135 (1988)
67. A. Brizard: J. Plasma Phys. **41**, 541 (1989)
68. B. Brosowski, R. Wegmann: Meth. Verf. Math. Phys. **8**, 125 (1973)
69. F. Brunel, T. Tajima: Phys. Fluids **26**, 535 (1983)
70. K.V. Brushlinskii, A.I. Morozow: 'Calculation of Two-dimensional Plasma Flow'. In: *Reviews of Plasma Physics, Vol. 8*, ed. by M.A. Leontovich (Atomizdat, Moscow 1974; Consultants Bureau, New York 1980) p. 105
71. S.V. Bulanov, P.V. Sasorov: Sov. Plasma Phys. **4**, 746 (1978)
72. S.V. Bulanov, I.V. Sokolov: Sov. Astron. J. **28**, 515 (1984)
73. O. Buneman: J. Comput. Phys. **1**(4), 517 (1967)
74. O. Buneman: 'TRISTAN'. In: *Computer Simulation of Space Plasma*, ed. by H. Matsumoto, Y. Omura (Terra Scientific, Tokyo 1993) p. 145
75. D. Burgess: J. Geophys. Res. **92**, 1119 (1987)
76. D. Burgess: Ann. Geophys. **5**, 133 (1987)
77. D. Burgess, S.J. Schwartz: J. Geophys. Res. **93**(10), 11327 (1988)
78. D. Burgess: 'Simulation of quasiparallel shocks'. In: *Proceedings of the 4th International School for Space Simulation, Kyoto-Nara, Japan, 1991*, ed. by Y. Omura, H. Matsumoto (Terra Scientific, Tokyo 1991) p. 29
79. S.Z. Burstein, A. Mirin: J. Comput. Phys. **5**, 547 (1968)

80. J. Busnardo-Neto, P.L. Pritchett, A.T. Lin, J.M. Dawson: J. Comput. Phys. **23**, 300 (1977)
81. J.C. Butcher: Math. Comp. **18**, 50 (1964)
82. J.A. Byers: 'Noise Suppression Techniques in Macroparticle Models of Collisionless Plasma'. In: *Proc. 4th Conf. Numer. Simul. Plasmas 1970*, ed. by J.P. Boros, R.A. Shanny (NRL, Washington, D.C. 1970) p. 496
83. J.A. Byers, B.I. Cohen, W.C. Condit, J.D. Hanson: J. Comput. Phys. **27**, 363 (1978)
84. J.A. Byers, J. Killeen: 'Finite-difference methods for collisionless plasma models'. In: *Methods in Computational Physics. Plasma Physics, Vol. 9*, ed. by B. Alder, S. Fernbach, M. Rotenberg (Academic Press, New York, London 1970) p. 259
85. D. Cai, L.R.O. Storey, T. Neubert: Phys. Fluids B **2**, 75 (1990)
86. H.J. Cai, D.Q. Ding, L.C. Lee: J. Geophys. Res. **99**, 35 (1994)
87. H.J. Cai, J.B. Scuder: EOS Trans. AGU **77**(46), 615 (1996)
88. C. Canuto, M.Y. Hussaini, A. Quarteroni, T.A. Zang: *Spectral Methods in Fluid Dynamics* (Springer-Verlag, New York 1987)
89. P.J. Cargill, T.E. Eastman: J. Geophys. Res. **96**, 13763 (1991)
90. P.J. Cargill, C.C. Googrich: Phys. Fluids **30**, 2504 (1987)
91. M.H. Carpenter, D. Gottlier, S. Abarbanel: J. Comput. Phys. **111**, 220 (1994)
92. P.J. Catto: Astrophys. Space Sci. **26**(1), 47 (1974)
93. P.J. Catto: Plasma Phys. **20**, 719 (1978)
94. C.L. Chang, A.S. Lipatov, A.T. Drobot, K. Papadopoulos, P. Satya-Narayana: Geophys. Res. Lett. **21**, 1015 (1994)
95. S. Chapman, V.C.A. Ferraro: J. Geophys. Res. **36**, 171 (1931)
96. C.Z. Cheng: J. Geophys. Res. **96**, 21159 (1991)
97. R. Chodura: Nucl. Fusion **15**, 55 (1975)
98. J.Y. Chou, D.H. Leventhal, B.C. Weinberg: Comput. Math. Appl. **25**, 105 (1993)
99. P.C. Chu, C. Fan: J. Comput. Phys. **140**, 370 (1998)
100. P. Cloutier: 'Solar-wind interaction with planetary ionospheres'. In: *Solar-wind Interaction with the Planets Mercury, Venus, and Mars. The Proceedings of a bilateral seminar of the US-USSR Joint Working group on Near-Earth Space, the Moon, and Planets, held at the Space Research Institute of Academy of Sciences of the USSR, Moscow, November 17–21, 1975*, ed. by N.F. Ness (NASA, SP-397, Washington, D.C. 1976) p. 111
101. P. Cloutier, C.C. Law, D.H. Crider, et al.: Geophys. Res. Lett. **26**(17), 2685 (1999)
102. B.I. Cohen: 'Orbit Averaging and Subcycling in Particle Simulation of Plasmas'. In: *Multiple Time Scales. Computational Techniques, Vol. 3*, ed. by J.U. Brackbill, B.I. Cohen (Academic Press, New York 1985) p. 311
103. B.I. Cohen, T.A. Brengle, D.B. Conley, R.P. Freis: J. Comput. Phys. **38**, 45 (1980)
104. B.I. Cohen, R.P. Freis: J. Comput. Phys. **45**, 367 (1982)
105. B.I. Cohen, R.P. Freis, V. Thomas: J. Comput. Phys. **45**, 345 (1982)
106. B.I. Cohen, A.B. Langdon, A. Friedman: J. Comput. Phys. **46**, 15 (1982)
107. B.I. Cohen, A.B. Langdon, D.W. Hewett, R.J. Procassini: J. Comput. Phys. **81**, 151 (1989)
108. B.I. Cohen, T.J. Williams: J. Comput. Phys. **107**, 282 (1993)
109. W.J. Coirier, K.G. Powell: J. Comput. Phys. **117**, 121 (1995)
110. P. Colella: J. Comput. Phys. **87**, 171 (1990)
111. P. Colella, M.R. Dorr, D.D. Wake: J. Comput. Phys. **149**, 168 (1999)
112. P. Colella, M.R. Dorr, D.D. Wake: J. Comput. Phys. **152**, 550 (1999)

113. L. Collatz: *The Numerical Treatment of Differential Equations* (Springer-Verlag, New York 1966) p. 538
114. P. Concus, G.H. Golub: SIAM J. Number Anal. **10**, 1103 (1973)
115. A.W. Cook: J. Comput. Phys. **154**, 117 (1999)
116. B. Coppi, G. Laval, R. Pellat: Phys. Rev. Lett. **16**, 1207 (1966)
117. F.V. Coronity: Phys. Rev. Lett. **38**, 1355 (1977)
118. J.M. Coyle, J. E. Flaherty, R. Ludwig: J. Comput. Phys. **62**, 26 (1986)
119. R. Courant, E. Issacson, M. Rees: Commun. Pure Appl. Math. **5**, 243 (1952)
120. G. Dahlquist, A. Bjorck: *Numerical Methods*, translated by N. Andersen (Prentice-Hall, Englewood Cliffs 1974) p. 269
121. W. Dai, P.R. Woodward: J. Geophys. Res. **100**, 14843 (1995)
122. C.G. Darwin: Philos. Mag. **39**, 537 (1920)
123. J.M. Dawson: 'The electrostatic sheet model for a plasma and its modification to finite-size particles'. In: *Method in Computational Physics. Plasma Physics, Vol. 9*, ed. by B. Alder, S. Fernbach, M. Rotenberg (Academic Press, New York, London 1970) p. 1
124. R.B. Decker: Space Sci. Rev. **48**, 195 (1988)
125. R.B. Decker, L. Vlahos: Astrophys. J. **306**, 710 (1986)
126. P.A. Delamere, D.W. Swift, H.C. Stenbaek-Nielsen: Geophys. Res. Lett. **26**, 2837 (1999)
127. J. Denavit: J. Comput. Phys. **9**, 75 (1972)
128. J. Denavit: J. Comput. Phys. **42**, 337 (1981)
129. R.E. Denton, M. Kotschenreuther: J. Comput. Phys. **119**, 283 (1995)
130. D. De Zeeuw, K.G. Powell: J. Comput. Phys. **104**, 56 (1993)
131. D.O. Dickman, R.L. Morse, C.W. Nielson: Phys. Fluids **12**, 1708 (1969)
132. A.M. Dimits, W.W. Lee: J. Comput. Phys. **107**, 309 (1993)
133. M. Dobrovolny: Nuovo Cimento **55b**, 427 (1968)
134. S.S. Dolginov, E.G. Eroshenko, L.N. Zhyzkov: Cosmic. Res. **13**, 108 (1975)
135. S.S. Dolginov, E.G. Eroshenko, L.N. Zhyzkov et al.: Sov. Astron. J. Lett. **2**, 88 (1976)
136. D.G. Dritschel: J. Comput. Phys. **77**, 240 (1988)
137. L.O'C. Drury, H.J. Volk: Astrophys. J. **248**, 344 (1981)
138. D.H.E. Dubin, J.A. Krommes, C. Oberman, W.W. Lee: Phys. Fluids **26**(12), 3524 (1983)
139. T.E. Eastman, E.W. Hones, S.J. Bame, J.R. Asbridge: Geophys. Res. Lett. **3**, 685 (1976)
140. T.E. Eastman, E.A. Greene, S.P. Christon, G. Gloeckler, D.C. Hamilton, F.M. Ipavich, G. Kremser, B. Wilken: Geophys. Res. Lett. **17**, 2031 (1990)
141. T.E. Eastman, G. Rostoker, L. Frank, C. Huang, D. Mitchell: J. Geophys. Res. **93**, 14411 (1988)
142. J.W. Eastwood, W. Arter, N.J. Brealey, R.W. Hockney: Comput. Phys. Comm. **87**, 155 (1995)
143. J.P. Edmiston, C.F. Kennel: J. Plasma Phys. **32**, 429 (1984)
144. D.C. Ellison, D. Eichler: Astrophys. J. **286**, 691 (1984)
145. V.G. Eselevich, A.G. Eskov, R.Kh. Kurtmulaev, A.I. Malytin: Sov. Phys. JETP **33**, 1120 (1971)
146. V.G. Eselevich: Planet. Space Sci. **31**, 615 (1983)
147. H.J. Fahr: Astron. Astrophys. **241**, 251 (1991)
148. H.J. Fahr: Space Sci. Rev. **78**, 199 (1996)
149. H.J. Fahr, R. Osterbart, D. Rucinski: Astron. Astrophys. **294**, 587 (1995)
150. D.H. Fairfield, W.C. Feldman: J. Geophys. Res. **80**, 515 (1975)
151. W.M. Farrell et al.: Geophys. Res. Lett. **23**, 1271 (1996)
152. W.M. Farrell, M.L. Kaiser, J.T. Steinberg: Geophys. Res. Lett. **24**, 1135 (1997)

153. W.M. Farrell, M.L. Kaiser, J.T. Steinberg, S.D. Bale: J. Geophys. Res. **103**(A10), 23653 (1998)
154. R.P. Fedorenko: USSR Comp. Maths. Math. Phys. **1**, 1092 (1962)
155. J.M. Finn, R.N. Sudan: Nucl. Fusion **22**, 1443 (1982)
156. W.L. Fite, A.C.H. Smith, R.F. Stebbings: Proc. R. Soc. Lond. Ser. A. **268**, 527 (1962)
157. C.A.J. Fletcher: *Computational Techniques for Fluid Dynamics I. Fundamental and General Techniques* (Springer-Verlag, Berlin, Heidelberg 1988)
158. E.A. Foote, R.M. Kulsrudlund: Phys. Fluids **24**, 1532 (1981)
159. D.W. Forslund: Space Sci. Rev. **42**, 3 (1985)
160. D.W. Forslund, J.P. Freidberg: Phys. Rev. Lett. **27**, 1189 (1971)
161. D.W. Forslund, K.B. Quest, J.U. Brackbill, K. Lee: J. Geophys. Res. **89**, 2142 (1984)
162. S.P. Frankel: 'Some Qualitative Comments on Stability Considerations in Partial Difference Equations'. In: *Proceeding of Sixth Symposia in Applied Mathematics, Vol. 6 - Numerical Analysis* (AMS, 1956) p. 73–75
163. A. Friedman, A.B. Langdon, B.I. Cohen: Comments Plasma Phys. Controlled Fusion **6**, 225 (1981)
164. M. Fujimoto, M.S. Nakamura, T. Nagai, T. Mukai, T. Yamamoto, S. Kokuban: Geophys. Res. Lett. **23**, 2533 (1996)
165. H.P. Furth, J. Kileen, M.N. Rosenbluth: Phys. Fluids **6**, 459 (1963)
166. A.A. Galeev: 'Collisionless shocks'. In: *Physics of Solar Planetary Environment*, ed. by D.J. Williams (AGU, Washington, D.C. 1976) p. 484
167. A.A. Galeev: Space Sci. Rev. **23**, 411 (1979)
168. A.A. Galeev, F.V. Coronity, M. Ashour-Abdalla: Geophys. Res. Lett. **5**, 707 (1978)
169. A.A. Galeev, B.E. Gribov, T.I. Gombosi, et al.: Geophys. Res. Lett. **13**, 841 (1986)
170. A.A. Galeev, C.F. Kennel, V.V. Krasnoselskikh: 'Quasi-Perpendicular Collisionless High Mach Number Shocks'. In: *Proceeding of Workshop on Plasma Astrophysics*, ed. by T.D. Guyenne, J.J. Hunt (ESA SP0285, Paris 1988) p. 173
171. A.A. Galeev, C.F. Kennel, V.V. Krasnoselskikh, V.V. Lobzin: 'The Role of Whistler Oscillations in the Formation of the Structure of High Mach Number Collisionless Shock'. In: *Proceeding of Workshop on Plasma Astrophysics*, ed. by T.D. Guyenne, J.J. Hunt (ESA SP0285, Paris 1988) p. 165
172. A.A. Galeev, A.S. Lipatov: Adv. Space Res. **4**(9), 229 (1984)
173. A.A. Galeev, A.S. Lipatov, V.A. Lobachev: 'Numerical simulation of whistler formation and electron acceleration at the front of the Collisionless shocks'. In: *Program of the Cornelius Lanczos International Centenary Conference (Computational Mathematics, Theoretical Physics, Astrophysics) at North Carolina State Univ., December 12–17, 1993* (North Carolina State Univ., Raleigh 1993) p. 113
174. A.A. Galeev, A.S. Lipatov, A.A. Malgichev: Sov. J. Plasma Phys. **17**(10), 701 (1991)
175. A.A. Galeev, A.S. Lipatov, R.Z. Sagdeev: Sov. Phys. JETP **62**(5), 866 (1985)
176. A.A. Galeev, A.S. Lipatov, R.Z. Sagdeev: Sov. J. Plasma Phys. **13**(5), 323 (1987)
177. A.A. Galeev, R.Z. Sagdeev: Sov. Phys. JETP **30**, 571 (1970)
178. A.A. Galeev, L.M. Zelenyi: Sov. Phys. JETP **70**, 2133 (1976)
179. A.L. Garcia, J.B. Bell, W.Y. Crutchfield, B.J. Alder: J. Comput. Phys. **154**, 134 (1999)

180. S. Gartenhaus: *Elements of plasma physics* (Holt, Rinehart and Winston, Orlando 1964)
181. S.P. Gary, C.D. Madland, D. Schriver, D. Winske: J. Geophys. Res. **91**, 4188 (1986)
182. S.P. Gary, C.D. Madland, N. Omidi, D. Winske: J. Geophys. Res. **93**(9), 9584 (1988)
183. S.P. Gary, S.W. Smith, M.A. Lee, M.L. Goldstein, D.W. Forslund: Phys. Fluids **84**, 1852 (1984)
184. C.W. Gear: *Numerical Initial Value Problems in Ordinary Differential Equations* (Prentice Hall, Englewood Cliffs 1971)
185. J.L. Geary, T. Tajima, J.N. Leboeuf, E.G. Zaidman, J.H. Han: Comp. Phys. Commun. **42**, 313 (1986)
186. R.A. Gentry, R.E. Martin, B.J. Daly: J. Comput. Phys. **1**, 87 (1966)
187. N.K. Ghaddar, G.E. Kariadakis, A.T. Patera: Numer. Heat Transfer **9**, 277 (1986)
188. A. Ghizzo, P. Bertrand, M.M. Shoucri, T.W. Johnston, E. Fijalkow, M.R. Feix: J. Comput. Phys. **90**, 431 (1990)
189. A. Ghizzo, P. Bertrand, M.M. Shoucri, E. Fijalkow, M.R. Feix: J. Comput. Phys. **108**, 105 (1993)
190. J. Giacalone, J.R. Jokipii, J. Kota: J. Geophys. Res. **99**(10), 19 351 (1994)
191. D. Givoli: J. Comput. Phys. **94**, 1 (1991)
192. N.T. Gladd, J.D. Huba: Phys. Fluids **22**, 911 (1979)
193. G. Gloecker, J. Geiss, E.C. Roelof, L.A. Fisk, F.M. Ipavich, K.W. Ogilvie, L.J. Lanzerotti, R. von Steiger, B. Wilken: J. Geophys. Res. **99**, 17637 (1994)
194. B.B. Godfrey: J. Comp. Phys. **15**, 504 (1974)
195. B.B. Godfrey: J. Comp. Phys. **20**, 251 (1976)
196. W.J. Goedheer, J.H.H.M. Potters: J. Comp. Phys. **61**, 269 (1985)
197. K.I. Golden, L.M. Linson, S.A. Mani: Phys. Fluids **16**, 2319 (1973)
198. H. Goldstein: Die Ionen-Tearing Instabilitat Zweidimensionaler Schichgleichgewichte und ihre Bedeutung für die Dynamik der Magnetosphäre. Dissertation zur Erlandung des Grades eines Doctors der Naturwissenschaften, Ruhr-Universitat, Bochum (1981)
199. M.L. Goldstein: 'Magnetohydrodynamics turbulence and its relationship to interplanetary fluctuation'. In: *Space Plasmas: Coupling Between Small and Medium Scale Processes, Geophysical Monograph Series, Vol. 86*, ed. by M. Ashour-Abdalla, T. Chang, P. Dusenbery (AGU, Washington, D.C. 1995) p. 7
200. C.C. Goodrich, P.J. Cargill: Geophys. Res. Lett. **18**, 65 (1991)
201. A.V. Gordeev, L.I. Rudakov: Sov. Phys. JETP **55**, 2310 (1968)
202. J.T. Gosling, M.F. Thomsen, S.J. Bame, W.C. Feldman, G. Paschmann, N. Scopke: Geophys. Res. Lett. **9**, 1333 (1982)
203. D. Gottlieb, S.A. Orzag: *Numerical Analysis of Spectral Methods* (SIAM, Philadelphia 1977)
204. K.I. Gringauz, V.V. Bezrukikh, T.K. Breus et al.: Sov. Astron. J. Lett. **2**, 82 (1976)
205. K.I. Gringauz, V.V. Bezrukikh, M.I. Verigin, A.P. Remizov: Cosmic. Res. **13**, 123 (1975)
206. V.T. Grudnitskii, Yu.A. Prokhorchuk: Doklady Acad. Nauk USSR **224**(6), 1249 (1977)
207. P.N. Guzdar, J.F. Drake, D. McCarthy, A.B. Hassam, C.S. Liu: Phys. Fluids B **5**(10), 3712 (1993)
208. B. Gustaffson, H.O. Kreiss, J. Oliger: *Time Dependent Problems and Difference Methods* (Wiley, New York 1994)

209. B. Gustaffson, P. Olsson: J. Comput. Phys. **117**, 300 (1995)
210. T. Hada, N. Omidi, M. Ashour-Abdalla: EOS Trans. AGU **67**, 334 (1986)
211. G. Haerendel, G. Paschmann, N. Sckopke, H. Rosenbauer, P.C. Hedgecock: J. Geophys. Res. **83**, 3195 (1978)
212. T.S. Hahm, W.W. Lee, A. Brizard: Phys. Fluids **31**, 1940 (1988)
213. D.T. Hall: Ultraviolet resonance radiation and the structure of the heliosphere. Ph.D. Thesis, Univ. of Arizona, Tucson (1992)
214. S. Hamasaki, N.A. Krall, C.E. Wagner, R.N. Byrne: Phys. Fluids **20**, 65 (1977)
215. J.M. Hamilton: The Numerical Solution of Ampere's Equation on a 2-Dimensional Mesh. Part I – Theoretical Discussion. Computer Science Report, RCS-49 (Reading University 1976)
216. D.S. Harned: private communication.
217. D.S. Harned: J. Comput. Phys. **47**, 452 (1982)
218. A. Harten, B. Enquist, S. Osher, R. Chakravarthy: J. Comput. Phys. **71**, 231 (1987)
219. A. Hasegawa, L. Chen: Phys. Fluids **19**, 1924 (1976)
220. A. Hasegawa, M. Wakatani: Phys. Fluids **26**, 2770 (1983)
221. A.B. Hassam, J.D. Huba: Phys. Fluids **31**, 318 (1988)
222. G.W. Hedstrom, G.H. Rodrigue, M. Berger, J. Oliger: *Adaptive mesh refinement for 1-dimensional gas dynamics, in IMACS* (North-Holland, Amsterdam 1983) p. 43
223. P. Hellinger, A. Mangeney: J. Geophys. Res. **104**, 4669 (1999)
224. M. Hesse, J. Birn: J. Geophys. Res. **97**, 3965 (1992)
225. M. Hesse, D. Winske: J. Geophys. Res. **99**, 11 177 (1994)
226. M. Hesse, D. Winske, M.M. Kuznetsova: J. Geophys. Res. **100**, 21 815 (1995)
227. D.W. Hewett: J. Comput. Phys. **38**, 378 (1980)
228. D.W. Hewett, J.K. Boyd: J. Comput. Phys. **70**, 166 (1987)
229. D.W. Hewett, G.E. Frances, C.E. Max: Phys. Rev. Lett. **61**(7), 893 (1988)
230. D.W. Hewett, A.B. Langdon: J. Comput. Phys. **72**, 121 (1987)
231. D.W. Hewett, C.A. Nielson: J. Comput. Phys. **29**, 219 (1978)
232. R.L. Higdon: Siam J. Numer. Anal. **27**(4), 831 (1990)
233. T.W. Hill: J. Geophys. Res. **80**, 4689 (1975)
234. F.L. Hinton, R.D. Hazeltine: Rev. Mod. Phys. **48**, 239 (1976)
235. R.S. Hirsh: J. Comput. Phys. **19**, 90 (1975)
236. R.W. Hockney, J.W. Eastwood: *Computer simulation using particles* (McGraw-Hill, New York 1981; Adam Hilger, IOP, Bristol 1988)
237. R. Horiuchi, T. Sato: Phys. Plasmas **4**, 277 (1997)
238. E.J. Horowitz, D. Shumaker, P. Anderson: J. Comput. Phys. **84**, 279 (1989)
239. W.T. Horton, T. Tajima, T. Kamimura: Phys. Fluids **30**, 3485 (1987)
240. M. Hoshino: J. Geophys. Res. **92**, 7368 (1987)
241. M. Hoshino, T. Terasawa: J. Geophys. Res. **90**(A1), 57 (1985)
242. G. Hu, J. Krommes: Phys. Plasmas **1**(4), 863 (1994)
243. V.P. Iljin: *Numerical methods for electrooptics* (Nauka Press, Novosibirsk 1974) p. 204
244. V.S. Imshennik: 'Nonhydrodynamic model of plasma focus'. In: *Two-Dimensional Models of Plasmas*, ed. by K.V. Brushlinskii (Keldysh Institute of Applied Mathematics, Moscow 1979) p. 120
245. Z.M. Ioffe: Sov. Astron. J. **11**, 1044 (1968)
246. W.H. Ip, W.I. Axford: 'Cometary plasma processes'. In: *Venus*, ed. by L.L. Wilkening (Univ. of Arizona Press, Tucson 1982) p. 588
247. K.P. Ivanov, O.A. Ladyzhenskaia, V.Ia. Rivkind: Matem. Mekh. Astron. Vest. Leningrad. Univ. **25**, 37 (1970)

248. V. Izmodenov, Yu.G. Malama, R. Lallement: Astron. Astrophys. **317**, 193 (1997)
249. B. Izrar, A. Ghizzo, P. Bertrand, E. Fijalkov, M. Feix: Comput. Phys. Comm. **52**(3), 375 (1989)
250. J.D. Jackson: *Classical Electrodynamics* (Wiley, New York 1962) p. 409
251. T.C. Johns, A.H. Nelson: Comput. Phys. Comm. **37**, 329 (1985)
252. J.R. Johnson, C.Z. Cheng: Geophys. Rev. Lett. **24**(11), 1423 (1997)
253. I.W. Johnson, A.J. Wathen, M.J. Baines: J. Comput. Phys. **79**, 270 (1988)
254. D.E. Jones, E.J. Smith, J.A. Slavin, B.T. Tsurutani, G.L. Siscoe, D.A. Mendis: Geophys. Res. Lett. **13**, 243 (1986)
255. G. Joyce, R. Hubbard, M. Lampe, S. Slinker: J. Comput. Phys. **81**, 193 (1989)
256. G. Joyce, M. Lampe, S. Slinker, W.M. Manheimer: J. Comput. Phys. **138**, 540 (1997)
257. B.B. Kadomtsev: *Plasma Turbulence* (Academic Press, New York 1965) p. 78
258. T. Kamimura, E. Montalvo, D.C. Barnes, J.N. Leboeuf, T. Tajima: J. Comput. Phys. **100**, 77 (1992)
259. J.R. Kan, D.W. Swift: J. Geophys. Res. **88**(9), 6919 (1983)
260. A. Kantorowitz, H.E. Petschek: 'MHD characteristics and shock waves'. In: *Plasma Physics in Theory and Application*, ed. by W. Kinkel (McGraw-Hill, New York 1966) p. 148
261. H. Karimabadi, D. Krauss-Varban, N. Omidi, H.X. Vu: J. Geophys. Res. **104**, 12313 (1999)
262. S.R. Karmesin, P.C. Liewer, J.U. Brackbill: Geophys. Res. Lett. **22**, 1153 (1995)
263. G.E. Karniadakis: Appl. Numer. Math. **5**, 1 (1989)
264. T. Katsouleas, J.M. Dawson: Phys. Rev. Lett. **51**, 392 (1983)
265. F. Kazeminezhad, J.M. Dawson, J.N. Leboeuf, R. Sydora, D. Holland: J. Comput. Phys. **102**, 277 (1992)
266. A.N. Kaufman, P.S. Rostler: Phys. Fluids **14**, 446 (1971)
267. K. Kecskemety, T.E. Cravens: Geophys. Res. Lett. **20**, 543 (1993)
268. C.F. Kennel, R.Z. Sagdeev: J. Geophys. Res. **72**, 3303 (1967)
269. C.F. Kennel, J.P. Edmistov, T.A. Hada: 'A quarter century of collisionless shock research'. In: *Collisionless Shocks in the Heliosphere: A Tutorial Review, Geophysical Monograph Series, Vol. 34*, ed. by B.T. Tsurutani, R.G. Stone (AGU, Washington, D.C. 1985) p. 1
270. J. Killeen, S.L. Rompel: J. Comp. Phys. **1**, 29 (1966)
271. M.G. Kivelson, Z. Wang, S. Joy, K.K. Khurana, C. Polansky, D.J. Southwood, R.J. Walker: Adv. Space Sci. **16**(4), 59 (1995)
272. A.J. Klimas, W.M. Farrell: J. Comput. Phys. **110**, 150 (1994)
273. G. Knorr: Phys. Fluids **11**, 885 (1968)
274. M.N. Kogan: *The Rarefied Gas Dynamics* (Nauka Press, Moscow 1967; Plenum Press, New York 1969)
275. A.E. Koniges, G.G. Craddock, D.D. Schnack, H.R. Strauss (Eds.): *Proceedings of the Workshop on Adaptive Grid Methods for Fusion Plasma*, Technical Report CONF-941279 (Lawrence Livermore National Laboratory, Berkeley July 1995)
276. G.A. Kotova, M.I. Verigin, A.P. Remizov, H. Rosenbauer, S. Livi, W. Riedler, K. Schwingenschun, M. Tatrallyay, K. Szegö, I. Apathy: Earth Planet. Sci. **52**(8), 501 (2000)
277. M. Kotschenreuther: Bull. Am. Phys. Soc. **34**, 2107 (1988)
278. V.M. Kovenya, N.N. Yanenko: *Splitting methods in gas dynamics problems* (Nauka, Novosibirsk 1981) (in Russian)

279. V.M. Kovenya, G.A. Tarnavsky, S.G. Cherny: *Application of the split method in the aerodynamics problems* (Nauka Press, Novosibirsk 1990) (in Russian)
280. D. Krauss-Varban, N. Omidi: Geophys. Res. Lett. **22**, 3271 (1995)
281. H.O. Kreiss, S.A. Orszag, M. Israeli: Annu. Rev. Fluid Mech. **6**, 281 (1974)
282. W.L. Kruer, J.M. Dawson, B. Rosen: J. Comput. Phys. **13**, 114 (1973)
283. H. Kucharek, M. Scholer: J. Geophys. Res. **100**, 1745 (1995)
284. W. Kutta: Z. Math. Phys. **46**, 435 (1901)
285. M.M. Kuznetsova, M. Hesse, D. Winske: J. Geophys. Res. **103**(A1), 199 (1998)
286. A.L. LaBelle-Hamer, Z.F. Fu, L.C. Lee: Geophys. Res. Lett. **15**, 152 (1988)
287. L.D. Landau, E.M. Lifshitz: *Field Theory* (Pergamon, Oxford 1951)
288. L.D. Landau, E.M. Lifshitz: *Electrodynamics of Continuous Media* (Pergamon, Oxford 1959) p. 216
289. A.B. Langdon: J. Comput. Phys. **30**, 202 (1979)
290. A.B. Langdon, D.C. Barnes: 'Direct Implicit Plasma Simulation'. In: *Multiple Time Scales, Computational Techniques, Vol. 3*, ed. by J.U. Brackbill, B.I. Cohen (Academic Press, New York 1985) p. 336
291. A.B. Langdon, B.I. Cohen, A. Friedman: J. Comput. Phys. **51**, 107 (1983)
292. A.B. Langdon, B.F. Lasinski: 'Electromagnetic and Relativistic Plasma Simulation Models'. In: *Methods in Computational Physics, Vol. 16. Controlled Fusion*, ed. by J. Killeen (Academic Press, New York, San Francisco, London 1976) p. 327–366
293. G. Lapenta, J.U. Brackbill: 'Three Dimensional Stability of Thin Current Sheets'. In: *Americal Geophysical Fall Meeting Abstract, San Francisco* (AGU, Washington, D.C. 1997) p. 576
294. W.W. Lee: Phys. Fluids **26**, 556 (1983)
295. L.C. Lee: 'A review of magnetic reconnection: MHD models'. In: *Physics of the Magnetosphere, Geophysical Monograph Series, Vol. 90*, ed. by P. Song, B.U.O. Sonnerup, M.F. Thomsen (AGU, Washington, D.C. 1995) p. 139
296. L.C. Lee, J.R. Kan: J. Geophys. Res. **84**, 6417 (1979)
297. M.A. Lee, V.D. Shapiro, R.Z. Sagdeev: J. Geophys. Res. **101**, 4777 (1996)
298. C.E. Leith: 'Numerical Simulation of the Earth's Atmospheres'. In: *Methods in Computational Physics, Vol. 4*, ed. by B. Alder, S. Fernbach, M. Rotenberg (Academic Press, New York, London 1965) p. 1–28
299. S.K. Lele: J. Comp. Phys. **103**(1), 16 (1992)
300. B. Lembege: Stabilite d'un modele bidimensionnel de la couche quasi-neutre de la queue magnetospherique terrestre vis a vis du mode de cisaillement (tearing-mode) lineare. Theses du Doctorat 3e cycle, L'Universite de Paris, XI (1976)
301. B. Lembege, J.M. Dawson: Phys. Rev. Lett. **62**, 2683 (1989)
302. D.S. Lemons, D. Winske, S.P. Gary: J. Plasma Phys. **21**, 287 (1979)
303. M.M. Leroy, D. Winske: Ann. Geophys. **1**(6), 527 (1983)
304. M.M. Leroy, D. Winske, C.C. Goodrich, C.S. Wu, K. Papadopoulos: J. Geophys. Res. **87**(7), 5081, (1982)
305. H.R. Lewis: 'Application of Hamilton's Principle to the Numerical Analysis of Vlasov Plasmas'. In: *Method in Computational Physics. Plasma Physics, Vol. 9*, ed. by B. Alder, S. Fernbach, M. Rotenberg (Academic Press, New York, London 1970) p. 307
306. P.C. Liewer, B.E. Goldstein, N. Omidi: J. Geophys. Res. **98**, 15211 (1993)
307. P.C. Liewer, S.R. Karmesin, J.U. Brackbill: J. Geophys. Res. **101**, 17119 (1996)
308. P.C. Liewer, S. Rath, B.E. Goldstein: J. Geophys. Res. **100**, 19809 (1995)
309. D.K. Lilly: Monthly Weather Review (U.S. Weather Bureau) **93**(1), 11 (1965)

310. T. Linde, T.I. Gombosi, P.L. Roe, K.G. Powell, D.L. DeZeeuw: J. Geophys. Res. **103**, 1889 (1998)
311. Y. Lin, D.W. Swift: J. Geophys. Res. **100**, 19859 (1996)
312. A.S. Lipatov: Propagation of the Hydromagnetic Waves in the Three Dimensional Magnetosphere. Ph.D. Thesis, Moscow Institute of Physics and Technology, Moscow (1972)
313. A.S. Lipatov: "Guiding Center in Cell" Method in the Three-Dimensional Nonstationary Problem of the Interaction of the Solar Wind with the Conducting Model of the Moon. Preprint No. 196 (Space Research Institute, Moscow 1974)
314. A.S. Lipatov: Cosmic Res. (Sov. J. Kosmich. Issled.) **14**(1), 103 (1976)
315. A.S. Lipatov: 'Numerical simulation of the effects of the magnetic field induced by plasma flow past nonmagnetic planets'. In: *Solar-wind Interaction with the Planets Mercury, Venus, and Mars. The Proceedings of a bilateral seminar of the US-USSR Joint Working group on Near-Earth Space, the Moon, and Planets, held at the Space Research Institute of Academy of Sciences of the USSR, Moscow, November 17–21, 1975*, ed. by N.F. Ness (NASA, SP-397 Washington, D.C. 1976) p. 151
316. A.S. Lipatov: 'Numerical Simulation of the Interaction of the Solar Wind with the Moon, Mars and Venus'. In: *Mathematical Models of the Nearest Space*, ed. by V.G. Pivovarov (Nauka Press, Novosibirsk 1977) p. 89
317. A.S. Lipatov: Numerical Simulation of Induced Magnetosphere under the Interaction of the Solar Wind with Venus's Ionosphere. Preprint No. 351 (Space Research Institute, Moscow 1977)
318. A.S. Lipatov: Cosmic Res. (Sov. J. Kosmich. Issled.) **16**(3), 346 (1978)
319. A.S. Lipatov: Cosmic Res. (Sov. J. Kosmich. Issled.) **23**(1), 135 (1985)
320. A.S. Lipatov: 'Numerical Simulation of the Explosion Growth of Magnetic Island and Particle Acceleration'. In: *The Physics of the Solar Flares, Proceedings of the Second Seminar for Solar Flare Problems, Riga, December 1983*, ed. by V.V. Fomichev (IZMIRAN, Moscow 1985) p. 134
321. A.S. Lipatov: Sov. J. Plasma Phys. **13**(8), 545 (1987)
322. A.S. Lipatov: Numerical Simulation of Plasma Processes under the Interaction of the Solar Wind with Magnetosphere of Planets. D.Sc. Thesis, Space Research Institute (IKI), Moscow (1988)
323. A.S. Lipatov: Numerical Simulation of Cometary Shocks Using the Hybrid Models. In: *Mathematical Models of the Space. Proceeding of USSR Meeting, held in November 1988, Moscow*, ed. by A.A. Galeev (IZMIRAN, Moscow 1988) p. 115
324. A.S. Lipatov: 'Numerical simulation of the cometary shocks waves'. In: *Nonlinear Phenomena in Vlasov Plasmas, Proceedings of the International Workshop, Cargese, France, July 1988*, ed. by F. Doveil (Les presses de l'imprimerie de l'Ecole Polytechnique, Palaiseau 1988) p. 207
325. A.S. Lipatov: 'Numerical simulation of the interaction of solar wind with comets'. Monograph. In: *Plasma Processes in Space, Vol. 1*, ed. by R.Z. Sagdeev (Itogy Nauki I Techniki, Vsesojuznyi Inst. Nauch. Tech. Inform. [VINITI], Moscow 1989) p. 49–136 (in Russian)
326. A.S. Lipatov: 'Numerical simulation of the structure of collisionless supercritical shocks'. In: *Solar Plasma Phenomena, Proceeding of the International Workshop, Cargese, July 1989*, ed. by M.A. Dubois, F. Bely-dubau, D. Gresillon (Editions de Physique, Orsay 1990) p. 57
327. A.S. Lipatov: 'Numerical simulation of the cometary shocks'. In: *Physics of the Outer Heliosphere, Proceeding of the COSPAR Colloquium, Warsaw, Poland,*

September 1989, ed. by S. Grzedzielski, D.E. Page (Pergamon Press, New York 1990) p. 395
328. A.S. Lipatov: 'Hybrid simulation of the structure of the Earth's shock including electron inertia'. In: *Proceedings of the 4th International School for Space Simulation, Kyoto-Nara, Japan, 1991*, ed. by Y. Omura, H. Matsumoto (Terra Scientific, Tokyo 1991) p. 190
329. A.S. Lipatov: Cosmic Res. (Sov. J. Kosmich. Issled.) **29**(5), 705 (1991)
330. A.S. Lipatov: 'Particle heating and acceleration by collisionless shocks'. In: *Collective Acceleration Processes in Collisionless Plasma, Proceeding of the International Workshop held at IESC, Corsica June 9–15, 1991*, ed. by D. Le Queau, A. Roux, D. Gresillon (Les Editions de Physique, Les Ulis 1991) p. 247
331. A.S. Lipatov: STEP SIMPO Newsl. **1**, 20 (1996)
332. A.S. Lipatov: 'Numerical simulation of the bow shocks near comets and planets'. In: *Plasma environments of non-magnetic planets, COSPAR Colloquia Vol. 4*, ed. by T.I. Gombosi (Pergamon Press, Oxford, New York, Seoul, Tokyo 1993) p. 203
333. A.S. Lipatov: Cosmic Res. (Sov. J. Kosmich. Issled.) **32**(1), 72 (1994)
334. A.S. Lipatov: STEP SIMPO Newsl. **5**(16), 11 (1996)
335. A.S. Lipatov, J. Buechner: Nonlinear Proc. Geophys. (submitted) (2000)
336. A.S. Lipatov, V.A. Lobatchov: Cosmic Res. (Sov. J. Kosmich. Issled.) **34**(5), 420 (1996)
337. A.S. Lipatov, G.M. Milikh, A.S. Sharma: 'Dusty Plasma at the Interaction of the Magnetospheres of Comet Shoemaker-Levy 9 and Jupiter'. In: *American Geophysical Union Fall Meeting Program* (AGU, San Francisco 1996) p. 199
338. A.S. Lipatov, U. Motschmann, T. Bagdonat: Planet. Space Sci. (2001) (accepted for publication)
339. A.S. Lipatov, K. Sauer, K. Baumgärtel: Adv. Space Res. **20**(2), 279 (1997)
340. A.S. Lipatov, A.S. Sharma, K. Papadopoulos: 'Simulation of comet Shoemaker-Levy 9 encounter with Jupiter'. In: *American Geophysical Union Fall Meeting Program* (AGU, San Francisco 1993) p. 95
341. A.S. Lipatov, A.S. Sharma: Geophys. Res. Lett. **21**(11), 1059 (1994)
342. A.S. Lipatov, A.S. Sharma, K. Papadopoulos: 'One- and Two-Dimensional Hybrid Simulations of the Tangential Discontinuities'. In: *Proceeding of American Physical Society-Division of Plasma Physics, Minnesota Meeting, 7-11 November*, (AIP, Minnesota 1994)
343. A.S. Lipatov, I.N. Syrovatskii: 'Numerical simulation of quasi-parallel cometary shocks'. In: *Collisionless Shocks, Proceeding of Intenational Symposium, Balatonfüred, Hungary, June 1987*, ed. by K. Szegö (CRIP, Budapest 1987) p. 112
344. A.S. Lipatov, I.N. Syrovatskii: Cosmic Res. (Sov. J. Kosmich. Issled.) **25**(6), 750 (1987)
345. A.S. Lipatov, G.P. Zank: Phys. Rev. Lett. **82**(18), 3609 (1999)
346. A.S. Lipatov, G.P. Zank: Solar System Res., translated from Astronomicheskii Vestnik **34**(2), 183 (2000)
347. A.S. Lipatov, G.P. Zank, H.L. Pauls: J. Geophys. Res. **103**, 20636 (1998)
348. A.S. Lipatov, G.P. Zank, H.L. Pauls: J. Geophys. Res. **103**, 29679 (1998)
349. A.S. Lipatov, L.M. Zelenyi: Sov. J. Plasma Phys. **5**(4), 525 (1979)
350. A.S. Lipatov, L.M. Zelenyi: The study of the processes of spontaneous magnetic field reconnection in collisionless and weakly collisionless plasmas. Preprint No. 592 (Space Research Institute, Moscow 1980)
351. A.S. Lipatov, L.M. Zelenyi: Plasma Phys. **24**(9), 1065 (1982)
352. R.G. Littlejohn: J. Plasma Phys. **29**, 111 (1983)

353. R.F. Lottermoser, M. Scholer, A.P. Matthews: J. Geophys. Res. **103**(A3), 4547 (1998)
354. J.G. Luhmann, J.U. Kozyra: J. Geophys. Res. **96**, 5457 (1991)
355. A.T. Lui, S.M. Krimigis: J. Geophys. Res. **86**, 11173 (1981)
356. P.M. Lyster, J.N. Leboeuf: J. Comput. Phys. **102**, 180 (1992)
357. K. Mahesh: J. Comput. Phys. **145**, 332 (1998)
358. Yu.G. Malama: Astrophys. Space Sci. **176**, 21 (1991)
359. M.E. Mandt, R.E. Denton, J.F. Drake: Geophys. Res. Lett. **21**, 73 (1994)
360. M.E. Mandt, J.R. Kan: J. Geophys. Res. **90**(1), 115 (1985)
361. A. Mankofsky, R. Sudan, J. Denavit: J. Comput. Phys. **70**, 89 (1987)
362. A. Mankofsky, T.M. Antonsen, A.T. Drobot: 'An implicit field algorithm for quasineutral hybrid plasma simulation'. In: *Plasma Technology Notes* (Plasma Technology Division, SAIC, McLean 1988)
363. N.N. Mansour, P. Moin, W.C. Reynolds, J.H. Ferziger: *Turbulent Shear Flows I* (Springer-Verlag, New York, Berlin 1977) p. 386
364. G.I. Marchuk: *Methods of Numerical Mathematics* (Springer-Verlag, New York, Heidelberg, Berlin 1982)
365. L.S. Marochnik: Sov. Astron. J. **6**, 828 (1963)
366. R. Mason: J. Comp. Phys. **41**, 233 (1981)
367. A.P. Matthews: J. Comput. Phys. **112**, 102 (1994)
368. S.F. McCormick: *Multilevel Adaptive Methods for Partial Differential Equations, Frontiers in Applied Mathematics, Vol. 6* (SIAM, Philadelphia, 1989)
369. M.E. McKean, N. Omidi, D. Krauss-Varban: J. Geophys. Res. **100**, 3427 (1995)
370. D.A. Mendis: 'Exploration of Halley's Comet: symposium summary'. In: *Exploration of Halley's Comet*, ed. by M. Grewing, F. Praderie, R. Reinhard (Springer-Verlag, Berlin, Heidelberg 1988) p. 939
371. F.C. Mitchel: Planet. Space Sci. **19**, 1580 (1971)
372. A.B. Mikhailovskii: *Theory of Plasma Instabilities, Vols. 1 and 2* (Consultant Bureau, New York 1974)
373. A. Miura: J. Geophys. Res. **89**, 801 (1984)
374. A. Miura: J. Geophys. Res. **92**, 3195 (1987)
375. A. Miura: Geophys. Res. Lett. **17**, 749 (1990)
376. A. Miura: J. Geophys. Res. **97**, 10655 (1992)
377. K. Miyamoto: *Plasma Physics for Nuclear Fusion* (MIT Press, Cambridge 1976)
378. S.S. Moiseev, R.Z. Sagdeev: Dokl. Akad. Nauk. USSR **146**, 329 (1962)
379. M.D. Montgomery, S.P. Gary, W.C. Feldman, D.W. Forslund: J. Geophys. Res. **81**, 2743 (1976)
380. C.S. Morawetz: Phys. Fluids **4**, 988 (1961)
381. R.L. Morse: 'Multidimensional Plasma Simulation by the Particle-in-Cell Method'. In: *Methods in Computational Physics. Plasma Physics, Vol. 9*, ed. by B. Alder, S. Fernbach, M. Rotenberg (Academic Press, New York, London, 1970) p. 213
382. R.L. Morse, C.W. Nielson: Phys. Fluids **14**, 830 (1971)
383. H.R. Müller, G.P. Zank, A.S. Lipatov: J. Geophys. Res. **105**, 27419 (2000)
384. T. Mukai, W. Miyake, T. Terasawa, M. Kitayma, K. Hirao: Nature **321**, 299 (1986)
385. T. Mukai, W. Miyake, T. Terasawa, M. Kitayma, K. Hirao: Geophys. Res. Lett. **13**(8), 829 (1986)
386. M.S. Nakamura, M. Fujimoto, K. Maezawa: J. Geophys. Res. **103**, 4531 (1998)

387. K. Nambu: 'Theoretical Basis of the Direct Simulation Monte Carlo Method'. In: *Proceeding of the Fifteenth International Symposium on Rarefied Gas Dynamics, Vol. 1*, ed. by V. Boffi, C. Cercignani (Teubner, Stuttgart 1986) p. 369–383
388. H. Naitou, S. Tokuda, T. Kamimura: J. Comput. Phys. **33**, 86 (1979)
389. N.F. Ness: Rev. Geophys. **7**, 97 (1969)
390. N.F. Ness, K.W. Behamon, H.E. Taylor, Y.C. Whang: J. Geophys. Res. **73**, 3421 (1968)
391. M. Neugebauer: Phys. Rev. Lett. **4**(1), 6 (1960)
392. C.W. Nielson, H.R. Lewis: 'Particle-Code Models in the Nonradiative Limit'. In: *Methods in Computational Physics. Controlled Fusion, Vol. 16*, ed. by B. Alder, S. Fernbach, M. Rotenberg (Academic Press, New York, London 1976) p. 367–388
393. C.W. Nielson, E.L. Lindman: 'An Implicit, Two-Dimensional, Electromagnetic Plasma Simulation Code'. In: *Proceedings 6th Conf. on Numerical Simulation of Plasmas, Paper E3* (Lawrence Livermore National Laboratory, Livermore 1973), p. 148
394. K.I. Nishikawa, G. Ganguli, Y.C. Lee, P.J. Palmadesso: J. Geophys. Res. **95**, 1029 (1990)
395. M.L. Norman, J.R. Wilson, R. Barton: Astrophys. J. **239**, 968 (1980)
396. T.G. Northrop: *The Adiabatic Motion of Charged Particles* (Interscience, New York 1963)
397. D. Nunn: J. Comput. Phys. **108**, 180 (1993)
398. J.T. Oden: *Finite Element of Non-linear Continua* (McGraw–Hill Book, New York 1972)
399. T. Ogino: 'Two-Dimensional MHD Code'. In: *Computer Simulation of Space Plasma*, ed. by H. Matsumoto, Y. Omura (Terra Scientific, Tokyo 1993) p. 161–209
400. Y. Ohsawa: Phys. Fluids **28**(7), 2130 (1985)
401. H. Okuda: Space Sci. Rev. **42**, 41 (1985)
402. H. Okuda: J. Geophys. Res. **98**, 3953 (1993)
403. H. Okuda, J.M. Dawson, A.T. Lin, C.C. Lin: Phys. Fluids **21**, 476 (1978)
404. Y.A. Omelchenko, R.N. Sudan: J. Comput. Phys. **133**, 146 (1997)
405. N. Omidi, D. Winske, C.S. Wu: ICARUS **66**, 165 (1986)
406. N. Omidi, D. Winske: J. Geophys. Res. **92**(12), 13409 (1987)
407. N. Omidi, D. Winske: 'Theory and simulation of cometary shocks'. In: *Cometary Plasma Processes*, ed. by A.D. Johnstone (AGU, Washington, D.C. 1991) p. 37
408. N. Omidi, D. Winske: J. Geophys. Res. **100**, 11935 (1995)
409. R. Osterbart, H.J. Fahr: Astron. Astrophys. **264**, 260 (1992)
410. K. Papadopoulos: 'Microinstabilities and Anomalous Transport'. In: *Collisionless Shocks in the Heliosphere: A Tutorial Review, Geophysical Monograph Series, Vol. 34*, ed. by B.T. Tsurutani, R.G. Stone (AGU, Washington, D.C. 1985) p. 59
411. K. Papadopoulos, J.D. Huba, A.T.Y. Lui: J. Geophys. Res. **92**(1), 47 (1987)
412. K. Papadopoulos, A. Mankofsky, R.C. Davidson, A.T. Drobot: Phys. Fluids B **3**, 1075 (1991)
413. K. Papadopoulos, A.S. Lipatov, A.S. Sharma: 'Numerical Simulation of the Structure of Tangential Discontinuities'. In: *American Geophysical Union Spring Meeting Program* (AGU, Baltimore 1994) p. 312
414. I. Papamastorakis, G. Paschmann, N. Sckopke, S.J. Bame, J. Berchem: J. Geophys. Res. **89**, 127 (1984)

415. W. Park, S. Parker, H. Biglari, M. Chance, et al.: Phys. Fluids B **4**, 2033 (1992)
416. E.N. Parker: *Interplanetary Dynamical Processes* (Interscience Publishers, John Wiley & Sons, New York, London 1963)
417. S.E. Parker, A.B. Langdon: J. Comput. Phys. **97**, 91 (1991)
418. G. Paschmann, G. Haerendal, J. Papamastorakis, N. Sckopke, S.J. Bame, J.T. Gosling, C.T. Russell: J. Geophys. Res. **4**, 2159 (1982)
419. A.T. Patera: J. Comput. Phys. **54**, 468 (1984)
420. H.L. Pauls, G.P. Zank, L.L. Williams: J. Geophys. Res. **100**(11), 21595 (1995)
421. R. Pellat: About Reconnection in a Collisionless Plasma. PPG-370 (Univ. of California, Los Angeles 1978)
422. F.W. Perkins: Phys. Fluids **19**, 1012 (1976)
423. H.E. Petschek: 'Magnetic field annihilation'. In: *AAS-NASA Symposium on the Physics of Solar Flares*, Spec. Publ. SP-50, ed. by W.N. Hess (NASA, Washington, D.C. 1964) p. 425
424. P.G. Petropoulos, L. Zhao, A.C. Cangellaris: J. Comput. Phys. **139**, 184 (1998)
425. A. Phavorskii et al.: About numerical dissipators for fluid equations. Preprint N78 (Keldysh Institute of Applied Mathematics, Moscow 1980)
426. I.M. Podgorny, E.M. Dubinin, Yu.M. Potanin, S.I. Shkolnihova: Astrophys. Space Sci. **61**(2), 369 (1979)
427. O.A. Pohotelov: Geomagn. Aeronom. **12**, 693 (1972)
428. J.W. Poukey, J.R. Freeman, G. Yonas: J. Vac. Sci. Technol. **10**(6), 954 (1973)
429. K.G. Powell, P.L. Roe, T.J. Linde, T.I. Gombosi, D.L. De Zeew: J. Comput. Phys. **154**, 284 (1999)
430. P.L. Pritchett, F.V. Coroniti: J. Geophys. Res. **89**, 168 (1984)
431. P.L. Pritchett, F.V. Coroniti, V.K. Decyk: J. Geophys. Res. **101**, 27413 (1996)
432. P.L. Pritchett, F.V. Coroniti, R. Pellat, H. Karimabadi: J. Geophys. Res. **96**, 11523 (1991)
433. K.B. Quest: 'Simulations of Quasi-Parallel Collisionless Shocks'. In: *Collisionless Shocks in the Heliosphere: Reviews of Current Research, Geophysical Monograph Series, Vol. 35*, ed. by B.T. Tsurutani, R.G. Stone (AGU, Washington, D.C. 1985) p. 185
434. K.B. Quest: Phys. Rev. Lett. **54**, 1872 (1985)
435. K.B. Quest: J. Geophys. Res. **91**, 8805 (1986)
436. K.B. Quest: 'Particle injection and cosmic ray acceleration at collisionless parallel shocks'. In: *Proceeding of Sixth International Solar Wind Conference, Vol. 2*, ed by. V.J. Pizzo, T. Holzer, D.G. Sime (National Center for Atmospheric Research, Boulder 1988)
437. K.B. Quest: J. Geophys. Res. **93**(9), 9649 (1988)
438. K.B. Quest: 'Hybrid Simulation'. In: *Tutorial Courses: Third International School for Space Simulation*, ed. by B. Lembege, J.M. Eastwood (Cepadues Ed., Toulouse 1989) p. 177
439. K.B. Quest, D.W. Forslund, J.U. Bracbill, K. Lee: Geophys. Res. Lett. **10**(6), 471 (1983)
440. M.M. Rai, P. Moin: AIAA-89-0369 (AIAA, Reno 1989)
441. P.W. Rambo: J. Comput. Phys. **118**, 152 (1995)
442. K.S. Ravichandran: J. Comput. Phys. **130**, 161 (1997)
443. H.L. Ripkin, H.J. Fahr: Astron. Astrophys. **122**, 181 (1983)
444. S. Rjasanow, W. Wagner: J. Comput. Phys. **124**, 243 (1996)
445. P.J. Roache: *Computational Fluid Dynamics* (Hermosa Publishers, Albuquerque 1970)
446. P.J. Roache, T.J. Mueller: AIAA J. **8**(3), 530 (1970)
447. K.V. Roberts, N.O. Weiss: Math. Comp. **20**, 272 (1966)

448. A.M. Roma, C.S. Peskin, M.J. Berger: J. Comp. Phys. **153**, 509 (1999)
449. M.N. Rosenbluth, M.N. Bussac: Nucl. Fusion **19**, 489 (1979)
450. A.S. Roshail, V.A. Leitan: Numer. Meth. Mech. Contin. Medium **7**(7), 63 (1976)
451. B.L. Rozdestvenskii, N.N. Yanenko: *The systems of quasi-linear equations and their applications to gas dynamics* (Nauka Press, Moscow 1979) p. 592
452. S.G. Rubin, P.K. Khosla: J. Comput. Phys. **24**, 217 (1977)
453. D. Rucinski, M. Bzowski: Astron. Astrophys. **296**, 248 (1995)
454. C. Runge: Math. Ann. (Ger.) **46**, 167 (1995)
455. V.V. Rusanov: Dolkady Acad. Nauk USSR **180**(6), 1303 (1968)
456. C.T. Russell, R.C. Elphic: Geophys. Res. Lett. **6**, 33 (1979)
457. W.M. Ruyten: J. Comput. Phys. **105**, 224 (1993)
458. R.Z. Sagdeev: Sov. Phys. Tech. Phys. **6**, 867 (1962)
459. R.Z. Sagdeev: 'Cooperative phenomena and shock waves in collisionless plasmas'. In: *Reviews of Plasma Physics, Vol. 4*, ed. by M.A. Leontovich (Consultants Bureau, New York 1966) p. 23
460. R.Z. Sagdeev, V.D. Shapiro, V.I. Shevchenko, K. Szegö: Geophys. Res. Lett. **13**, 85 (1986)
461. R.Z. Sagdeev, V.D. Shapiro: Sov. JETP Lett. **17**, 279 (1973)
462. J. Saltzman: J. Comput. Phys. **115**, 153 (1994)
463. A.A. Samarskii: *Introduction to the Theory of Difference Schemes* (Nauka Press, Moscow 1971) (in Russian)
464. A.A. Samarskii: 'About the conservative difference schemes'. In: *The Problems of Applied Mathematics and Mechanics*, ed. by A.A. Samarskii (Nauka Press, Moscow 1971) p. 129–136
465. E.T. Sarris, S.M. Krimigis, A.T.Y. Lui, K.L. Ackerson, L.A. Frank: Geophys. Res. Lett. **8**, 349 (1981)
466. T. Sato: J. Geophys. Res. **84**, 7177 (1979)
467. K. Sauer, A. Bogdanov, K. Baumgärtel, E. Dubinin: Planet. Space Sci. **44**(7), 715 (1996)
468. K. Sauer, A.S. Lipatov, K. Baumgärtel, E. Dubinin: Adv. Space Res. **20**(2), 295 (1997)
469. M.A. Saunders: Geophys. Res. Lett. **16**, 1031 (1989)
470. P. Savoini, M. Scholer, M. Fujimoto: J. Geophys. Res. **99**, 19377 (1994)
471. K. Schindler: J. Geophys. Res. **79**, 2803 (1974)
472. H.U. Schmidt, R. Wegmann: 'Flow and magnetic field in comets'. In: *Venus*, ed. by L.L. Wilkening (Univ. of Arizona Press, Tucson 1982) p. 538
473. M. Scholer: 'Diffusive Acceleration'. In: *Collisionless Shocks in the Heliosphere: Reviews of Current Research, Geophysical Monograph Series, Vol. 35*, ed. by B.T. Tsurutani, R.G. Stone (AGU, Washington, D.C. 1985) p. 287
474. M. Scholer: J. Geophys. Res. **94**, 8805 (1989)
475. M. Scholer: 'Models of flux transfer events'. In: *Physics of the Magnetosphere, Geophysical Monograph Series, Vol. 90*, ed. by P. Song, B.U.O. Sonnerup, M.F. Thomsen (AGU, Washington, D.C. 1995) p. 167
476. M. Scholer, T. Terasawa: Geophys. Res. Lett. **17**, 119 (1990)
477. G. Schybert, C. Sonett, K. Schwartz: J. Geophys. Res. **78**, 2094 (1973)
478. N. Sckopke: Adv. Space Res. **15**, 261 (1995)
479. J.D. Scudder, A. Mangeney, C. Lacombe, C.C. Harvey, T.L. Aggson, R.R. Anderson, J.T. Gosling, G. Paschmann, C.T. Russel: J. Geophys. Res. **91**(10), 11019 (1986)
480. A.G. Sgro, C.W. Nielson: Phys. Fluids **19**, 126 (1976)
481. J.S. Shang: J. Comput. Phys. **118**, 109 (1995)
482. J.S. Shang: J. Comput. Phys. **153**, 312 (1999)

483. M.A. Shay, J.F. Drake: Geophys. Res. Lett. **25**(20), 3759 (1998)
484. M.A. Shay, J.F. Drake, R.E. Denton, D. Biskamp: J. Geophys. Res. **103**, 9165 (1998)
485. M.A. Shay, J.F. Drake, B.N. Rogers, R.E. Denton: Geophys. Res. Lett. **26**(14), 2163 (1999)
486. D.E. Shumaker, P.V. Anderson, G.F. Simonson: 3D and 2D Simulation of an Expanding Plasma Using a Darwin Hybrid Particle Code, Preprint, UCRL-ID-109852 (Lawrence Livermore National Laboratory, Livermore 1992)
487. M.M. Shoucri, R.R.J. Gagne: J. Comput. Phys. **23**, 242 (1977)
488. Yu.S. Sigov, Yu.V. Khodyrev: About the theory of discrete models of plasma. Preprint No. 83 (Keldysh Institute of Applied Mathematics, Moscow 1975)
489. G.L. Siscoe, E.F. Lyon, J.H. Binsack, H.S. Bridge: J. Geophys. Res. **74**, 59 (1969)
490. D.V. Sivukhin: 'Motion of Charge Particles in the Electromagnetic Fields'. In: *Reviews of Plasma Physics, Vol. 1*, ed. by M.A.Leontovich (Consultants Bureau, New York 1965) p. 1
491. C.V. Solodyna, J.W. Sari, J.W. Belcher: J. Geophys. Res. **82**, 10 (1977)
492. C.P. Sonett, D.S. Colburn, P. Dyal, et al.: Nature **230**, 359 (1971)
493. E. Sonnendrücker, J.J. Ambrosiand, S.T. Brandon: J. Comput. Phys. **121**, 281 (1995)
494. B.U.O. Sonnerup, I. Papamastorakis, G. Paschmann, H. Luhr: J. Geophys. Res. **95**, 10541 (1990)
495. M.R. Spiegel: *Vector Analysis* (Schaum, New York 1959)
496. J.R. Spreiter et al.: Cosmic Electrodyn. **1**(1), 5 (1970)
497. J.R. Spreiter, S.S. Stahara: J. Geophys. Res. **85**, 7715 (1980)
498. J.R. Spreiter, S.S. Stahara: 'Magnetohydrodynamic and Gas Dynamic Theories for Planetary Bow Waves'. In: *Collisionless Shocks in the Heliosphere: Reviews of Current Research, Geophysical Monograph Series, Vol. 35*, ed. by B.T. Tsurutani, R.G. Stone (AGU, Washington, D.C. 1985) p. 85
499. R.F. Stebbings, A.C.H. Smith, H. Ehrhardt: J. Geophys. Res. **69**, 2349 (1964)
500. O. Steiner, M. Knolker, M. Schussler: *Solar Surface Magnetism* (Kluwer Academic, Dordrecht/Norwell 1994) p. 441
501. R.S. Steinolfson, V.J. Pizzo, T. Holzer: Geophys. Res. Lett. **21**, 245 (1994)
502. H.L. Stone, P.L.T. Brian: Am. Inst. Chem. Eng. J. **9**, 681 (1963)
503. J.McL. Stone: Numerical simulations of mass outflows from star forming regions. Ph.D. Thesis, Univ. of Illinois, Urbana-Champaign (1990)
504. W. Strang: SIAM J. Numer. Anal. **5**, 506 (1968)
505. P.A. Sweet: 'The neutral point theory of solar flares'. In: *Proc. Int. Astr. Union Symp. on Electromagnetic Phenomena in Cosmic Physics, No. 6*, ed. by B. Lehnert (Cambridge Univ. Press, New York 1958) p. 123
506. D.W. Swift: J. Geophys. Res. **91**, 219 (1986)
507. S.I. Syrovatsky, A.M. Zaborov, K.V. Brushlinsky: Numerical two-dimensional plasma flow simulation near the magnetic neutral line. Preprint No. 61 (Keldysh Institute of Applied Mathematics, Moscow 1977)
508. T. Tajima, J.N. Leboeuf, J.M. Dawson: J. Comput. Phys. **38**, 237 (1980)
509. T. Tajima: *Computational Plasma Physics, Application to Fusion and Astrophysics, Frontiers in physics. Lecture Note Series. V.72* (Addison-Wesley, New York, Bonn, Tokyo 1989)
510. T. Tajima, W. Horton, P.J. Morrison, J. Schutkeker, T. Kamimura, K. Mima, Y. Abe: Phys. Fluids B **3**, 938 (1991)
511. Z. Tan, P.L. Varghese: J. Comput. Phys. **110**, 327 (1994)

512. M. Tanaka: 'MACRO-EM'. In: *Computer Space Plasma Physics: Simulation Techniques and Software*, ed. by H. Matsumoto, Y.Omura (Terra Scientific, Tokyo 1993) p. 85
513. M. Tanaka: J. Comput. Phys. **107**, 124 (1993)
514. T. Terasawa: J. Geophys. Res. **86**, 9007 (1981)
515. T. Terasawa, M. Hoshino, J.I. Sakai, T. Hada: J. Geophys. Res. **91**, 4171 (1986)
516. T. Terasawa, M. Fujimoto, H. Karimabadi, N. Omidi: Phys. Rev. Lett. **68**, 2778 (1992)
517. V.A. Thomas: J. Geophys. Res. **94**(9), 12009 (1989)
518. V.A. Thomas: J. Geophys. Res. **94**, 13579 (1989)
519. V.A. Thomas: J. Geophys. Res. **100**, 19429 (1995)
520. V.A. Thomas, S.H. Brecht: Phys. Fluids **29**(8), 2444 (1986)
521. V.A. Thomas, S.H. Brecht: Phys. Fluids **29**(10), 3398 (1986)
522. V.A. Thomas, S.H. Brecht: J. Geophys. Res. **93**, 11341 (1988)
523. V.A. Thomas, D. Winske, N. Omidi: J. Geophys. Res. **95**, 18809 (1990)
524. V.A. Thomas, D. Winske: J. Geophys. Res. **98**, 11425 (1993)
525. D.A. Tidman, N.A. Krall: *Shock waves in collisionless plasmas* (Wiley-Interscience, New York 1971)
526. A.N. Tikhonov, A.A. Samarskii: *Equations of the mathematical physics* (Nauka Press, Moscow 1972) (in Russian)
527. R.L. Tokar, C.H. Aldrich, D.W. Forslund, K.B. Quest: Phys. Rev. Lett. **56**, 1909 (1986)
528. A.I. Tolstykh: *High Accuracy Non-centered Compact Difference Schemes for Fluid Dynamics Applications* (World Scientific, Singapore 1994)
529. A.I. Tolstykh, M.V. Lipavskii: J. Comput. Phys. **140**, 205 (1998)
530. B.T. Tsurutani: 'Comets: A laboratory for plasmawaves and instabilities'. In: *Cometary Plasma Processes*, ed. by A. D. Johnstone (AGU, Washington, D.C. 1991) p. 189
531. B.T. Tsurutani, E.J. Smith: Geophys. Res. Lett. **13**, 259 (1986)
532. A.A. Tusheva, Yu.N. Shokin, N.N. Yanenko: 'About the design of higher order difference schemes on basis of differential consequences'. In: *Some Problems of Computational and Applied Mathematics* (Nauka Press, Novosibirsk 1975) p. 184 (in Russian)
533. O.L. Vaisberg, A.V. Bogdanov, V.N. Smirnov, C.I. Romanov: Cosmic Res. **13**, 129 (1975)
534. O.L. Vaisberg, A.A. Galeev, C.I. Klimov, M.N. Nozdrachev, A.N. Omelcheko, R.Z. Sagdeev: Sov. Phys. JETP Lett. **35**(1), 25 (1982)
535. L.L. Vanyan, I.V. Egorov: Cosmic Res. **11**(6), 913 (1973)
536. L.L. Vanyan, A.S. Lipatov: Geomagn. Aeron. **14**, 417 (1974)
537. L.L. Vanyan, A.S. Lipatov: Geomagn. Aeron. **14**, 714 (1974)
538. O.V. Vasilyev, S. Paolucci: J. Comput. Phys. **138**, 16 (1997)
539. V.M. Vasyliunas, G.L. Siscoe: J. Geophys. Res. **81**, 1247 (1976)
540. R. Vichnevetsky: Math. Comput. Simul. **21**, 170 (1979)
541. J. Villasenor, O. Buneman: Computer Phys. Comm. **69**, 306 (1992)
542. V.A. Vshivkov, G.I. Dudnikova, Yu.P. Zaharov, A.M. Orishich: Generation of Plasma Disturbancies under Collisionless Interaction of the SuperAlfvén Beams. Preprint No. 20-87 (Institute of Theoretical and Applied Mechanics, Novosibirsk 1987)
543. V.A. Vshivkov, G.I. Dudnikova: Simul. Mech. **4**(21), 1 (1990) (in Russian).
544. H.X. Vu, J.U. Brackbill: J. Comput. Phys. **116**, 384 (1995)
545. M.K. Wallis: Cosmic Electrodyn. **3**, 45 (1972)
546. M.K. Wallis: Planet. Space Sci. **23**, 1647 (1973)

547. M.K. Wallis, M. Dryer: Astrophys. J. **205**, 895 (1976)
548. A. Wambecq: Computing **20**, 33 (1978)
549. R. Warming, P. Kutler, H. Lomas: AIAA J. **11**(2), 189 (1973)
550. K. Watanabe, T. Sato: 'High-Precision MHD Simulation'. In: *Computer Simulation of Space Plasma*, ed. by H. Matsumoto, Y. Omura (Terra Scientific, Tokyo 1993) p. 209-216
551. Y.C. Whang: Phys. Fluids **11**(8), 1713 (1968)
552. Y.C. Whang: Phys. Rev. **186**(1), 143 (1969)
553. Y.C. Whang: Space Sci. Rev. **78**, 387 (1996)
554. Y.C. Whang, N.F. Ness: J. Geophys. Res. **75**(31), 6002 (1970)
555. R.B. White, D.A. Monticello, M.N. Rosenbluth, B.V. Waddell: Plasma Phys. Contr. Nucl. Fusion Res. (IAEA Vienna) **1**, 569 (1977)
556. M. Wilcoxson, V. Manousiouthakis: J. Comput. Phys. **115**, 376 (1994)
557. L.L. Williams, D.T. Hall, H.L. Pauls, G.P. Zank: Astrophys. J. **476**, 3766 (1997)
558. J.H. Williamson: J. Comp. Phys. **8**, 258 (1971)
559. D. Winske: Phys. Fluids **24**(6), 1069 (1981)
560. D. Winske: 'Microtheory of Collisionless Shock Current Layers'. In: *Collisionless Shocks in the Heliosphere: Reviews of Current Research, Geophysical Monograph Series, Vol. 35*, ed. by B.T. Tsurutani, R.G. Stone (AGU, Washington, D.C. 1985) p. 225
561. D. Winske: J. Geophys. Res. **93**, 2539 (1988)
562. D. Winske, S.P. Gary: J. Geophys. Res. **91**, 6825 (1986)
563. D. Winske, S.P. Gary, M.E. Jones, M. Rosenberg, V.W. Chow, D.A. Mendis: Geophys. Res. Lett. **22**(15), 2069 (1995)
564. D. Winske, S.P. Gary, D.S. Lemons: 'Diffusive transport at the magnetopause'. In: *Physics of Space Plasmas, 1990 SPI Conf. Proc. Reprint Ser., Vol. 10*, ed. by T. Chang, G.B. Crew, J.R. Jasperse (Scientific, Cambridge 1991) p. 397
565. D. Winske, M.M. Leroy: 'Hybrid Simulation Techniques Applied to the Earth's Bow Shock'. In: *Computer Simulation of Space Plasma*, ed. by H. Matsumoto, T. Sato (Terra Scientific, Tokyo 1984) p. 256
566. D. Winske, M.M. Leroy: J. Geophys. Res. **89**, 2673 (1984)
567. D. Winske, N. Omidi: 'Hybrid Codes'. In: *Computer Simulation of Space Plasma*, ed. by H. Matsumoto, Y. Omura (Terra Scientific, Tokyo 1993) p. 145
568. D. Winske, N. Omidi: J. Geophys. Res. **100**, 11923 (1995)
569. D. Winske, N. Omidi: J. Geophys. Res. **101**(8), 17287 (1996)
570. D. Winske, N. Omidi, K. Quest, V.A. Thomas: J. Geophys. Res. **95**, 18821 (1990)
571. D. Winske, K. Quest: J. Geophys. Res. **91**(8), 8789 (1986)
572. D. Winske, K. Quest: J. Geophys. Res. **93**(9), 9681 (1988)
573. D. Winske, M. Tanaka, C.S. Wu, K.B. Quest: J. Geophys. Res. **90**(1), 123 (1985)
574. D. Winske, C.S. Wu, Y.Y. Li, Z.Z. Mou, S.Y. Guo: J. Geophys. Res. **90**(3), 2713 (1985)
575. M. Witte, H. Rosenbauer, M. Banaszkiewicz, H.J. Fahr: Adv. Space Res. **13**(6), 121 (1993)
576. R.R. Woodward, P. Collela: J. Comput. Phys. **54**, 115 (1984)
577. R.A. Wolf: J. Geophys. Res. **73**(A13), 4281 (1968)
578. C.C. Wu: J. Geophys. Res. **91**, 3042 (1986)
579. C.S. Wu, R.C. Davidson: J. Geophys. Res. **77**, 5399 (1972)
580. C.S. Wu: Space Sci. Rev. **32**, 83 (1982)

581. N.N. Yanenko: *The Fractional Step Methods for Solving Multi-Dimensional Problems of Mathematical Physics* (Nauka, Novosibirsk 1967; Springer-Verlag, New York 1971)
582. N.N. Yanenko, N.G. Danaev, V.D. Liseikin: Numer. Meth. Mechan. Contin. Medium **8**, 157 (1977)
583. S.P. Yu, G.P. Kooyers, O. Buneman: J. Appl. Phys. **36**(8), 2550 (1965)
584. A.L. Zachary, B.I. Cohen: J. Comput. Phys. **66**, 469 (1986)
585. S.T. Zalesak: J. Comp. Phys. **31**, 363 (1979)
586. S.T. Zalesak: J. Comp. Phys. **40**, 497 (1981)
587. G.P. Zank, A.S. Lipatov, H. Müller: 'The Interaction of Heavy Interstellar Atoms with the Heliosphere'. In: *Proceedings of the Ninth International Solar Wind Conference, Nantucket, Massachustts*, ed. S.R. Habbal, R. Esser, J.V. Hollweg, P.A. Isenberg, Conference Proceedings 471 (AIP Press, College Park 1999) p. 811
588. G.P. Zank, H.L. Pauls, I.H. Cairns, G.M. Webb: J. Geophys. Res. **101**, 457 (1996)
589. G.P. Zank, H.L. Pauls, A.S. Lipatov: 'Interaction of the Solar Wind with the Local Interstellar Medium'. In: *Proceedings of the 25th ICRC held 30 July–6 August 1997, Durban, South Africa, Invited, Rapporteur, and Highlight Papers, Vol. 8*, ed. by M.S. Potgieter, B.C. Raubenheimer, D.J. van der Walt (World Scientific, Singapore, New Jersey, London, Hong Kong 1998) p. 333
590. G.P. Zank, H.L. Pauls, L.L. Williams, D.T. Hall: 'Multidimensional modeling of the solar wind–LISM interaction including neutrals: A Boltzmann approach'. In: *Solar Wind Eight, Conf. Proc, 382*, ed. by D. Winterhalter et al. (AIP Press, College Park 1996) p. 654
591. G.P. Zank, H.L. Pauls, L.L. Williams, D.T. Hall: J. Geophys. Res. **101**, 21639 (1996)
592. L.M. Zelenyi, A.S. Lipatov, A.L. Taktakishvili: Cosmic Res. (Sov. J. Kosmich. Issled.) **20**, 714 (1982)
593. L.M. Zelenyi, A.S. Lipatov, D.G. Lominadze, A.L. Taktakishvili: Energetic proton bursts during magnetic island formation in the Earth's magnetotail. Preprint No. 697 (Space Research Institute, Moscow 1982)
594. L.M. Zelenyi, A.S. Lipatov, D.G. Lominadze, A.L. Taktakishvili: Planet. Space Sci. **32**(1), 313 (1984)
595. J. Zhang: J. Comput. Phys. **143**, 449 (1998)
596. V.P. Zhukov: Comput. Maths. Math. Phys. **34**(8/9), 1155 (1994)
597. D. Zilbersher, M. Gedalin: Planet. Space Sci. **45**, 693 (1997)
598. N.M. Zueva, V.S. Imshennik, O.B. Lokutsievskii, M.S. Mikhailov: The model of nonhydrodynamic stage of plasma focus. Preprint No. 73 (Keldysh Institute of Applied Mathematics, Moscow 1975)

Index

Active magnetospheric particle tracer explorers 14, 320
Adaptive mesh and particle refinement 158, 161
Advection equation 131
Alfvén Mach number 192, 317
Alfvén turbulence 222
Alfvén wave 14
Ampere's plasma model 27
Analytical integration of particle motion equations 74
Anomalous cosmic rays 224
Anomalous resistivity 247
Area weighting 90
Assignment function 83

Background ions 36
Boltzmann equation 344
Boris's scheme 73
Boundary layer 7
Bow shock 347
Buneman's scheme 74

C1 class schemes 78
Cell-centered mesh 129
Chapman–Ferraro problem 14
Charge exchange processes 56, 180, 344
Cloud shape 84
Cloud-in-cell model 85
Collisionless shock 12
Combined model 37
Comet 10, 309
Cometary material 319
Compact finite-difference schemes 126
Composite grids 159
Computational experiment 4
Computational simulation 3
Conducting wall 185
Conservation laws 92
Conventional hybrid codes 143
Conventional hybrid models 26

Cosine scheme 86
Courant–Fridrich–Levy number 370, 373
Cross-field diffusion 245
Current and neutral sheets 12

D1 class schemes 79
Darwin model 26
Delta-F method 42
Differencing errors 130
Diffusion equation 131
Direct Vlasov methods 17
Dissipator 134
Drift-kinetic equation 40, 55
Dust grains 29
Dust-grain acceleration 14

Effects of anomalous resistivity 275
Electromagnetic ion-beam-driven instability 203
Electromagnetic potentials 35, 135
Electron (ion) pressure 30
Electron effects 268
Electron inertia 270
Electron pressure anisotropy 270
Electron pressure equation 140
Electron pressure tensor 273
Electron sound Mach number 192
Electrostatic model 26
Energy conservation 98
Evaporation rate 315
Explosive regime 264
Extremely ultraviolet sources 12, 62

Field absorption 182
Field radiation 182
Field-aligned effects 302
Filaments 302
Filtering of spurious oscillations 134
Finite-difference methods 19
Finite-size beam 13, 301
Fluid ion background 38

Index

Foot 227
Force interpolation 90
Four-fluid simulation 348
Fourier transform 18
Fourier–Hermite transform method 18
Full particle models 50
Fusion plasma 14

Gaussian particle shape 86
General hybrid codes 15
Generalized electromagnetic field 138
Geomagnetic tail 8, 255
Guiding center 46
Gyrocenter equation 38
Gyrokinetic equation 41

Heavy cometary ions 295
Heliopause 12, 347
Heliosphere 12
Hierarchy of time and space scales 15
High-Mach-number parallel shocks 206
High-order schemes 124

Implicit calculation of the electric field 114
Implicit method 71
Induced magnetosphere 317
Interplanetary shocks 13
Inversion of the cumulative density 165
Ion (electron) kinetic energy tensor 30, 136
Ion acceleration 14, 264
Ion acoustic mode 227
Ion cyclotron emission 147
Ion tearing instability 256

Kelvin–Helmholtz instability 13, 237, 248
Klimontovich representation 43
Knudsen number 57, 344

Large-scale hybrid simulations 255
Leapfrog method 71
Leapfrog schemes 61, 76, 109
Local interstellar medium 344
Low-Mach-number parallel shock 203
Lower hybrid waves 246
Lower-hybrid drift mode 227
Lunar wake 8

Magnetic barrier 316, 336

Magnetic field reconnection 8, 255
Magnetic flux 273
Magnetic Reynolds number 190
Magnetopause 7, 237, 302
Magnetosheath 7, 237
Magnetosonic Mach number 317
Magnetosphere 7
Mass conservation 92
Mass loading 9, 291
Mathematical modeling 3
Maxwellian distribution flux 174
Maxwellian velocity distribution 166
MHD turbulence 222
Modified two-stream instability 227
Moment method 145
Momentum conservation 94
Monte-Carlo technique 34, 61, 344
Multiple reflection 226
Multiple-space/time-scale methods 156
Multiple-time-scale methods 147
Multipoint stencil schemes 124
Multipole expansion method 102

Nearest grid point interpolation 85
Non-full modeling 23
Nongyrotropic quasiviscous effects 273
Nonneutral models 49
Nonradiative limit 31

Oblique cometary shock 215
Oblique shocks 198
Operator splitting method 73, 119
Orbit averaging 25, 154
Overshoot 201

Padé operator 126
Particle acceleration 224
Particle electrons 48
Particle injection 174
Particle ion elements 37
Particle loading 165
Particle motion equation 67
Particle splitting procedure 350
Particle weighting 89
Particle-mesh (PM) model 22
Photo-ionization 56, 344
Pickup ion acceleration 224
Pickup ions 13
Plasma cloud 13, 14, 301
Plasma transport across the tangential discontinuity 251
Plasma–vacuum interface 181

Pluto   329
Positron acceleration   14
Positrons   29
Predictor–corrector method   107

Quadratic-spline interpolation scheme   86
Quasineutral model   25
Quasiparallel cometary shock   219
Quasiparallel shock   202
Quasiperpendicular cometary shock   210
Quasiperpendicular shocks   192

Ramp   227
Rankine–Hugoniot relation   194, 214
Relativistic particle motion   81
Resistive diffusion length   227
Resonant charge exchange   344
Richardson extrapolation   147
Riemann problem   109
Ring distribution flux   176
Ring velocity distribution   167
Runge–Kutta algorithm   80, 145

Shell distribution flux   178
Shell velocity distribution   173
Shock surfing   224
Shocklets   218
Shot noise effects   259
Solar wind   6
Solar-wind–Earth interaction   7
Solar-wind–Mars interaction   8
Solar-wind–Moon interaction   8, 337
Solar-wind–Venus interaction   8, 333
Splitting of operator method   120

Splitting of particle   158
Standing whistler   194
Subcycling   25
Surface waves   302
Synchronization   162

Tail splitting structure   328
Tangential discontinuity   237, 302
Termination shock   12, 347
Test particle regime   321
Thermoforce   33
Time integration   67
Tokamak   15, 147
Tongue-like distribution   351
Transform method   18
Transportive property   121
Trapping of the beam ions   286
Triangular-shaped density cloud scheme   85
Two-phase flow   316
Two-step Lax–Wendroff scheme   111

Uniform distribution   166
Upstream-differencing method   123
Upwind method   108

Variational methods   157
Very weak comet   321
Vlasov hybrid simulation method   21

Water-bag method   21
Wave steepening   218
Weak comet   320, 329
Weighting function   87
Whistler   14